LASER SOURCES AND APPLICATIONS

LASER SOURCES AND APPLICATIONS

Proceedings of the Forty Seventh Scottish
Universities Summer School in Physics,
St Andrews, June 1995.

A NATO Advanced Study Institute.

Edited by

A Miller –University of St Andrews
D M Finlayson –University of St Andrews

Series Editor

P Osborne – University of Edinburgh

CRC Press
Taylor & Francis Group
Boca Raton London New York

CRC Press is an imprint of the
Taylor & Francis Group, an **informa** business

CRC Press
Taylor & Francis Group
6000 Broken Sound Parkway NW, Suite 300
Boca Raton, FL 33487-2742

First issued in hardback 2018

© 1996 by Taylor & Francis Group, LLC
CRC Press is an imprint of Taylor & Francis Group, an Informa business

No claim to original U.S. Government works

ISBN 13: 978-0-7503-0444-3 (pbk)
ISBN 13: 978-1-138-45585-6 (hbk)

Visit the Taylor & Francis Web site at
http://www.taylorandfrancis.com

and the CRC Press Web site at
http://www.crcpress.com

SUSSP Proceedings

/continued

SUSSP Proceedings (continued)

1. Virgilijus Vaicaitis
2. Hans Lange
3. Gianpaolo Barozzi
4. Antonio Lucianetti
5. Georg Sommerer
6. James Fraser
7. Chris Howle
8. Mathias Hilpert
9. Milan Sejka
10. Mike Hasselbeck
11. Sean Ross
12. Valdas Pasiskevicius
13. Linas Giniunas
14. Paul Gunning
15. Chris Dorman
16. J.F. Ripoche
17. Dmitry Mashkovsky
18. John Collier
19. Chris King
20. Briggs Atherton
21. Tiejun Xia
22. Janet Noon
23. Philip Riblet
24. Thomas Graf
25. Jason Sutherland
26. Carlo Boutton
27. Eric Buckland
28. Per Olof Hedekvist

29. Alasdair Cameron
30. Gunnar Haeffler
31. Paul McNamara
32. Brett D. Guenther
33. Steve Lane
34. Phil Gorman
35. Thorsten Schweizer
36. Matthias Jaeger
37. Patrick Moebert
38. Nicholas Brooks
39. Jan Valenta
40. Stuart Butterworth
41. Graham Friel
42. Stratis Georgiou
43. Monika Pietrzyk
44. Ana Sacedón
45. J.L. Sanchez-Rojas
46. Arlene Gordon
47. Tracy Stevenson
48. Simone Hartung
49. Thomas Jensen
50. David Matthews
51. Tim Kellner
52. Scott Diddams
53. Livio Fornasiero
54. Teunis Tukker
55. Alexander Novikov
56. Dom Withers

57. Richard Hainzl
58. Bengt-Erik Olsson
59. Majid Ebrahimzadeh
60. Roger Edwin
61. André Mysyrowicz
62. Eric Van Stryland
63. John Reintjes
64. Alan Miller
65. Bruce Sinclair
66. Gunnar Bjork
67. Gunther Huber
68. Bob Boyd
69. Brett Patterson
70. Ian Lindsay
71. Christopher Bollig
72. David Coppeta
73. Alejandro Rodriguez
74. Christian Rahlff
75. Andrei Fotiadi
76. Magnus Olson
77. A. Pelin Aksoy
78. Ali Çetin
79. Peter Henriksson
80. Gary Morrison
81. Peggy Perozzo
82. Mike Sundheimer
83. David Armstrong

Executive Committee

Prof Alan Miller	University of St Andrews	*Director and Co-Editor*
Dr Bruce D Sinclair	University of St Andrews	*Secretary*
Dr Roger P Edwin	University of St Andrews	*Treasurer*
Dr Miles J Padgett	University of St Andrews	*Social Secretary*
Dr Majid Ebrahimzadeh	University of St Andrews	
Dr David M Finlayson	University of St Andrews	*Co-Editor*
Prof Willie J Firth	University of Strathclyde	

International Advisory Committee

Prof Eric W Van Stryland	University of Central Florida, USA
Prof Algis Piskarskas	Vilnius University, Lithuania
Prof Wilson Sibbett	University of St Andrews, Scotland
Prof Malcolm Dunn	University of St Andrews, Scotland
Prof Evgeny M Dianov	General Physics Institute, Moscow, Russia
Prof Yoshihisa Yamamoto	Stanford University, USA

Lecturers

Gunnar Björk	Stanford University
Robert W Boyd	University of Rochester
Malcolm H Dunn	University of St Andrews
Majid Ebrahimzadeh	University of St Andrews
T Y Fan	Lincoln Laboratory, MIT
David C Hanna	University of Southampton
Andreas Hemmerich	University of Munich
Günter Huber	University of Hamburg
Pavel Mamyshev	AT and T Bell Laboratories
André Mysyrowicz	École Polytechnique
Miles J Padgett	University of St Andrews
John F Reintjes	Naval Research Laboratory
Wilson Sibbett	University of St Andrews
Eric W Van Stryland	University of Central Florida
Ian White	University of Bath

Preface

Lasers have stealthily progressed from being principally research tools in laboratories to becoming ubiquitous in our daily lives by incorporation into CD players, bar code readers, printers, communications systems etc. Further exploitation of these coherent optical emitters can be expected based on some remarkable advances in laser technology during the past few years. Compact, efficient, solid state, fibre and semiconductor lasers are allowing yet more applications of lasers to be conceived. In addition, new nonlinear optical materials have extended the available coherent sources to a wide range of wavelengths, bandwidths and pulse widths.

This book is the Proceedings of the 47th Scottish Universities Summer School in Physics, a NATO Advanced Studies Institute on New Perspectives in Laser Sources and Applications. The chapters are written by an outstanding group of internationally renowned experts in the field of lasers and their applications. The intention of the book, as it was with the school, is to present a relatively broad coverage of the basic principles of new solid state laser systems and their applications, highlighting the most exciting recent developments. The coverage can be loosely divided into three areas, (i) materials, (ii) lasers, and (iii) applications. The chapters have been written as tutorial-style introductions to the various topics at post-graduate level, suitable for research students and others with a basic knowledge of lasers and nonlinear optics. The topics cover both physics and engineering aspects of the field.

The 47th SUSSP was held in the School of Physics and Astronomy and John Burnet Hall at the University of St Andrews between 25th June and 8th July 1995. The 73 participants from 19 countries attended 43 lectures presented by 15 lecturers over 10 working days. In addition, a number of successful discussion sessions were held. Participants were given a chance to present their own work in 8 short oral and 57 poster papers in several very lively sessions.

A busy social programme and the numerous St Andrews pubs kept everybody well occupied outside of the formal sessions. The weather was outstanding during the school, so that golfing on the famous golf courses, walking and even swimming in the North Sea from the adjacent West Sands Beach were popular. Participants were able to sample the Scottish mountains, some Highland castles, the ship Discovery, a Secret Bunker and a Witches tour of St Andrews. They were introduced to malt whisky and Scottish dancing in a final night ceilidh. Two receptions hosted by the University of St Andrews and North East Fife District Council were greatly appreciated.

The organisers are grateful for the help of many organisations and individuals in the success of the school. The principal sponsors were NATO, Euroconferences, the National

Science Foundation and SUSSP. The director and all participants greatly appreciated the unstinting efforts of the local organising team, Dr Bruce Sinclair as school secretary, Dr Roger Edwin as treasurer, Dr Miles Padgett as social secretary, Dr Majid Ebrahimzadeh for arranging the lecture notes, and Dr David Finlayson who provided much needed advice and experience for the organisation of a successful summer school.

Special thanks go to Carol Anders for all her efforts in ensuring the smooth running of the school and Wendy Webster for her assistance in producing this volume on time. The staff of John Burnet Hall, headed by Jackie Mathews, provided a high quality and very friendly residential service in spite of the short preparation time in the newly renovated building. Thanks are also due to Professor Malcolm Dunn, the staff and students of the School of Physics and Astronomy for so willingly assisting in countless ways.

The editors would like to thank the authors for their excellent and timely contributions to this volume.

Alan Miller and David M Finlayson

St Andrews, December 1995

Contents

The Nature of the Nonlinear Optical Susceptibility

Robert W Boyd

University of Rochester, New York, USA

1   Introduction

This chapter presents some ideas regarding methods for calculating the nonlinear optical susceptibility that differ from the standard approach commonly used by most workers. The alternative approaches described here are not really new; the ideas behind these approaches date back to the 1960's. However, because these approaches lie outside of conventional methods, they provide different insights into the nature of the nonlinear optical susceptibility, and perhaps can provide increased understanding of the nature of nonlinear optical interactions.

The standard approach for calculating the nonlinear optical susceptibility entails assuming that a time varying field of the form $E(t) = E_0 \exp(-i\omega t)$+c.c. is applied to an atomic system. The atomic response is then calculated by finding how the state of the system is modified by the applied field. Time-dependent, quantum mechanical perturbation theory is used. Under many circumstances, it is adequate to determine how the atomic wavefunction is modified. Under more general circumstances, where damping effects play an important role in determining the atomic response, the density matrix formalism of quantum mechanics must be used. In either case, one calculates the quantum mechanical expectation value of the induced dipole moment per atom. From this result, polarisabilities and hyperpolarisabilities are extracted.

This procedure is well known and leads to predictions that are usually accepted to be reliable. The calculational procedure is described in standard textbooks on nonlinear

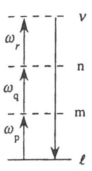

Figure 1. *The nonlinear optical interaction represented by Equation 1.*

optics (see, for example, Boyd (1992)). This method is especially useful for the case of atomic vapours because, in this case, the resonance frequencies and dipole matrix elements that are needed to evaluate the expression for the nonlinear susceptibility are often known to high accuracy. This method has been successfully applied to experimental situations, for example, by Miles and Harris (1973). Even subtle aspects of the nonlinear optical response, such as the existence of pressure-induced resonances, which are a manifestation of quantum interference effects, are well described by the nonlinear optical susceptibility obtained by the density matrix version of the theory (Rothberg 1987, Bloembergen *et al.* 1978).

Nonetheless, this 'standard approach' has some limitations. For example, the expression for the nonlinear optical susceptibility $\chi^{(3)}(\omega_p+\omega_q+\omega_r,\omega_r,\omega_q,\omega_p)$ derived using the density matrix theory consists of 48 different terms that have to be summed to obtain quantitative predictions. Performing such a summation is computationally tedious. A typical term in this sum has the form

$$\frac{N}{\hbar^3}\sum_{\nu nml}\frac{\mu_{l\nu}\mu_{\nu n}\mu_{nm}\mu_{ml}}{(\omega_{\nu l}-\omega_p-\omega_q-\omega_r-i\gamma_{\nu l})(\omega_{nl}-\omega_p-\omega_q-i\gamma_{nl})(\omega_{ml}-\omega_p-i\gamma_{ml})} \tag{1}$$

and can be represented symbolically by the interaction illustrated in Figure 1. Note that the numerator has the form of the product of four dipole matrix elements and that the denominator has terms representing one-, two-, and three-photon resonances. To see another limitation of the standard approach, note that for $n = l$ and for $\omega_p = -\omega_q$ the two-photon resonance factor vanishes, leading to a divergence in this contribution to $\chi^{(3)}$. This divergence is an awkward feature of the standard quantum mechanical expression for $\chi^{(3)}$. In fact, there is no divergence in the complete expression for $\chi^{(3)}$, but a complicated rearrangement of the complete expression must be performed to demonstrate this fact, as first pointed out by Orr and Ward (1971). For these reasons, it would be desirable to formulate a simpler theory of the quantum mechanical susceptibility. Such a theory is formulated in the next section.

2 Nonlinear optical susceptibility in the static limit

As noted above, one usually calculates χ^{NL} using time-dependent perturbation theory. But time-independent perturbation theory is much easier to apply, and should yield the same answer in the limit of static fields, or in fact under any circumstance in which the applied field frequency is much smaller than any resonance frequency.

We now formulate such a theory, following the procedure of Jha and Bloembergen (1968). Since we conventionally write

$$P = \chi^{(1)}E + \chi^{(2)}E^2 + \chi^{(3)}E^3 + \dots \tag{2}$$

where P is the polarisation (dipole moment per unit volume), the energy stored in polarising the medium is given by

$$
\begin{aligned}
W = -\int_0^E P(E')\,dE' &= -\frac{1}{2}\chi^{(1)}E^2 - \frac{1}{3}\chi^{(2)}E^3 - \frac{1}{4}\chi^{(3)}E^4 - \dots \\
&\equiv W^{(2)} + W^{(3)} + W^{(4)} + \dots
\end{aligned}
\tag{3}
$$

Thus if we know $W(E)$ (*e.g.* from Stark shift measurements or calculations), we can determine the (static) nonlinear susceptibilities by simple differentiation as follows:

$$\chi^{(1)} = -\frac{d^2 W^{(2)}}{dE^2}, \tag{4}$$

$$\chi^{(2)} = -\frac{1}{2}\frac{d^3 W^{(3)}}{dE^3}, \tag{5}$$

$$\chi^{(3)} = -\frac{1}{6}\frac{d^4 W^{(4)}}{dE^4}, \qquad etc. \tag{6}$$

We do not necessarily need to know $W(E)$ as a power series in E to use this method. For example, we can write $\chi^{(3)}$ as

$$\chi^{(3)} = -\frac{1}{6}\frac{d^4 W}{dE^4}\bigg|_{E\to 0}. \tag{7}$$

We can see from Equations 4-6 or 7 that there is a direct relation between the Stark effect and the (static) nonlinear susceptibility. Note that the present model is not quite as idealised (*i.e.* irrelevant) as one might think. One usually wants nonlinear optical materials that are largely lossless, and this usually means working at frequencies much lower than any resonance frequency. For certain model systems, $W(E)$ can be calculated directly: two examples are presented below. Later in this chapter we will see how to apply this method in general.

2.1 Hydrogen atom

For the hydrogen atom, for example, Sewell (1949) has shown that in the presence of a static electric field F, the ground state energy can be expressed as a power series expansion of the form

$$E = E_0 + E_1 F + E_2 F^2 + \dots \tag{8}$$

Here and throughout this chapter we use the notation of the original sources in so far as possible. Solutions to the time-independent Schrödinger equation give explicit predictions for the coefficients that appear in this expansion. One finds that the odd terms vanish and that the first several even coefficients are given by

$$E_0 = -\frac{1}{2} \qquad E_2 = -\frac{9}{4} \qquad E_4 = -\frac{3555}{64} \ldots \tag{9}$$

where 'atomic units' are used. When converted to conventional (*i.e.* Gaussian) units, one obtains

$$\chi^{(1)} = N\alpha \qquad \alpha = \frac{9}{2} a_0^3 \tag{10}$$

$$\chi^{(3)} = N\gamma \qquad \gamma = \frac{3555}{16} \frac{a_0^7}{e^6} \tag{11}$$

where a_0 denotes the Bohr radius. Although these results are specific to hydrogen, they display the correct scaling laws for any non-resonant atomic system:

$$\alpha \propto (\text{atomic dimension})^3 \propto V$$
$$\gamma \propto (\text{atomic dimension})^7 \propto V^{7/3} \tag{12}$$

Since $N \approx 1/V$ (for condensed matter), we see that

$$\chi^{(1)} \text{ is independent of atomic volume} \tag{13}$$
$$\chi^{(3)} \propto V^{4/3} \propto (\text{atomic dimension})^4 \tag{14}$$

2.2   Square well potential

Another problem that can be solved by this method is that of a square well potential. This model is often applied to the response of delocalised π-electrons in a conjugated polymer, as illustrated in Figure 2. The chemical structure of polymers are described in standard organic chemistry textbooks, for example, Kemp (1980). Since the π-electrons are largely delocalised, it makes sense to model this system as a collection of nearly free electrons constrained to move in a square well potential. Recent work has shown

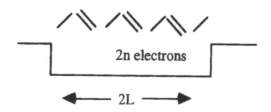

Figure 2. *Model of the potential energy function representing a conjugated polymer.*

that conjugated chain-like polymers can possess very large nonlinear optical responses (Prasad and Williams 1991). For example, for polydiacetylene $\chi^{(3)} = 2.5 \times 10^{-10}$ esu (Blau *et al.* 1992) whereas for CS_2 $\chi^{(3)} = 1.9 \times 10^{-12}$ esu.

Let us consider why the π-electrons in conjugated structures can be treated as being delocalised. By way of notation, we note that a polymer is said to be conjugated if it contains alternating single and double (or triple) bonds. A saturated polymer, in contrast, contains no double bonds.

We consider a chain of the sort

Here the single line represents a single bond and the double line represents a double bond. A single bond is always a σ-bond. A σ-bond consists of one electron from each atom. It has cylindrical symmetry about the internuclear axis. A double bond consists of a σ-bond and a π-bond. A π-bond is comprised of the overlap of two p-orbitals, one for each atom.

The optical properties of σ-electrons and π-electrons (*i.e.* the electrons contained within a σ-bond or π-bond, respectively) are very different. σ-electrons tend to be localised in space. In general, each σ electron makes a certain contribution to the linear and hyper polarisabilities, and the total response of a number of σ-electrons is simply the sum of the individual contributions. π-electrons, on the other hand, tend to be delocalised in the sense that the electron can be found anywhere on the polymer chain. This occurs first because the chain can co-exist in either of the two following forms or in a linear combination of the two (this in fact is why these structures are said to be conjugated).

For this reason, it is somewhat arbitrary which bond we call the double bond and which we call the single bond. Secondly, π-electrons are located away from the molecular axis. Even though the structure is typically written as

$$= C - C = C - C = C -$$

in fact the π-orbitals extend equal amounts to the left and right of the carbon atom. It is only because the single bond is slightly longer, that the double bond is said to occur at the position indicated.

Rustagi and Ducuing (1974) treated the quantum mechanical problem of the motion of free electrons in a square well potential. They found that to a good approximation (*i.e.* lowest order in $1/n$), the linear and nonlinear optical properties of conjugated polymers are described by

$$\alpha = \frac{8\,L^4}{3a_0\pi^2 n} \tag{15}$$

$$\gamma = \frac{256\,L^{10}}{45a_0^3 e^2 \pi^6 n^5} \tag{16}$$

Since n is proportional to L, we see that

$$\alpha \propto L^3 \tag{17}$$

and

$$\gamma \propto L^5 \quad \text{or as} \quad \alpha^{5/3}. \tag{18}$$

Since $N \propto 1/L^3$, we expect that

$$\chi^{(1)} \text{ is independent of } L \tag{19}$$

and

$$\chi^{(3)} \propto L^2 \tag{20}$$

Some experimental results relevant to these predictions have been tabulated by Ducuing (1977). These results show that for a specific set of conjugated polymers, γ increases initially as the fifth power of the number of double bonds and then begins to saturate as the number of double bonds exceeds roughly ten. In contrast, for saturated molecules (*i.e.* for molecules containing no double bonds) γ increases linearly with the number of carbon atoms, both for the alkanes and the cyclo-alkanes. The deviation from the L^5 law for large L for conjugated molecules is likely due to the fact that the potential function experienced by the *pi*-electrons is not really a square well. Bond length alternation, for example, will give a periodic component to the potential. When one solves Schrödinger's equation for such a potential well, somewhat different results are found. Finally, it should be noted that the fullerenes C_{60} and C_{70} (see, for example, Kajzar *et al.* 1994) can be modelled as a square well potential for which the electrons are constrained to move on the surface of a sphere.

3   Static nonlinear optical susceptibility

Let us now see how to implement this method in general. The quantum mechanical expression for the ground state energy of a material system is given by

$$W = N\,E_0 \tag{21}$$

where N is the number density of atoms and where the ground state energy of an atom can be represented as

$$E_0 = E_0^{(0)} + E_0^{(1)} + E_0^{(2)} + \dots \tag{22}$$

The unperturbed energies of the various energy eigenstates are given by the time-independent Schrödinger equation

$$H_0\,|m\rangle = E_m^{(0)}\,|m\rangle \tag{23}$$

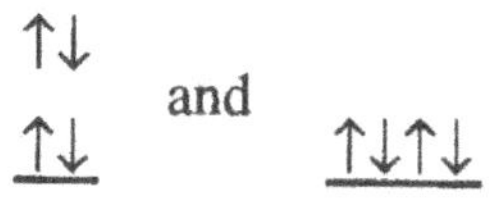

$$\uparrow\downarrow \atop \underline{\uparrow\downarrow} \quad \text{and} \quad \underline{\uparrow\downarrow\uparrow\downarrow}$$

Figure 3. *Illustration of two-photon- and one-photon-resonant contributions to γ, which represent the first and second summations of Equation 29.*

By explicit application of time-independent perturbation theory, one finds that the first four lowest order corrections to the ground state energy are given by

$$E_0^{(1)} = e\mathcal{E}\langle 0|\,x\,|0\rangle \quad \text{(we assume that this term vanishes)} \tag{24}$$

$$E_0^{(2)} = -e^2\mathcal{E}^2 \sum_s{}' \frac{\langle 0|\,x\,|s\rangle\,\langle s|\,x\,|0\rangle}{E_s^{(0)} - E_0^{(0)}} \tag{25}$$

$$E_0^{(3)} = e^3\mathcal{E}^3 \sum_{st}{}' \frac{\langle 0|\,x\,|s\rangle\,\langle s|\,x\,|t\rangle\,\langle t|\,x\,|0\rangle}{\left(E_s^{(0)} - E_0^{(0)}\right)\left(E_t^{(0)} - E_0^{(0)}\right)} \tag{26}$$

$$E_0^{(4)} = e^4\mathcal{E}^4 \sum_{stu}{}' \frac{\langle 0|\,x\,|s\rangle\,\langle s|\,x\,|t\rangle\,\langle t|\,x\,|u\rangle\,\langle u|\,x\,|0\rangle}{\left(E_s^{(0)} - E_0^{(0)}\right)\left(E_t^{(0)} - E_0^{(0)}\right)\left(E_u^{(0)} - E_0^{(0)}\right)}$$
$$- e^2\mathcal{E}^2 E_0^{(2)} \sum_u{}' \frac{\langle 0|\,x\,|s\rangle\,\langle s|\,x\,|0\rangle}{\left(E_u^{(0)} - E_0^{(0)}\right)^2} \tag{27}$$

In these expressions, the prime on the summation symbol indicates that the ground state is to be omitted from the summation, and we are now using the symbol $\mathcal{E}$ to denote the amplitude of the applied electric field. The fact that $E_0^{(4)}$ decomposes into the sum of two terms is a standard, if not often appreciated, aspect of time-independent perturbation theory. By introducing these expressions into Equations 4–6, one finds by differentiation that

$$\alpha \equiv \alpha_{xx} = \frac{2e^2}{\hbar} \sum_{i\neq 0} \frac{x_{0i}x_{0i}}{\omega_{i0}} \tag{28}$$

and

$$\gamma \equiv \gamma_{xxxx} = \frac{4e^4}{\hbar^3} \left[\sum_{ijk\neq 0} \frac{x_{0i}x_{ij}x_{jk}x_{k0}}{\omega_{i0}\omega_{j0}\omega_{k0}} - \sum_{ij\neq 0} \frac{x_{0i}x_{i0}x_{0j}x_{j0}}{\omega_{i0}^2\omega_{j0}} \right] \tag{29}$$

Note that when using this procedure the expression for γ neatly decomposes into the sum of two terms, without the need for the algebraic contortions mentioned above. These two terms describe processes which, in the context of time-dependent perturbation theory, would be represented by the two diagrams shown in Figure 3.

3.1 Nonlinear susceptibility in the Unsold approximation

These expressions can be simplified further through use of what is known as the *Unsold approximation*. First, we assume that each of the resonance frequencies ω_{i0} can be replaced by some average frequency ω_0 to obtain, for example,

$$\alpha = \frac{2e^2}{\hbar\omega_0} \sum_i{}' \langle 0| x |i\rangle \langle i| x |0\rangle = \frac{2e^2}{\hbar\omega_0} \langle 0| x\hat{O}x |0\rangle \tag{30}$$

where we have introduced the operator

$$\hat{O} = \sum_i{}' |i\rangle \langle i| \approx \sum_i |i\rangle \langle i| = \hat{I} \tag{31}$$

The approximate equality expresses the assumption that the summation is not expected to change dramatically by adding one additional state (the ground state) to the summation. But then by the closure relation of quantum mechanics, this summation is just equal to the identity operator $\hat{I}$. We thus find that

$$\alpha = \frac{2e^2}{\hbar\omega_0} \langle x^2 \rangle . \tag{32}$$

We similarly find that

$$\beta = \frac{-3e^3}{\hbar^2\omega_0^2} \langle x^3 \rangle \tag{33}$$

and

$$\gamma = \frac{4e^4}{\hbar^3\omega_0^3} \left[\langle x^4 \rangle - 2\langle x^2 \rangle^2 \right] . \tag{34}$$

These results show that in the static limit χ^{NL} *can be thought of in terms of moments of the ground state electron density distribution.* This view is very different from the standard view which represents the nonlinear susceptibility in terms of a summation over all of the excited states of the molecule.

We can carry this analysis one step further by using the *Thomas-Reiche-Kuhn sum rule*:

$$\frac{2m}{\hbar} \sum_k \omega_{k0} |x_{k0}|^2 = Z \tag{35}$$

where Z denotes the number of electrons. Again replacing ω_{k0} by a constant ('the average transition frequency'), we obtain a prediction for ω_0:

$$\omega_0 = \frac{Z\hbar}{2m\langle x^2 \rangle} \tag{36}$$

Through use of this result we obtain:

$$\alpha \approx \frac{4e^2}{\hbar^2} \frac{\langle x^2 \rangle^2}{Z} \qquad \propto L^3$$

$$\gamma \approx \frac{2^5 e^4 m^3}{\hbar^6 Z^3} \langle x^2 \rangle^3 \left[\langle x^4 \rangle - 2\langle x^2 \rangle^2 \right] \propto L^7 \tag{37}$$

where L now represents some characteristic linear dimension of the system and where we have assumed that Z scales linearly with L. Note that we recover the familiar scaling law that the linear polarisability scales with the system volume and we predict that the hyperpolarisability scales as the seventh power of the linear dimensions of the system.

4 Miller's rule and Wang's rule

Miller's rule (Miller 1964) was initially derived for $\chi^{(2)}$ using the classical anharmonic oscillator model of optical nonlinearities. It states that

$$\chi^{(2)}(\omega_3 ; \omega_2, \omega_1) = \Delta\, \chi^{(1)}(\omega_1)\, \chi^{(1)}(\omega_2)\, \chi^{(1)}(\omega_3) \tag{38}$$

where Δ is a constant that is independent of frequency and nearly the same for all materials. The validity of Miller's rule is well established for a variety of second-order nonlinear optical materials (see, for example, the data presented in Table 16.2 of Yariv (1975)).

Despite early success in generalising Miller's rule to third-order nonlinear optical interactions for certain optical materials such as ionic crystals (Wynne 1969), it appears that the third-order generalisation is not universally valid.

Wang (1970) has shown that a different relationship seems to be more generally valid than Miller's rule. Wang's rule states that, in the static limit $\omega \to 0$,

$$\chi^{(3)} = Q' \left(\chi^{(1)}\right)^2 \quad \text{where} \quad Q' = g/N_{\text{eff}}\hbar\omega_0 . \tag{39}$$

Here g is a dimensionless parameter of the order of unity, N_{eff} is the product of the molecular number density with the oscillator strength, and ω_0 is the mean absorption frequency. Wang showed empirically that this new rule is valid both for low-pressure gases (where Miller's rule fails) and for ionic crystals (where Miller's rule is also valid).

We can understand the origin of Wang's rule on the basis of Jha and Bloembergen's treatment of static optical nonlinearities in terms of the moments of the ground state electron distribution. Recall that Equations 32 and 34 provide empirical predictions for the quantities α and γ. These equations can be rearranged to show that

$$\gamma = Q\alpha^2 \quad \text{where} \quad Q = \frac{g}{\hbar\omega_0}, \quad g = \left[\frac{\langle x^4 \rangle}{\langle x^2 \rangle^2} - 2\right] \tag{40}$$

Note that g is a dimensionless quantity that provides a measure of the normalised fourth moment (or *kurtosis*) of the ground state electron distribution. The validity of Wang's rule, which is well established empirically, thus implies that g has nearly the same value for all optical materials. There does not appear to be any simple physical argument to explain why this should be so.

5 Formula of Boling, Glass, and Owyoung

Wang's rule can be used to derive a well known empirical relationship that allows one to predict the nonlinear refractive index of a material from measurements of linear optical properties (Boling, Glass, and Owyoung 1978). Here we review the derivation of this result, using the notation of the original reference. We assume that the linear optical properties of the material are described by the Lorentz-Lorenz equation in the form

$$\frac{n^2 - 1}{n^2 + 2} = \frac{4\pi}{3} \sum_i N_i \alpha_i \tag{41}$$

The atomic polarisability is assumed to be given by the standard Lorentz oscillator result

$$\alpha_i = \frac{s_i e^2/m}{\omega_i^2 - \omega^2} \tag{42}$$

where s_i denotes the oscillator strength of transition i. The summation is to include all of the molecular species that constitute the sample. We further assume that the nonlinear optical properties are described by the system of equations

$$n_2 = \frac{3\pi}{n}\chi^{(3)} \qquad \chi^{(3)} = f^4 \sum_i N_i \gamma_i \qquad f = \frac{n^2 + 2}{3} \qquad \gamma_i = Q\alpha_i^2 \tag{43}$$

We now assume that the resonance transition of only one constituent dominates the two sums that appear in these equations. For the case of silicate glass, this constituent is accepted to be oxygen. With the sums reduced to single terms, a unique relation between n_2 and n can be derived. It is

$$n_2 = \frac{(gs)(n^2 + 2)^2(n^2 - 1)^2}{48\pi n\hbar\omega_0(Ns)} \tag{44}$$

with

$$\frac{n^2 - 1}{n^2 + 2} = \frac{4\pi}{3}\frac{(Ns)(e^2/m)}{\omega_0^2 - \omega^2} \tag{45}$$

The equation for n_2 is written in terms of the product Ns because, as is clear from the second equation, only the product can be deduced from linear response data. The theory then contains a single free parameter, the product gs, which is assumed to be a constant for various optical materials. The value $gs=3$ is found to give good agreement with the experimental data. The other parameters needed to determine n_2 (*i.e.* Ns and ω_0) can be found by measuring n at two different frequencies.

Boling's equation is known to be highly accurate in its ability to predict the nonlinear refractive index of optical materials. Some data from Adair *et al.*(1989) comparing the measured value of n_2 to that obtained using Boling's equation are shown in Figure 4.

6   Discussion and summary

In this chapter, we have seen how to formulate a theory of the nonlinear optical susceptibility which is based on the application of time-independent perturbation theory and which is valid for low-frequency (strictly speaking, for static) optical fields. The resulting expression (Equation 29) for the nonlinear optical susceptibility (Jha and Bloembergen 1968) is considerably simpler than that resulting from the application of time-dependent perturbation theory.

The expression for the static susceptibility can be further simplified by making the Unsold approximation, that is, by replacing the transition frequencies ω_{i0} that appear in this expression by a suitably defined mean transition frequency ω_0. The resulting expressions show that the linear and nonlinear susceptibilities are actually measures of the moments of the electron distribution in the atomic ground state. The *linear polarisability* α is a measure of the quadrupole moment of the electron distribution, the

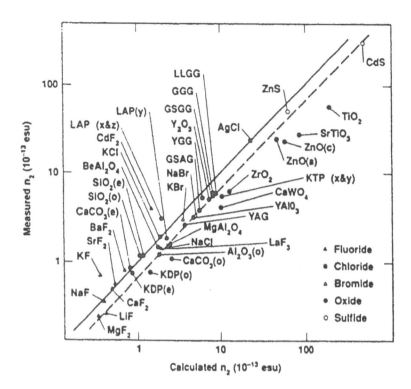

Figure 4. *Measured value of the nonlinear refractive index compared to that predicted by the formula of Boling, Glass, and Owyoung for a variety of optical materials (after Adair et al. 1989).*

first hyperpolarisability β is a measure of the octupole moment, and the *second hyperpolarisability* γ is a measure of the normalised hexadecimal-pole moment. This static model of the nonlinear susceptibility within the Unsold approximation is thus quite different from the time-dependent perturbation theory model, which ascribes the origin of the nonlinear response to the mixing of excited states into the atomic wavefunction. These two views thus provide different and complementary insights into the nature of the nonlinear optical response. Of course, there is no incompatibility between these two views. The shape of the ground state electron distribution depends on the form of the potential energy function $V(r)$ which appears in Schrödinger's equation, and which is also the function that determines the parameters that appear in the time-dependent perturbation theory model, such as the energies of the excited states and the dipole matrix elements.

Throughout the history of nonlinear optics, various attempts have been made to correlate the nonlinear optical properties of materials with their linear properties. The empirical rule of Wang (Equation 39) which relates the static value of the third-order susceptibility to the linear susceptibility is believed to be more generally valid than the third-order version of Miller's rule. The validity of Wang's rule seems to hinge on the constancy over a broad class of optical materials of the parameter g (see Equation 40), which is related to the kurtosis of the ground state electron distribution. Starting from Wang's rule, Boling, Glass, and Owyoung (1978) derived an empirical formula relating the nonlinear coefficient n_2 to the linear refractive index and to its dispersion. This formula seems to have broad validity in its ability to predict the nonlinear response of optical materials, as illustrated for example by the good level of agreement between the measured and calculated values of n_2 shown in Figure 4. However, a number of interesting and important questions involving the applicability of this model remain unresolved. One such question is what the limits of validity of this model are - that is, for what types of materials is the model valid? For example, one would not expect this model to be valid for conjugated polymers because of the dependence of α and γ on L, as predicted by Equations 17 and 18. Another question is why the model works as well as it does for the cases presented in Figure 4. Certainly the model is somewhat crude, being based on a long chain of approximations as described in the text. Especially puzzling in this regard is the assumption that the parameter g has the same value for different optical materials. For these reasons, it is somewhat surprising that the model works as well as it does. Still another open question is how the model of Boling, Glass and Owyoung relates to other models of the nonlinear optical response, such as that of Wherrett (1984) and of Sheik-Bahae, Hagan, and Van Stryland (1990). Perhaps questions of this sort can provide directions for future studies.

Acknowledgements

Discussions of the material presented in this chapter with G. S. Agarwal, N. F. Borrelli, E. Buckland, A. J. Glass, A. Owyoung, E. Van Stryland and D. Weidman are gratefully acknowledged. This work was supported by the United States National Science Foundation and the Army Research Office.

References

Adair R, Chase L L, and Payne S A, 1989, *Phys Rev B* **39** 3337.
Blau W J, Byrne H J, and Cardin D J, 1991, *Phys Rev Lett* **67** 1423;*ibid.*, 1992, **68** 2704.
Bloembergen N, Lotem H, and Lynch R T, 1978, *Indian Journal of Pure and Applied Physics* **16** 151.
Boling N L, Glass A J, and Owyoung A, 1978, *IEEE J Quantum Electron* **14** 601.
Boyd R W, 1992, *Nonlinear Optics* (Academic, Boston).
Ducuing F, 1977, in *Proceedings of the International School of Physics 'Enrico Fermi', Course LXIV*, ed Bloembergen N (North Holland, Amsterdam).
Jha S S and Bloembergen N, 1968, *Phys Rev* **171** 891.
Kajzar F, Taliani C, Danieli R, Rossini S, and Zamboni R, 1994, *Phys Rev Lett* **73** 1617.
Kemp D S, 1980, *Organic Chemistry* (Worth Publishers, New York).
Miles R B and Harris S E, 1973, *IEEE J Quantum Electron* **9** 470.
Miller R C, 1964, *Appl Phys Lett* **5** 17.

Orr B J and Ward J F, 1971, *Molecular Physics* **20** 513.

Prasad P N and Williams D J, 1991, *Introduction to Nonlinear Optical Effects in Molecules and Polymers* (Wiley, New York).

Rothberg L, 1987, in *Progress in Optics XXIV*, ed Wolf E (Elsevier, Amsterdam).

Rustagi K C and Ducuing J, 1974, *Opt Commun* **10** 258.

Sewell G L, 1949, *Proc Cam Phil Soc* **45** 678.

Sheik-Bahae M, Hagan D J, and Van Stryland E W, 1990, *Phys Rev Lett* **65** 96; see also, 1991, *IEEE J Quantum Electron* **27** 1296.

Wang C C, 1970, *Phys Rev B* **2** 2045.

Wherrett B S, 1984, *J Opt Soc Am B* **1** 67.

Wynne J J, 1969, *Phys Rev* **178** 1295.

Yariv A, 1975, *Quantum Electronics* (Wiley, New York).

Third-Order and Cascaded Nonlinearities

Eric W Van Stryland

Center for Research and Education in Optics and Lasers (CREOL)
University of Central Florida, USA

1 Introduction

Nonlinear optics (NLO) refers to any change in the optical properties of a material induced by light (see, for example, Harper and Wherrett 1977, Shen 1984 and Boyd 1992). A common example is the increased absorption seen in photochromic sunglasses when exposed to sunlight (this results from a reversible chemical reaction). Besides irradiance (I) dependent changes in the absorption coefficient, $\alpha(I)$, there can also be changes in index of refraction, $n(I)$, or even changes in scattering. Another common example is the index change resulting from thermal expansion induced by linear absorption. This effect is readily observed by dissolving a small amount of organic dye (e.g. food colouring) in water or other solvent and shining a HeNe laser beam through it. By blocking and unblocking the light beam, the transmitted beam observed at several feet away is seen to change size because of the thermally induced 'lens' formed in the solution. That is, the index at the centre of the beam, where the power per unit area (irradiance) is highest, is changed more than in the wings of the beam and a lens is formed due to the optical path length gradient. Subsequent free-space propagation of this phase distorted wavefront results in a change in the spatial irradiance distribution.

In this chapter we discuss a number of mechanisms that result in *nonlinear absorption* (NLA) or *nonlinear refraction* (NLR) or a combination of the two. Nonlinear materials have been studied for over 30 years. However, to date, there are few commercial or military applications other than for frequency conversion nonlinearities, which we only discuss here in terms of their contributions to NLA and NLR. One reason is that in general the materials have not performed up to requirements. In addition, there has been a lack of a reliable database for nonlinear coefficients. More characterisation has been needed to determine trends (*e.g.* from one material to another) and to determine scaling rules. Proper characterisation can lead to an understanding upon which theoretical models can be based and they, in turn, can lead to an ability to predict the nonlinear response of classes of materials. Feedback of this information to the materials scientists can help to optimise material parameters. Close cooperation between experiment, theory, and materials synthesis is needed for this field to progress.

The difficulty for a designer of devices based on nonlinear optical effects is exemplified by the data of Figure 1 which shows a plot of the two-photon absorption (2PA) coefficient, $\beta(\text{cm/GW})$ (an ultrafast nonlinearity), for GaAs on a semilogarithmic scale as a function of year published in the literature (Van Stryland and Chase 1994). The large differences shown in Figure 1 are not due to differences in the materials, but are due to experimental problems and errors in interpretation of data (Van Stryland *et al.* 1993). We will discuss the reasons for the discrepancies after introducing a formalism for discussing NLA and NLR. Clearly, there are a great number of problems that researchers may encounter in experimental measurements and modelling of nonlinear optical coefficients. This makes it even more difficult to determine trends from one material to another.

We divide this chapter into two main sections. The first concentrates on ultrafast third-order bound-electronic NLA and NLR and temporally cumulative free-carrier or excited-state nonlinearities. These excited-state nonlinearities can be understood in terms of a sequential cascading of lower order nonlinearities. In the second main section, we describe how frequency conversion nonlinearities can also lead to NLA and NLR through another cascading process, in this case, a *cascaded second-order nonlinearity*.

The first section also reviews linear dispersion relations and outlines the application of causality to NLO. We will see how NLR is intimately tied to NLA through Kramers-Krönig relations (Sheik-Bahae *et al.* 1991a, Bassini and Scandolo 1991, Hutchings *et al.* 1992). We will also discuss the difficulties in determining the dominant nonlinearities in a given experiment (Van Stryland *et al.* 1993). We will find that this depends on many parameters including wavelength, λ, pulsewidth, t_p, and focusing geometry, *e.g.* spot size, w_0. In general, several experiments are needed, varying these parameters, to determine nonlinear coefficients.

2 Odd order nonlinearities

We will see below that odd order nonlinearities lead directly to self interactions, in that light input at frequency ω leads to changes in absorption and changes in refraction at ω. Even order nonlinearities, in general, lead to new frequencies. We will see, however, in Section 3 that cascading of even order nonlinearities can also lead to self-action.

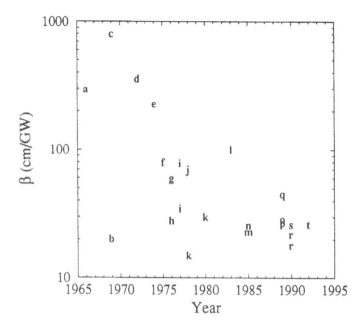

Figure 1. *The two-photon absorption coefficient β as a function of year published for GaAs. The letters refer to references as given in Van Stryland and Chase (1994).*

2.1 Formalism for 'third-order' nonlinearities

The discussions in this chapter deal with *nonlinear self-action*, *i.e.* NLA and NLR. Here, it will normally be assumed that we input a nearly monochromatic laser beam into a material and look for loss from NLA or lensing from NLR. As defined in most nonlinear optics texts, the *total material polarisation*, **P**, that drives the wave equation for the electric field, **E**, is (ignoring nonlocality) (Flytzanis 1975):

$$\mathbf{P}(t) = \varepsilon_0 \int_{-\infty}^{+\infty} \chi^{(1)}(t - t_1)\mathbf{E}(t_1)\,dt_1 + \int_{-\infty}^{+\infty}\int_{-\infty}^{+\infty} \chi^{(2)}(t - t_1, t - t_2)\mathbf{E}(t_1)\mathbf{E}(t_2)\,dt_1dt_2$$

$$+ \int_{-\infty}^{+\infty}\int_{-\infty}^{+\infty}\int_{-\infty}^{+\infty} \chi^{(3)}(t - t_1, t - t_2, t - t_3)\mathbf{E}(t_1)\mathbf{E}(t_2)\mathbf{E}(t_3)\,dt_1dt_2dt_3 + \dots \quad (1)$$

where $\chi^{(n)}$ is defined as the *n-th order time dependent response function* or *time dependent susceptibility*. Thus, the nonlinear polarisation is electric field dependent. Here, we take **E** as

$$|\mathbf{E}(t)| = \mathrm{Re}\{E(t)\,e^{i(kz-\omega t)}\} \quad (2)$$

with E a complex, slowly varying function of time. As an example, for harmonic generation (2nd harmonic from $\chi^{(2)}$ or 3rd from $\chi^{(3)}$) the nonlinearity, by necessity, follows the rapidly varying field. The only material response capable of this is the ultrafast bound-electronic response, *i.e.* the so-called 'instantaneous' response. Self-action effects, on the other hand, can be caused by a variety of nonlinearities with response times from 'instantaneous' for bound-electronic NLR in silica fibres, to seconds as for some photochromic effects (*e.g.* in photodarkening sunglasses). We will first look at the ultrafast nonlinearities and then examine slow or cumulative nonlinearities. A cumulative nonlinearity has a relatively slow response (*i.e.* memory) such that the effect builds to larger values during a laser pulse. Terms that can lead to self-action are the odd order susceptibilities since only these terms lead to oscillation at the input frequency. The lowest order nonlinear process comes from the third-order susceptibility, $\chi^{(3)}$.

As is usual, Equation 1 is Fourier transformed to give frequency dependent functions where the electric fields (now, strictly speaking, only functions of frequency, taken to be nearly monochromatic) are still allowed to be slowly varying functions of time (see, for example, Harper and Wherrett 1977, Shen 1984 and Boyd 1992). Similarly the polarisation is allowed to slowly follow the field envelope. For a single frequency input at ω, and looking only at self-action, this leads to a *slowly varying* polarisation at ω of

$$P(t) = \varepsilon_0 \chi E(t) = \varepsilon_0 \left[\chi^{(1)} + \chi^{(3)} \frac{|E|^2}{2} + \dots \right] E. \tag{3}$$

Here we are ignoring the usual and confusing degeneracy factors and the tensor and polarisation properties of these nonlinearities. Thus, for example, $\chi^{(3)} = \chi^{(3)}(\omega)$ is an effective nonlinear susceptibility, $\chi_{\text{eff}}^{(3)}$. Including such degeneracy factors, using the conventions of Hutchings *et al.* (1992), and looking only at self-action for linearly polarised light along x;

$$\chi_{\text{eff},x}^{(3)} = \frac{3}{2} \chi_{xxxx}^{(3)}(\omega\,;-\omega,\omega,\omega). \tag{4}$$

The frequency arguments of $\chi^{(3)}$ are written in a convention to show the input field frequencies leading to a polarisation at the first frequency argument, *i.e.* self-action.

In order to demonstrate how this nonlinearity leads to NLA and NLR, we return to the rapidly varying field and polarisation and insert this nonlinearity into the wave equation neglecting diffraction effects. The neglect of diffraction effects is equivalent to working in the 'external self-action' regime as discussed in the appendix to this chapter: see Kaplan(1969). The wave equation is then given by

$$\frac{d^2 E}{dz^2} - \frac{1}{c^2} \frac{d^2 E}{dt^2} = \mu_0 \frac{d^2 P}{dt^2}. \tag{5}$$

Keeping only the first and third-order susceptibilities and making the slowly varying envelope approximation on the amplitudes $E(z,t)$ and $P(z,t)$ (*e.g.* ignoring second order derivatives), leads to (Meystre and Sargent 1991)

$$\frac{dE}{dz} + \frac{1}{c} \frac{dE}{dt} = i \frac{\omega}{2\varepsilon_0 c} P, \tag{6}$$

where the amplitudes E and P can now be considered the same as the Fourier transformed, slowly varying amplitudes for nearly monochromatic inputs. Making a transformation of coordinates to travel with the wave, $\tau = t - z/c$ and $z' = z$ leads to the simplified equation (Meystre and Sargent 1991)

$$\frac{dE}{dz'} = i\frac{\omega}{2\varepsilon_0 c} P = i\frac{\omega}{2c}\left[\chi^{(1)} + \chi^{(3)}\frac{|E|^2}{2}\right]E, \tag{7}$$

where E and P are now functions of z' and τ. Defining the magnitude and phase of E by

$$E = E_0\, e^{i\phi}, \tag{8}$$

leads to

$$\frac{dE_0}{dz'} = -\frac{\omega}{2c}\left[\text{Im}\{\chi^{(1)}\} + \text{Im}\{\chi^{(3)}\}\frac{E_0^2}{2}\right]E_0 \tag{9}$$

and

$$\frac{d\phi}{dz'} = \frac{\omega}{2c}\left[\text{Re}\{\chi^{(1)}\} + \text{Re}\{\chi^{(3)}\}\frac{E_0^2}{2}\right]. \tag{10}$$

Given that the irradiance I is proportional to E_0^2, Equation 9 and 10 clearly show how the real and imaginary parts of $\chi^{(3)}$ lead to irradiance dependent phase shifts and loss respectively.

The 'i' in Equation 7 is important. Looking back at the wave equation (Equation 5), it shows the $\pi/2$ phase shift between polarisation and field. Thus, the polarisation can be viewed as having a real part which is in phase with the driving electric field leading to a change of field phase (Equation 10) and an imaginary part which leads to a change in the field amplitude (Equation 9) (index and absorption respectively).

Rewriting Equation 9 and 10 in terms of the irradiance, considering only the nonlinearly induced phase change Φ and including the linear phase change in the linear index n, results in,

$$\frac{dI}{dz'} = -\left[\frac{\omega}{2nc}\text{Im}\{\chi^{(1)}\}\right]I - \left[\frac{\omega}{n^2 c^2 \varepsilon_0}\text{Im}\{\chi^{(3)}\}\right]I^2 = -\alpha I - \beta I^2, \tag{11}$$

and with $k_0 = \omega/c$,

$$\frac{d\Phi}{dz'} = k_0\left[\frac{1}{2n^2 c\varepsilon_0}\text{Re}\{\chi^{(3)}\}\right]I = k_0 n_2 I, \tag{12}$$

where β is defined as the *two-photon absorption* (2PA) *coefficient* (in m/W) and n_2 is defined as the *nonlinear refractive index* (in m²/W). In the literature, n_2 is often used to discuss everything from thermal and reorientational nonlinearities (*e.g.* for CS$_2$), to changes in index due to saturation of absorption, to ultrafast $\chi^{(3)}$ nonlinearities. Here we restrict the use of n_2 to describe the ultrafast index change. Gaussian units (esu) are often used for n_2 and a useful relation is

$$n_2(\text{esu}) = \frac{cn}{40\pi}\, n_2(\text{SI}) \tag{13}$$

where the right hand side is all in MKS (SI).

Another way to remember how $\chi^{(3)}$ gives loss and/or phase shifts is to consider the linear index of refraction $n \simeq 1 + \text{Re}\{\chi^{(1)}\}/2$. Replacing $\chi^{(1)}$ by $\chi^{(1)} + \chi^{(3)}|E|^2/2$ gives the 'new' index. Similarly for loss, the linear absorption coefficient is given by $\alpha = k_0/2\,\text{Im}\{\chi^{(1)}\}$ which, with the same replacement of $\chi^{(1)}$, gives 2PA.

Another important piece of information is contained in this pair of equations (*i.e.* Equation 11 and 12). Namely it is known that the real and imaginary parts of the linear susceptibilities are connected through causality by the Kramers-Krönig relations and we should expect that there should be an analogous connection between the real and imaginary parts of the nonlinear susceptibility. We discuss these relations and the associated physics next.

2.2   Kramers-Krönig relations

Kramers-Krönig (KK) relations are dispersion relations relating the frequency dependent refraction, $n(\omega)$ to an integral over all frequencies of the absorption $\alpha(\omega)$ and vice-versa. An interesting way of viewing the necessity of these dispersion relations was given by Toll (1956) as shown in Figure 2a. A wave train (a), consisting of a superposition of many frequencies, arrives at an absorbing medium. If one frequency component (b) is completely absorbed we could naively expect that the output should be given by the difference between (a) and (b) as shown in (c). However, it can be seen that such an output would violate causality with an output signal occurring at times before the incident wave train arrives. In order for causality to be satisfied, the absorption of one frequency component must be accompanied by phase shifts of all of the remaining components in just such a fashion that, when the components are summed, zero output results for times before the arrival of the wave train. Such phase shifts result from the index of refraction and its dispersion.

The most frequent expression of the dispersion relation in optics relates the refractive index, n, to the absorption coefficient, α, integrated over all frequencies,

$$n(\omega) - 1 = \frac{c}{\pi} \, \wp \int\limits_{0}^{+\infty} \frac{\alpha(\Omega)}{\Omega^2 - \omega^2} \, d\Omega \tag{14}$$

where $\wp$ denotes the Cauchy principal value. We drop the $\wp$ notation in what follows for simplicity, although it should always be understood. It is this relation that is most commonly referred to when one speaks of KK relations in optics and is the original form of the relation as given by Kramers and Krönig.

Although dispersion relations for linear optics (*i.e.* KK relations) are well understood, confusion has existed about their applications to nonlinear optics. Clearly causality holds for nonlinear as well as linear systems. The question is, what form do the resulting dispersion relations take? For self-action NLO effects, a nondegenerate form of dispersion relation is needed where the nonlinearity of two frequency arguments (for example, the index change at ω due to the presence of a strong perturbing field at ω_e) is related to an integral over Ω of the NLA at Ω due to the presence of the same perturbing field at ω_e. Thus, both the NLA and the NLR are equivalent to pump-probe spectra with a fixed pump frequency and variable probe frequency. We will look at this relation in more detail after a description of the linear KK relations.

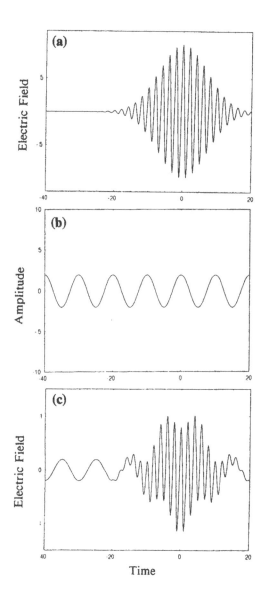

Figure 2. *Illustration of the connection between causality and dispersion relations. (a) temporal input, (b) single frequency absorption, (c) curves (a-b) = output as calculated without dispersion. This result violates causality.*

Linear Kramers-Krönig relations

In a dielectric medium, the linear optical polarisation, $\mathbf{P}(t)$ can be obtained from Equation 1 as,

$$\mathbf{P}(t) = \varepsilon_0 \int\limits_{-\infty}^{+\infty} \chi(\tau)\,\mathbf{E}(t - \tau)\,d\tau \tag{15}$$

The *response function*, $\chi(\tau)$, is equivalent to a Green's function, as it gives the response (polarisation) resulting from a delta function input (electric field). This equation is often stated in terms of its Fourier transform, where the convolution in Equation 15 is transformed into a product

$$P(\omega) = \chi(\omega)\,E(\omega) \tag{16}$$

where $\chi(\omega)$ is the susceptibility defined in terms of the response function as

$$\chi(\omega) = \int\limits_{-\infty}^{+\infty} \chi(\tau)\,e^{-i\omega\tau}\,d\tau\,. \tag{17}$$

(Note that, as defined, $\chi(\omega)$ and $\chi(\tau)$ are not exactly a Fourier transform pair because of a missing 2π). Causality states that the effect cannot precede the cause. In the above case, this requires that $\mathbf{E}(t - \tau)$ cannot contribute to $\mathbf{P}(t)$ for $t < (t - \tau)$. Therefore, in order to satisfy causality, $\chi(\tau) = 0$ for $\tau < 0$. An easy way to see this is to consider the response to a delta function $E(T) = E_0\,\delta(T)$, where the polarisation would then follow $\chi(t)$. This has important consequences for the relation between the susceptibility, $\chi(\omega)$, and the response function, $\chi(t)$, since the integration need be performed only for positive times. Therefore, the lower limit in the integral in Equation 15 can in general be replaced by zero.

The usual method for deriving the KK relation from this point is to consider a Cauchy integral in the complex frequency plane. However, in the Cauchy integral method, the physical principle from which dispersion relations result (namely causality) is not obvious. The principle of causality can be stated mathematically as

$$\chi(T) = \chi(T)\,\theta(T) \tag{18}$$

i.e. the response to an impulse at $t = 0$ must be zero for $t < 0$. Here $\theta(T)$ is the *Heaviside step function* defined as $\theta(T) = 1$ for $T > 0$ and $\theta(T) = 0$ for $T < 0$. (It is also possible to use the 'sign' function at this point or any other function that requires $\chi(T) = 0$ for $T < 0$.) On Fourier transforming this equation, the product in the time domain becomes a convolution in frequency space

$$
\begin{aligned}
\chi(\omega) &= \chi(\omega) * \left[\frac{\delta(\omega)}{2} + \frac{i}{2\pi\omega}\right] \\
&= \chi(\omega) + \frac{1}{2\pi}\int\limits_{-\infty}^{+\infty} \frac{\chi(\Omega)}{\omega - \Omega}\,d\Omega \\
&= \frac{1}{i\pi}\int\limits_{-\infty}^{+\infty} \frac{\chi(\Omega)}{\Omega - \omega}\,d\Omega\,,
\end{aligned}
\tag{19}
$$

which is the KK relation for the linear optical susceptibility. Thus, the KK relation is simply a restatement of the causality condition (Equation 18) in the frequency domain. Taking the real part with $\chi = \chi' + i\chi''$, we have,

$$\chi'(\omega) = \frac{1}{\pi} \int_{-\infty}^{+\infty} \frac{\chi''(\Omega)}{\Omega - \omega} d\Omega.$$

It is more usual to write the optical dispersion relations in terms of the more familiar quantities of refractive index, $n(\omega)$, and absorption coefficient, $\alpha(\omega)$. These relations are derived in Hutchings *et al.* (1992). If we assume dilute media with small absorption and indices, we can simply set $n - 1 = \chi'/2$ and $\alpha = \omega\chi''/c$. This results in

$$n(\omega) - 1 = \frac{c}{2\pi} \int_{-\infty}^{+\infty} \frac{\alpha(\Omega)}{\Omega - \omega} \frac{d\Omega}{\Omega}. \tag{20}$$

Since $\mathbf{E}(t)$ and $\mathbf{P}(t)$ are real, $n(-\omega) = n(\omega)$ and $\alpha(-\omega) = \alpha(\omega)$ which, when transforming the integral in Equation 20 from zero to infinity, gives the final result of Equation 14.

Nonlinear Kramers-Krönig formalism

Here we discuss the Kramers-Krönig relation used to calculate the change in refractive index from the change in absorption due to some external perturbation. The linear Kramers-Krönig relation can be applied both in the presence and in the absence of a perturbation, and the difference taken between the two cases. Doing this, we can write down a modified form of KK relation (which we also derive below specifically for an optical perturbation),

$$\Delta n(\omega\,;\zeta) = \frac{c}{\pi} \int_{0}^{+\infty} \frac{\Delta\alpha(\Omega\,;\zeta)}{\Omega^2 - \omega^2} d\Omega \tag{21}$$

where ζ denotes the perturbation. An equivalent relation also exists whereby the change in absorption coefficient can be calculated from the change in the refractive index, but this is rarely used for reasons described below. Note that it is essential that the perturbation be independent of the frequency of observation, Ω, in the integral (*i.e.* the excitation must be held constant).

This form of calculation of the refractive index for nonlinear optics is often more useful than the analogous linear optics relation, as absorption changes (which can be either calculated or measured) usually occur only over a limited frequency range and, thus, the integral in Equation 21 need be calculated only over this finite frequency range. In comparison, for the linear KK calculation, absorption spectra tend to cover a very large frequency range and it is necessary to take account of this full range in order to obtain a quantitative fit for the dispersion, although a qualitative fit to the dispersion can often be obtained using a limited frequency range. Unfortunately, the converse is not true, as refractive index changes are usually quite extensive in frequency, so a calculation of absorption changes from refractive index changes is seldom performed.

The nonlinear susceptibility can be determined by integration over positive times only

$$\chi^{(n)}(\omega_1, \omega_2, \ldots, \omega_n) = \int\limits_0^{+\infty} d\tau_1 \ldots \int\limits_0^{+\infty} d\tau_n \, \chi^{(n)}(\tau_1, \ldots, \tau_n) \, e^{-i(\omega_1 \tau_1 + \omega_2 \tau_2 + \ldots + \omega_n \tau_n)} \quad (22)$$

It is now possible to use the method used earlier for the linear susceptibility in order to derive a dispersion relation for the nonlinear susceptibility. For example, we can write

$$\chi^{(n)}(\tau_1, \tau_2, \ldots, \tau_n) = \chi^{(n)}(\tau_1, \tau_2, \ldots, \tau_n) \, \theta(\tau_j) \quad (23)$$

and then calculate the Fourier transform of this equation. Here j can apply to any one of the indices $1, 2, \ldots, n$. We could also use any number and combination of step functions; however, the simplest result is obtained by taking just one.

We can thus obtain the generalised nonlinear KK relation for a non-degenerate nonlinear susceptibility

$$\chi^{(n)}(\omega_1, \omega_2, \ldots, \omega_n) = \frac{1}{i\pi} \int\limits_{-\infty}^{+\infty} \frac{\chi^{(n)}(\omega_1, \omega_2, \ldots, \Omega, \ldots, \omega_n)}{\omega_j - \Omega} \, d\Omega. \quad (24)$$

Specifically, for the third-order susceptibility, $\chi^{(3)}$, and choosing $j=1$,

$$\chi^{(3)}(\omega_1, \omega_2, \omega_3) = \frac{1}{i\pi} \int\limits_{-\infty}^{+\infty} \frac{\chi^{(3)}(\Omega, \omega_2, \omega_3)}{\omega_1 - \Omega} \, d\Omega. \quad (25)$$

Note here that the integral is over only one frequency argument, Ω: all other frequencies are held constant. Thus, we cannot obtain any relationship between the degenerate Kerr coefficient, $n_2(\omega)$, and the degenerate two-photon absorption coefficient, $\beta(\omega)$.

Using an analogous definition for the nondegenerate n_2 and β, defined by Equations 11 and 12, leads to;

$$n_2(\omega\,;\omega_e) = \frac{1}{n^2 c \varepsilon_0} \operatorname{Re}\{\chi^{(3)}(\omega\,; -\omega_e, \omega_e, \omega)\} \quad (26)$$

and

$$\beta(\omega\,;\omega_e) = \frac{2\omega}{c^2 n^2 \varepsilon_0} \operatorname{Im}\{\chi^{(3)}(\omega\,; -\omega_e, \omega_e, \omega)\}. \quad (27)$$

As discussed in Sheik-Bahae et al. (1992, 1994), there is a factor of 2 difference between the degenerate and nondegenerate definitions. This accounts for cross-modulation or 'beating' terms that result for the nondegenerate case for nonlinearities that can respond fast enough. This is sometimes referred to as *weak-wave retardation* (Van Stryland et al. 1982).

In addition, we must mention that the definition of the nondegenerate β now includes all possible nonlinear mechanisms including 2PA, Raman and AC-Stark effects, as discussed in Sheik-Bahae et al. (1991a). In most situations of interest to us, the 2PA dominates both in NLA and in its contribution to n_2.

We can relate these quantities by choosing $\omega_1 = -\omega_2$ with $\omega_2 = \omega_e$ to be the perturbing field frequency (or excitation field) in Equation 25 and convert to an integral from zero to infinity to give

$$\text{Re}\{\chi^{(3)}(\omega\,;-\omega_e,\omega_e,\omega)\} = \frac{2}{\pi} \int\limits_0^{+\infty} \frac{\Omega\,\text{Im}\{\chi^{(3)}(\Omega\,;-\omega_e,\omega_e,\Omega)\}}{\Omega^2 - \omega^2}\,d\Omega\,, \tag{28}$$

which, with the definitions of Equations 26 and 27, leads to

$$n_2(\omega\,;\omega_e) = \frac{c}{\pi} \int\limits_0^{+\infty} \frac{\beta(\Omega\,;\omega_e)}{\Omega^2 - \omega^2}\,d\Omega\,. \tag{29}$$

Although the calculation as illustrated above gives the nondegenerate NLR for a specific pair of frequencies, in most cases we set $\omega = \omega_e$ (after the integral is performed) and consider self-refraction. This gives the degenerate n_2, *i.e.* what is commonly referred to as *the* n_2 (we note again that there is a factor of 2 difference related to 'weak-wave retardation').

Now we can interpret the external perturbation in Equation 21 to be an electromagnetic field. The index change Δn is $n_2 I_e$ and the absorption change is βI_e, where I_e is the irradiance of the excitation beam, *i.e.* the beam that alters the optical properties of the material. Note that this perturbation is constant on both sides of the relation and must not vary as the integral is computed. Thus, a linear KK calculation is being performed on a new system consisting of the material plus an electromagnetic field of fixed frequency and irradiance. This is a significant step, as we have extended the use of the modified linear relation, as given in Equation 21, for resonant optical nonlinearities to the nonresonant case where there is no real excited intermediate state. However, by viewing the nonlinear optical process in terms of virtual excited states, this step is clear.

These nondegenerate forms of the NLR and NLA are equivalent to pump-probe spectra, in that $\Delta n(\omega\,;\omega_e)$ or $n_2(\omega\,;\omega_e)$ and $\Delta\alpha(\omega\,;\omega_e)$ or $\beta(\omega\,;\omega_e)$ describe the change in refractive index and absorption coefficient, respectively, for a weak optical probe of frequency ω when a strong pump of fixed frequency ω_e is applied.

The cause need not be of optical origin, but can be any external perturbation. For example, this method has been used to calculate the refractive index change resulting from an excited electron-hole plasma (Miller *et al.* 1981) and a thermal shift of the band edge (Wherrett *et al.* 1986). For cases where an electron-hole plasma is injected, the subsequent change of absorption gives the plasma contribution to the refractive index. In this case, the ζ parameter in Equation 21 is taken as the change in plasma density, regardless of the mechanism of generation of the plasma or the pump frequency. Van Vechten and Aspnes (1969) obtained the low frequency limit of n_2 from a similar KK transformation of the *Franz-Keldysh electro-absorption effect* where, in this case, ζ is the DC field.

It is important to note that the integral with $\omega_e = \Omega$ (*i.e.* the degenerate case) does not obey causality. This has been a source of confusion in considering saturation of a two-level atomic system, as discussed below (Meystre and Sargent 1991, Yariv 1975).

Example 1: Two-level atom

The 'two-level atom' problem is a familiar one and is illustrative of how the nonlinear Kramers-Krönig relation can be applied. The absorption spectrum is given by (Meystre and Sargent 1991);

$$
\begin{aligned}
\alpha(\omega) &= \alpha_0 \frac{\gamma^2}{(\omega - \omega_0)^2 + \gamma^2} \frac{1}{1 + \frac{I}{I_s}\frac{\gamma^2}{(\omega-\omega_0)^2+\gamma^2}} \\
&= \alpha_0 \frac{\gamma^2}{(\omega - \omega_0)^2 + \gamma^2(1 + I/I_s)},
\end{aligned}
\tag{30}
$$

and is shown in Figure 3a. As noted, for example in Yariv's text (Yariv 1975), this does not obey causality! The problem is that as ω is tuned, the excitation is tuned, and the saturation changed, for this *degenerate* form of the absorption. What is needed is the nondegenerate, or pump-probe, absorption spectrum shown in Figure 3b;

$$
\begin{aligned}
\alpha(\omega\,;\Omega) &= \alpha_0 \frac{\gamma^2}{(\omega - \omega_0)^2 + \gamma^2} \frac{1}{1 + \frac{I}{I_s}\frac{\gamma^2}{(\Omega-\omega_0)^2+\gamma^2}} \\
&= \alpha_0(\omega) \frac{1}{1 + \frac{I}{I_s}\frac{\gamma^2}{(\Omega-\omega_0)^2+\gamma^2}}.
\end{aligned}
\tag{31}
$$

Here Ω is fixed and α clearly obeys causality since now the absorption spectrum is identical in shape to the linear absorption, but simply reduced in amplitude by homogeneous saturation. (Note that we have ignored population pulsations and other more exotic effects in this oversimplified description of the 2-level atom) (Meystre and Sargent 1991).

Example 2: NLA and NLR in semiconductors

Here we examine data taken on a variety of materials, primarily semiconductors, and compare it with the predictions of a relatively simple 2-band model. It would be best to make this comparison with the nonlinear spectra for a given material. Unfortunately, there are few materials for which nonlinear spectra are known. One reason for this is that tuneable sources with the required irradiance, pulsewidth and beam quality are not typically available. Instead, we use simple scaling relations predicted by the 2-band model to scale the material dependence. Wherrett (1984) has shown that the third-order nonlinear susceptibiltiy $\chi^{(3)}$ in inorganic solids should scale as

$$
\chi^{(3)} \simeq \frac{1}{E_g^4} f(\hbar\omega/E_g),
\tag{32}
$$

where the complex function f depends only on the ratio $\hbar\omega/E_g$ (*i.e.* upon which states are optically coupled). Thus;

$$
\beta(\hbar\omega/E_g) \propto \frac{\hbar\omega}{n_0^2} \operatorname{Im}\{\chi^{(3)}\} \propto \frac{1}{n_0^2 E_g^3} \frac{\hbar\omega}{E_g} \operatorname{Im}\{f(\hbar\omega/E_g)\} \propto \frac{1}{n_0^2 E_g^3} F(\hbar\omega/E_g)
\tag{33}
$$

and

$$
n_2(\hbar\omega/E_g) \propto \frac{1}{n_0^2} \operatorname{Re}\{\chi^{(3)}\} \propto \frac{1}{n_0^2 E_g^4} \operatorname{Re}\{f(\hbar\omega/E_g)\} = \frac{1}{n_0^2 E_g^4} G(\hbar\omega/E_g)
\tag{34}
$$

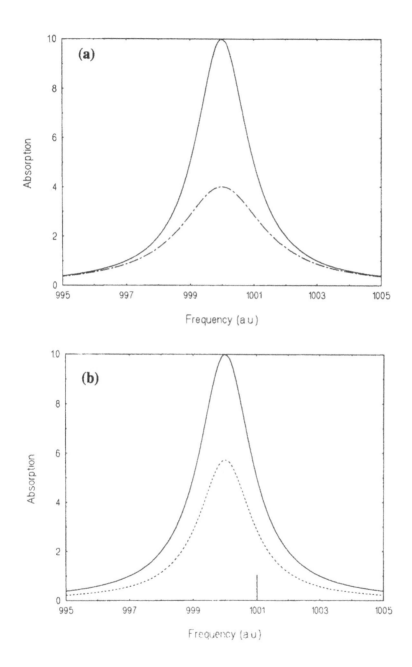

Figure 3. *(a) (Solid line) The normalised low irradiance absorption spectrum of a 2-level atom. (Dot-dash line) The absorption as seen by a tuneable saturating laser source. (b) (Solid line) same as in (a). (Dashed line) The absorption as seen by a weak tuneable probe beam in the presence of a saturating beam fixed at frequency 1001. In both (a) and (b) $I/I_{sat} = 1.5$ and the FWHM of the transition is 1.*

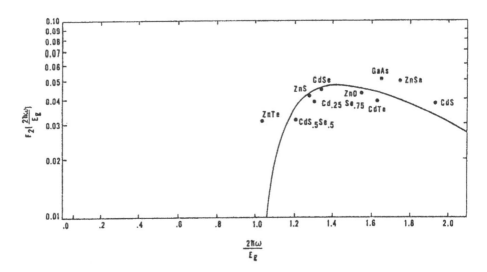

Figure 4. *The solid line is the two-parabolic band prediction for the function F plotted as a function of $2\hbar\omega/E_g$ using $K=3100$ in Equation 35. The data, scaled according to Equation 35, are from Van Stryland et al. (1985,1988).*

where the defined functions F and G are band structure dependent. Thus, F gives the 2PA spectrum and G gives the dispersion of n_2. One method to test the above scaling relations is to scale the experimental data to obtain the experimental functions

$$F^{\text{expt}}(\hbar\omega/E_g) = \frac{1}{K\sqrt{E_p}}\, n_0^2\, E_g^3\, \beta^{\text{expt}} \tag{35}$$

and

$$G^{\text{expt}}(\hbar\omega/E_g) = \frac{1}{K'\sqrt{E_p}}\, n_0\, E_g^4\, n_2^{\text{expt}} \tag{36}$$

where β^{expt} and n_2^{expt} are experimental values of β and n_2 and K and K' are proportionality constants. Here E_p is the *Kane energy* and is nearly material independent with a value near 21eV (Wherrett 1984). Figures 4 and 5 plot these scaled data versus $\hbar\omega/E_g$, along with the predicted dependence from a two-parabolic band model using a value of $K=3100$ in units such that E_p and E_g are in eV and β is in cm/GW (Van Stryland et al. 1985). The value of $K'=0.94\times10^{-8}$ is determined from the Kramers-Krönig integral of the NLA spectrum using the above value for K (Sheik-Bahae et al. 1991a).

Figure 5 shows a small, positive, nearly dispersionless n_2 for $\hbar\omega/E_g$ much less than E_g, reaching a peak near $E_g/2$ (where 2PA turns on) and then decreases, reaching negative values as $\hbar\omega$ approaches the band edge. This curve is reminiscent of the behaviour of the linear index in a solid which has its peak value at the band edge, where linear absorption turns on, and then rapidly turns down toward smaller values as $\hbar\omega$ increases. Just as the linear index n is related to the linear absorption through Kramers-Krönig relations, so the nonlinear index is related to the NLA.

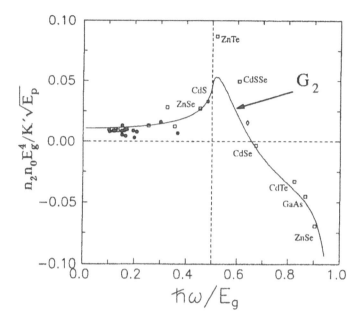

Figure 5. *A plot of n_2 data scaled according to Equation 36. The circles are measurements in Adair et al. (1987), the diamond is from Ross et al. (1990), and the squares are Z-scan measurements from Sheik-Bahae et al. (1991a). We have labelled the semiconductor data. The solid line is the function $G(\hbar\omega/E_g)$ derived here for a two-band model of a semiconductor using the 2PA data to give K'.*

We find that the general trends in the data displayed in Figures 4 and 5 are well described using the simplest possible band structure, *i.e.* two-parabolic bands. The solid line in Figure 4 comes from a calculation of the transition rate for 2PA using such a band structure. Performing a Kramers-Krönig transformation on the NLA calculated using this band structure gives the solid line of Figure 5. While there are deviations from these curves of up to factors of 3, in general there is surprisingly good agreement considering the range of materials and differences in band-gap energies (from 0.2 to 10eV). Using the calculated spectral responses, we can compare the range of values of β and n_2 for the different materials studied by replotting the scaled data on a log-log plot versus E_g as in Figures 6 and 7 (*i.e.* dividing out the respective frequency dependences of the nonlinearities). This shows the E_g^{-3} dependence of 2PA in Figure 6 and the E_g^{-4} dependence of n_2 in Figure 7, revealing more than four orders-of-magnitude change in n_2.

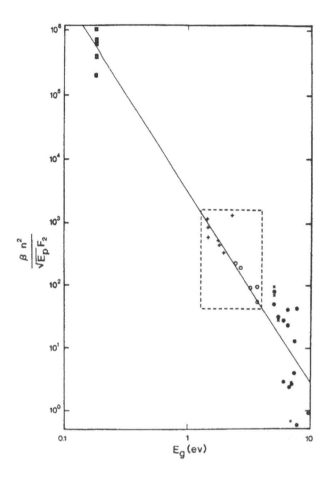

Figure 6. *A log-log plot of the scaled* 2PA *coefficient* β *as a function of the band-gap energy* E_g *(in eV) (data scaled from Equation 35). The straight line is a fit to the data within the dashed box from Van Stryland et al. (1985) for a line of fixed slope -3. (Figure taken from Van Stryland et al. 1988.)*

2.3 Difficulties in Determining Nonlinearities

It is illustrative to look briefly at why there exists such large discrepancies for 2PA coefficients of GaAs. The following list gives some of the problems.

1. Poor laser output properties and/or characterisation.
2. Too long pulsewidths used, such that other nonlinearities dominate.
3. Nonlinear refraction always accompanies nonlinear absorption.

The first problem was true for early work, but at least for 1.06μm Nd:YAG lasers,

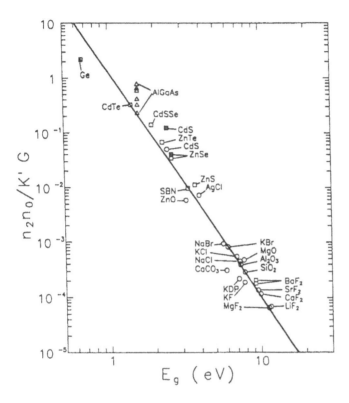

Figure 7. *A log-log plot of scaled n_2 data as a function of E_g (data scaled from Equation 36). The data (circles) are from Adair et al. (1987) using nearly degenerate three-wave mixing, (squares) are Z-scan measurements from Sheik-Bahae et al. (1991a) (solid squares at 0.532 μm, open squares at 1.06 μm and half-shaded squares at 10.6 μm), and the AlGaAs data (triangles) are taken from LaGasse et al. (1990). These data are now scaled by the dispersion function G. The solid straight line has a slope of -4.*

these problems have been minimised in recent years. Problems 2 and 3 are related. In semiconductors such as GaAs, the 2PA process creates electrons and holes which can subsequently linearly absorb light. This free-carrier absorption actually dominates the original 2PA for pulses of nanosecond duration in GaAs. In addition, these carriers alter the linear refractive index, lowering it for excitation below the band edge - as is the case for 2PA experiments. The lowered refractive index leads to self-defocusing of the input beams which usually have a near Gaussian spatial profile. In very thick samples, this can lead to a reduction of irradiance; however, most researchers have used relatively thin samples such that the defocusing only occurs outside the sample (the *external self-*

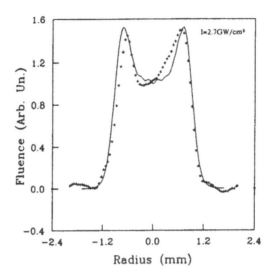

Figure 8. *The transverse beam profile in the near field of an initially Gaussian-shaped picosecond 532 nm pulse after transmission through a ZnSe sample showing self-defocusing.*

action regime—see the appendix at the end of this chapter). What actually happens in experiments, is that it becomes extremely difficult to collect all the transmitted energy after the sample (*i.e.* the energy can simply be missed by the detector if its area is too small). Figure 8 shows the beginnings of such defocusing for picosecond pulses in ZnSe. This leads to an apparent large loss and an overestimate of the 2PA coefficient as seen for most of the data in Figure 1. This free-carrier defocusing is greatly enhanced for longer pulses for the same reasons that free-carrier absorption is enhanced for long pulses. The greater energy for a given irradiance creates more carriers. The statement of the third problem that NLR *always* accompanies NLA concerns more than just free-carrier absorption. As mentioned earlier, NLR and NLA are related through causality.

The solution to the three problems mentioned after they have been identified is relatively simple; use smooth beam profiles, *e.g.* TEM$_{0,0}$, carefully characterise the output, use short pulses (for most semiconductors in the near infrared and visible spectral regions, 30ps is short enough to nearly eliminate free-carrier absorption effects, and the carrier defocusing is reduced to a manageable level), carefully collect all the transmitted light (*e.g.* place a large area detector directly at the back of the sample), and be sure to use samples short enough and irradiance low enough to be in the external self-action regime.

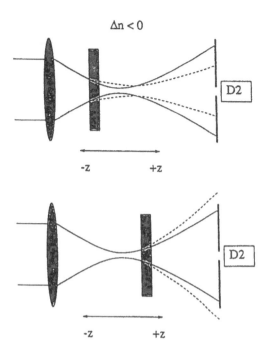

Figure 9. *The Z-scan (aperture) experimental apparatus showing linear (solid line) and nonlinear (dashed line) propagation for a self-defocusing nonlinearity. The EZ-scan apparatus replaces the apertures with disks and uses large area detectors.*

2.4 Z-Scan techniques

In order to describe the competing free-carrier nonlinearites we first describe experimental techniques for measuring NLA and NLR and then describe how to interpret such measurements. This interpretation is valid for other techniques that measure absorptive changes and index changes. We first describe the use of these techniques for measuring NLR. We then describe their use for measuring NLA, and finally describe how NLR can be measured in the presence of NLA. Using a single Gaussian laser beam in a tight focus geometry, as depicted in Figure 9, we measure the transmittance of a nonlinear medium through a finite aperture (Z-scan (Sheik-Bahae *et al.* 1989a,b)) or around an obscuration disk (EZ-scan (Xia *et al.* 1994, Van Stryland *et al.* 1994)), both positioned in the far field, as a function of the sample position Z measured with respect to the focal plane. The following example qualitatively describes how such data (Z-scan or EZ-scan) are related to the NLR of the sample.

Assume, for example, a material with a positive nonlinear refractive index. Starting the scan from a distance far away from the focus (negative Z), the beam irradiance is low and negligible NLR occurs; hence, the transmittance remains relatively constant. The transmittance here is normalised to unity as shown in Figure 10. As the sample is brought closer to focus, the beam irradiance increases leading to self-focusing in the sample. This positive NLR moves the focal point closer to the lens leading to a

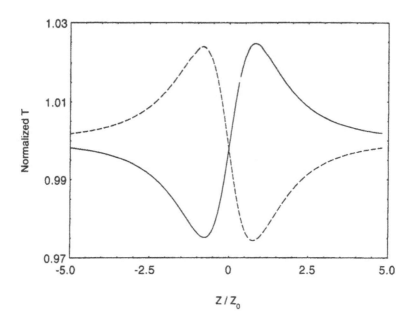

Figure 10. *Z-scan signal (i.e. normalised output transmittance versus Z for a self-focusing medium (solid line) and self-defocusing medium (dashed line)) for $\Delta\Phi_0 = \pm 0.25$.*

larger divergence in the far field. Thus, the transmittance is reduced. Moving the sample to behind the focus, the self-focusing helps to collimate the beam increasing the transmittance of the aperture. Scanning the sample farther toward the detector returns the normalised transmittance to unity. Thus, the valley followed by peak signal shown in Figure 10 is indicative of positive NLR, while a peak followed by valley shows self-defocusing.

The EZ-scan can be described in nearly identical terms except that we monitor the complementary information of what light leaks past the obscuration disk, or eclipsing disk. Since, in the far field, the largest fractional changes in irradiance occur in the wings of a Gaussian beam, the EZ-scan can be more than an order-of-magnitude more sensitive than the Z-scan. Figure 11 demonstrates the remarkable, interferometric sensitivity of this method by showing a signal of peak optical path length change $\lambda/2200$ ($\Delta\Phi_0 = 2\pi/2200$, where $\Delta\Phi_0$ is defined as the *temporally integrated peak-on-axis phase shift*). The EZ-scan signal shows self-focusing in BK-7 glass at 532nm with 30ps pulses of approximately 50nJ per pulse (energy turned down to display the signal-to-noise ratio). Note the inverted signature for self-focusing since what is transmitted by an aperture is blocked by a disk.

It is an extremely useful feature of the Z-scan (or EZ-scan) method that the sign of the nonlinear index is immediately obvious from the data. In addition, the methods

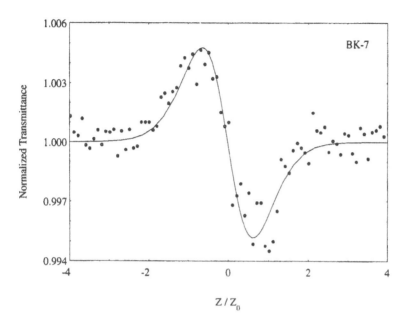

Figure 11. *EZ-scan of a 1.5 mm thick BK-7 sample performed with a frequency-doubled psec Nd:YAG laser. The solid line is a fit to the data, indicating a peak wavefront distortion of $\lambda/2200$. (Figure taken from Xia et al. 1994.)*

are sensitive and simple single beam techniques. We can define an easily measurable quantity ΔT_{pv} as the difference between the normalised peak and valley transmittance: $T_p - T_v$. The variation of ΔT_{pv} is found to be linearly dependent on $\Delta \Phi_0$ (for a bound-electronic n_2, $\Delta \Phi_0$ involves a temporal integral of Equation 12 (Sheik-Bahae *et al.* 1989b). For example, in a Z-scan using an aperture with a transmittance of $S{=}40\%$;

$$\Delta T_{pv} \simeq 0.36 \, |\Delta \Phi_0| \qquad \text{for} \quad \Delta T_{pv} < 1. \qquad (37)$$

With experimental apparatus and data acquisition systems capable of resolving transmission changes $\Delta T_{pv} \simeq 1\%$, Z-scan is sensitive to less than $\lambda/225$ wavefront distortion (*i.e.* $\Delta \Phi_0 = 2\pi/225$). The Z-scan has a demonstrated sensitivity to a nonlinearly induced optical path length change of nearly $\lambda/10^3$, while the EZ-scan has shown a sensitivity of $\lambda/10^4$.

In the above picture we assumed a purely refractive nonlinearity with no absorptive nonlinearities (such as multiphoton or saturation of absorption). Qualitatively, multiphoton absorption suppresses the peak and enhances the valley, while saturation produces the opposite effect. If NLA and NLR are simultaneously present, a numerical fit to the data can extract both the nonlinear refractive and absorptive coefficients. The NLA leads to a symmetric response about Z=0, while the NLR leads to an asymmetric response (if ΔT_{pv} is not too large), so that the fitting is unambiguous. In addition, noting that the sensitivity to NLR in a Z-scan is entirely due to the aperture, removal of the aperture completely eliminates the effect. In this case, the Z-scan is only sensitive to NLA. Nonlinear absorption coefficients can be extracted from such 'open' aperture experiments. A further division of the apertured Z-scan (referred to as 'closed aperture' Z-scan) data by the open aperture Z-scan data gives a curve that, for small nonlinearities, is purely refractive in nature (Sheik-Bahae et al. 1989b). In this way we can have separate measurements of the absorptive and refractive nonlinearities without the need of computer fits with the Z-scan. Figure 12 shows such a set of Z-scans for ZnSe. Separation of these effects without numerical fitting for the EZ-scan is more complicated.

The single beam Z-scan can be modified to give nondegenerate nonlinearities by focusing two collinear beams of different frequencies into the material and monitoring only one of the frequencies (different polarisations can be used for degenerate frequencies) (Sheik-Bahae et al. 1992). This '2-colour Z-scan' can separately time resolve NLR and NLA by introducing a temporal delay in the path of one of the input beams. This method is particularly useful to separate the competing effects of ultrafast and cumulative nonlinearities.

2.5   Higher order nonlinearities in semiconductors

Discrepancies in measured values of 2PA coefficients in GaAs (see Figure 1) and other semiconductors can be understood better by looking at the irradiance dependence of NLR for picosecond pulses using the Z-scan technique. Figure 13 shows the index change divided by the input irradiance, I_0, in ZnSe at 532nm, where it exhibits 2PA, as a function of I_0. The index change is calculated from the measured $\Delta \Phi_0$. For a purely third-order response, $\Delta n = n_2 I_0$ and this figure would show a horizontal line. The slope of the line shown in Figure 13 shows a fifth-order response, while the intercept gives n_2, in this case the negative (i.e. defocusing) bound-electronic n_2 at 532nm.

The physical interpretation of this fifth-order response is the defocusing caused by the carriers generated by 2PA. This is described well by combining this effect with the ultrafast n_2 in Equation 12 to give

$$\frac{\mathrm{d}\Phi}{\mathrm{d}z'} = kn_2 I + k\sigma_r N \,, \tag{38}$$

the second term showing NLR from photogenerated carriers of density N (the k in the second term is sometimes dropped to give the refractive cross section σ_r in units of cm^2 rather than cm^3 as we do later for molecular NLR). Assuming 2PA is the only mechanism for generating carriers, and neglecting decay within the pulse, the *carrier generation rate* is given by,

$$\frac{\mathrm{d}N}{\mathrm{d}t} = \frac{\beta I^2}{2\hbar\omega} \,. \tag{39}$$

This interpretation is also consistent with degenerate four-wave mixing measurements (Canto-Said et al. 1991) and time-resolved 2-colour Z-scan measurements shown in Figure 14 (Wang et al. 1994). This figure shows the NLR as well as the NLA of ZnSe when excited by 532nm light and probed at $1.06\mu m$. The 2PA generated carriers both defocus and absorb the $1\mu m$ light long after the pump pulse is over. The signals near zero delay also have large contributions from the ultrafast nonlinearities. This is a good

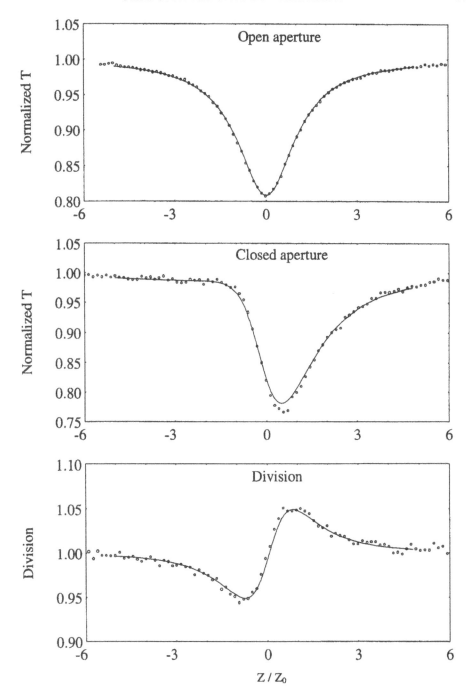

Figure 12. *Normalised Z-scan transmittance of ZnSe measured using picosecond pulses at λ=532nm. (a) Open aperture data and fit (solid line) (b) 40% aperture data and fit (solid line) and (c) the result of the division of the Z-scans of (b) by (a).*

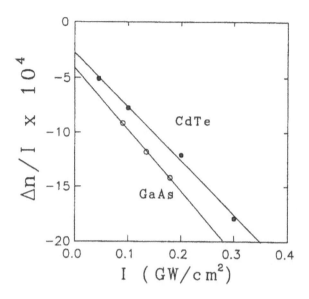

Figure 13. $\Delta n/I_0$ *directly derived from* ΔT_{pv} *plotted as a function of* I_0 *for GaAs (open circles) and CdTe (filled circles). (Figure taken from Said et al. 1992.)*

example which shows the importance of performing several different types of experiment on a given material. Analysis of only the single beam Z-scan measurements is open to different interpretations. Other semiconductors show similar behaviour for excitation above the 2PA edge.

Organic materials can also show similar behaviour (Said *et al.* 1994). Figure 15 shows $\Delta n/I_0$ versus I_0 for BBTDOT along with a plot of $\Delta\alpha/I_0$ versus I_0. BBTDOT is a solution of a bisbenzethiozole-substituted thiophene compound (see inset for chemical structure). Here the corresponding nonlinearities to the free-carrier effects are excited-state refraction (ESR) and excited-state absorption (ESA), where the excited states are created via 2PA.

These higher order nonlinearities observed in semiconductor and organic materials give some indication of the importance of the careful characterisation needed to interpret the measured nonlinear loss and phase. An even more striking caution for interpreting third-order nonlinearities is examined in the next section.

2.6   Third-order nonlinearities: ultrafast or cumulative

As discussed in the previous section, carriers or excited states can lead to NLA and NLR. How they are generated is unimportant. If the carriers or excited states are created by 2PA, the resulting nonlinearities are fifth order, *i.e.* an effective $\chi^{(5)}$ (Said *et al.* 1992). Depending on the absorption spectra, these states can also be created by *linear*

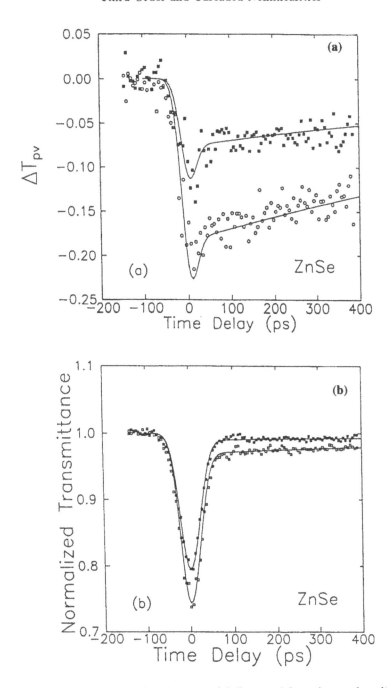

Figure 14. *Temporally resolved Z-scans. (a) Temporal dependence of nonlinear absorption as determined from open aperture Z-scan. (b) Temporal dependence of nonlinear refraction as determined from closed aperture Z-scan divided by open aperture data. (Figure taken from Wang et al. 1994.)*

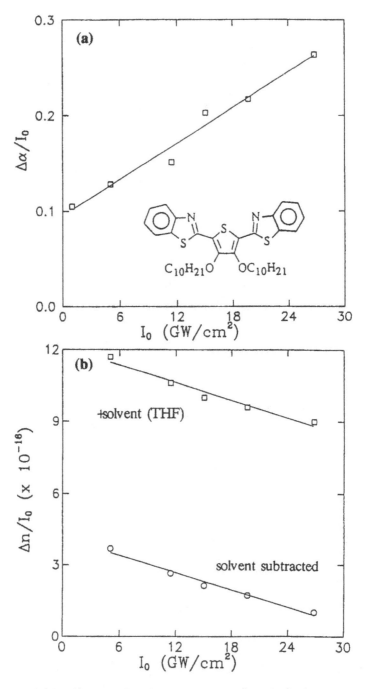

Figure 15. *(a) $\Delta\alpha/I_0$ vs I_0 for the* BBTDOT *sample with the least squares fit (solid line), (b) $\Delta n/I_0$ vs I_0 with solvent n_2 (and glass cell n_2) present and subtracted along with fits (solid lines). (Figure taken from Said et al. 1994.)*

absorption where, neglecting decay within the pulse (Wei *et al.* 1992),

$$\frac{dN}{dt} = \frac{\alpha I}{\hbar \omega}. \qquad (40)$$

Concentrating on molecular nonlinearities, we refer to these nonlinearities as excited-state nonlinearities, ESA and ESR, as defined above. Assuming that depletion of the ground state can be ignored (*i.e.* no saturation),

$$\frac{dI}{dz'} = -\alpha I - \sigma N I. \qquad (41)$$

By temporal integration of Equations 40 and 41, we find,

$$\frac{dF}{dz'} = -\alpha F - \frac{\alpha \sigma}{2\hbar \omega} F^2, \qquad (42)$$

where F is the *fluence* (*i.e.* energy per unit area). This equation is exactly analogous to Equation 11 which describes 2PA, except that the irradiance is replaced by the fluence and the 2PA coefficient, β, is replaced by $\alpha \sigma / 2\hbar \omega$. Thus experiments such as Z-scan will monitor a third-order nonlinear response that could easily be mistaken for 2PA. However, there must be some linear absorption present, however small, for ESA to take place. Two-photon absorption does not require linear loss. Unfortunately this is not enough to differentiate the processes as there can be linear absorption present in 2PA materials unrelated to the NLA process, *e.g.* from impurities or other absorbing levels. A temporally resolved measurement, such as degenerate four-wave mixing (DFWM) or time-resolved Z-scan, would also show the excited-state lifetime assuming the pulsewidth was short compared to the excited state lifetime. Another way to determine the mechanism is to measure the nonlinear response for different input pulsewidths, again assuming the pulses can be made shorter than the excited state lifetime. Figure 16 shows this measurement performed on a solution containing chloro-aluminium phthalocyanine (CAP). While the energy in the pulses was held fixed, the irradiance was changed by a factor of two by changing the pulsewidth. The nonlinear transmittance remained the same in the open aperture Z-scans clearly indicating that the NLA is fluence rather than irradiance dependent and, therefore must be described by a real state population, *i.e.* ESA. In CAP, at a wavelength of 532nm, the ESA cross section σ is considerably larger than the ground-state cross section leading to α. This type of absorber is referred to as a *reverse-saturable absorber* since the absorption increases with increasing input. Such effects are useful in optical limiting (Hagan *et al.* 1993). For large inputs the ground state can become depleted reducing the overall NLA.

Associated with the ESA is ESR, as given by the second term in Equation 38, which is simply due to the redistribution of population from ground to excited state. This is analogous to the index change in a laser from gain saturation which leads to frequency pulling of the cavity modes (Meystre and Sargent 1991). Ground state absorbers are being removed and excited state absorbers are being added. Depending on the spectral position of the input with respect to the peak linear and peak excited-state absorption, the NLR can be of either sign. For reverse saturable absorbing materials, the NLR is most likely controlled by the addition of excited-state absorbers and their spectrum, since the cross section is larger. Thus N is determined by Equation 40. Figure 16 also shows the NLR in CAP for two different pulsewidths demonstrating that it is also fluence dependent and thus, dependent on real state populations.

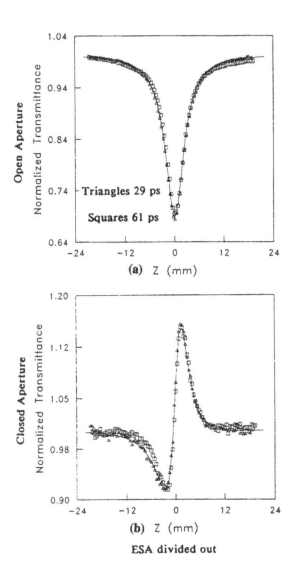

Figure 16. (a) Open aperture Z-scans for 29ps (squares) and 61ps (triangles) pulsewidths at an incident energy of 1.16μJ in CAP. (b) The results of the division of the closed aperture Z-scan data by the open aperture Z-scan data of (a) for 29ps (squares) and 61ps (triangles) pulsewidths. (Figure taken from Wei et al. 1992.)

2.7 Interpretation

It should now be clear that the interpretation of NLA and NLR measurements is fraught with many pitfalls. Great care must be taken. In extensive studies of a wide variety of materials, we have found that there is seldom a single nonlinear process occurring. Often several processes occur simultaneously, sometimes in unison, sometimes competing. It is necessary to experimentally distinguish and separate these processes in order to understand and model the interaction. There are a variety of methods and techniques for determining the nonlinear optical response, each with its own weaknesses and advantages. In general, it is advisable to use as many complementary techniques as possible over a broad spectral range in order to unambiguously determine the active nonlinearities.

Numerous techniques are known for measurements of NLR and NLA in condensed matter. Nonlinear interferometry, degenerate four-wave mixing (DFWM), nearly-degenerate three-wave mixing, ellipse rotation, beam distortion, beam deflection, and third-harmonic generation, are among the techniques frequently reported for direct or indirect determination of NLR (see Chase and Van Stryland 1994). Z-scan is capable of separately measuring NLA and NLR (Sheik-Bahae *et al.* 1989b). Other techniques for measuring NLA include transmittance, calorimetry, photoacoustic, and excite-probe methods (see Van Stryland and Chase 1994).

2.8 Cascading of odd-order nonlinearities

We started our discussion of nonlinear optics with an expansion of the polarisation to various orders of the electric field. We found that the lowest order self nonlinearities come from the third-order term of proportionality constant $\chi^{(3)}$. This term well describes two-photon absorption and ultrafast NLR (described by n_2). However, we also found that a linear absorption process followed by a second linear absorption (or an analogous index of refraction change) also leads to a third-order response. The question here is, are both processes, ultrafast and cumulative, described well by $\chi^{(3)}$.

For the bound-electronic response, two optical pulses with the same electric field, but different pulsewidths, result in the same polarisation P (see, for example, Equation 3). This statement is also approximately true if the pulsewidths of both pulses are much longer than the material response time contained in the susceptibilities in Equation 1. However, the time integrals allow for memory. Thus, slower responding materials can have their nonlinear response build-up with time. For example, in CS_2, the molecules can be reoriented due to the difference in polarisability along the long and short axes of the molecule which changes the macroscopic index as seen by the incoming light. For pulses shorter than the $\simeq 2ps$ reorientational response time, this build-up can be measured. Or, for a thermal nonlinearity, the light heats the material through linear absorption ($\mathrm{Im}\{\chi^{(1)}\}$) which changes the linear index ($\mathrm{Re}\{\chi^{(1)}\}$).

Fourier transformation of Equation 1 results in the usually quoted frequency dependent susceptibility $\chi^{(n)}(\omega_1, \omega_2 \ldots \omega_3)$. Memory, which was previously explicitly included in the response function, is lost in the dispersion. Thus, irradiance, I, and fluence, F, dependences are treated equally. This can lead to confusion as the two processes (ultrafast and cumulative) are indistinguishable for pulses long compared to relevant relaxation processes.

Such processes as reorientational, electrostrictive, thermal, saturation and excited-state nonlinearities can also be thought of as two step processes, or cascaded $\chi^{(1)} : \chi^{(1)}$ nonlinearities (see thermal nonlinearity above). For example, for ESA, light first induces a transition to the excited state (an $\text{Im}\{\chi^{(1)}\}$ process) and then the excited state absorbs (a second $\text{Im}\{\chi^{(1)}\}$ process), $i.e.$ two $linear$ absorption processes.

For these types of slow cumulative nonlinearities, the irradiance (or field) may no longer be the important input parameter. For ESA nonlinearities, 1GW/cm^2 for a picosecond pulse could give the same $\Delta\Phi_0$ as 1kW/cm^2 for a microsecond pulse. In any case, quoting β or n_2 without a clear explanation of the physical mechanisms can be misleading. It makes sense to describe the ultrafast process of 2PA by $\chi^{(3)}$ or β (sometimes α_2 is also used in analogy to n_2), and the cumulative process of ESA by an absorption cross section. An analogous statement applies to the corresponding refractive processes.

2.9   Applications

For applications, knowing whether a quoted β is due to 2PA or ESA can be extremely important. Below we briefly discuss two application areas, optical limiting and all-optical switching. We specifically give an example for optical limiting showing how this differentiation is a necessity.

Optical limiting

Effective optical limiting devices have high transmittance for low inputs and an output that becomes constant for high inputs (Leite et $al.$ 1967, Soileau et $al.$ 1983). Such devices can be useful for sensor protection. Let us look at a hypothetical example where, for simplicity, no spatial or temporal integrations are performed. We are given a 'β' in the literature of 100cm/GW and a linear absorption coefficient of 0.1cm^{-1}. The measurements were performed with 10 ns laser pulses and the authors quote the energy and irradiance so that we can calculate the fluence and the length. If this is a 2PA coefficient, we expect a transmittance of $T_{\text{NL}}=0.83$ at $I=1\text{MW/cm}^2$ - where the linear transmittance is $T_{\text{L}}=0.90$. If on the other hand, the loss is from ESA, with a fluence of 10^{-2}J/cm^2, we find $\sigma_e = 2.0 \times 10^{-17}\text{cm}^2$ assuming $\sigma_g = 1.0 \times 10^{-18}\text{cm}^2$, $N_g = 10^{17}\text{cm}^{-3}$ and $\hbar\omega = 10^{-19}\text{J}$ ($i.e.$ for the above values, the losses from 2PA or from ESA are assumed the same). However, if the experiment is repeated for $1\mu\text{s}$ pulses with the same irradiance, we would see the same loss for 2PA, but low transmittance ($\simeq 0.14$) for ESA - since the fluence is now 1J/cm^2. Similarly, if 100ps pulses are used with the same irradiance, the ESA loss becomes negligible ($F=10^{-4}\text{J/cm}^2$). Simply stated, 2PA is irradiance dependent and ESA is fluence dependent.

From these arguments, we see that if we want to perform optical limiting of picosecond pulses, 2PA may be more useful (assuming normal values of β from 1–100cm/GW). If we need to limit longer pulses from nanoseconds to microseconds, ESA may be more useful. Short pulses can be quite effectively limited by 2PA and combinations of 2PA and carrier absorption or ESA and carrier refraction or ESR. Such processes in semiconductors have been shown to work well (Van Stryland *et al.* 1985,1988).

All-optical switching

Following the work of Stegeman (Mizrahi *et al.* 1989, DeLong *et al.* 1989), a useful figure of merit (FOM) can be given for all-optical switching devices. The basic requirement is to obtain a significant phase shift for devices such as interferometers to switch from high to low transmittance or vice-versa, before there is significant loss. If no NLA is present, and looking only at instantaneous nonlinearities, the FOM is

$$\text{FOM} = \frac{k_0 \Delta n}{\alpha} = k_0 \frac{n_2 I}{\alpha}. \tag{43}$$

If 2PA is present the FOM becomes

$$\text{FOM} = \frac{k_0 \Delta n}{\Delta \alpha} = k_0 \frac{n_2}{\beta}. \tag{44}$$

For certain devices both these FOM's need to be larger than $\simeq 2\pi$. Figure 17 shows the FOM for semiconductors in the 2PA regime. Note that the ratio of two experimental nonlinear coefficients is compared to the ratio as determined from a two-band model and using KK relations, and the agreement is excellent. Unfortunately, this figure shows that semiconductors are not very useful within the frequency range where they exhibit 2PA. This has led to the development of devices operating just below the 2PA edge (Al-hemyari *et al.* 1993).

3 Introduction to second-order cascading

The previous section discussed cascading of odd-order nonlinearities to obtain NLA and NLR. Here we discuss a cascading of $\chi^{(2)} : \chi^{(2)}$ to also obtain effects similar to NLA and NLR. However, while this cascading can lead to an effective $\chi^{(3)}$ for low irradiance, at high irradiance the nonlinearities deviate strongly from third order. Previously it had been recognised that $\chi^{(2)}$ could contribute to third-order nonlinearities (Armstrong *et al.* 1962), and this effect was included in many of the expansions for the nonlinear polarisation (Flytzani and Bloembergen 1976, Karamzin and Sukhorukov 1975, Belan-shenkov *et al.* 1989). However, it was not predicted that this would lead to large effects. With significant advances in $\chi^{(2)}$ materials (see, for example, Marder *et al.* 1994) as well as the development of a deeper understanding of this nonlinearity, it is now established that this effect can lead to record large effective nonlinearities, larger than the best available from non-resonant $\chi^{(3)}$'s (Stegeman *et al.* 1993).

We first describe the regime where this cascaded effect resembles a third-order response by looking at the 'small depletion limit'. We then look at the more interesting new effects when significant fundamental beam depletion occurs (large SHG efficiency) by examining numerical solutions to the SHG equations derived below.

The physical process whereby second-order effects can lead to NLA and NLR are straightforward, and the analogy to third-order nonlinearities clear. Here we present arguments only for second-harmonic generation (SHG), but in general any parametric

Eric W Van Stryland

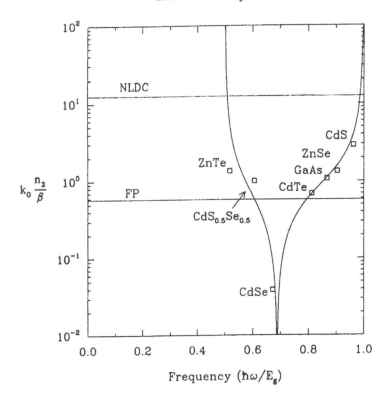

Figure 17. *The theoretical figure-of-merit, $k_0|n_2|/\beta$ (along with data for the materials indicated) as a function of frequency. Also shown by horizontal lines are the minimum limits needed for all-optical switching in a Fabry-Perot device or a Nonlinear Directional Coupler. (Taken from Sheik-Bahae et al. 1991a.)*

process can lead to similar effects. The nonlinear loss for SHG is analogous to two-photon absorption, as two fundamental photons are annihilated to produce a single second-harmonic photon. For 2PA, two fundamental photons are annihilated to produce excited states which eventually lead to fluorescence and/or heat. The end result for the fundamental is the same. An important difference for the SHG process is that it is reversible, *i.e.* downconversion from the second harmonic back to the fundamental is possible. The NLR on the fundamental comes about from the combination of SHG and downconversion. For example, for an input fundamental beam tuned off phase matching, the energy in the fundamental oscillates between fundamental and second harmonic every coherence length (see oscillations of the fundamental in Figure 21). However, while in the form of second harmonic it travels at a different phase velocity (since the indices are not matched). Thus, when it is down converted to the fundamental it has a shifted phase. This phase shift is analogous to the shift of phase from a third order nonlinearity described by n_2.

3.1 Small depletion limit

The above statements can be put on a mathematical footing by looking at the equations for SHG derived below which follow closely the derivation of third-order nonlinearities in Section 2.1.

In analogy to Equation 2, we define the field for the fundamental (at ω) and second harmonic (at 2ω) as,

$$
\begin{aligned}
|\mathbf{E}_\omega(t)| &= \mathrm{Re}\{E_\omega(t)\exp[i(k_\omega z - \omega t)]\} \\
|\mathbf{E}_{2\omega}(t)| &= \mathrm{Re}\{E_{2\omega}(t)\exp[i(k_{2\omega} z - 2\omega t)]\}
\end{aligned}
\tag{45}
$$

and the polarisations as,

$$
\begin{aligned}
|\mathbf{P}_\omega(t)| &= \mathrm{Re}\{P_\omega(t)\exp[i(k_\omega^{\mathrm{NL}} z - \omega t)]\} \\
|\mathbf{P}_{2\omega}(t)| &= \mathrm{Re}\{P_{2\omega}(t)\exp[i(k_{2\omega}^{\mathrm{NL}} z - 2\omega t)]\}
\end{aligned}
\tag{46}
$$

where k^{NL} is determined from the nonlinear interaction. For the quasi- monochromatic, slowly varying polarisations, we have

$$
\begin{aligned}
P_{2\omega}(t) &= \frac{\varepsilon_0}{2}\chi^{(2)}(2\omega;\omega,\omega)E_\omega^2 && \text{and} \quad k_{2\omega}^{\mathrm{NL}} = 2k_\omega\,, \\
P_\omega(t) &= \frac{\varepsilon_0}{2}\chi^{(2)}(\omega;-\omega,2\omega)2E_{2\omega}E_\omega^* && \text{and} \quad k_\omega^{\mathrm{NL}} = k_{2\omega} - k_\omega\,,
\end{aligned}
\tag{47}
$$

where $\chi^{(2)}(\omega;-\omega,2\omega) = \chi^{(2)}(2\omega;\omega,\omega)$ for real $\chi^{(2)}$ by the Manley Rowe relations (energy conservation between the ω and 2ω pulses).

We substitute these definitions for $\mathbf{E}$ and $\mathbf{P}$ into the wave equation below including linear index n_ω for the fundamental and $n_{2\omega}$ for the second harmonic and ignore linear absorption;

$$
\frac{d^2\mathbf{E}}{dz^2} - \frac{n^2}{c^2}\frac{d^2\mathbf{E}}{dt^2} = \mu_0\frac{d^2\mathbf{P}}{dt^2}\,.
\tag{48}
$$

After making the slowly varying envelope approximation (*e.g.* ignoring second derivatives) and transforming to a coordinate system moving with the pulse, this gives,

$$
\begin{aligned}
\frac{dE_{2\omega}}{dz'} &= i\frac{2\omega}{2n_{2\omega}\varepsilon_0 c}P_{2\omega}\exp[i(k_{2\omega}^{\mathrm{NL}} - k_{2\omega})z'] \\
&= i\frac{\omega}{2n_{2\omega}c}\chi^{(2)}(2\omega;\omega,\omega)E_\omega^2\exp[-i\Delta k z']\,,
\end{aligned}
\tag{49}
$$

$$
\begin{aligned}
\frac{dE_\omega}{dz'} &= i\frac{\omega}{2n_\omega\varepsilon_0 c}P_\omega\exp[i(k_\omega^{\mathrm{NL}} - k_\omega)z'] \\
&= i\frac{\omega}{2n_\omega c}\chi^{(2)}(\omega;-\omega,2\omega)E_{2\omega}E_\omega^*\exp[i\Delta k z']\,,
\end{aligned}
\tag{50}
$$

where $\Delta k = k_{2\omega} - 2k_\omega = 2\omega[n_{2\omega} - n_\omega]/c$ is the wave vector mismatch. Here we will not be concerned with the tensorial nature of $\chi^{(2)}$ and will limit the discussion to type I phase matching (*i.e.* the fundamental is polarised along a principle axis). In terms of the commonly used $d_{\mathrm{eff}} = \chi^{(2)}/2$, and dropping the prime on z,

$$
\frac{dE_{2\omega}}{dz} = i\frac{\omega}{n_{2\omega}c}d_{\mathrm{eff}}E_\omega^2\exp[-i\Delta k z]\,,
\tag{51}
$$

$$
\frac{dE_\omega}{dz} = i\frac{\omega}{n_\omega c}d_{\mathrm{eff}}E_{2\omega}E_\omega^*\exp[i\Delta k z]\,.
\tag{52}
$$

Eric W Van Stryland

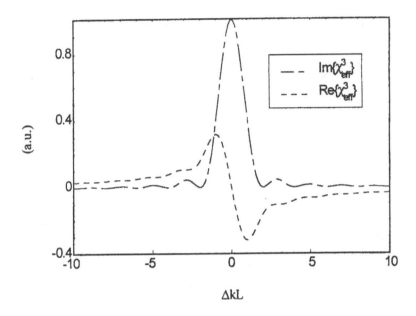

Figure 18. *Real and imaginary parts of the third-order susceptibility as calculated in the small depletion limit.*

We can show how these equations lead to loss and phase shifts by solving them in the small depletion limit (*i.e.* the energy at 2ω is small compared to the energy at ω). Integrating Equation 51 in this limit, assuming no initial second harmonic, gives

$$E_{2\omega}(z) = i\frac{\omega}{n_{2\omega}c} d_{\text{eff}} \exp\left[-i\Delta kz/2\right] \text{sinc}\left(\frac{\Delta kz}{2}\right) z\, E_{\omega}^2. \tag{53}$$

Substituting this into Equation 52 gives for the fundamental;

$$\frac{dE_{\omega}(z)}{dz} = -\frac{\omega^2 d_{\text{eff}}^2}{n_{\omega}n_{2\omega}c^2} \exp\left[i\Delta kz/2\right] \text{sinc}\left(\frac{\Delta kz}{2}\right) z\, |E_{\omega}|^2\, E_{\omega}, \tag{54}$$

which when rewritten in terms of real and imaginary parts yields,

$$\frac{dE_{\omega}(z)}{dz} = i\frac{\omega^2 d_{\text{eff}}^2}{n_{\omega}n_{2\omega}c^2} \text{sinc}\left(\frac{\Delta kz}{2}\right)\left[-\sin\left(\frac{\Delta kz}{2}\right) + i\cos\left(\frac{\Delta kz}{2}\right)\right] z\, |E_{\omega}|^2\, E_{\omega}. \tag{55}$$

Remembering the equation for a third-order, $\chi^{(3)}$ nonlinearity (see Equation 7),

$$\frac{dE}{dz} = i\frac{\omega}{4nc}\chi^{(3)}|E|^2\, E \tag{56}$$

shows the direct analogy of the cascaded second-order nonlinear effect to that of the third-order nonlinearity. However, as the coefficient of $|E_{\omega}|^2 E_{\omega}$ depends on z, we must integrate along z to define an effective $\chi^{(3)}$. Of course we do the analogous operation for the true third-order nonlinearity with the same assumption that $|E_\omega|^2 E_\omega$ is constant. This integration is important in that it shows the phase building up with z and no oscillatory behaviour. We can thus define an effective $\chi^{(3)}$ as

$$\chi_{\text{eff}}^{(3)} = \frac{2\omega d_{\text{eff}}^2}{n_{2\omega}c} \left\{ \frac{2}{\Delta k} \left[\text{sinc}\,(\Delta kz) - 1 \right] + i \left[z\,\text{sinc}^2(\Delta kz/2) \right] \right\}. \tag{57}$$

The imaginary part is related to loss and the real part shows a phase shift as discussed for $\chi^{(3)}$ nonlinearities (see Equations 11 and 12). Figure 18 shows the real and imaginary parts of $\chi_{\text{eff}}^{(3)}$. This gives an effective 2PA coefficient and effective n_2 for a sample of length L as;

$$\beta_{\text{eff}} = \frac{\omega}{n_\omega^2 c^2 \varepsilon_0} \text{Im}\{\chi^{(3)}\} = \frac{2\omega^2 d_{\text{eff}}^2}{n_\omega^2 n_{2\omega} c^3 \varepsilon_0} \text{sinc}^2\left(\frac{\Delta kL}{2}\right) L \tag{58}$$

$$n_2^{\text{eff}} = \frac{1}{2n_\omega^2 c \varepsilon_0} \text{Re}\{\chi^{(3)}\} = \frac{2\omega d_{\text{eff}}^2}{n_\omega^2 n_{2\omega} c^2 \varepsilon_0} \frac{[\text{sinc}\,(\Delta kL) - 1]}{\Delta kL} L \tag{59}$$

from Equations 11 and 12 respectively. Note that these processes are proportional to the usual figure-of-merit for $\chi^{(2)}$ materials, namely d_{eff}^2/n^3. Also, for a given value of phase mismatch, longer samples will give larger effective nonlinearities.

Here it is important to distinguish between phase shift, Φ, and index change. There is not an index change associated with the phase shifts from second-order nonlinearities since another beam, passing through the sample at the same position at the same time, will not in general be affected. It would need to couple to the other beam via a nonzero $\chi^{(2)}$ tensor component as well as exchange energy (*i.e.* the light must travel some distance as a 2ω beam in order to accumulate a phase shift). This usually will require very special circumstances. Note that either sign of phase shift can occur (*i.e.* either self-focusing or self-defocusing) depending on the sign of the phase mismatch.

Figure 19 shows the loss and phase shift induced on a 1.06μm beam by a KTP crystal as a function of the phase mismatch for a fixed input irradiance (DeSalvo *et al.* 1992). The depletion curve, Figure 19a, shows the typical sinc squared 'Maker fringe' pattern for the transmitted pulse energy. The fit is a numerical solution to Equations 51 and 52 including averages over the temporal Gaussian picosecond pulses and spatially Gaussian shaped beam profile. Figure 19b shows the induced phase shift as measured by Z-scan (*i.e.* $\Delta\Phi_0 \propto \Delta T_{\text{pv}}$). These two curves qualitatively agree with the shapes of the curves in Figure 18 in the small depletion limit. Both sets of curves are reminiscent of the loss and phase shift expected from a two-photon absorbing state as a function of frequency (here $k = n\omega/c$) as determined by Kramers-Krönig relations discussed previously. Here the 2PA resonance is replaced by the energy exchange to the second harmonic which is maximised at $\Delta kL = 0$.

Physically the loss and phase shift can be looked at as the cascading of $\chi^{(2)}(\omega\,; -\omega, 2\omega) : \chi^{(2)}(2\omega\,; \omega, \omega)$ corresponding to the d_{eff}^2 term in Equation 55, *i.e.* ω is first upconverted (doubled) to 2ω and then downconverted to ω with a phase that depends on ΔkL. It is instructive to look at how the phase of this downconverted second-harmonic light affects the fundamental. On phase match, $\Delta kL=0$ and the i in each of Equations 51 and 52 corresponds to a $\pi/2$ phase shift, thus upconversion

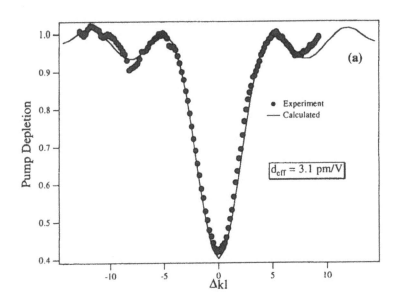

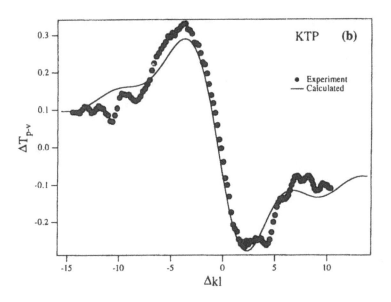

Figure 19. *(a) The normalised fundamental pump depletion and (b) the phase shift (proportional to ΔT_{pv}) induced on a 1.06μm beam by a* KTP *crystal as a function of the phase mismatch for a fixed input irradiance. (Figure taken from DeSalvo et al. 1992.)*

followed by downconversion results in a total phase shift of π, corresponding to loss of the fundamental (a minus sign upon downconversion). Overall these $\pi/2$ phase shifts lead to energy conservation of the combination of ω and 2ω light. Off phase match the picture gets more complex. With $\Delta k L > 0$ such that $n_{2\omega} > n_\omega$, there is an additional phase shift of the downconverted light from the loss term at π which leads to an in-quadrature component. This component causes self-focusing for $\Delta k L < 0$ and self-defocusing for $\Delta k L > 0$. Note that this phase shift is not the nonlinear phase change, Φ, which is also determined by *how much* '2ω' is downconverted. The larger the depletion of the fundamental, the more energy there is to be downconverted with a phase shift, and the more the possible net phase shift induced on the fundamental beam. Thus, both energy conversion and phase mismatch are needed in this case (no initial second harmonic) for self-refraction effects.

Away from perfect phase matching ($\Delta k L \neq 0$), the 2ω irradiance oscillates every two 'coherence' lengths $2L_c = 2\pi/\Delta k$ (*i.e.* the ω and 2ω beams add an extra π phase difference due to linear dispersion every L_c). When the second harmonic (developed by $\chi^{(2)}$) is allowed to act back on E_ω the cascading process is introduced, and $\chi^{(3)}$-like effects are seen on the fundamental (see Figure 21 for the numerical results for the fundamental depletion versus z). As the second harmonic periodically goes to zero with propagation along z, the interaction can be terminated with no loss. This is very different from $\chi^{(3)}$ materials exhibiting 2PA.

The nonlinear refraction is maximised in the small depletion limit for $\Delta k L \simeq \pi$ where the loss to second harmonic is small. This gives a maximum n_2^{eff} of

$$n_2^{\text{eff}} \simeq -\frac{2\omega d_{\text{eff}}^2 L}{\pi n_\omega^2 n_{2\omega} c^2 \varepsilon_0} \tag{60}$$

showing how n_2^{eff} is proportional to the material length L leading to an accumulated nonlinear phase shift proportional to L^2. For high $\chi^{(2)}$ materials this nonlinear refractive index can be very large. For example, N-(4-Nitrophenyl)-(L)-prolinol, (NPP) with a $d_{\text{eff}} \simeq 80 \text{pM/V}$, has a maximum $n_2^{\text{eff}} \simeq 10^{-10} \text{cm}^2/\text{W}$ ($\simeq 10^{-7}$ esu) for a 1cm length. Clearly keeping the beam together for long distances will lead to larger nonlinearities so that waveguiding becomes important. It is with waveguided NLO devices where it is expected that this NLO process will find the most applications.

3.2 Numerical results

The simple effective third-order description of cascading breaks down as the small depletion limit is violated. While there are more sophisticated analytical models that can faithfully describe the nonlinearities for significantly higher inputs, numerical solutions are relatively quick to perform and can easily incorporate the spatial and temporal averaging usually needed to interpret experimental results. We rely on these in the remainder of this chapter. Examination of these full numerical solutions shows some other very interesting features and demonstrates why, in general, the cascaded process cannot simply be described as a third-order response.

Figure 20 shows the effects of increasing irradiance on the phase mismatch dependence of the fundamental loss and phase shift. As the irradiance is increased, the

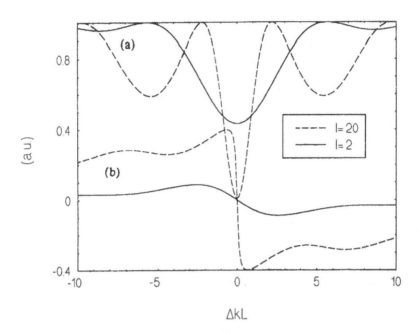

Figure 20. ΔkL *dependence of the (a) depletion and (b) phase shift for two different values of irradiance.*

phase shift effectively 'detunes' the SHG process farther from phase match so that the Maker fringe pattern squeezes in like an accordion. This in turn sharpens up the phase shift curve near $\Delta kL=0$ to a nearly step function-like behaviour. Looking at the z dependence of the loss and phase shift, we see that Φ exhibits 'jumps' as large as $\pi/2$ near $\Delta kL=0$ as shown in Figure 21. These jumps occur when the energy is contained primarily in the 2ω beam where large downconversion is taking place into a strongly depleted fundamental. The behaviour appears similar in a plot of the loss and phase shift versus irradiance, but an apparent saturation of the nonlinear phase shift occurs at very high irradiance as shown in Figure 22. For the limit of high irradiance, the variation of the phase shift with input goes as $\sqrt{I}$ or as the input electric field. Such a dependence can be obtained analytically as in the following example given by Kaplan (1993). Here it is assumed that there is an initial $E_{2\omega}(z=0)$ and solutions are sought where the irradiances at ω and 2ω are constant, *i.e.* an eigensolution of Equations 51 and 52. By defining $E_\omega = \sqrt{I_\omega}e^{i\phi_1}$ and $E_{2\omega} = \sqrt{I_{2\omega}}e^{i\phi_2}$, the result is,

$$I_\omega = 2I_{2\omega} \quad \text{and} \quad \phi_2(z) - \phi_1(z) = 0$$

$$\frac{d\Phi}{dz} = -\text{constant} \times |E_\omega|\,, \tag{61}$$

i.e. a field dependent phase shift rather than the I dependence of a third-order effect. This solution is obtained for $\Delta kL = 0$!

Because SHG is a coherent phenomenon, *i.e.* depends on phase mismatch, as opposed

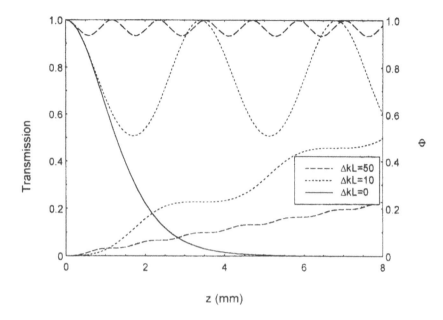

Figure 21. *The z dependence of the depletion and phase shift as calculated by numerical integration of Equations 51 and 52 for three values of ΔkL. For $\Delta kL = 0$, $\Phi = 0$.*

to third-order self-action nonlinear effects which are automatically phase matched, different effects can be seen. It is important to remember that the processes of up and downconversion are continuous processes occurring simultaneously along z. It is the net effect which gives the fields at any z. Thus, the nonlinear phase shift, Φ, can continually grow with propagation length. This is illustrated by the above example. In general the effects of 'seeding' the SHG process (*i.e.* starting with $E_2(z=0) \neq 0$) lead to a new means of producing nonlinear phase shifts. This problem was first examined in the original work of Armstrong *et al.* (1962).

Figure 23 shows the results of an experiment where 1% of the energy of the fundamental was converted to 2ω (in a separate SHG crystal) to seed the SHG process in KTP (3% of the fundamental irradiance) (Hagan *et al.* 1994). The relative phase of the seed was varied and the change in fundamental transmittance for KTP was maximised by tuning the phase mismatch to obtain these results. At an input of $20\mathrm{GW/cm^2}$ the transmittance change upon a π phase shift of the seed was maximised for $\Delta kL \simeq -1.3$. Note that as the process depends on d_{eff}^2, changing from KTP $d_{\mathrm{eff}}=3.1\mathrm{pM/V}$ to NPP will lower this irradiance to $\simeq 30\mathrm{MW/cm^2}$.

How the phase of the seed beam changes the SHG interaction is seen by comparison of Figures 24 and 25 where we plot the transmittance and nonlinear phase shift versus z for $\Delta\phi_0 = 0$, $\pi/2$ and π in Figure 24 and for $\Delta kL = -1.1$, 0 and 1.1 in Figure 25. Apparently the effects of the seed are analogous to changing the phase-mismatch condition.

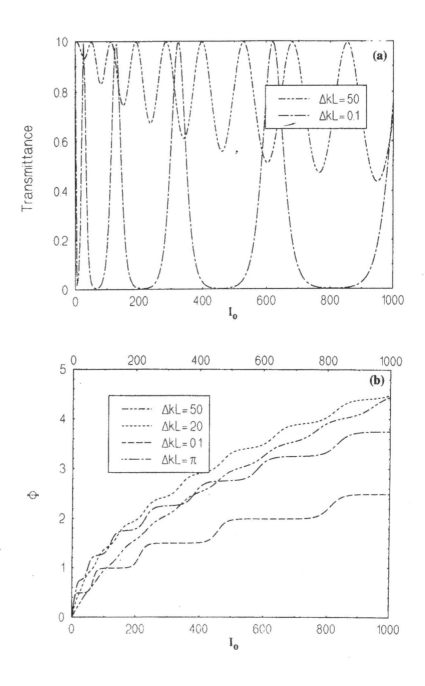

Figure 22. *The irradiance dependence of the (a) depletion and (b) phase shift for different values of ΔkL.*

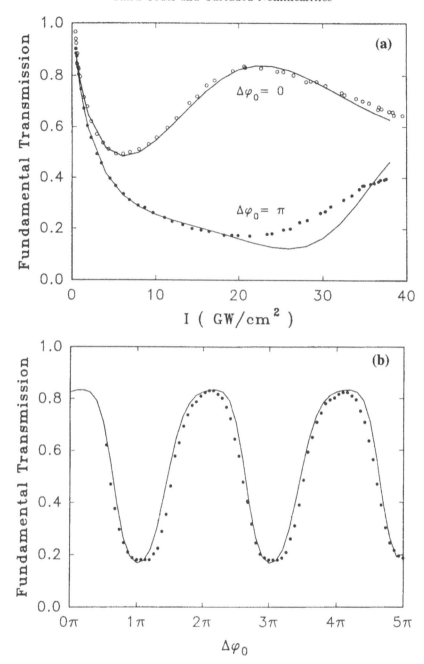

Figure 23. *The results of a seeded* SHG *experiment using a 1 mm thick* KTP *sample with 30ps 1.06μm pulses and 1% by energy, 532nm pulses at* $\Delta k L = -1.3$. *(a) fundamental transmittance versus irradiance, (b) fundamental transmittance versus relative phase of the seed for a fixed input fundamental irradiance of 20GW/cm². (Figure taken from Hagan et al. 1994.)*

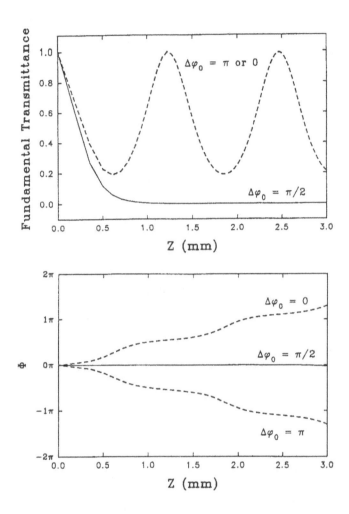

Figure 24. *Plot of calculated transmittance (top) and nonlinear phase shift (bottom) versus z, for a relative seed phase $\Delta\phi_0$ of 0, $\pi/2$ and π, at a fundamental input irradiance of $20GW/cm^2$ in* KTP. *The seed irradiance is 1% of the fundamental's.*

Note that the $\Delta\phi_0 = \pi/2$ case corresponds to $\Delta kL=0$, as the second harmonic starts at $z=0$ with a $\pi/2$ phase shift. This analogy is even more dramatically demonstrated by looking at the ΔkL dependence for different relative seed phases as shown in Figure 26. To first order, the initial phase of the seed is seen to simply shift the phase mismatch curves to one side of $\Delta kL=0$ or the other. A similar shift of the Φ versus ΔkL curves occurs, so that the seed phase effectively changes the phase mismatch condition.

A very important example of these cascaded nonlinear interactions is the existence of spatial solitary wave solutions. These have been predicted for some time (Karamzin and Sukhorukov 1975), and both one (Schiek *et al.* 1995) and two-dimensional (Torruellas

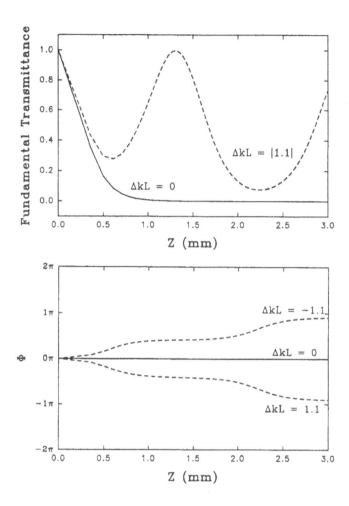

Figure 25. *Plot of calculated fundamental transmittance (top) and nonlinear phase shift (bottom) for ΔkL=-1.1, 0 and 1.1 versus z, at a fundamental input irradiance of $20GW/cm^2$ in* KTP.

et al. 1995) spatial solitary waves have been recently demonstrated. The 2-D experiment was demonstrated by imaging the fundamental output of a 1cm long KTP crystal onto a CCD detector array (Torruellas *et al.* 1995). Above a threshold value the normal diffraction is overcome and the beam collapses to a z independent axially symmetric beam. Interestingly, while the threshold for this collapse increases when the phase mismatch is changed to give an initial self-defocusing rather than self-focusing effect, the spatial solitary solution still exists (both theoretically and experimentally). Thus, the solution can be interpreted more by a 'parametric gain guiding' type of interaction than by a self-lensing phenomenon. These solitary waves can be generated with or

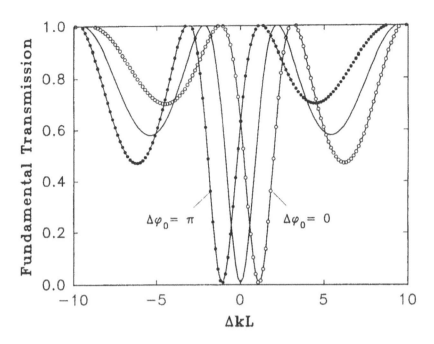

Figure 26. *Fundamental depletion versus phase mismatch for a 1% (by irradiance) 2ω seed beam for relative input phases of 0, π/2 and π.*

without a seed. With only the ω beam present the 2ω is generated within the crystal to give the z independent solution. The threshold occurs near where the parametric gain length becomes shorter than a diffraction length.

4   Conclusion

We have reviewed the basic aspects of self-action effects in materials that lead to non-linear absorption and nonlinear refraction. We have found that NLR and NLA are intimately connected by causality through Kramers-Krönig relations. Examining the lowest order term of the polarisation that leads to NLA and NLR, we found considerable ambiguity. We have seen that $\chi^{(1)}$, $\chi^{(2)}$, and $\chi^{(3)}$ effects can all lead to 'third-order' nonlinear processes. Cascaded first-order nonlinearities lead to third-order effects by changing the optical properties through linear absorption, $\chi^{(2)}$ processes lead to third-order effects through energy conversion to another frequency and back conversion, and $\chi^{(3)}$ processes are, by definition, purely third-order. The similarities of these can some-times make experimental interpretation difficult. For this reason, as well as the fact that in many experiments self-lensing effects lead to spatial amplitude changes and higher order nonlinearities can be present, several experiments are usually needed to determine underlying physical processes. Each process has its own unique uses and the next several years should show the practicality of any of these processes for devices.

The least explored of these processes for applications (and currently the most interesting) is the cascaded $\chi^{(2)} : \chi^{(2)}$ nonlinearity. The possibility of experimentally choosing the sign and magnitude of the nonlinearity as well as the relative magnitudes of loss and phase shift gives considerable flexibility to the device designer. This has prompted considerable interest as evidenced by the large number of publications in the past two years. In addition, since there is no intrinsic absorption there is no heating. However, while the $\chi^{(2)}$ process is 'instantaneous', the nonlinear response is confined by phase matching considerations to a finite bandwidth which is now a function of irradiance. This limits the ability to use ultrashort pulses and complicates the analysis for devices.

Appendix: External self-action regime

An important simplification for interpreting and modeling nonlinear experiments is to work in the 'external self-action' limit as described below (Kaplan 1969). Nonlinear absorption directly affects the amplitude of the propagating electric field while NLR directly affects the phase. However, during propagation, phase changes propagate to give spatial (and temporal) amplitude changes. This can be seen as the coupling of phase and amplitude in the differential equation describing this propagation (*i.e.* the wave equation). A great simplification results by making the 'thin sample approximation' which is also referred to as the external self-action limit. In this approximation we can separate the wave equation into an equation for the phase ϕ, and an equation for the irradiance, I, as a function of the depth z within the sample, as was done in Equations 11 and 12 in the text. This approximation is valid when the sample is thin compared to the depth of focus of the beam (*i.e.* the diffraction length $Z_0 = n_0 \pi w_0^2 / \lambda$ where w_0 is the half-width at the e^{-2} of maximum, HW1/e^2M, of the irradiance distribution, n_0 is the linear index and λ is the wavelength in air). In addition, irradiances must be used that give integrated phase shifts less than approximately π so that nonlinear induced phase shifts do not propagate far enough to lead to irradiance distribution changes within the sample.

If this assumption is not valid, the full wave equation must be solved numerically including both spatial and temporal beam characteristics. This often requires a supercomputer. Throughout this paper, we assume the thin sample approximation to be valid. For a discussion of 'thick' samples the reader is referred to Sheik-Bahae *et al.* (1991b) and Hermann and McDuff (1993).

Acknowledgements

We gratefully acknowledge the support of the National Science Foundation ECS-9320308 and the Naval Air Warfare Center Joint Service Agile Program Contract N66269-C-93-0256. The work presented in this paper represents many years of effort involving many former and current students as well as post-doctoral fellows, research scientists and faculty. We thank all those involved and acknowledge their contributions through the various referenced publications. We explicitly thank Gaetano Assanto, Edesly J. Canto-Said, J. Richard DeSalvo, David J. Hagan, David C. Hutchings, Alan

Miller, Ali A. Said, M. J. Soileau, George Stegeman, William Torruellas, Hermann Vanherzeele, Tai H. Wei, Brian Wherrett, Milton A. Woodall, Jiangwei Wang, Zuo Wang, and Yuen-Yen Wu for their many contributions. Special acknowledgement goes to Mansoor Sheik-Bahae who not only developed the Z-scan technique but determined the appropriate Kramers-Krönig relations for β and n_2.

References

Adiar R, Chase L L, and Payne S A, 1987, Nonlinear refractive index of optical crystals, *Phys Rev B* **39** 3377.

Al-hemyari K, Villeneuve A, Kang J, Aitchison J, Ironside C, and Stegeman G, 1993, Ultra-fast all-optical switching in GaAlAs directional couplers at 1.55μm without multiphoton absorption, *Appl Phys Lett* **63** 3562.

Armstrong J A, Bloembergen N, Ducuing J, and Pershan P S, 1962, Interaction between light waves in a nonlinear dielectric, *Phys Rev* **127** 1918.

Bassani F and Scandolo S, 1991, Dispersion relations in nonlinear optics, *Phys Rev B* **44** 8446.

Belashenkov N R, Gagarskii S V, and Inochkin M V, 1989, Nonlinear refraction of light on second-harmonic generation, *Opt Spectrosc* **66** 806.

Boyd R W, 1992, *Nonlinear Optics* (Academic Press, San Diego).

Canto-Said E, Hagan D J, Young J, and Van Stryland E W, 1991, Degenerate four-wave mixing measurements of high-order nonlinearities in semiconductors, *IEEE J Quantum Electron* **27** 2274.

Chase L and Van Stryland E, 1994, Nonlinear refractive index: Inorganic materials, in *Handbook of Laser Science and Technology; Supplement 2: Optical Materials* **8** 269, ed Weber M (CRC Press).

DeLong K, Rochford K, and Stegeman G, 1989, Effect of two-photon absorption on all-optical waveguide devices, *Appl Phys Lett* **55** 1823.

DeSalvo J R, Hagan D J, Sheik-Bahae M, Stegeman G, and Van Stryland E W, 1992, Self-focusing and self-defocusing by cascaded second-order effects in KTP, *Opt Lett* **17** 28.

Flytzanis C, 1975, Theory of nonlinear optical susceptibilities, in *Quantum Electronics, Nonlinear Optics* Vol 1 Part A, eds Rabin H and Tang C L (Academic Press, New York).

Flytzanis C and Bloembergen N, 1976, Infrared dispersion of third-order susceptibilities in dielectrics: Retardation effects, *Quantum Electron* **4** 271; see also, Flytzanis C, 1975, Theory of nonlinear optical susceptibilities, in *Quantum Electronics, Nonlinear Optics* Vol 1 Part A, eds Rabin H and Tang C L (Academic Press, New York).

Hagan D J, Xia T, Said A A, Wei T H, and Van Stryland E W, 1993, High dynamic range passive optical limiters, *Int J Nonlinear Opt Phys* **2** 483.

Hagan D J, Sheik-Bahae M, Wang Z, Stegeman G, Van Stryland E W, and Assanto G, 1994, Phase-controlled transistor action by cascading of second-order nonlinearities in KTP, *Opt Lett* **19** 1305.

Harper P G and Wherrett B S, 1977, *Nonlinear Optics* (Academic Press, London).

Hermann J A and McDuff R G, 1993, Analysis of spatial scanning with thick optically nonlinear media, *J Opt Soc Am B* **10** 2056.

Hutchings D C, Sheik-Bahae M, Hagan D J, and Van Stryland E W, 1992, Kramers-Krönig relations in nonlinear optics, *Opt Quantum Electron* **24** 1.

Kaplan A E, 1969, External self-focusing of light by a nonlinear layer, *Radiophys Quantum Electron* **12** 692.

Kaplan A E, 1993, Eigenmodes of $\chi^{(2)}$ wave mixings: Cross-induced second-order nonlinear refraction, *Opt Lett* **18** 1223.

Karamzin Y N and Sukhorukov A P, 1975, Mutual focusing of high-power light beams in media with quadratic nonlinearity, *Pis'ma Zh Eksp Teor Fiz* **68** 834 (1976, *Sov Phys JETP* **41** 414).

LaGasse M J, Anderson K K, Wang C A, Haus H A, and Fujimoto J G, 1990, Femtosecond measurements of the nonresonant nonlinear index in AlGaAs, *Appl Phys Lett* **56** 417.

Leite R C C, Porto S P S and Damen P C, 1967, The thermal lens effect as a power-limiting device, *Appl Phys Lett* **10** 100.

Marder S R, Perry J W, and Yakymyshyn C R, 1994, Organic salts with large second-order nonlinearities, *Chem Mater* **6** 1137.

Meystre P and Sargent III M, 1991, *Elements of Quantum Electronics*, 2nd ed (Springer Verlag, Berlin).

Miller D A B, Seaton C T, Prise M E, and Smith S D, 1981, Band-gap-resonant nonlinear refraction in III–IV semiconductors, *Phys Rev Lett* **47** 197.

Mizrahi V, DeLong K, Stegeman G, Saifi M, and Andrejco M, 1989, Two-photon absorption as a limitation to all-optical switching, *Opt Lett* **14** 1140.

Ross I N, Toner W T, Hooker C J, Barr J R M, and Coffey I, 1990, Nonlinear properties of silica and air for picosecond ultraviolet pulses, *J Mod Opt* **37** 555.

Said A A, Sheik-Bahae M, Hagan D J, Wei T H, Wang J, Young J, and Van Stryland E W, 1992, Determination of bound-electronic and free-carrier nonlinearities in ZnSe, GaAs, CdTe, and ZnTe, *J Opt Soc Am B* **9** 405.

Said A A, Wamsley C, Hagan D J, Van Stryland E W, Reinherdt B A, Roderer P, and Dillard A G, 1994, Third- and fifth-order optical nonlinearities in organic materials, *Chem Phys Lett* **228** 646.

Schiek R, Baek Y, and Stegeman G I, 1995, One-dimensional spatial solitons due to cascaded second-order nonlinearities in planar waveguides, submitted to *Phys Rev Lett*.

Sheik-Bahae M, Said A A, and Van Stryland E W, 1989a, High sensitivity, single beam n_2 measurements, *Opt Lett* **14** 955.

Sheik-Bahae M, Said A A, Wei T H, Hagan D J, and Van Stryland E W, 1989b, Sensitive measurement of optical nonlinearities using a single beam, *IEEE J Quantum Electron* **26** 760.

Sheik-Bahae M, Hutchings D C, Hagan D J, and Van Stryland E W, 1991a, Dispersion of bound electronic nonlinear refraction in solids, *IEEE J Quantum Electron* **27** 1296.

Sheik-Bahae M, Said A A, Hagan D J, Soileau M J and Van Stryland E W, 1991b, Nonlinear refraction and optical limiting in thick media, *Opt Eng* **30** 1288.

Sheik-Bahae M, Wang J, DeSalvo J R, Hagan D J, and Van Stryland E W, 1992, Measurement of nondegenerate nonlinearities using a 2-colour Z-scan, *Opt Lett* **17** 258.

Sheik-Bahae M, Wang J, and Van Stryland E W, 1994, Nondegenerate optical Kerr effect in semiconductors, *IEEE J Quantum Electron* **30** 249.

Shen Y R, 1984, *The Principles of Nonlinear Optics* (John Wiley and Sons, New York).

Soileau M J, Williams W E, and Van Stryland E W, 1983, Optical power-limiter with picosecond response time, *IEEE J Quantum Electron* **19** 731.

Stegeman G I, Sheik-Bahae M, Van Stryland E W, and Assanto G, 1993, Large nonlinear phase-shifts in second-order nonlinear optical processes, *Opt Lett* **18** 13.

Toll J S, 1956, Causality and the dispersion relation: Logical foundations, *Phys Rev* **104** 1760.

Torruellas W E, Wang Z, Hagan D J, Van Stryland E W, Stegeman G I, Torner L and Menyuk C R, 1995, Observation of two-dimensional spatial solitary waves in a quadratic medium, *Phys Rev Lett* (in press).

Van Stryland E W, Smirl A L, Boggess T F, Soileau M J, Wherrett B S, and Hopf F, 1982, Weak-wave retardation and phase-conjugate self-defocusing in Si, in *Picosecond Phenomena III*, eds Eisenthal K B, Hochstrasser R M, Kaiser W, and Laubereau A (Springer-Verlag, Berlin) p368.

Van Stryland E W, Vanherzeele H, Woodall M A, Soileau M J, Smirl A L, Guha S, and Boggess T F, 1985, Two-photon absorption, nonlinear refraction, and optical limiting in semiconductors, *Opt Eng* **24** 613.

Van Stryland E W, Wu Y Y, Hagan D J, Soileau M J, and Mansour K, 1988, Optical-limiting with semiconductors, *J Opt Soc Am B* **5** 1980.

Van Stryland E W, Sheik-Bahae M, Said A A, and Hagan D J, 1993, Characterisation of nonlinear optical absorption and refraction, *Journal of Progress in Crystal Growth and Characterisation* **27** 279.

Van Stryland E and Chase L, 1994, Two-photon absorption: Inorganic materials, in *Handbook of Laser Science and Technology; Supplement 2: Optical Materials* **8** 299, ed Weber M (CRC Press).

Van Stryland E W, Sheik-Bahae M, Xia T, Wamsley C, Wang Z, Said A A, and Hagan D J, 1994, Z-scan and EZ-scan measurements of optical nonlinearities, *Int J Nonlinear Opt Phys* **3** no 4 489.

Van Vechten J A and Aspnes D E, 1969, Franz-Keldysh contribution to third-order optical susceptibilities, *Phys Lett* **30A** 346.

Wang J, Sheik-Bahae M, Said A A, Hagan D J, and Van Stryland E W, 1994, Time-resolved Z-scan measurements of optical nonlinearities, *J Opt Soc Am B* **11** 1009.

Wei T H, Hagan D J, Spence M J, Van Stryland E W, Perry J W, and Coulter D R, 1992, Direct measurements of nonlinear absorption and refraction in solutions of phthalocyanines, *Appl Phys B* **54** 46.

Wherrett B S, 1984, Scaling rules for multiphoton interband absorption in semiconductors, *J Opt Soc Am B* **1** 67.

Wherrett B S, Hutchings D C, and Russel D, 1986, Optically bistable interference filters: Optimisation considerations, *J Opt Soc Am B* **3** 351.

Xia T, Hagan D J, Sheik-Bahae M, and Van Stryland E W, 1994, Eclipsing Z-scan measurement of $\lambda/10^4$ wavefront distortion, *Opt Lett* **19** 317.

Yariv A, 1975, *Quantum Electronics*, 2nd ed (Wiley, New York).

Picosecond Optical Pulse Generation Using Semiconductor Lasers

I H White

University of Bath, England.

1 Introduction

Semiconductor lasers have attracted considerable interest over the last thirty years, in particular because of their key applications in high capacity telecommunication networks (Ikegama 1992). Increasing demand from telephone, entertainment channels and computer links has resulted in the widespread implementation of optical fibre in terrestrial trunk links and in submarine systems. The compactness, low cost, efficiency, reliability and direct modulation capability of diode lasers has made them very robust for these applications. However their ease of integration and high power operation has lead to additional interest in a range of applications including optical storage, ranging, sensing, sampling, signal processing, microwave signal distribution, and high power pumping of other linear and nonlinear optical systems, such as frequency doubling and parametric oscillator systems. Many of these applications rely on the ability of laser diodes to generate short optical pulses at high repetition rates and methods of achieving this have therefore become important.

The design of lasers for such applications depends on a number of system determined pulse characteristics. Control over the pulse profile, duration, peak power, repetition rate, and stability are critical to sampling and communication applications. Tolerances in fibre communication applications are further complicated as the wavelength shift or chirp during the pulse determines the pulse propagation characteristics in fibre. For the pumping of nonlinear components, the beam profile limits the achievable peak intensity and the absolute wavelength becomes more critical. In this chapter, methods of picosecond pulse generation will be considered in the context of potential pulse specifications.

2   Picosecond pulse generation techniques

The unique ability of the diode laser to generate light on direct injection of electrical carriers has led to a wide range of techniques for picosecond optical pulse generation, each typically leading to the development of different laser structures for optimised pulse generation. Of the techniques which have been studied to date, there are three major classes of pulse generation; gain-switching, Q-switching and mode-locking. In general, gain-switching is the simplest technique for generating pulses on demand with typical durations of tens of picoseconds. However the optical spectrum of the gain switched pulses is typically chirped through high carrier depletion during pulse generation, and as a result, compensatory techniques to achieve Fourier limited pulse generation have been studied. Alternatively Q-switching allows pulse generation with higher peak pulse powers, though typically, passive Q-switching techniques have been found to generate pulses with worse noise and jitter in addition to chirp. Recently therefore, actively Q-switched schemes in laser diodes have been studied in order to reduce these limitations. Finally mode-locking, including active, passive and colliding pulse forms, has been used to generate high power pulse trains with excellent quality (*i.e.* with low jitter and chirp), though confined in repetition rate to narrow operating regimes. This has been achieved in diode lasers using both external and integrated optical cavities, some with wavelength selective grating elements. This chapter will proceed to review each of these techniques in detail.

2.1   Gain-switching

The simplest of the pulse generation techniques, gain switching, was observed in the early days of the laser diode, but became of particular interest following the development of the double heterostructure laser which made possible high repetition rate pulse generation at high duty cycles and moderate drive currents (Ito *et al.* 1979). Using this technique, large amplitude short current pulses are applied directly to the laser. As the current is injected, the carrier concentration increases rapidly in the laser. Pulses are then formed from the transient oscillation in the photon output as the laser turns on. There is strong interaction between the populations of photons and injected electrons. The delay between the increased rate of photon generation due to carrier density increases and the build up of the photon population (Figure 1) in part enables the first spike to be much shorter than one might expect from small signal operation.

The generated photon population rapidly depletes the electron concentration to prevent further lasing so that a single resonance spike is generated. The current is abruptly terminated after the charge carrier concentration has exceeded threshold to ensure that no further relaxation oscillations occur. Pulses generated in this manner are found to be typically a few tens of picoseconds wide and typically have peak powers from a few tens

(a)

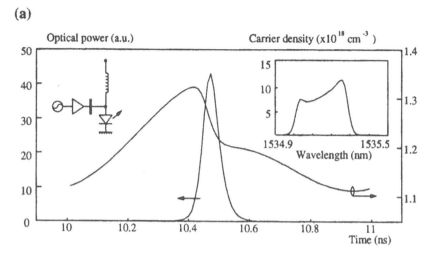

(b)

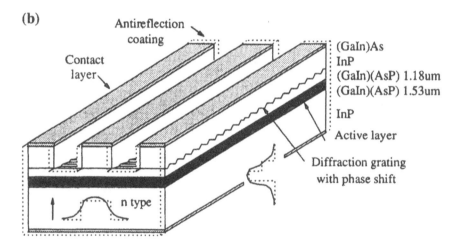

Figure 1. *(a) Simulation of gain-switching in a bulk multiple phase shift distributed feedback laser, showing reducing carrier levels during pulse generation with the time averaged red shifted optical spectrum. (b) Schematic for a ridge waveguide InGaAsP distributed feedback laser often used to generate short pulses. The inset plots show the typical horizontal and vertical guide profiles (dotted lines) and optical distributions (solid lines) of the beam in the region of the active layer.*

to a few hundreds of milliwatts. To achieve short pulses of the highest possible power, a high gain condition is achieved before pulse turn-on by driving the laser using large signal microwave modulation (Torphammar and Eng 1980), or with subnanosecond high current pulses from a comb generator (Lin *et al.* 1980) or an avalanche transistor pulse generator (Bimberg *et al.* 1984). These cause carrier densities to increase, enhancing

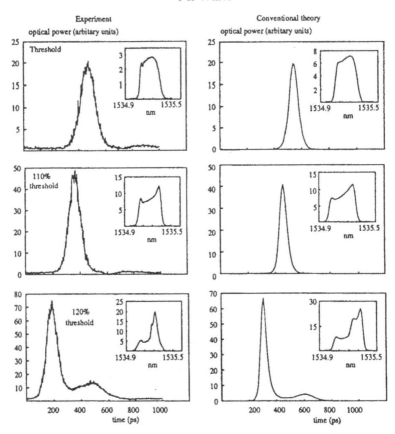

Figure 2. *Comparison of experimental gain-switched performance with simulation for varied DC bias currents and 15dBm, 1GHz RF power in bulk multiple phase shift distributed feedback lasers. Temporal performance with spectra inset.*

gain before optical turn-on. Typical examples of the generated pulses are shown in Figure 2.

In detail, the mechanisms involved in gain switching can be understood by studying the electron and photon rate equations which describe the carrier-photon interaction in terms of the photon density, S, in the cavity and the carrier density, n, in the active region. The rate equations can be written in different forms, one of which is as follows (Vasil'ev 1993)

$$\frac{dn}{dt} = \frac{j}{ed} - g_0(n - n_t)S - \frac{n}{\tau_s} \tag{1}$$

$$\frac{dS}{dt} = \Gamma g_0(N - n_t)S - \frac{S}{\tau_{ph}} + \beta \Gamma \frac{n}{\tau_s} \tag{2}$$

where n_t, is the transparency density, g_0, is the differential gain coefficient, β is the spontaneous coupling factor, e is the electron charge, d is the active layer thickness, j is the time varying current density, Γ is the optical confinement factor, and τ_s and τ_{ph} are the carrier and photon lifetimes. In Equations 1 and 2 the gain is assumed to be proportional to the carrier concentration. These simple equations are single-mode, that is, they do not take into account the distribution of S between cavity modes. It is assumed also that the laser operates in a single lumped spatial mode.

The photon lifetime τ_{ph} in Equation 2 is given as

$$\tau_{ph} = \frac{1}{v_g(\alpha_m + \alpha_i)} \tag{3}$$

where v_g is the group velocity and α_m and α_i is the mirror and internal loss, respectively. If one assumes that the current density consists of short electrical pulses, the carrier concentration within the laser can be approximately written as,

$$n(t) = n_t - n_1 \tanh(\gamma t)$$

where n_1 is a constant. Neglecting spontaneous emission coupling, the photon rate equation may be solved so that

$$S(t) = S_0 \text{sech}^2(\gamma t)$$

where $\gamma = -gn_1/2$.

The strong depletion of the carriers therefore causes a short pulse to be generated, the width of which does not depend directly on the current drive, but rather is a function of the gain of the laser material and the carrier population at pulse turn-on. As the carrier concentration is linked to the photon lifetime, devices with short lifetimes are preferred (Lin and Bowers 1985). The shape of the pulse depends upon the laser device structure and material. A larger cavity allows higher total carrier injection to the active layer before pulse turn on and therefore higher optical pulse energies. However, longer pulse durations result from the increased, cavity length dependent, photon lifetime and so there is no significant peak power improvement.

The change in carrier density does however lead to a large refractive index modulation and a red shift in the lasing wavelength. This in turn leads to duration bandwidth products which can be over five times greater than the Fourier limit for single mode lasers (Lin and Bowers 1985). Fabry Perot diode lasers under high current driving are prone to multi-longitudinal-mode pulse generation which usually is most susceptible to mode partitioning (Van der Ziel 1979). A range of techniques have therefore been used to achieve single mode gain switched pulse generation. Short external cavities have been used to enhance the gain within a single longitudinal mode and successful monomode operation has been achieved even under high current injection. Injection locking of a pulsed laser by a CW driven diode laser has also been successful (Hughes *et al.* 1994).

The use of distributed feedback (DFB) and distributed Bragg reflector devices has proved most effective in ensuring that a single mode is generated. Onodera *et al.* (1984) first reported generation of bandwidth-limited pulses from a $1.3\mu m$ gain-switched DFB InGaAsP diode laser. Here, a DFB diode laser was driven with a strong RF modulation superimposed on the d.c. bias current. The minimum pulsewidth was 26ps at a 500MHz repetition rate. Since then DFB gain-switched lasers have become the most widely used devices for generating single-frequency low-chirp picosecond pulses with duration in the range of 25-30ps.

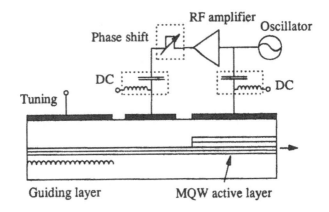

Figure 3. *A scheme for the generation of transform limited gain-switched pulse generation as demonstrated by Lourtioz et al. (1992).*

A triple contact DBR diode laser has been used to generate single-frequency picosecond pulses at 1.56μm with reduced chirp, Figure 3(Lourtiouz *et al.* 1992). The laser consisted of two gain and phase-control sections. Gain switching was realized by application of a strong (over 30dBm) microwave signal to the end gain sections. The chirp of the gain-switched pulses was compensated by applying a fraction of the microwave signal (1.32GHz) to the central phase-control section. This resulted in a frequency shift of the light travelling through the phase-control section which is opposite to that in the gain section. A time-bandwidth product of 0.35 was obtained for the 100–200ps pulses.

White *et al.* (1992) demonstrated linewidth reduction and generation of shorter bandwidth-limited gain-switched picosecond pulses by injection locking a triple contact 1.5μm DFB laser (Figure 4). Here, the end sections of the triple contact laser were gain-

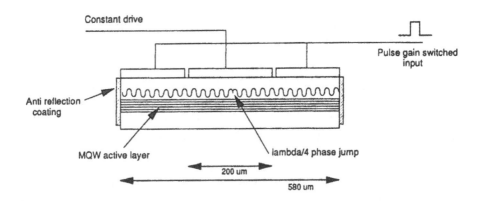

Figure 4. *A scheme for the generation of transform limited gain-switched pulse generation as demonstrated by White et al. (1992).*

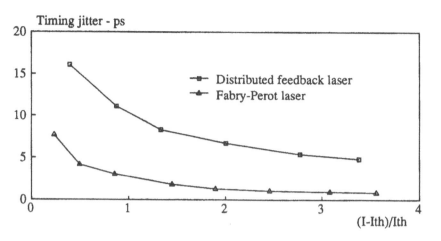

Figure 5. *Timing jitter in multimode and single mode gain-switched lasers for varied DC bias conditions. After Weber et al. (1992).*

switched using a strong RF signal while the central section is d.c. biased. The latter provided an effective optical injection locking mechanism in the laser. Adjustment of the current amplitude I_c allowed variations of the carrier density in active region. As a result, the central section at appropriate current amplitudes caused locking of the wavelength of the gain-switched pulse output. This was experimentally observed over a wide range of the bias currents I_c. Locking action of the central contact also resulted in significant reduction of frequency chirp of the output pulses. The chirp as low as 0.06nm for 32ps gain-switched pulses was demonstrated, allowing time-bandwidth products of 0.26.

Spectral filtering also has been used to cause near transform limited pulse generation through the use of semiconductor amplifiers (Sundersan and Wickens 1990), diffraction gratings (Iwatsuki *et al.* 1991) and pulse compression in optical fibre (Takada *et al.* 1986). Chirp compensation is achieved by using a medium with the opposite sign of dispersion which leads to pulse width compression. The intrinsically red-shifted spectrum of a picosecond pulsed diode laser experiences a relative retardation of the shorter wavelength components at the leading edge of the pulse in anomalous dispersion fibre, a grating cavity or combinations thereof.

As potential applications for high repetition rate picosecond pulses have been explored, it has become increasingly apparent that pulse to pulse amplitude and repetition rate fluctuations can be important for certain diode laser applications. In terms of gain-switching, (Böttcher *et al.* 1988), the spontaneous emission noise leads to a random turn-on time described hereon as timing jitter. As the spontaneous emission levels increase with decreased bias current, timing jitter has been observed to decrease with increasing current in gain-switched lasers. Figure 5 illustrates such jitter reduction with bias for one monomode distributed feedback laser and one multimode laser as determined in a study by Weber *et al.* (1992). The differing pulse statistics resulting

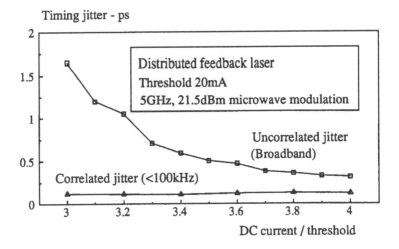

Figure 6. *Timing jitter for gain-switched distributed feedback lasers at 5GHz after filtering and compression. After Jinno (1993).*

from reduced levels of spontaneous coupling explain the increased jitter for the DFB.

No data is given on the pulse profile, although a tail is typically observed with such high current, long duration excitation pulses. Such pulse profile limitations have been overcome for distributed feedback lasers in a study by Jinno (1993). While bias schemes involving above threshold bias current are used, subpicosecond jitter is achieved by direct microwave modulation at 5GHz, subsequently filtering the nonlinear chirp and compressing the pulses (Figure 6).

Here the picosecond pulse train is generated tail-free at 5GHz, irrespective of bias condition as the nonlinear chirp characteristic of pulse tails has been removed before compression. Jitter levels as low as 0.5ps are achieved. The pulse compression itself is not expected to affect measured timing jitter. It is worth noting however, that while offering no noise improvements over mode-locking (Burns *et al.* 1990), the scheme has become more complex.

In order to reduce jitter in conventional gain switched structures, optical feedback techniques have been introduced. Optical feedback requires either an external mirror (Solgaard and Lau 1993), grating (Huhse *et al.* 1994) or fibre loop to feed back light into the cavity. Such self-seeded schemes do not involve laser facet coatings, therefore leading to two cavities as opposed to extending the laser cavity in external cavity mode-locking. The cavity provides a second round trip time synchronised to the modulation rate in order to assist optically subsequent pulse generation with stimulated emission. The reduced dependence on amplified spontaneous noise at turn on reduces timing jitter in the pulse train. Total jitter ultimately relies on the quality of electrical modulation, the stability of the cavity and reliance on spontaneous emission for pulse generation. A reduction in jitter with increased feedback is ultimately limited by pulse profile deterioration as pulse turn off is prevented.

Hybrid feedback techniques have also been studied. These use the ability to control gain switching pulse generation by feeding back information about the optical pulse train incident on a photodetector. The amplified photocurrent directly modulates the laser. After a number of modulation periods the signal stabilises to generate a microwave signal defined by the frequency response of the feedback loop. Pulse widths decrease from 60ps to 35ps as the repetition rate is tuned from 230–500MHz under the regeneratively gain-switched scheme. The original self-pulsing scheme employing the same optoelectronic feedback however allows the generation of 20ps pulses at 1GHz, decreasing in width to 10ps as the repetition rate increases to 5GHz.

Electronic feedback techniques have also been used to reduce jitter (Williams 1995). Here the forward voltage of the laser device is monitored and fed back into the device. The jitter of the Fabry-Perot laser shows significantly reduced jitter using the scheme. However, poorer pulse profiles are noted when compared with gain-switched Fabry-Perot lasers. The back of the pulse shows an exponential fall off at the higher bias conditions. Without any filter in the feedback loop, self-pulsation occurs at 3GHz with 4ps timing jitter. The lack of frequency discrimination in the feedback loop results in the amplification of broadband noise responsible for much of the jitter. The inclusion of the narrow band filter reduces jitter to 0.8ps with a bias dependence as outlined in Figure 7. Jitter levels for the Fabry Perot lasers are not significantly better than observed for distributed feedback lasers however, indicating that a limit is being reached.

2.2 Q-switching

At an early stage in the development of the laser, trains of pulses or self-pulsations were observed, these being later found to be due to the presence of regions of saturable absorption in the laser devices. Rapidly it was recognised that such pulsations could have use, and hence research was initiated into the use of saturable absorbers in laser diodes to generate optical pulses by Q-switching. Here, it was recognised that if additional loss could be placed within the cavity prior to pulse generation, either by providing multicontact sections with loss or gain or by specially introducing loss by irradiation or damage, then according to the theoretical arguments described above for gain switching, the carrier density before lasing could be increased, indeed, by as much as 4.5 times (Vasil'ev 1993). If the loss was then reduced, the pulses generated could typically be expected to exhibit shorter pulse widths and higher pulse powers than in the gain switched case.

To illustrate this, a schematic comparison is presented in Figure 8, showing the variation in the carrier density and optical power during gain and Q-switching. The Q-switching is represented for a multicontact device with separate gain and loss regions. The carrier densities in both the gain and loss section are given for the Q-switched example. The pulse turn-on is now additionally initiated by absorption in the loss section. The enhanced pulse powers and reduced durations of the optical pulses result from enhanced carrier levels as a result of the delayed pulse turn on.

For a passively Q-switched laser, the irradiated absorber experiences the excitation of carriers to higher energy states and a resultant shift in the absorption edge until the material is bleached. With the significantly higher cross-section in the loss regime (below transparency), a significantly reduced perturbation of the carrier concentration

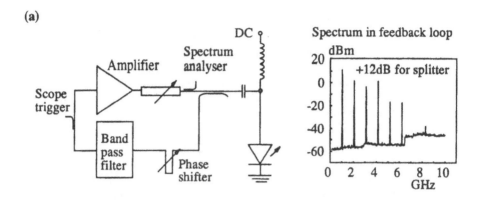

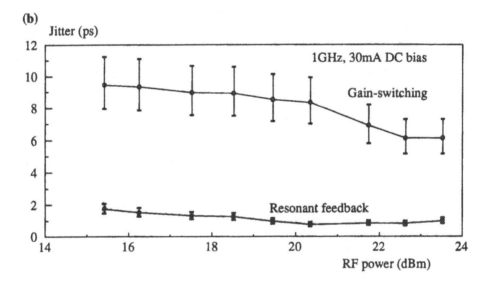

Figure 7. *(a) Resonant electrical feedback in a single contact diode laser for picosecond pulse generation at gigahertz repetition rates. (b) Timing jitter in a single contact distributed feedback laser under gain-switching and resonant electrical feedback.*

in the loss section is required to change the gain of the cavity by as much as four times more than that expected for gain-switching. For the passive generation of Q-switched pulse trains therefore, the differential absorption should be greater than the differential gain and the gain section carrier lifetime should be greater than the carrier lifetime in the absorber section (Ueno and Lang 1985).

The most common forms of saturable absorption have been formed using proton bombardment of the facets (Volkov *et al.* 1989), or with a segmented electrode multi-contact structure (Tsang and Walpole 1983). A centrally positioned absorber has been recommended by Vasil'ev to increase the rate of excited carriers transferring from the loss to the amplifying sections (Vasil'ev 1988). Transferred carriers should be recycled

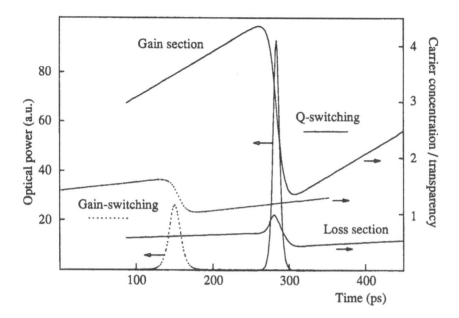

Figure 8. *Comparison of carrier dynamics and optical intensity variation during (i) gain-switching and (ii) Q-switching.*

to take part in the stimulated emission rather than vanish through spontaneous recombination. The upper limit of pulse repetition frequency is not limited by the absorber spontaneous recombination lifetime but by the population transportation rate. The Q-switching efficiency thus increases by the extra pumping of the carrier drift from the absorber as shown theoretically by Vasil'ev (1993). Here a reverse biased absorber is modelled by assuming a differential absorption eight to ten times greater than the differential gain and a 50–100ps carrier lifetime in the absorber. Such modelling allows 2–10ps duration pulses for 50–500ps recombination times. Pulses of 10ps duration with 200–700mW peak power have been generated by Q-switched InGaAsP lasers at pulse generator limited 140MHz rates with evidence of near Fourier-limited pulses (Vasil'ev *et al.* 1993b). Self-Q-switching has also been observed in aged lasers and as a result of dark line defects (Paoli 1977).

Passively Q-switched laser diodes thus can generate very high optical power pulse trains whose repetition rates can be controlled over a significant range by the d.c. drive current. However they do suffer from increased jitter and noise. To overcome this limitation, actively or forced-Q-switched techniques can be used where the loss section is electrically modulated to define a non-recovery time dependent repetition rate (Figure 9). This has been shown to reduce the jitter of the generated pulses by modulating the stimulated absorption processes and allow control of pulse shape without affecting necessarily the peak power. Table 1 shows a survey of Q-switched GaAs lasers to indicate the wide variation in attainable performance levels.

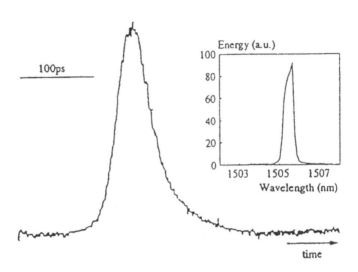

Figure 9. *The optical pulse profile and spectrum for a 1.55 micron multiquantum well distributed feedback laser under Q-switching.*

Ref	Rate	Q-switching scheme	FWHM ps	Power W
a	500Hz	Optically Q-switched broad area laser	21	6
b	kHz	Transverse mode switched broad area laser (Single heterostructure)	40	140
c	40MHz	Passive Q-switching in a grating cavity (Discrete absorber)	20	0.75
d	400MHz	Passive Q-switched broad area laser	40	7.5
e	1.5GHz	Actively Q-switched broad area laser (Quantum well)	24	1.3
f	2.5GHz	Radiofrequency locked self Q-switching	17	1
g	18GHz	Self Q-switching (multilobe far field)	5	10

Table 1. *Comparison of state of the art Q-switched narrow-stripe GaAs lasers. Although there have been significantly fewer reported results, long wavelength devices typically appear to operate at lower peak powers, often less than 1 W. References: (a) Thedrez et al. 1993; (b) Volpe et al. 1994; (c) Bouchoule et al. 1993; (d) Gavrilovic et al. 1995; (e) O'Gormann et al. 1991; (f) Vasil'ev et al. 1993a; (g) Vasil'ev 1988.*

Q-switching is thus now established as a technique for generating high power, high energy pulses from narrow stripe lasers, and much work has also been carried out on broad area lasers with the highest reported powers (140W) (Volpe *et al.* 1994) for large optical volume single heterojunction lasers. This often results in multiple transverse mode operation and unstable near and far field patterns. High frequency pulsations have also been observed recently in distributed feedback lasers. Both single and multimode pulsations have been observed at modulation rates up to 80GHz (Marcenac and Carroll 1994), though in some cases at low peak powers. Electrical repetition rate tuning is not consistently observed.

As jitter in Q-switched lasers has been of considerable importance, even in active Q-switching, optoelectronic feedback schemes have also been studied (Figure 10). These have proved successful, the uncorrelated jitter being reduced substantially. Here, rather than using photodiode detection, shot noise from a dc-biased gain section is coupled into the feedback loop through a microwave tee. This causes repetitive pulse generation at frequencies determined predominantly by the electrical bandwidth of the feedback loop. With no microwave source or critical narrow-band filtering in the feedback loop system, costs are kept to a minimum. The signal is amplified and injected into a subthreshold biased absorber section.

For example, using quantum well InGaAsP/InP distributed feedback lasers, stable self Q-switching at 1GHz has been demonstrated with reduced jitter. As the amplification in the loop is increased, a uniform increase in noise across the frequency spectrum occurs, indicating the existence of higher levels of amplitude noise. Such increases in amplitude noise with feedback loop gain are also reported for continuous wave feedback linewidth reduction schemes. Feedback has also been shown to lead to a reduction in the jitter observed for diode lasers in Fabry-Perot cavity configurations to 0.8ps at 1GHz repetition rate, the jitter being relatively insensitive to electrical feedback phase. Further improvements regarding the long term repetition rate stability are expected with the inclusion of narrow band filters and low noise amplifiers in the feedback loop.

2.3 Mode-locking.

While unlocked modes in a laser diode randomly related in phase result in irregular irradiance fluctuations on a steady dc optical level, locked modes initiate an optical pulse to circulate back and forth inside the laser cavity (Vasil'ev 1995). A mode-locked laser therefore emits pulses separated by a time equal to the round trip time of the laser cavity. The phase relation may be enforced by modulating loss or gain in the laser cavity at a frequency equal to the intermode frequency separation. This corresponds to the reciprocal of the round trip time as shown in Figure 11.

Under steady state operation an optical pulse arrives at the gain section slightly after the peak of the current pulse. The leading portion of the pulse receives a small amount of amplification as the electrical pulse has just increased the segment gain. The middle portion receives the largest amount of gain. The trailing edge of the pulse receives reduced amplification as the earlier portions of the pulse deplete the gain. The output pulse is therefore shortened on successive round trips to become much narrower than the electrical current pulses. The final optical pulse width is achieved when the gain mod-

(a)

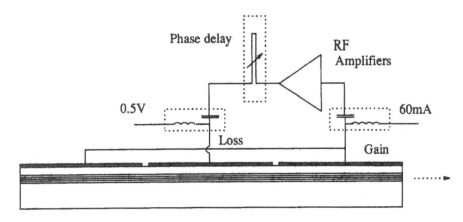

(b)

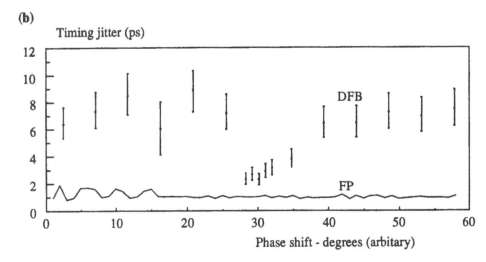

Figure 10. *(a) Schematic to show the implementation of optoelectronic feedback on a single chip for a multicontact laser diode. (b) Uncorrelated timing jitter in multimode and single mode self-Q-switched laser with optoelectronic feedback.*

ulation section's pulse shortening is counterbalanced by cavity dispersion. Extended cavities are required for actively mode-locked lasers to allow modulation at electrically achievable frequencies. For GaAs lasers mode-locked at 200MHz in an external grating cavity, 15ps duration, 1W peak power pulses are typically reported (Vasil'ev *et al.* 1995). However active mode-locked lasers with single gain elements in extended cavities are known to be very susceptible to multiple pulse formation (Schell *et al.* 1991). The undesirable secondary pulsations are initiated by reflections from imperfect antireflection coatings on the diode laser facet. The reflected pulse is then amplified because the

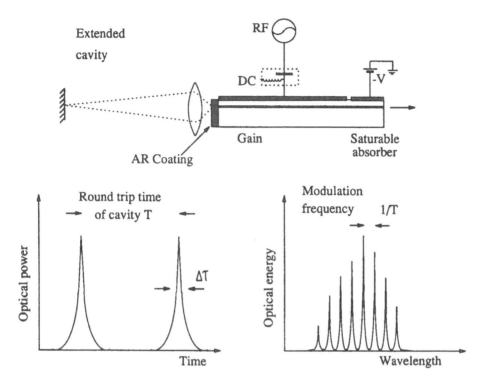

Figure 11. *Mode-locking in an external cavity. Schematic spectra and pulse profiles are included for clarification.*

main pulse does not fully deplete the gain and the current drive to the segment may still be creating new carriers. Multiple pulse formation for reflectivities as low as 10^{-5} have been suggested by Schell *et al.* (1991). A saturable absorber section within the laser diode has therefore been used to achieve single pulse operation and limit secondary pulsations at higher bias levels. Using these techniques, 2.8ps duration, 25mW peak power pulses have been obtained by Derickson *et al.* (1992). Such hybrids of active and passive mode-locking schemes have been implemented with ion implantation at the emitting facet. The use of saturable absorption also causes reduction of pulse widths. Finally as external cavities can be difficult to implement in a robust manner, the use of fibre mirror and grating schemes has been studied (Morton *et al.* 1994).

For ultrahigh repetition rate pulse trains, in excess of around 100GHz, purely passive mode-locking schemes using either a saturable absorbing section (Vasil'ev and Sergeev 1989) or bombarded facets (Deryagin *et al.* 1994) have been found useful for mode-locking. These typically operate on the straightforward principle that as the pulse enters the highly attenuating absorber section, the leading edge is absorbed with the creation of excitons. Saturation of the absorption reduces the attenuation for the following part of the pulse. Efficient carrier sweepout resets the absorption for improved mode-locked operation. Pulsation can thus occur at repetition rates essentially unlimited by the laser parasitics and damping mechanisms. Using these techniques repetition rates of a few

hundred GHz have been observed with pulse widths of less than 1ps. Average powers
are typically a few milliwatts. In addition recent work has shown that the linewidth
enhancement factors of the loss regions can be used to compensate for chirp caused by
the gain regions.

For a centrally placed absorber, the passively mode-locked laser can simultaneously
support two counter-rotating pulses in the cavity. Known as colliding pulse modelocking (Chen and Wu 1992, Eisenstein *et al.* 1986) the pulses collide in the absorber to
doubly saturate for further pulse narrowing, while occupying different regions of gain.
This results in shorter duration pulses at twice the repetition rate. A comparison of
state of the art mode-locked InGaAsP lasers is given in Table 2 to identify the roles of
the different schemes.

Ref	Rate GHz	Cavity and scheme	Duration ps	Power(mW) peak(av)	$d\nu d\tau$
	2	Active with fibre cavity			
		Comb generator	4.8^1	56(0.54)	
b		Radiofrequency	7.7^2	39(0.6)	0.56
b	2.5	Hybrid fibre grating cavity	23	137(7.8)	0.88
		HR/AR coated laser			
c	4.4	Active monolithic	9	95(4)	15.6
d	5	Active grating cavity	10	2(0.1)	0.36
e	20	Active fibre cavity	5^1	18(1.8)	0.44
		HR/AR coated laser			
f	40	CPM monolithic	1.1^1	1(0.05)	0.34
a	80	CPM monolithic	1.28^2	5(0.5)	0.34
g	104	Ion bombarded facets	0.64^2	20(1.3)	0.51
h	1540	Passive DBR laser	0.26^3	40	0.52

Table 2. *State of the art mode-locking in (m) InGaAsP lasers. HR/AR coatings
indicate emission from an anti reflection coated facet, with high reflectivity coatings at
the often inaccessible second facet. The superscripts indicate the assumed pulse profiles
used in the width calculation: 1. Single sided exponential profile, 2. sech-squared profile,
3. Gaussian profile. References:(a) Eisenstein et al. 1986; (b) Morton et al. 1994; (c,
Raybon et al. 1992; (d) Bird et al. 1990; (e) Tucker et al. 1989; (f) Chen et al. 1992,
(g) Deryagin et al. 1994; (h) Arahira et al. 1994.*

3   High power laser structures

There has been considerable interest in enhancing the peak powers generated by short
pulse diode lasers. Various forms of broad area and array laser diodes have been studied
and success has been achieved in generating high powers at very low repetition rates.
However, high repetition rate operation is difficult due to thermal effects and the pulses
generated typically do not have well defined beams. Recently therefore much interest
has focused on tapered waveguide devices.

The development of tapered-waveguide lasers results from demonstrations of single pass semiconductor tapered amplifiers (Walpole *et al.* 1992). The tapered amplifier maximises its efficiency by expanding the gain volume along the amplifier length in order to maintain a near uniform power density and hence reduced gain saturation. Since the amplified beam is diverging, the extent of the spatial overlap of the reflected backward travelling wave and the forward input wave is reduced relative to nontapered structures. Lasers are thus readily implemented (Walpole *et al.* 1994) by say, allowing an uncoated 30% facet reflectivity at the narrow facet of a 2mm long device. Powers of up to 2W have been achieved in a near-diffraction-limited (0.23°) lobe. Straight waveguide distributed Bragg reflector lasers have also been integrated with the tapered amplifiers leading to single mode powers of 1.3W pulsed. Continuous wave powers have subsequently increased to 2W (Parke *et al.* 1993) and more recently 3W. External cavity tapered lasers have also demonstrated high power(1W) diffraction limited sources with a 50nm wavelength tunability (Mehuys *et al.* 1993).

Much of the short pulsed operation of tapered-waveguide laser systems has focused on mode-locked operation in the 1–5GHz range (Mar *et al.* 1994). Cavity design involving a mode-locked source as the master-oscillator and a tapered-waveguide amplifier as the power-amplifier has lead to actively mode-locked pulse trains with pulses as short as 4.2ps with powers as high as 28W. Passively mode-locked 4.1ps duration 5W peak power pulses have also demonstrated using a tapered-waveguide postamplifier configuration.

Also tapered devices have been used alone in a grating cavity to generate picosecond pulses. In a direct comparison between a narrow-stripe and tapered-waveguide, Helkey has demonstrated 4.2ps pulses with 1W power at 5GHz repetition rate for the tapered source, an energy enhancement of 2.3 over the straight waveguide control.

Other work however has considered the use of monolithic devices for short pulse generation. Here much attention has focused on the bow-tie laser. This combines advantages of the enhanced optical power performance of broad area lasers with the mode control associated with narrow ridge waveguide lasers by incorporating a narrow central waveguide for mode control between two tapered waveguide sections (Figure 12).

Comparisons between the bow-tie laser and non-tapered multicontact lasers of otherwise identical dimensions indicate an order of magnitude increase in power and energy (Williams *et al.* 1994), with the Q-switched devices having achieved peak powers in excess of 8W or pulse energies in excess of 100pJ (Figure 13). Despite these high powers, narrow far field operation is achieved.

High power pulsed and CW mode-locking has also been achieved in InGaAs/GaAs multiquantum well bow-tie lasers. Stable generation of 750fs pulses with average powers of up to 35mW at frequencies of 132GHz are observed. Both bulk active layer AlGaAs/GaAs and quantum well InGaAs/GaAs bow-tie lasers are shown to mode-lock. For the AlGaAs/GaAs lasers, average powers of up to 50mW have been achieved under pulsed conditions. However, modulation is sinusoidal with an estimated modulation depth of 95%. At a repetition rate of 100GHz, a time bandwidth product of 0.9 has been estimated from both grating spectrometer and first order interferometric spectral measurement. Preliminary work by Summers *et al.* (1995) on the InGaAs/GaAs laser has indicated continuous wave mode-locked operation with a time bandwidth product of 0.61. While average powers of around 17mW are readily generated for mode-locked

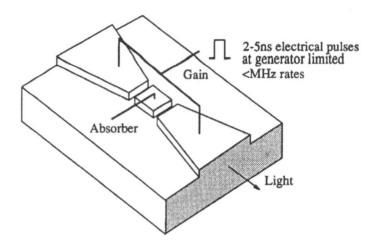

Figure 12. *Schematic for a double tapered waveguide Bow-tie laser with patterned p-side electrodes for picosecond pulsed operation.*

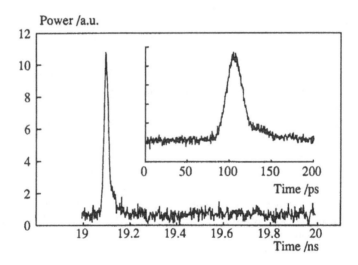

Figure 13. *Q-switched pulse profile measured with a 32GHz photodetector and 50GHz sampling oscilloscope. Tail-free pulsation is noted. Inset is an expanded trace to indicate profile detail.*

pulse trains under a pulsed electrical modulation, increased electrical bias to the taper regions allows average powers of up to 35mW. Single contact InGaAs/GaAs tapered waveguide lasers have also been shown to mode lock at promising power levels (Williams *et al.* 1995)(Figure 14).

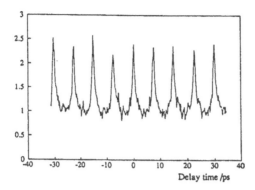

Figure 14. *Second harmonic autocorrelation trace indicating mode-locking in a In-GaAs/GaAs multiquantum well single taper laser. The repetition rate is 116GHz and the pulse width is 850fs.*

4 Conclusions

In conclusion therefore, diode lasers have excellent picosecond pulse generation properties in allowing the direct generation of pulses with significant powers on demand either by gain switching or Q-switching at repetition rates of up to more than 10GHz. A range of techniques has been devised to control spectral properties and noise. Mode-locking techniques are able to generate pulses of high quality at ultrahigh repetition rates. As a result, a wide range of applications can now be fulfilled by these devices. However much work remains to be carried out in this area, particularly in terms of increasing pulse powers and reducing pulse widths to levels currently attained by other laser systems. This work is likely to benefit much from the new laser diode architectures now becoming important, such as the surface emitting systems.

References

Arahira S, Oshiba S, Matsui Y, Kuni T, and Ogawa Y, 1994, Terrahertz rate optical pulse generation from a passively mode-locked semiconductor laser diode, *Opt Lett*, **19** 834.

Bimberg D, Ketterer K, Schöll H E, and Vollmer H P, 1984, Generation of 4ps light pulses from directly modulated V-groove lasers, *Electron Lett* **20** 343.

Bird D M, Fatah R M, Cox M K, Constantine P D, Regnault J C, and Cameron K H, 1990, Miniature packaged actively mode-locked semiconductor laser with tunable 20ps transform limited pulses, *Electron Lett,* **26** 207.

Böttcher E H, Ketterer K, and Bimberg D, 1988, Turn-on delay time fluctuations in gain-switched AlGaAs/GaAs multiple quantum well lasers, *J Appl Phys* **63** 2469.

Bouchoule S, Stelmakh N, Cavelier M, and Lourtioz J M, 1993, Highly attenuating external cavity for picosecond tunable pulse generation from gain/Q-switched laser diodes, *IEEE J Quantum Electron,* **29** 1594.

Burns D, Finch A, Sleat W and Sibbett W, 1990, Noise characterisation of a mode-locked InGaAsP semiconductor diode laser, *IEEE J Quantum Electron*, 26 1860.

Buus J, 1985, Dynamic line broadening of semiconductor lasers modulated at high frequencies, *Electron Lett*, 21 129.

Chen Y K, and Wu M C, 1992, Monolithic colliding pulse mode-locked quantum well lasers, *IEEE J Quantum Electron*, 28 2176.

Derickson D J, Helkley R J, Mar A, Karin J R, Bowers J E, and Thornton R L, 1992, Suppression of multiple pulse formation in external cavity mode-locked semiconductor lasers using interwaveguide saturable absorbers, *IEEE Photon Tech Lett*, 4 333.

Deryagin A G, Kuksenkov D V, Kuchinskii V I, Portnoi E L, and Khrushchev I Yu, 1994, Generation of 110GHz train of subpicosecond pulses in 1.535μm spectral region by passively modelocked InGaAsP/InP laser diodes, *Electron Lett* 30 309.

Eisenstein G, Tucker R S, Koren U, and Korotky S K, 1986, Active mode-locking characteristics of InGaAsP single mode fibre composite cavity lasers,*IEEE J Quantum Electron*, 22 142.

Gavrilovic P, Stelmakh N, Zarrabi J H, and Beyea D M,1995, High energy CW Q-switched operation of multicontact semiconductor laser, *Electron Lett*, 31 1154.

Hughes D M, Burns D, Sibbett W, Williams K A, and White I H, 1994, Ultrashort pulse semiconductor lasers with improved timing jitter, *Proc. CLEO, (Anaheim), paper CWN4*.

Huhse D, Schell M, Utz W, Kässner J, and Bimberg D, 1994, Generation of low jitter (210fs) single mode pulses from a 1.3μm Fabry-Perot laser diode by self-seeding, *Proc. ECOC, paper Tu.P.24, Florence,*

Ikegama T, 1992, Survey of telecommunications: Application of quantum electronics—Progress with fibre communications, *IEEE Proceedings*, 80 411.

Ito H, Yokohama H, Murata S, and Inaba H,1979, Picosecond optical pulse generation from an r.f. modulated AlGaAs DH laser diode,*Electron Lett*, 15 738

Iwatsuki K, Suzuki K, and Nishi S, 1991, Generation of transform limited gain-switched DFB-LD pulses <6ps with linear fibre compression and spectral window, *Electron Lett*, 27 1981

Jinno M, 1993, Correlated and uncorrelated timing jitter in gain-switched laser diodes, *IEEE Photon Tech Lett*, 5 1140.

Lin C, Liu P L, Damen T C, Eilenberger D J, and Hartman R L, 1980, Simple picosecond pulse generation scheme with injection lasers, *Electron Lett*, 16 600.

Lin C, and Bowers J E, 1985, Measurement of 1.3 and 1.5μm gain-switched semiconductor pulses with picosecond IR streak camera and a high speed InGaAsa PIN photodiode, *Electron Lett*, 21 1200.

Lourtiouz J-M, Chusseau L, Brun E, Hamaide J-P, Lesterlin D, and Leblond, 1992, Fourier-transform limited pulses from gain-switched distributed Bragg reflector laser using simultaneous modulation of the gain and phase sections, *Electron Lett*, 28 1499.

Mar A, Helkey R, Bowers J E, Mehuys D, and Welch D,1994, Mode-locked operation of a master oscillator power amplifier,*IEEE Photon Tech Lett*, 6 1067.

Marcenac D D, and Carroll J E, 1994, Comparison of self-pulsation mechanisms in DFB laser diodes, *LEOS'94, paper SL8.5.*

Mehuys D, Welch D F, and Scifres D R,1993, High power diffraction limited external cavity tunable diode lasers,*LEOS'93, paper SCL7.3.*

Morton P A, Mizrahi V, Tanbun-ek T, Logan R A, Lemaire P, Erdogan T, Sciortino P F, Sergent A M, and Wecht K W, 1994, High-powermode-locked hybrid pulse source using two-section laser diodes, *Opt Lett*, 19 725.

O'Gormann J, Level A F J, and Hobson W S, 1991, High power switching of multielectrode broad area lasers, *Electron Lett*, 27 13.

Onodera N, Ito H, and Inaba H, 1984, Fourier-transform-limited single-mode picosecond optical pulse generation by a distributed feedback laser, *Appl Phys Lett,* **45** 843.

Paoli T L, 1977, Changes in the optical properties of CW (AlGa)As junction lasers during accelerated aging,*IEEE J Quantum Electron,* **13** 351.

Parke R, Welch D F, Hardy A, Lang R J, Mehuys D, O'Brien S, Dzurko K M, and Scrifes D R, 1993, 2.0W cw diffraction-limited operation of a monolithically integrated master oscillator power amplifiers,*IEEE Photon Tech Lett,* **5** 297.

Raybon G, Hansen P B, Koren U, Miller B I, Young M G, Newkirk M, Iannone P P, Burrus C A, Centanni J C, and Zirngibl M, 1992, Two contact 1cm long monolithic extended cavity actively mode-locked at 4.4GHz,*Electron Lett,* **28** 2220.

Schell M, Weber A, Schol E, and Bimberg D, 1991, Fundamental limits of subpicosecond pulse generation by active mode-locking of semiconductor lasers : The spectral gain width and the facet reflectivities, *IEEE J Quantum Electron,* **27** 1661.

Solgaard O, and Lau K Y, 1993, Optical feedback stabilization of the intensity oscillations in ultrahigh-frequency passively mode-locked monolithic quantum well lasers, *IEEE Photon Tech Lett,* **5** 1264.

Summers H D, White I H, Laughton F R, Ralston J D, Penty R V, Williams K A, Sarma J, Middlemast I, and Ryan T, 1995, Passive mode-locking in p-doped MQW bow-tie lasers, *Proc CLEO, Baltimore.*

Sundersan H, and Wickens G E, 1990, Very high amplitude, minimal chirp optical pulse generation at 1.55μm using multicontact DFBs and an Erbium doped fibre amplifier, *Electron Lett,* **26** 725.

Takada A, Sugie T, and Saruwatari M, 1986, Transform limited 5.6ps optical pulse generation at 12GHz repetition rate from gain-switched distributed feedback laser diode by employing pulse compression technique, *Electron Lett,* **22** 1347.

Thedrez B J, Saddow S E, Liu Y Q, Wood C, Wilson R, and Lee C H, 1993, Experimental and theoretical investigation of large output power Q-switched AlGaAs semiconductor lasers, *IEEE Photon Tech Lett,* **5** 19.

Torphammer P and Eng S T, 1980, Picosecond pulse generation in semiconductor lasers using resonance oscillation, *Electron Lett,* **16** 587

Tsang D Z, and Walpole J N, 1983, Q-switched semiconductor diode lasers, *IEEE J Quantum Electron,* **19** 145.

Tucker R S, Koren U, Raybon G, Burrus C A, Miller B I, Koch T L, and Eisenstein G, 1989, 40GHz active mode-locking in a 1.5μm monolithic extended cavity laser,*Electron Lett,* **25** 621.

Ueno M, and Lang R, 1985, Conditions for self-sustained pulsation and bistability in semiconductor lasers, *J Appl Phys* **58** 1689.

Van der Ziel J P, 1979, Spectral broadening of pulsating AlGaAs double heterostructure lasers, *IEEE J Quantum Electron,* **15** 1277.

Vasil'ev P P, 1988, Picosecond injection laser : a new technique for ultrafast Q-switching,*IEEE J. Quantum Electronics,* **24,** 2386.

Vasil'ev P P, 1993, High power high frequency picosecond pulse generation by passively Q-switched 1.55μm diode lasers, *IEEE Journal of Quantum Electron,* **29** 1687 .

Vasil'ev P P, 1995, Ultrafast diode lasers : Principles and applications,(Artech House).

Vasil'ev P P, and Sergeev A B, 1989, Generation of bandwidth limited 2 ps pulses with 100GHz repetition rate from multisegmented injection laser, *Electron Lett,* **25** 1049.

Vasil'ev P P, White I H, Burns D, and Sibbett W,1993a, High power low jitter encoded picosecond pulse generation using an RF-locked self Q-switched multicontact GaAs/GaAlAs diode laser, *Electron Lett,* **29** 1594.

Vasil'ev P P, White I H, and Fice M J, 1993b, Narrow line high power picosecond pulse generation in a multicontact distributed feedback laser using modified Q switching, *Elect. Lett.*, **29**, 561.

Volkov L A, Guriev A L, Danil'chenko V G, Deryagin A G, Kuksenkov D V, Kuchinskii V I, Portnoi E L, and Smirnitskii V B,1989, Generation and detection of picosecond optical pulses in InGaAsP/InP lasers ($\lambda = 1.5 - -1.6\mu$m) with passive Q-switching, *Sov Technol Lett*, **15** 497.

Volpe F P, Gorfinkel V, Sola J, Kompa G, 1994, 140W, 40ps single optical pulses for laser sensor application, *Proc. CLEO,(Anaheim) paper CWC4.*

Walpole J N, Kinzter E S, Chinn S R, Livas J C, Wang C A, Missaggia L J, Woodhouse J D, 1994, High power monolithic tapered semiconductor oscillators, *Proc. CLEO,(Anaheim) invited paper CMA1,*

Walpole J N, Kinzter E S, Chinn S R, Wang C A, and Missaggia L J, 1992 High-power strained layer InGaAs/AlGaAs tapered travelling wave amplifier, *App Phys Lett*, **61** 740.

Weber A G, Ronghan W, Böttcher E H, Schell M, and Bimberg D, 1992, Measurement and simulation of the turn-on delay time jitter in gain-switched semiconductor lasers, *IEEE J Quantum Electron*, **28** 441.

White I H, Griffin P S, Fice M J, and Whiteaway J E A, 1992, Line narrowed picosecond optical pulse generation using a three contact InGaAsP/InP multiquantum well distributed feedback laser under gain-switching, *Electron Lett*, **28** 1257.

Williams K A, 1995, Q-switched Diode Lasers, *Univ of Bath, Ph.D. thesis.*

Williams K A, Sarma J, White I H, Penty R V, Middlemast I, Ryan T, Laughton F R, and Roberts J S, 1994, Q-switched bow-tie lasers for high energy picosecond pulse generation, *Electron Lett*, **30** 320.

Williams K A, Summers H D, Abd Rahman M K, Muller J, Laughton F R, White I H, Penty R V, Jiang Z, Sarma J, Middlemast I, Ralston J D, 1995, Monolithic passive mode-locking in single contact tapered lasers, *Proc. LEOS'95, (San Francisco) paper SCL9.1*

Microcavity Lasers and QED

Gunnar Björk

Stanford University, USA

1   Introduction

Radiative emission processes always come about through the coupling between an excited atom, an exciton or some other excited particle and one or more modes of the electromagnetic field. In many textbooks, even today, this coupling is treated as if it was independent of the boundary conditions. Of course this is not true in any dynamic process—the boundary conditions often play as important a role in determining the system dynamics as the equation of motion. The success of classical physics largely comes from the fact that, for a system with a large number of internal degrees of freedom (a large number of modes), the mean equation of motion will always give a good approximation of the system evolution. However, when the system becomes small and the internal degrees of freedom small, the radiative decay can be radically modified.

When an excited atom interacts with a large number of modes the atom excitation decays exponentially with a fixed rate. Purcell (1946) was the first to suggest that if the atom was only interacting with a small number of modes in a resonant cavity, the spontaneous emission decay rate would be modified. The subject of spontaneous emission control remained largely a theoretical subject. Dicke(1954) treated the case of radiative decay of a collection of atoms prepared in a collective state. Jaynes and Cummings (1963) advanced the field further by showing that an atom coupling to a single mode would not decay at all, the excitation would instead oscillate periodically between the atom and the field. During the early 1970s material technology matured to allow the fabrication of mono-layer films. In a series of pioneering works, Drexhage (1974) experimentally demonstrated modifications of spontaneous emission patterns by a reflector coated with a mono-layer of atoms. However, it was not until the 1980s that the technology was mature enough for a device demonstration. In 1985 the first experiment showing a clear change of spontaneous emission rate was performed (Hulet *et al.* 1985). Soon thereafter, the first wavelength-size microlaser was demonstrated (De Martini *et al.* 1987). This marked a breakthrough in the microlaser field since the experiment clearly demonstrated the benefits to be gained from spontaneous emission engineering in the form of a reduced laser threshold pump-rate.

It was not long thereafter that experimentalists also entered the strong coupling regime predicted by Jaynes and Cummings. Raizen *et al.* (1992) were the first to observe the effect of cavity dressing on a collection of atoms. Weisbuch(1992) and co-workers demonstrated the same effect in a semiconductor microlaser.

Today the microcavity laser field is vigorously pursued by researchers from at least three continents. This chapter, first presented as a lecture series, intends to give a pedagogical overview of the field as it stands at the time of writing in mid 1995. We shall devote the first section to the radiative processes of a single atom and of single atom spontaneous emission engineering. In the following section we shall see how laser devices and light emitting diodes stand to benefit from spontaneous emission engineering. In the subsequent section we shall see how the picture radically changes when there is more than one atom in the cavity. As shall be seen, the many atom (or exciton) assumption will make things more complicated but also better in many respects (*e.g.* fast decay and directional emission). However, dephasing effects (such as phonon scattering in semiconductors) will quickly destroy the spatial and temporal coherence of the atom ensemble and bring the radiative decay physics back to the single atom case. In the final section we shall present a brief summary of the chapter.

2 Fundamental light-atom interaction

2.1 A single electromagnetic mode and a single two-level atom

The simplest light-atom interaction one can think of is the coupling between a single atom and a single mode of the electromagnetic field. This system provides considerable insight into the fundamental physics of the light-atom coupling but the reader is warned that the more general many mode-many atom interactions are much richer in their dynamics.

The interaction Hamiltonian, in the dipole approximation, between the electromagnetic mode and the two-level atom is

$$H_I = -q\,\mathbf{r}\cdot\mathbf{E}, \tag{1}$$

where q is the elementary charge, $q\mathbf{r}$ is the atom's dipole moment operator and $\mathbf{E}$ is the electric field operator. We will use a state representation $|a, b\rangle$ where the index a refers to the atom excitation and b to the electric field excitation. Hence for $|0, 0\rangle$ both the atom and the field are in their ground states, for $|1, 0\rangle$ the atom is excited but the field-mode is in the vacuum state, and for $|0, 1\rangle$ the atom is in its ground state but the field mode has one quantum of excitation. It is convenient to express the dipole operator using Pauli spin-flip operators, σ^+, σ^- and σ_z. These operate on the states according to

$$
\begin{aligned}
\sigma^+|0,b\rangle &= |1,b\rangle & \sigma^+|1,b\rangle &= 0 \\
\sigma^-|0,b\rangle &= 0 & \sigma^-|1,b\rangle &= |0,b\rangle \\
\sigma_z|0,b\rangle &= -|0,b\rangle & \sigma_z|1,b\rangle &= |1,b\rangle.
\end{aligned}
$$

For the field we use the standard creation and annihilation operators $a^\dagger$ and a. The interaction Hamiltonian becomes

$$
H_I = \hbar(a + a^\dagger)(g\sigma^+ + g^*\sigma^-), \tag{2}
$$

where

$$
g = q\left(\frac{\omega_a}{2\hbar\epsilon}\right)^{1/2} \Psi(\mathbf{r}_a) \cdot \int d^3r\, \Phi_e^*(\mathbf{r})\, \mathbf{r}\, \Phi_g(\mathbf{r}), \tag{3}
$$

where the ω_a is the atom transition (angular) frequency, $\epsilon = \epsilon_r\epsilon_0$ is the dielectric constant, $\Psi(\mathbf{r}_a)$ is the mode-function of the field at the location of the atom, and $\Phi_e^*(\mathbf{r})$ and $\Phi_g(\mathbf{r})$ are the mode-functions of the excited and ground state atom respectively. It is important to note that the atom's position enters explicitly in the coupling between the coupling coefficient. If, for example, $\Psi(\mathbf{r}_a) = 0$, that is the atom is located in a node-position of the cavity field, then the atom and field do not couple. In Equation 3 we have assumed that g is real. If no external reference 'clock' is available (such as phase-information from a laser that was used to invert the atom at $t = 0$) we can do so without loss of generality. In spatially extended multi-atom cases, each atom has its own coupling coefficient with the field, and in this case the coupling coefficient phase is important.

In the rotating wave approximation the total Hamiltonian of the system is

$$
H = \frac{\hbar\omega_a}{2}(1 + \sigma_z) + \hbar\omega_f a^\dagger a + \hbar g(a\sigma^+ + a^\dagger\sigma^-). \tag{4}
$$

where ω_f is the angular frequency of the field mode. One notes that the Hamiltonian couples the state $|1,n\rangle$ only to the state $|0,n+1\rangle$, where $n = 0,1,2,\ldots\infty$, and to no other state. In any given excitation manifold these states are called the bare states, since they are eigen-states of the non-coupled system Hamiltonian (with $g = 0$). Due to the rotating wave approximation the Hamiltonian conserves the total excitation number. The Hamiltonian matrix $H_{ij} \equiv \langle i|H|j\rangle$, where $i, j = 1, 2$ in the state basis $|1\rangle = |1,n\rangle$, $|2\rangle = |0,n+1\rangle$ is

$$
H = \hbar\begin{pmatrix} \omega_a + n\omega_f & g\sqrt{n+1} \\ g\sqrt{n+1} & (n+1)\omega_f \end{pmatrix}. \tag{5}
$$

Diagonalising the matrix we find that the eigen-energies are

$$
\omega_- = \frac{\omega_a + \omega_f}{2} + n\omega_f - \frac{1}{2}\left(4(n+1)g^2 + (\omega_a - \omega_f)^2\right)^{1/2} \tag{6}
$$

$$
\omega_+ = \frac{\omega_a + \omega_f}{2} + n\omega_f + \frac{1}{2}\left(4(n+1)g^2 + (\omega_a - \omega_f)^2\right)^{1/2}, \tag{7}
$$

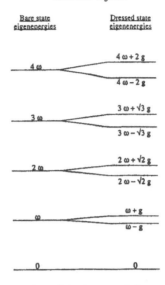

Figure 1. *Schematic representation of the bare and dressed state eigen-energies for the first few excitation manifolds.*

and the corresponding eigen-states are

$$|-,n+1\rangle = \frac{1}{N_-}\left[\frac{\omega_a - \omega_f - (4(n+1)g^2 + (\omega_a - \omega_f)^2)^{1/2}}{2\sqrt{n+1}g}|1,n\rangle + |0,n+1\rangle\right] \quad (8)$$

$$|+,n+1\rangle = \frac{1}{N_+}\left[\frac{\omega_a - \omega_f + (4(n+1)g^2 + (\omega_a - \omega_f)^2)^{1/2}}{2\sqrt{n+1}g}|1,n\rangle + |0,n+1\rangle\right], \quad (9)$$

where $N_\pm$ is a (relatively uninteresting) normalisation factor. The states $|-,n\rangle$ and $|+,n\rangle$ are called the dressed system states. We see that the atom-field coupling splits the eigen-frequencies of the two eigen-modes by $\Omega_{\text{Rabi}} = \sqrt{4(n+1)g^2 + (\omega_a - \omega_f)^2}$. We see that even in the $N = 1$ manifold (spanned by the bare states $|1,0\rangle$ and $|0,1\rangle$) the eigen-energies are separated by the energy $2\hbar g$ (when $\omega_a = \omega_f$), so even an excited atom sitting in a empty cavity gets 'dressed'. A standard interpretation is that the atom gets dressed by the vacuum fluctuations in the cavity, because even though the expectation value for the electric field $E(\mathbf{r}) = (\hbar\omega_f/2\epsilon)\Psi(\mathbf{r})(a^\dagger + a)$ is zero since $\langle a,0|(a^\dagger + a)|a,0\rangle = 0$, the expectation value of $E^2 \propto (a^\dagger + a)^2$ is non-zero, since $\langle a,0|(a^\dagger + a)^2|a,0\rangle = 1$. The splitting of the atom transition frequencies was first derived by Rabi. However, he considered a quantised atom in a classical, un-quantised field. He found virtually the same formula for the eigen-frequency difference, but with a factor of n instead of $n + 1$. The additional factor of '+1' in $n + 1$ is usually attributed to the zero-point energy of the field and it is a true quantum effect. It is the same effect and has the same origin as spontaneous emission which will be treated in the next two subsections. In the case of zero detuning, $\omega_a = \omega_f = \omega_0$, the expressions simplify considerably. The eigen-frequency splitting becomes $\Omega_{\text{Rabi}} = 2g\sqrt{n+1} = 2g\sqrt{N}$. The dressed state energies in the manifold ladder is depicted in Figure 1. The two eigen-modes in the $N = n$ manifold case are equal superpositions of the $|1,n-1\rangle$ and the $|0,n\rangle$ states.

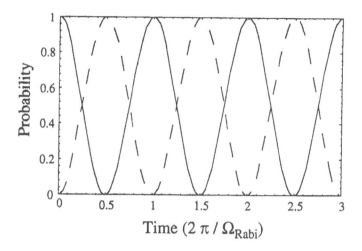

Figure 2. *Probabilities of finding the system in the respective bare states as a function of time. The solid line is $P_{|0,n\rangle}$ and the dashed line is $P_{|1,n-1\rangle}$.*

Hence,

$$|-,n\rangle = \frac{1}{\sqrt{2}}(|1,n-1\rangle - |0,n\rangle) \tag{10}$$

$$|+,n\rangle = \frac{1}{\sqrt{2}}(|1,n-1\rangle + |0,n\rangle) \tag{11}$$

Since these are eigen-states of the Hamiltonian they evolve in time according to $|\pm(t)\rangle = \exp(-i\omega_{\pm}t)|\pm\rangle$. Hence, if the system is started with n photons in the cavity and the atom in the ground state, $|\psi(t_0)\rangle = |0,n\rangle = (|+,n\rangle - |-,n\rangle)/\sqrt{2}$, the time evolution of the system is simply

$$|\psi(t)\rangle = \{\cos(\Omega_{\text{Rabi}}t/2)|0,n\rangle + i\sin(\Omega_{\text{Rabi}}t/2)|1,n-1\rangle\}\exp[-i\omega_0(t-t_0)]. \tag{12}$$

The two eigen-modes will beat against each other and one unit of excitation will be periodically transferred from one bare state to the other. The probabilities of finding the system in the two respective bare states are given by

$$P_{|0,n\rangle}(t) = |\langle 0,n|\psi(t)\rangle|^2 = \cos^2(\Omega_{\text{Rabi}}t/2) \tag{13}$$

$$P_{|1,n-1\rangle}(t) = |\langle 1,n-1|\psi(t)\rangle|^2 = \sin^2(\Omega_{\text{Rabi}}t/2). \tag{14}$$

The time evolution of these states is plotted in Figure 2. In an isolated system energy cannot be dissipated. This is why the emission of the excited atom is eventually transferred back to the atom. The atom-field coupling is coherent and therefore the emission becomes reversible. To investigate the spontaneous emission process we need to consider an open system. This is the topic of the next subsection.

2.2   Weisskopf-Wigner theory of spontaneous emission

An atom in free space couples not only to one, but to a continuum of modes. As we have seen in the previous section, the coupling shifts the transition eigen-frequencies slightly from the bare state eigen-frequencies. When the time evolution of the coupling between the atom and the field modes proceed at detuned frequencies, the respective atom to field energy transfer will mutually decohere and, if the number of modes is infinite, will lead to irreversible transfer of the excitation from an excited atom to the field modes.

The interaction Hamiltonian in the case of a large number of field modes can be written

$$H = \frac{\hbar\omega_a}{2}\left(1 + \sigma_z\right) + \sum_j \left(\hbar\omega_{fj}a_j^\dagger a_j + \hbar(g_j a_j \sigma^+ + g_j^* a_j^\dagger \sigma^-)\right), \tag{15}$$

where the creation and annihilation operators $a_j^\dagger$ and a_j operate on the jth field mode. Note that in this case, with multiple field modes we can no longer assume that the coupling coefficient g is real for all modes. Since the Hamiltonian only couples states like $|1, 0, 0, \ldots, 0\rangle$ and $|0, 0, \ldots, 0, 1, 0, \ldots, 0\rangle \equiv |0, 1_j\rangle$, where j is the index of the excited field mode after the interaction, the state at any time t can be expressed

$$|\psi(t)\rangle = c_a(t)|1, 0, \ldots, 0\rangle \exp(-i\omega_a t) + \sum_j c_j(t)|0, 1_j\rangle \exp(-i\omega_{fj}t). \tag{16}$$

Inserting this state expansion in the Schrödinger equation

$$i\hbar\frac{d}{dt}|\psi(t)\rangle = H|\psi(t)\rangle,$$

we obtain the following coupled equations

$$\dot{c}_a(t) = -i\sum_j g_j c_j(t)\exp[-i(\omega_{fj} - \omega_a)t] \tag{17}$$

$$\dot{c}_j(t) = -ig_j^* c_a(t)\exp[i(\omega_{fj} - \omega_a)t]. \tag{18}$$

Formally integrating Equation 18 and inserting the expression for $c_j(t)$ in Equation 17 we obtain the following integro-differential equation for $c_a(t)$

$$\dot{c}_a(t) = -\sum_j |g_j|^2 \int_0^t dt' c_a(t')\exp[-i(\omega_{fj} - \omega_a)(t - t')]. \tag{19}$$

To solve this equation we let the field quantisation volume go to infinity. The sum $\sum_j$ over all the modes can then be approximated by an integral over the density of states. The number of plane wave modes of one polarisation per volume differential d^3k in k-space is $d^3k\, V/(2\pi)^3$. Using a spherical coordinate system so that we can express the k vector in the coordinates (k, θ, φ), the differential becomes $d^3k = k^2 \sin\theta\, dk\, d\theta\, d\varphi$. Using the transformation $k = \omega/c$, we arrive at the number of modes m for a fixed polarisation per (angular) frequency differential $d\omega$

$$m = \frac{V\omega^2}{(2\pi c)^3}\sin\theta d\omega d\theta d\varphi. \tag{20}$$

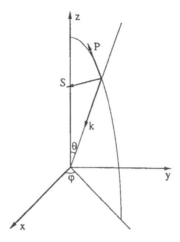

Figure 3. *Illustration of the dipole orientation relative to that of the field modes and the spherical coordinate system.*

Hence, a sum over a large number of modes of some k vector dependent function $f(\mathbf{k})$ can be transformed into an integral over frequency according to the transformation

$$\sum_j f(\mathbf{k}) \rightarrow \frac{V}{(2\pi c)^3} \int d\omega\, \omega^2 \int_0^\pi d\theta \sin\theta \int_0^{2\pi} d\varphi\, f(\omega/c, \theta, \varphi). \tag{21}$$

We shall use this transformation to evaluate the sum $\sum_j |g_j|^2$. We shall assume that the dipole-moment $\mathbf{p}$ of the atomic transition lies along the $\hat{\mathbf{e}}_z$ axis in the coordinate system in Figure 3. We see that only the plane waves with an electric field component in the $\hat{\mathbf{e}}_z$-k plane couple to the atom, the other half of the modes with $\mathbf{E}$ perpendicular to the $\hat{\mathbf{e}}_z$-k plane do not contribute to the decay. The coupling coefficient between the modes that do couple is proportional to $\sin\theta$. Furthermore the mode-functions of the plane electromagnetic field waves are $\Psi(\mathbf{r}) = \exp(i\mathbf{k}\cdot\mathbf{r})/V$. Hence

$$\sum_j |g_j|^2 = \frac{V}{(2\pi c)^3} \int d\omega_j\omega_j^2 \int_0^\pi d\theta \sin^3\theta \int_0^{2\pi} d\varphi \frac{\omega_a|M|^2}{2\hbar\epsilon V}, \tag{22}$$

where we have introduced the transition matrix element

$$M = q \int d^3r\, \Phi_e^*\, z\, \mathbf{r}\, \Phi_g(\mathbf{r}). \tag{23}$$

Since

$$\int_0^\pi d\theta \sin^3\theta = 4/3, \tag{24}$$

we arrive at the following expression for the equation of motion for $c_a(t)$:

$$\dot{c}_a(t) = -\frac{1}{6\pi^2\hbar c^3\epsilon} \int d\omega\, \omega^2\omega_a|M|^2 \int_0^t dt'c_a(t') \exp[-i(\omega - \omega_a)(t - t')]. \tag{25}$$

Before we try to simplify this expression we will briefly discuss the reason that the equation of motion of c_a is independent both of the position of the atom within the cavity and of the quantisation volume. We have assumed plane wave modes in our quantisation cavity. This is slightly unphysical, because in general, cavity modes have more complicated mode functions. In a rectangular box cavity the modes are standing sinusoidal waves and the field mode function Ψ, and hence the coupling coefficient g, involve a factor $\sqrt{2}\sin(k_x x)/L_x$ associated with the x direction; here L_x is the box length in the x-direction and the factor $\sqrt{2}$ is due to the normalisation of the mode-function. Hence, $|g_j|^2$ is zero if the atom is located in a standing wave node position and it is twice as large as the plane wave coupling coefficient (which is position independent) when the atom is located in a standing wave anti-node position. However, if the mode number is large, the atom will be located in the node of one mode, in the anti-node of another mode and somewhere in-between for a third mode etc. Therefore the large number of modes in the cavity will effectively probe the spatially averaged coupling coefficient. Hence, $\langle|\psi(\mathbf{r}_a)|^2\rangle = V^{-1}$. This is the same as for the plane wave modes. However, as the cavity gets larger, the density of states increases linearly with V, as can be seen in Equation 20. Therefore the total coupling of all the modes to the atom becomes cavity volume and geometry independent. In order for Equation 20 to be true the cavity volume must be much larger than the pertinent wavelength in all spatial dimensions. Tailoring of spontaneous emission decay can *only* be accomplished in cavities small enough for the concept of density of states to be invalid so that the behaviour of the system is not the spatially (and frequency) averaged response we are calculating here.

Also worth noticing from Equation 25 is that if the cavity is filled with some dielectric, then $c = c_0/\sqrt{\epsilon_r}$ and $\epsilon = \epsilon_0\epsilon_r$, where ϵ_r is the relative dielectric constant. Hence the spontaneous emission coupling factor is proportional to $\sqrt{\epsilon_r}$, or in other words, to the refractive index of the dielectric. In reality the story is a bit more complicated since the local field correction factor should also be included (Glauber and Lewenstein 1991, Björk et al. 1991), but for our purposes we need not include this factor in this overview. The reader can rest assured that the omission is without consequences in this case.

Now let us return to Equation 25. First we note that the integral over t' has an appreciable value only when $\omega_a \approx \omega_f$, Therefore it is justifiable to take the factor $\omega_a\omega^2|M|^2$ outside the integral and replace it by $\omega_a^3|M|^2$. Secondly we shall assume that $c_a(t')$ varies slowly compared to the exponential factor preceding it in the time integral so that we can remove it from the integral and set $c_a(t') \approx c_a(t)$. Finally we note that the remaining integral can be solved in the limit $t \to \infty$ where

$$\lim_{t\to\infty} \int_{t_0}^{t} dt' \exp[-i(\omega - \omega_a)(t - t')] = \pi\delta(\omega - \omega_a) - P\left\{\frac{i}{\omega - \omega_a}\right\} \qquad (26)$$

where P stands for the principal value. The first term (which is real) gives us the spontaneous emission decay rate, whereas the second term (imaginary) describes (a small) radiative shift. Ignoring the radiative shift, the equation of motion for $c_a(t)$ becomes

$$\dot{c}_a(t) = -\frac{\omega_a^3|M|^2 c_a(t)}{6\pi\hbar c^3\epsilon} = -\Gamma_0 c_a(t)/2, \qquad (27)$$

where $\Gamma_0 \equiv 1/\tau_{sp0}$ is the spontaneous radiative decay rate. The final expression for the radiative decay rate of a single atom is

$$\Gamma_0 = \frac{\omega_a^3 |M|^2}{3\pi \hbar c^3 \epsilon}. \tag{28}$$

The mathematics of the Weisskopf-Wigner theory of spontaneous emission tends to obscure the relatively simple physics of the emission process. Essentially the difference from the previous section is that when an excited atom interacts with a mode continuum, the decay becomes exponential and non-reversible in contrast to the atom-finite mode interaction, which features revival. The reason is that with a discrete number of system eigen-modes, there tends to be a time when the excited superposition of eigen-modes re-phases to the initial condition phases (being eigen-modes, their amplitudes are constants of motion). This is not the case when the atom couples to a mode continuum, re-phasing never occurs. The other important piece of physics to retain is that, for a given atom, the atom-field coupling depends on the field mode-function at the location of the atom. The mode function is a 'classical' function and is computed from Maxwell's equations and the boundary conditions. Therefore, to first order, spontaneous emission control does not involve quantum mechanics. We also saw that since we did not assume any particular cavity geometry in our discussion, the decay rate can only be influenced by the boundary conditions of a cavity if some of the assumptions we made above are not valid, *i.e.* if the cavity is small enough so that the field modes interacting with the atom are discrete and not (quasi)-continuous.

2.3 Fermi's Golden Rule

An alternative, shorter way of calculating the single atom spontaneous emission rate is to use Fermi's Golden rule, which is a prescription for calculating general transition rates to first order in perturbation theory. According to the Golden Rule, the transition rate is given by

$$\Gamma_0 = \frac{2\pi}{\hbar^2} \varrho(\omega_a) |\langle \psi_f | H_I | \psi_i \rangle|^2, \tag{29}$$

where ψ_f (ψ_i) is the final (initial) state of the system, H_I is the interaction picture Hamiltonian and any frequency dependent term in the interaction Hamiltonian is to be taken at the transition (angular) frequency ω_a. The density of states (per unit angular frequency for a particular polarisation) is $\varrho(\omega) = V\omega^2/(2\pi^2 c^3)$. Note that the modes polarised perpendicular to the $\hat{e}_z$-$\mathbf{k}$ plane do not couple to the dipole so we should use ϱ and not 2ϱ. Finally, the single atom coupling term (the bracket in equation 29) depends on the angle between the dipole vector and the mode $\mathbf{k}$-vector. Therefore the average value of the bracket squared is 2/3 of its maximum value obtained when the mode propagates perpendicular to the dipole moment. The final result for the spontaneous emission rate is

$$\Gamma_0 = \frac{2\pi}{\hbar^2} \frac{V\omega_a^2}{2\pi^2 c^3} \frac{2}{3} \frac{\hbar \omega_a |M|^2}{2V\epsilon} = \frac{\omega_a^3 |M|^2}{3\pi \hbar c^3 \epsilon}, \tag{30}$$

This is the same result as from the Weisskopf-Wigner theory, but with an important distinction in interpretation. We see that in the Fermi Golden Rule formula, the density of states explicitly enters. Therefore one may rightfully claim that in order to control spontaneous emission, the density of states should be modified. In the literature different authors take different viewpoints. Some attribute the spontaneous emission modification in small cavities to the modification of the density of states, some attribute it to the modification of the mode-function. Neither viewpoint is right or wrong, handled correctly they both lead to the same result. In this chapter we will adopt the viewpoint that the mode-functions are being modified. We will use a very large quantisation cavity and put a small microcavity inside it. If the microcavity is small enough, the cavity modes of the large cavity will be negligibly disturbed on the large scale, but locally the mode functions can be quite different in the presence of the micro-cavity.

2.4 Einstein's A and B coefficients

The spontaneous emission rate calculated in the previous two subsections is, by definition, the same as Einstein's A coefficient. The B_{21} coefficient is defined as the rate of stimulated emission per unit energy density (energy per unit volume) and per unit frequency bandwidth. Expressed in the entities above it can be written

$$B_{21} = \frac{\pi |M|^2}{3\hbar\epsilon}. \tag{31}$$

Hence the ratio between the A and B_{21} coefficients is

$$\frac{A}{B_{21}} = \frac{\hbar\omega_a^3}{\pi^2 c^3}. \tag{32}$$

The right hand side of Equation 32 can be factored in the (mean) energy density of one photon (at the transition frequency) in the cavity $\hbar\omega_a/V$ times the density of states per unit frequency bandwidth at the transition frequency $\varrho(\omega_a)$. This means that if we put a single photon in every cavity mode, the stimulated emission rate per unit emission frequency bandwidth would equal the spontaneous emission rate per unit emission frequency bandwidth. This result is general and it is independent of the size or shape of the cavity. Remember that Einstein invoked only the assumption of thermodynamic equilibrium when he derived the rates. If the result is true for any cavity of any shape, the result must also hold true for every mode in the cavity. Therefore the result above tells us that the spontaneous emission rate in any mode is equal to the stimulated emission rate per mode at the (mean) mode excitation of one photon. Hence, for a specific mode i we can write the emission rate per excited atom Γ_i as

$$\Gamma_i = g_i(n+1), \tag{33}$$

where g_i is the modal gain (per photon) of the mode. Equation 33 in all its simplicity is very important. It tells us that if we want to improve the rate of stimulated emission into a specific mode, the spontaneous emission will increase too. Conversely, it is sufficient to increase the spontaneous emission rate into a mode to assure that the stimulated emission rate will also increase.

It is often argued that spontaneous emission is emission stimulated by the vacuum field. If that was unconditionally true, one would expect a spontaneous emission rate equal to half that of the stimulated emission rate per photon since the zero-field energy is $\hbar\omega/2$. So why is the stimulated emission rate twice what one would expect? One interpretation is that the emitted field acts back on the emitting atom. The process is called radiation reaction and arises in quantum mechanical systems for the same reason that zero-field fluctuations arise, *i.e.* operators do not necessarily commute. The interested reader is referred to Milonni (1994). What is important for us is the fact that boundaries affect the vacuum field and the self-field (radiation reaction) in the same way. Hence we can simply see by what factor the spontaneous emission of the zero-point field induced decay is modified by a boundary and rest assured that the radiation reaction induced decay will be modified by the same factor, and hence the total decay process too.

2.5 Modification of spontaneous emission rate and pattern by the presence of boundaries

In this subsection we shall look at two simple problems where boundaries modify the spontaneous emission rate of an excited atom. Both examples have historical significance. In the first observation of modified spontaneous emission by Drexhage (1974), he used a thin molecular Langmuir-Blodgett film attached to a glass plate which acted as a weak mirror. We shall first examine how an ideal plane mirror affects the spontaneous decay rate of atoms put in close vicinity to it. The second example will be that of an ideal mirror planar cavity. This was the first configuration in which microlaser action was reported by De Martini and Jacobvitz (1988). We will conclude the subsection with some general comments on spontaneous emission control in other confined geometries.

Our viewpoint will be that we have a large enough system so that we can assume that the modes incident on our small system are unaltered by it. Again we will assume that the normal modes of the unaltered systems are plane waves and therefore we can use one polarisation index and the mode wavevector to specify the mode. However, it will prove convenient to work in a spherical coordinate system and we can use the notation (k, θ, ϱ) to specify the state. Since $k = \omega/c$, it is actually more convenient to use the notation $(\omega, \theta, \varrho)$. We have seen in Section 2.2 that the coupling with a dipole oriented in the z-direction is proportional to $\sin^2 \theta$ for the modes polarised in the $\hat{e}_z$-k plane (we shall call these waves P-modes), and zero for the modes polarised normal to that plane (the S-modes, see Figure 3). Assume now that the normalisation volume is roughly cubic and large enough so that we can assume that the modes are isotropically and continuously distributed in k-space. We shall also assume that the mode density is constant over the small frequency ranges $\Delta\omega$ we consider. This is certainly true if $\Delta\omega \ll \omega_a$. In this case we can write the spontaneous emission rate from the atom into P-polarised modes per unit solid angle and unit angular frequency

$$\gamma_{\rm sp}(\omega, \theta, \varrho) = C(\omega) \sin^2 \theta, \tag{34}$$

where C is a frequency dependent function chosen to get both the correct decay rate and line-shape of the atom. Integrating the emission over all of k-space we get

$$\Gamma = \int_0^\infty d\omega \, C(\omega) \int_0^\pi d\theta \sin^3 \theta \int_0^{2\pi} d\varrho = \frac{8\pi}{3} \int_0^\infty d\omega \, C(\omega). \tag{35}$$

Identifying this expression with the free space spontaneous emission rate Γ_0, where Γ_0

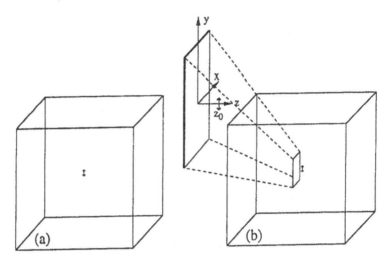

Figure 4. *The large cavity without (a) and with (b) a planar mirror. The inset in (b) shows a blowup of the mirror and the atomic dipole.*

is given by Equation 28, one finds that

$$\gamma_{sp}(\omega, \theta, \varrho) = \frac{3\Gamma_0 \sin^2 \theta f(\omega)}{8\pi}, \tag{36}$$

where f is a normalised function such that $\int d\omega f(\omega) = 1$. The reason we need this function is that due to homogeneous and inhomogeneous broadening, the atom-field coupling is distributed in frequency. As long as the spontaneous emission rates that we calculate in the presence of a microcavity are not vastly different from the free space rates, we can consider the function f as a fixed function describing the free space spontaneous line-shape. In most semiconductor systems the emission line-shape is dominated by inhomogeneous broadening and in this case f is quite independent of Γ. In most solid state systems it is a very good approximation to consider f as a fixed function. It is quite trivial to extend the discussion to dipoles oriented in the xy-plane. For an atom with its dipole moment oriented in the $\hat{\mathbf{e}}_y$-direction, the corresponding expressions are

$$P\text{-mode} \qquad \gamma_{sp}(\omega, \theta, \varrho) = \frac{3\Gamma_0 \cos^2 \theta \sin^2 \varphi f(\omega)}{8\pi}, \tag{37}$$

$$S\text{-mode} \qquad \gamma_{sp}(\omega, \theta, \varrho) = \frac{3\Gamma_0 \cos^2 \varphi f(\omega)}{8\pi}, . \tag{38}$$

The expression for the emission per solid angle for the $\hat{\mathbf{e}}_x$-oriented dipole is simply obtained by interchanging $\sin \varphi$ and $\cos \varphi$ in Equations 37 and 38.

Look now at what happens if we insert a perfect 'metallic' mirror just to the left of the dipole: see Figure 4. The mirror is very large on the scale of the atom and the wavelength, but is small on the scale of the quantisation volume. Therefore the mode-functions are going to be approximately the same in the absence or presence of the mirror. We shall assume that the mirror is a perfect conductor with unit reflectivity at all (relevant) wavelengths and angles. From subsection 2.2 we realise that the only change the mirror brings to the system is that the field mode-functions $\Psi(r_a)$ become modified. Therefore, if we insert the modification factors in Equations 37 and 38 and integrate over all modes (over all solid angles and all frequencies), we can calculate the modified spontaneous emission rates. Using classical field theory the modification factors are trivial to compute; they are $2\sin(k_z z_0) = 2\sin(k z_0 \cos\theta)$ for all modes of both polarisations and dipole moment orientations, but the P-mode coupling to the z-oriented dipole. Here z is the distance between the mirror and the atom in the mirror normal direction. For the P-mode coupling to the z-oriented dipole the modification factor is $2\cos(k_z z_0)$. In Figure 5 the spontaneous emission radiation patterns for free space and several different combinations of dipole orientations and positions are plotted. Integrating the emission over solid angle and frequency we get the following expressions for the spontaneous emission rates:

$$
\begin{aligned}
\Gamma &= \frac{3\Gamma_0}{8\pi} \int_0^\infty d\omega\, f(\omega) \int_0^{\pi/2} d\theta\, 4\cos^2(kz\cos\theta)\sin^3\theta \int_0^{2\pi} d\varphi \\
&\approx 3\Gamma_0 \left(\frac{1}{3} - \frac{\cos(2kz_0)}{4k^2 z_0^2} + \frac{\sin(2kz_0)}{8k^3 z_0^3} \right)
\end{aligned}
\tag{39}
$$

for a dipole oriented in the z-direction, and

$$
\begin{aligned}
\Gamma &= \frac{3\Gamma_0}{8\pi} \int_0^\infty d\omega\, f(\omega) \int_0^{\pi/2} d\theta\, 4\sin^2(kz\cos\theta)\sin\theta \int_0^{2\pi} d\varphi \left(\cos^2\varphi + \cos^2\theta\sin^2\varphi \right) \\
&\approx \frac{3\Gamma_0}{2} \left(\frac{2}{3} - \frac{\cos(2kz_0)}{4k^2 z_0^2} + \frac{(4k^2 z_0^2 - 1)\sin(2kz_0)}{8k^3 z_0^3} \right)
\end{aligned}
\tag{40}
$$

for a y-oriented dipole. Note that due to the mirror, only the modes impinging from the right can reach the dipole, therefore the θ integral goes only from 0 to $\pi/2$. The solutions of these integrals are particularly simple when $z_0 \gg \lambda$ or $z_0 \ll \lambda$ and when $f(\omega)$ is a sharply peaked function on the ω scale. In the first case the factors $\cos^2(2kz\cos\theta)$ and $\sin^2(2kz\cos\theta)$ are oscillating so rapidly as a function of θ that they both effectively can be taken to be $1/2$. In this case both expressions simplify to Γ_0. This is sensible because when the atom sits many many wavelengths away from the mirror wall it is effectively located in free space. Both Equations 39 and 40 also simplify when $z_0 = 0$. In this case the spontaneous emission rate of the z-oriented dipole is $2\Gamma_0$ and the rate for the y-oriented dipole is zero. These are the limiting decay rates for an atom located close to a mirror. The reason the factor of two (and zero) emerges is that the mirror changes the modes seen by the atom from plane waves to standing waves. Close to the mirror it is possible to find positions where almost all the impinging waves interfere constructively (destructively), leading to a doubling (quenching) of the local electric field density. Far from the mirror the atom sees the average electric field density, which is the same regardless of whether the modes are plane waves or standing waves. In Figure 6 the spontaneous emission rate is plotted as a function of the mirror to dipole spacing. It is seen that, as demonstrated above, for large distances the emission rate approaches that of free space. This is a general trait. One can say that if the atom to mirror spacing is large, then the atom actually is in the free space.

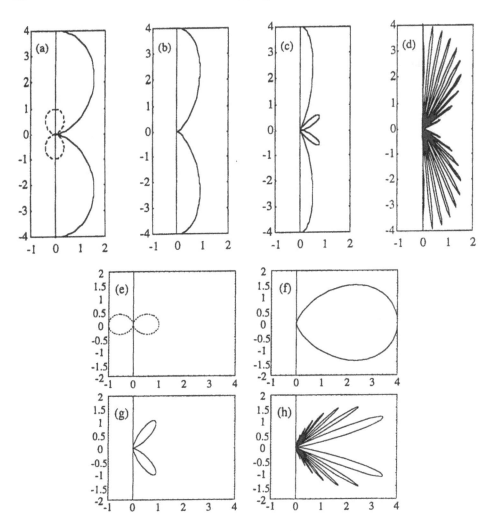

Figure 5. *(a)—(d) Emission patterns in the yz-plane from a z-oriented dipole in free space (dashed) and in front of a perfect mirror. The straight vertical line represents the mirror plane. In (a), (b), (c) and (d) the dipole is z-oriented and $z_0 = 0$, $\lambda/4$, $\lambda/2$ and 5λ, respectively. (e)—(h) Emission patterns in the yz-plane from a y-oriented dipole in free space (dashed) and in front of a perfect mirror. In (e), (f), (g) and (h), $z_0 = 0$, $\lambda/4$, $\lambda/2$ and 10λ, respectively.*

Before finishing this subsection we shall briefly look at a second example of spontaneous emission control which is analytically solvable. If two plane mirrors are held parallel to each other a planar cavity is formed, Figure 7. In this case too it is simple to derive the mode modification functions. For the waves impinging from the right they

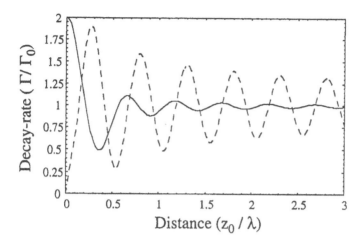

Figure 6. *Computed spontaneous emission rate, normalised to the free space decay rate, as a function of the mirror spacing z. The solid line is when the dipole moment points in the mirror normal direction and the dashed line is when the dipole moment lie in the mirror plane.*

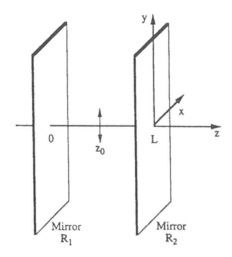

Figure 7. *The geometry of the planar cavity. In this figure the dipole-moment of the atom is oriented in the y-direction, parallel to the cavity mirrors.*

are

$$\frac{(1 - R_2)[1 + R_1 - 2\sqrt{R_1}\cos(2kz\cos(\theta))]}{(1 - \sqrt{R_1 R_2})^2 + 4\sqrt{R_1 R_2}\sin^2(kL\cos(\theta))} \tag{41}$$

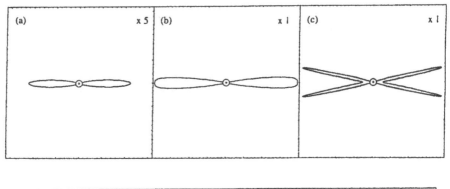

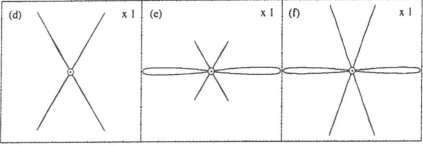

Figure 8. *The emission patterns in the xz-plane for an atom with dipole moment in the y-direction for different cavity configurations. In (a), (b) and (c) the atom is in the center of the cavity and $L = 0.49\lambda$, $\lambda/2$ and 0.51λ. In (d) and (e), $L = \lambda$ and in (f) $L = 3\lambda/2$. In (d) and (f) the atom is in the center of the cavity while in (e) $z_0 = L/4$. $R_1 = R_2 = 0.95$ in all plots.*

for the P- and S-polarised modes coupling to the in-plane dipole moment, and

$$\frac{(1 - R_2)[1 + R_1 + 2\sqrt{R_1}\cos(2kz\cos(\theta))]}{(1 - \sqrt{R_1 R_2})^2 + 4\sqrt{R_1 R_2}\sin^2(kL\cos(\theta))} \tag{42}$$

for the P-polarised modes coupling to the cavity mirror normal dipole moment, where R_1 and R_2 are the left and right mirror reflectivities. One can show, using time-reversibility arguments that the emission 'caused' by zero-point fluctuations impinging from the right of the cavity will also be emitted to the right. The importance of this is that if the cavity is small and the atom is not sitting in the centre, the emission spectra measured on the left and right hand side of the cavity need not be identical (Lei *et al.* 1993). To get the mode modification functions for the waves impinging from the left, make the substitution $z \to L - z$ and exchange R_1 and R_2. We can immediately see that if $R_1 = R_2 = 0$, the modification factors reduce to unity. In Figure 8 the modification of the emission patterns for several different configurations is shown.

Inserting the modification factors in Equations 36, 37 and 38, and numerically evaluating the subsequent integrals, the spontaneous emission lifetime can be computed. In Figure 9 the normalised lifetime is shown, assuming nearly perfect mirrors ($R_1 = R_2 = 0.99$) and a very narrow atom coupling function $f(\omega)$. It is seen that for

(a)

(b)

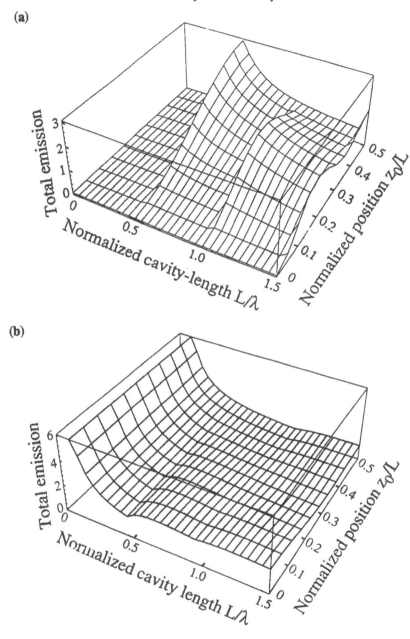

Figure 9. *(a) The spontaneous emission decay rate versus the normalised cavity length L/λ and dipole position z_0/L. The atomic dipole moment is oriented parallel to the cavity mirrors. (b) The spontaneous emission decay rate versus the normalised cavity length L/λ and dipole position z_0/L. The atomic dipole moment is oriented normal to the mirrors.*

a dipole oriented parallel to the mirrors, the spontaneous emission can be increased to $3\Gamma_0$ for a half wavelength long cavity with the atom located at its centre. If the cavity is shorter than half a wavelength the emission is strongly suppressed. For the normal dipole, the maximum spontaneous emission enhancement takes place in a short cavity. For very short cavities the decay rate approaches $2\Gamma_0(1 + 3/\pi^2)/(1 - R)$, (in theory). Due to surface plasmon effects this enhancement is not likely to be seen in the laboratory. For both configurations the decay rate Γ approaches Γ_0 if the cavity is many wavelengths long, except if the dipole sits very close to one mirror. In this case the geometry reduces to that treated above (single mirror in a large cavity) as it should, and the decay rates approach 0 and $2\Gamma_0$ for the parallel and perpendicular dipole, respectively. An ideal planar cavity can enhance the spontaneous emission rate from an atom with in-plane atomic dipole moment by a factor of three at most. This factor is sufficiently small that it is difficult to measure it. The factor can be decomposed into two parts. A factor of two results from placing the atom in the (quasi)-mode anti-node of the cavity standing wave. A collection of atoms randomly spaced in the cavity would have an (average) decay rate a factor of two smaller. Another factor of 3/2 comes from the fact that the modes allowed into the cavity are all propagating in the cavity normal direction so that the field-atomic dipole interaction is maximised. One can show that the on-resonance planar cavity only redistributes the total zero-point field density, the total field density remains constant (Björk 1994). In general cavity geometries with more confinement (in two and three dimensions) can achieve larger spontaneous emission enhancements (Brorson *et al.* 1990). Still, the simple planar geometry achieves an important goal in any spontaneous emission controlling structure, *it channels all of the emitted radiation into a single quasi-mode*, as can be seen from Figure 5(b). If the atomic dipole moments lie randomly oriented in the cavity mirror plane, and this is the situation in planar semiconductor microcavities using quantum well active material, the emission goes into two polarisation-orthogonal states. Still, the emission from such a light emitting diode would be 50%. However, when dielectric cavities are used, the simple analysis above breaks down, and in general the modification rates and single-mode emission coupling efficiency becomes quite low.

2.6 Dielectric cavities

In the previous subsection we examined how the emission characteristics could (in principle) be modified by mirrors. However, in reducing theory to practice one needs to take additional factors into account. One such factor is that metal mirrors are far from ideal at optical frequencies. First, the maximum reflectivity obtainable with thin-film metal mirrors is on the order of 95%. Worse yet is that high reflectivity metal film mirrors absorb a substantial fraction of the transmitted light. In order to have reasonable photon lifetimes in wavelength size cavities one needs reflectivities of the order of 99–99.9%. Such reflectivities are obtainable with dielectric Bragg stack mirrors. In addition, the transmission absorption of such mirrors can be kept small. However, the high reflectivity is bought at the expense of limited ranges for wavelength and angle of incidence. In practice the latter (at a fixed wavelength) is the more limiting. The range of angles over which a dielectric mirror is highly reflective depends on the Bragg layer refractive index difference. For semiconductor epitaxial materials (material

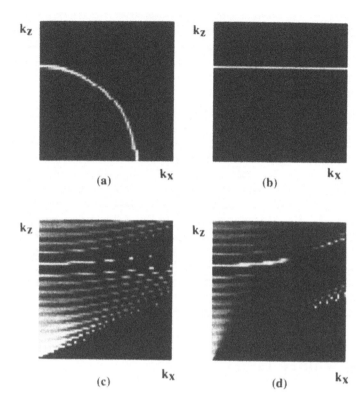

Figure 10. *The electric field density as a function of the k_x-k_z coordinates. In (a) the emission pattern into the S-polarised free-space modes for a y-oriented dipole is shown. In (b) the corresponding field density for an ideal, $L = \lambda/2$, planar metallic mirror cavity is shown and in (c) and (d) the field density for the S- and P-polarised modes of a half wavelength long planar dielectric cavity is shown.*

compositions with the same crystal lattice spacing), such as GaAlAsP, the mirror is only good at incidence angles up to about 25 degrees. At larger angles of incidence the mirror is essentially transparent and does little to modify the incident zero-point field modes. In Figure 10 the electric field densities in the centre of an ideal and a dielectric microcavity are shown. The white horizontal line in Figure 10(b) (indicating a large field density) shows that the k_z becomes quantised in the ideal planar cavity. The emission pattern of an atom placed in the cavity is essentially the product of the patterns in Figure 10 (a) and (b). Hence, if the cavity is tuned to the transition so that the emitted light $|k|$ is equal to k_z of the cavity quantised mode, then light will only be emitted into the $\hat{e}_z$ direction as indicated in Figure 10 (b). However, in the dielectric case, cavity modes are only suppressed and enhanced in a narrow band. The resonance is still seen as a slightly curved white line in the centre of a wider black band, the area in which the Bragg mirrors are highly reflecting. However, if we were to multiply (a) by (c) or (d) we find that lots of emission will be emitted at angles between 25 and 70 degrees from the normal. This is called the open window of the Bragg mirror. Careful analysis reveals that actually most of the emission leaks out through the open windows, and only a few percent is captured in the resonant mode (Björk *et al.* 1991; Yamamoto *et al.* 1992; Lin *et al.* 1994). In addition the spontaneous emission lifetime will be modified by marginal amounts in realistic planar structures (Björk *et al.* 1991; Lin *et al.* 1994; Suzuki *et al.* 1991).

Similar results hold for other microcavity geometries such as hemispherical cavities (Matinaga *et al.* 1993), microcylinders (Chu and Ho 1993), microdisc cavities (Levi *et al.* 1992; Chu *et al.* 1994; Chin *et al.* 1994) and vertical post microcavities (Baba *et al.* 1991; Baba *et al.* 1992). So far, the structure which most efficiently collects the emitted spontaneous emission into a single cavity quasi-mode is the microdisc laser, for which a spontaneous emission coupling factor of 15% has been claimed (Chu *et al.* 1994). In spite of being far from 100%, this result is significant, as will be discussed in the next section. In closing this section we would like to stress that the optimum device geometry has still not been found. Many candidates exist, possibly the vertical post structure looks most promising at the moment. The reason it is difficult to come up with an ideal structure is that realistic three-dimensional structures are difficult both to fabricate and to compute. Only the simplest geometries can be computed analytically. In addition, most of the promising structures are very difficult to fabricate due to the small dimensions and the tight fabrication tolerances.

3 Microcavity lasers

For device applications, probably the single most important figure of merit for micro-cavities is the spontaneous emission coupling ratio. The reason that this is so can be understood from the insights from the previous section. There we showed that the spontaneous and stimulated emission rate (per photon) into any given mode are identical. We also demonstrated that it is difficult to modify the spontaneous emission rate of an atom by any substantial amount in real cavities at optical frequencies. We can write the spontaneous emission rate as a sum over all modes,

$$\Gamma = \sum_i g_i \approx \Gamma_0. \tag{43}$$

The spontaneous emission rate into a given mode can be written formally as $\beta\Gamma$, where β is the spontaneous emission coupling ratio into the mode. Hence

$$\beta \equiv \frac{g_i}{\Gamma}. \tag{44}$$

Since g_i is also the stimulated emission gain (per photon) into the mode, it seems likely that it would be advantageous to have lasers with large values of β. It follows from the definition that $0 \leq \beta \leq 1$. In an ordinary semiconductor laser $\beta \approx 10^{-4}$-10^{-5}. Hence, in principle, it should be possible to improve β and consequently the stimulated emission gain per photon by several orders of magnitude. We shall see below that this will lead to a corresponding reduction of the laser threshold.

3.1   Micro-cavity rate equations

A rate equation analysis geared specifically towards microcavity devices was developed independently in Yokoyama and Brorson (1989) and Björk and Yamamoto (1991). The results and conclusions are similar, but the notations are slightly different. In this subsection we shall follow that of Björk and Yamamoto (1991).

The rate equations for the photon number n and the free carrier density N in any laser can be written

$$\frac{dN}{dt} = \frac{I}{qV} - \left(\frac{1-\beta}{\tau_{\text{sp}}} + \frac{\beta}{\tau_{\text{sp}}}\right) N - \frac{N}{\tau_{\text{nr}}} - \frac{gn}{V} \tag{45}$$

$$\frac{dn}{dt} = -(\gamma - g)n + \frac{\beta NV}{\tau_{\text{sp}}} \tag{46}$$

where I is the injection current, q is the elementary charge, V is the volume of the *active material*, τ_{nr} is the non-radiative decay rate, g is the modal gain per photon per unit time, and γ is the cavity decay rate per unit time. Note that g is not simply a material dependent constant, as seen in Sections 2.2 and 2.5, it actually depends on the cavity geometry and the exact location of the gain volume. In a macroscopically sized cavity, averaging in space and in frequency makes this constant material dependent only. In these equations we have used $1/\tau_{\text{sp}}$ instead of Γ, so as not to confuse the notation with the mode confinement factor Γ used in many rate equation models. Also note that we have assumed that the non-radiative decay rate is independent of N. This is not necessarily true in real systems, but it simplifies the analysis substantially and allows us to bring out the relevant physics.

In semiconductors, a simple and reasonably accurate gain model is to assume that

$$g = g'(N - N_0), \tag{47}$$

where N_0 is the transparency free carrier density. The stimulated emission gain per excited carrier and cavity photon is hence $g'N$, whereas the definition of β and τ_{sp} gives the total spontaneous emission rate into the mode $\beta NV/\tau_{\text{sp}}$. The factor NV is the total excited carrier number, and remember that τ_{sp} is the decay rate *per excited carrier*. We saw in Section 2.4 that the two rates equal each other for every mode and every excited atom. Therefore it must also hold true for a mutually incoherent collection of excited atoms (or for carriers in a semiconductor). This leads to the relation

$$g' = \frac{\beta V}{\tau_{\text{sp}}}. \tag{48}$$

Note that the spontaneous emission rate and the modal gain are interconnected by β which depends on the cavity geometry and especially the cavity volume.

3.2   The concept of threshold

With the equations in the previous subsection we are ready to look at one important characteristic of microlasers — the threshold current. Recently there has been some discussion in the literature about the definition of a threshold in microlasers. Essentially three viewpoints have been advocated. The first is that these small devices do not have a threshold, they are thresholdless (DeMartini and Jacobovitz 1988). The second is that the lasers have a threshold, but that this threshold should be defined differently from the ordinary macroscopic laser threshold (Björk and Yamamoto 1991). The third and most recent viewpoint is that the concept of a microcavity laser threshold is not well defined. In a microscopic (or rather mesoscopic) system one cannot expect to see effects associated with a macroscopic system size (Rice and Carmichael 1994). While the third viewpoint is probably the most correct from a strict axiomatic point of view, we shall stick to the second view in this section with the justification that the so defined threshold will at least serve as an operational definition. The device behaviour, say, ten times below the threshold defined in this way will be quite different from the behaviour, say, ten times above the threshold.

From an operational point of view, the 'laser' below threshold is a (linear) amplifier, while above the threshold it is a (non-linear) oscillator. The reason a laser can oscillate in a sustained and stable fashion is that there is a self-regulating mechanism built into the system, the gain saturation. Below threshold the gain saturation is negligible. The gain saturation is due to the stimulated emission which rapidly depletes the gain if the photon number in the cavity exceeds the equilibrium value. A reasonable way to define the threshold is to say that the threshold is reached when the stimulated emission overtakes the spontaneous emission as a generation source. Physically this occurs when the mean photon number in the mode equals unity [see Equation 33].

$$n_{\mathrm{th}} = 1. \tag{49}$$

Substituting this condition into the rate-equations above, the threshold current can be written (Björk and Yamamoto 1991)

$$I_{\mathrm{th}} = \frac{q\gamma}{2\beta}\left[1 + \beta + \frac{\tau_{\mathrm{sp}}}{\tau_{\mathrm{nr}}} + \xi\left(1 - \beta + \frac{\tau_{\mathrm{sp}}}{\tau_{\mathrm{nr}}}\right)\right], \tag{50}$$

where the dimensionless parameter $\xi \equiv (N_0\beta V)/(\gamma\tau_{\mathrm{sp}})$. In Figure 11 I_{th} is plotted for various cases.

Essentially Equation 50 has two contributions if the non-radiative decay rate is negligible ($\tau_{\mathrm{sp}} \ll \tau_{\mathrm{nr}}$). For any laser $1 < 1 + \beta < 2$ so the first term (approximately) reads $q\gamma/\beta$. This is the pump current needed to sustain spontaneous emission in all modes at the threshold free carrier density. Not unexpectedly the term is proportional to β^{-1}. Hence all improvements of the spontaneous emission coupling ratio will lead to a decrease of the threshold (assuming that all other parameters remain fixed). The other term will essentially read $q\gamma\xi/2\beta = qN_0V/2\tau_{\mathrm{sp}}$. This is the current needed to sustain the spontaneous emission at the transparency free-carrier density. It is seen from Equation 50 that if the non-radiative recombination is negligible and β approaches unity this term approaches zero anyhow, that is, the threshold can formally be achieved in an 'absorbing' (or non-inverted) laser medium. This is not as strange as it sounds, remember that our usual picture of absorption is associated with multi-mode decay. As seen in Section 2.1 above, the situation is different in a (nearly) single mode device (*i.e.* in a device with $\beta \approx 1$) with no other decay channel for the excited free carriers (or atoms) than radiative 'decay'. In such a device 'dissipation' does not exist, so

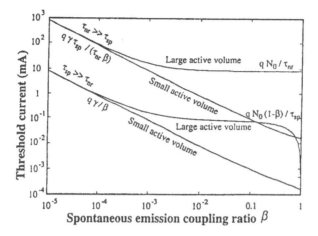

Figure 11. *The threshold currents as a function of the spontaneous emission coupling ratio β. The two top traces are the curves when non-radiative decay dominates the spontaneous emission decay. In the lower two curves the opposite has been assumed to be true. In the two curves denoted 'large active volume' ξ has been assumed to be $10^3\beta$. In the 'small active volume' curves, $\xi = \beta$ has been assumed. All curves coincide within a factor of two with the conventional definition except for the dramatic drop near $\beta = 1$ of the threshold current of a 'large active volume' laser in which non-radiative recombination dominates.*

the concept of 'transparency' loses meaning. Only in this regime does the proposed threshold definition differ substantially from the conventional one (Björk and Yamamoto 1991; Yamamoto and Björk 1991). We also see that in lasers with a small active region, for example in quantum well lasers, the threshold is essentially proportional to β^{-1} irrespective of whether the non-radiative or radiative recombination dominates. In Figure 12 we have plotted reported threshold currents for vertical cavity lasers as a function of the year it was reported. It can be seen that the threshold for the lowest threshold steadily decreases. It can also be seen that although the sub 100μA threshold level is already surpassed (Huffaker *et al.* 1994) and a few lasers come close to it (Numai *et al.* 1993), no laser comes close to the 1μA threshold that should be attainable in theory. One reason is that no laser is quite as small as it possibly could be. That the threshold reduction is directly coupled to the active volume size can be seen in Figure 13. In this figure the same data is plotted versus (estimated) cavity volume. The best devices all lie on a line proportional to V_{cav}. However, it is clear that there is still room for improvement. As micro- and nano-fabrication technology matures at least this author expects minimum threshold currents to drop by another order of magnitude.

3.3 Micro cavity laser characteristics

In most respects the device characteristics of microlasers will not differ from the characteristics of macroscopic lasers. Fundamentally the operational principle is the same.

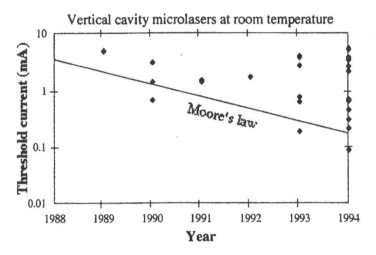

Figure 12. *The reported threshold currents for electrically pumped vertical cavity semiconductor lasers emitting at 800-1000 nm at room temperature, as a function of year of the report. It can be seen that the lowest reported threshold current steadily decreases as time proceeds. It can also be seen that the interest in VCSEL is steadily increasing.*

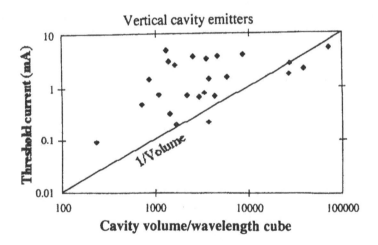

Figure 13. *The threshold currents for the same devices as in the previous figure, but now plotted versus size. The diagonal trend-line is the expected slope for good devices assuming that $\beta \propto V_{cav}^{-1}$.*

Output powers will quite naturally be lower, but linewidths, modulation bandwidths, beam quality etc. should in principle be comparable. As for the linewidth, the cold cavity bandwidth of a microlaser is of the same order of magnitude as cleaved facet Fabry-Perot semiconductor lasers. Since it should be possible to drive microlasers

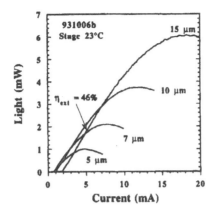

Figure 14. *The output power versus pump current for electrically pumped vertical cavity lasers operated at room temperature. All lasers, independent of size, shows a saturation behavior. The saturation current density is roughly device size independent. The saturation is most likely due to a combination of non-linear gain saturation and cavity mode/emission wavelength mismatch due to thermal effects. (Reproduced with permission from J W Scott, B J Thibault, D B Young, L A Coldren and F H Peters, IEEE Phot. Techn. Lett 6, 678 (1994). ©1994 IEEE.)*

as high above threshold (say ten times) as macroscopic lasers, the linewidths should be comparable. Of course, extremely narrow linewidth microlasers may be difficult to achieve since the high cold cavity Q-value (small cold cavity bandwidth) is often achieved by means of a physically large (long) cavity. Some investigations indicate that microcavity linewidths are intrinsically larger than the linewidths of macroscopic lasers (Jahnke *et al.* 1993; Mohideen *et al.* 1994). However, the evidence for this behaviour is still inconclusive.

The modulation bandwidth of microlasers should also be close to that of conventional semiconductor lasers, but at much smaller bias currents. The modulation speed increases with the stimulated emission rate, which in turn increases with increasing pumping above threshold, or equivalently, increasing differential gain. However, due to the non-linear gain effect, it is unlikely that the differential gain can be made higher in microlasers than in macroscopic lasers (Karlsson *et al.* 1994). In Figure 14 typical output power versus pump curves are shown for different size vertical cavity lasers. It is seen that the saturation current (and the corresponding output power) decreases roughly according to decreasing cavity size. Therefore it is unrealistic to believe that a μm sized laser will emit more than perhaps 100μW of optical power. In Figure 15 the calculated modulation bandwidth is plotted versus the injection current density. It turns out (Karlsson *et al.* 1994) that the threshold current densities in microlasers and macrolasers are very similar. Hence the plot is more or less independent of the semiconductor laser size. When a non-linear gain model is used (but device and driving circuit parasitic elements are neglected), the modulation frequency 3dB cutoff bandwidth saturates at a few tens of GHz as shown by the solid line in Figure 15. The calculated

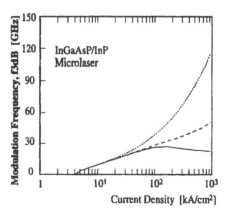

Figure 15. *The calculated intensity modulation 3dB cutoff bandwidth versus the injection current density in InGaAs/InP vertical cavity lasers. The figure is roughly independent of the device diameter. The dotted line represent the case where the gain is assumed to be strictly independent of the intracavity photon density, the solid and dashed lines are calculated for different gain-saturation models. The solid line represents the most commonly employed model where the gain has been assumed to be proportional to* $(N - N_0)/(1 + \delta n)$, *where δ is a material dependent constant.*

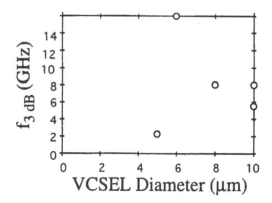

Figure 16. *Reported* VCSEL *semiconductor intensity modulation bandwidths versus the* VCSEL *cavity diameter.*

behaviour is consistent with measured cutoff values, which are plotted as a function of the VCSEL diameters in Figure 16. We see that the (few) reports (Michler *et al.* 1995; Shtengel *et al.* 1993) show no correlation between size and bandwidth, but that the bandwidth figures, between ten and twenty GHz, are consistent with theoretical predictions.

3.4   Microcavity light emitting diodes

Recently microcavities have been employed to enhance the performance not only of lasers, but also of LEDs. In most display applications the output power and mode quality of LEDs are sufficient, diode lasers are not needed. Stated differently, the number of applications where LEDs are used vastly outnumbers the diode laser applications. In all consumer electronics, and in many cars, LEDs are used as indicator lights. A possible future application for LEDs is colour flat panel displays where presently many different technologies compete and no clear winner has emerged.

For LED applications, the reason for incorporating microcavities is slightly different from that for laser applications. LEDs typically emit light over such a wide wavelength range that spontaneous emission lifetime engineering is difficult. What one can do, however, is to improve the coupling efficiency to specific mode(s), and enhance/inhibit spontaneous emission at particular wavelengths.

The former is employed in the Er-doped LEDs developed mainly by AT&T Bell Laboratories (Hunt *et al.* 1992). These are intended for moderately fast fibre optic communication over intermediate distances. For this application one has to worry both about the amount of LED power one can couple into the optical fibre and about chromatic dispersion. Both are improved by using a microcavity. As seen above, a well designed microcavity will redistribute the radiation both spatially and spectrally. The main reason one cannot employ macroscopic cavities for this application is that the large mode volume allows for too many modes within the emission band. A large cavity will therefore only 'modulate' the emission pattern and the emission spectrum. In contrast to lasers, LEDs are linear, and no mode-competition and gain-saturation can suppress the 'side-modes'. Figure 5 (d) and (h) illustrate quite well why macroscopic cavities cannot be used to channel more light into a specific mode when the numerical apertures involved are large.

A particular application in which the huge free spectral range of microcavities comes in handy is polymer light emitting diodes. These devices which are organic semiconductors (such as polyphenylene vinylene) are intrinsically white light emitters, since they emit light over the whole visible spectrum. However, in many applications such as flat panel colour displays, one needs sources of the three primary colours red, green and blue rather than white light sources. Enclosing the 'white-light' LED in a microcavity accomplishes the needed spectral filtering in a particularly simple fashion. The active layer is enclosed between two flat and reflecting metal electrodes. By adjusting the thickness of a (transparent) spacer layer sitting between for example the active polymer layer and the top electrode, the light from the same polymer film can be tuned to either red, green or blue. The polymer layer can be fabricated inexpensively by spin-coating, a micro-patterned spacer layer can subsequently be deposited between the film and the metal-film top electrode, resulting in a large array of LEDs emitting alternatively in red, green and blue (Dodabalapur *et al.* 1994).

Spontaneous emission 'control' can be achieved by other means than modifying the mode structure of the vacuum-fields. One way of increasing the quantum efficiency of light emitting diodes is to try to reduce the radiative loss of the unwanted spurious modes (without eliminating the modes, which is the objective when one encloses the LED by a microcavity). The simplest way to do this is to make all (or in reality, most) modes resonant in a low loss cavity. The cavity prevents the energy in the spurious modes being radiated into free space modes. The simplest such configuration is the metal clad dielectric cylindric rod cavity, although other more elaborate schemes have been tested (Gigase *et al.* 1991). The modes travelling along the rod symmetry axis couple strongly to the free space modes. The modes propagating in the rod normal direction, however, are trapped inside the metal cladding. If the absorption loss of the metal cladding is small enough, the energy in these modes will eventually be re-absorbed by the emitting atoms in the rod. In presence of phonon scattering, the absorbing atoms will dephase, and the absorbed radiation will be re-emitted in an arbitrary direction. Eventually, in an ideal structure, most radiation will leave the rod cavity along the rod axis, independent of the size of the metal cladding cavity (provided the length of the rod is much larger than the rod diameter).

One reason for the low quantum efficiency in semiconductor LEDs are their high refractive indices and corresponding small light escape cones. All light emitted in a planar LED propagating at an angle higher than $\arcsin(1/n)$, where n is the semiconductor index of refraction, will be totally internally reflected and trapped in the planar structure. In GaAs which has a refractive index of about 3.5, the top angle of the light escape cone is about 17 degrees, corresponding only to about 14% of the total 4π steradian solid angle. Most of the radiation (which is emitted more or less isotropically) is therefore trapped inside the planar structure. One way of overcoming this restriction is to pattern the device surface to make it irregular. This will 'frustrate' the propagation of the light such that the motion of the internally trapped photons become chaotic (Schnitzer *et al.* 1993). This way the probability of re-absorption and subsequent re-emission into the escape-cone will increase, leading to substantially improved quantum efficiencies (Gourley 1994; Schnitzer *et al.* 1993).

3.5   Microcavity geometries

So far we have talked in quite some detail about microcavity effects as such, but very little about the real cavity geometries. At present, there exist many different geometries, all with their respective advantages and disadvantages. The first requirement of a good microcavity geometry is that it must be possible to fabricate, preferably by semiconductor thin film fabrication techniques. If the device is to be pumped electrically one must incorporate conductive layers and bonding pads. (A complication with microlasers is that the bonding pads typically are much larger than the microcavity itself.) A third choice to be made is whether or not to make a surface emitter or an edge emitter. Today, the latter can often be made smaller, but in principle surface emitters could also be scaled to sub-wavelength size. The trend today seems to favour surface emitters for their ability to be tested on the chip, and for their ability to be incorporated in large two-dimensional laser arrays. In addition, many other choices must be made, the most obvious one being the emission wavelength. The selected wavelength severely restricts the range of fabrication materials, and hence the fabrication techniques.

One can roughly divide the fabricated optical microcavities into a few broad categories: planar, dielectric post, disc, ring, droplet and hemispherical. All of them are made from dielectrics for reasons stated above. We are already familiar with the planar cavity, which is the simplest cavity to fabricate. So far, it is the most widely employed

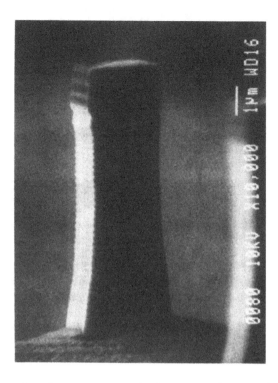

Figure 17. *A scanning electron micrograph of a $2 \times 2\,\mu m$ large, and $8\mu m$ high dielectric InGaAsP micropost laser with a Si/SiO_2 top mirror. The light bouncing between the top and bottom mirror is confined in the lateral direction by the high refractive index post. The microlaser was fabricated at the Department of Electronics, KTH, Stockholm, Sweden. Photo courtesy of Mr. Janos André and Dr. Klaus Streubel.*

cavity geometry (De Martini and Jacobovitz 1988; Yamamoto *et al.* 1992; Yamanishi 1992; Suzuki *et al.* 1991; De Neve *et al.* 1995). We have already seen its limitations; if the cavity loss is low, its cavity volume becomes rather large (Ujihara 1991). Therefore the spontaneous emission can only be engineered to a limited extent.

To make a dielectric post cavity, one patterns the surface of a planar cavity structure with photoresist and etches away everything but narrow dielectric pillars. Figure 17 shows a scanning electron micrograph of such a structure. The high refractive index of the pillar confines the light tightly to the post. This particular post has an epitaxially grown bottom mirror and a dielectric (amorphous) top mirror. The refractive index step between the different top mirror dielectrics (Si and SiO_2) is rather high (both layers are a quarter of a wavelength thick, but the low refractive index SiO_2 layers are much thicker than the high refractive index Si layers). Due to the large refractive index contrast, the mirror need only be a few layers thick. The disadvantage with the dielectric post cavity is that it requires etching very close to the active region. The etching typically induces surface defects that serve as recombination centres. Therefore, lasers like this typically feature high non-radiative recombination rates. An advantage with the structure is that the lasers are surface emitters and that the design is scalable

Figure 18. *A InGaAs/InGaAsP double-disc microdisc laser, 2µm in diameter. The bottom disk with InGaAs quantum wells is the actual laser. The top InGaAsP disk is a passive guiding disk which in principle can be used for electrical bonding. This laser has a threshold (optical) pump power of only 28µ W. (Copyright American Institute of Physics, 1994. Photo courtesy of Prof. S. T. Ho Northwestern University, Evanston, Illinois. Reproduced with permission)*

both in device size and in array size. An alternative to etching pillars is to passivate the layer closest to the active layer by ion implantation. The pump current is then channeled into small holes in the passive layer, defining the active regions. No etching needs to be done, the lasers are gain-guided. A disadvantage with this method is that the mode volume (and hence β) becomes larger than if dielectric pillars standing in free space are formed.

Disc cavities are formed by patterning a planar structure, etching down the top layers to form a thin circular disc, and under-etching the disc to make it sit, more or less free standing perched atop a thin stem (Levi *et al.* 1992; Chu *et al.* 1994). The finished device looks much like a mushroom. The disc is made sufficiently thin so that the modes bouncing between the top and the bottom of the disc all have higher resonant energies than the emission energy. The mode supported at the transition energy bounces around the perimeter of the disc, just like acoustic whispering gallery modes. The discs can be made rather small, disc diameters of 2 μm and disc thicknesses of a few 100nm have been reported (Chu *et al.* 1994). Likewise, the calculated and measured spontaneous emission collection efficiencies are as high as 15% (Levi *et al.* 1992, Chin *et al.* 1994). Bonding can be achieved by making a double disc structure, where the top disc serves as a bonding pad. Such a double-disc can be seen in Figure 18. By locating the bonding disc close to the laser disc, and removing a 'slice of the pie' from the top disc, the output can be made 'directional' (or at least not omni-directional). The main disadvantages with these structures are that they are fragile, and at present, edge-emitting.

Recently the first reported micro-ring cavity was reported. This is essentially a

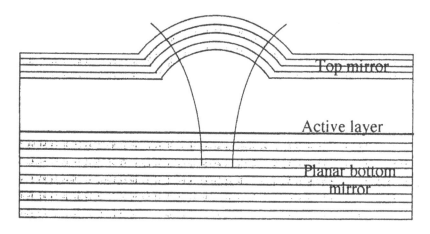

Figure 19. *A schematic drawing of a hemispherical microcavity.*

microdisc cavity with the central (unneeded) portion removed. Instead of sitting at the top of a stem, these cavities sit on top of a low refractive index film. While the reports are preliminary at present, spontaneous emission coupling efficiencies in excess of 0.5 are expected!

Microdroplet cavities are high quality dielectric spheres. Since the geometry is favoured by nature, the cavities can be made very nearly perfectly spherical leading to extremely high Q-values (Collot *et al.* 1993). The simplest microdroplet cavities are made from falling dye microdroplets (Lin and Hsieh 1991). The droplets are excited by a short laser-pulse while they fall, and their temporal and spectral characteristics are subsequently studied. More practical (from a device point of view) microdroplet cavities are formed by melting silica optical fiber to form a microdroplet attached to a small glass stem (Collot *et al.* 1993). Impurities in the melted fiber serve as emitters. The advantage with the cavities are their relative simplicity and high quality, the disadvantage is that they are incompatible with semiconductor thin film fabrication, and that it is difficult to see how such a cavity could be pumped electrically.

The hemispherical microcavity, finally, is fabricated essentially by thin film methods. A bottom Bragg mirror is grown, followed by a spacer region and an active region. On top of the active region a rather thick semiconductor layer is deposited. This top layer is patterned by small circular resist discs. The wafer is then heated which makes the resist discs melt and form small lens-like islands on the semiconductor surface (provided the surface is properly treated) (Matinaga *et al.* 1993). Subsequent ion milling will eat away resist and semiconductor film alike, slowly transferring the lens island shape from the resist onto the semiconductor surface. When the resist has been milled away, the process is stopped, and the lens array patterned surface is coated, by conventional means, with dielectric films forming a non-planar top Bragg mirror surface. Hemispherical cavities are formed between the (roughly) spherical top mirrors and the planar bottom mirrors, Figure 19. It turns out that it is hard to make small radius of curvature mirrors this way. This effectively prohibits very small cavities to be fabricated since the planar mirror should located near the focal point of the top mirror. An advantage with this design is that neither etching nor ion bombardment needs to be done anywhere near the active region, keeping the non-radiative recombination to a minimum.

3.6   Microcavity device applications

The most promising application for microlasers seems to be in short range optical communication. Historically, optical communication was first employed in transoceanic cables, followed by national trunk networks. Relatively quickly optical networks started to make economic sense even for relatively short distances. Today even local area networks and vehicular networks (such as that in the recently deployed fly-by-wire Boeing 777 passenger jetliner) are in use. The trend to communicate by optics over shorter and shorter distances will probably continue for some time yet. The new series of IBM mainframe computers have their motherboards connected by a fibre optical bus. It is likely that many other electronic systems also will incorporate fibre optics for board-to-board and even board interchip communication. There are even those that envisage microlaser intrachip communication (Iwata and Hayashi 1993). Some steps towards this have already been taken. On the market there exist VCSEL arrays specifically geared towards coupling to fibre-ribbons, which consist of ten or so optical fibres configured in a flat ribbon cable. What makes the microcavity lasers desirable (and feasible) in such applications are their small size and low threshold. Both space and power are limiting factors in electronics. Although present day macroscopic laser driving currents may seem sufficiently small, very low threshold current lasers will probably be in demand if issues such as cost, reliability etc. can be met (Iwata and Hayashi 1993; Cutrer and Lau 1995). The output power (maybe 100μW) is sufficient for communications over small distances. As mentioned above, modulation speeds of microcavity lasers are expected to be of the same order of magnitude as macroscopic lasers. To meet the bandwidth demands, the most likely way is to use parallel communication. To this end, surface emitters seem to have an advantage in that they are easily fabricated in arrays.

4   The QED regime, collective effects

4.1   Collective effects—two atoms coupled to one field mode

As mentioned in Section 2, the system dynamics gets substantially richer when there are many atoms (and many excitations) in the system. In this subsection we shall look at a simple system, namely two atoms coupled to one field mode only. In the rotating wave approximation the total Hamiltonian of the system can be written

$$H = \frac{\hbar\omega_a}{2}(2 + \sigma_{1z} + \sigma_{2z}) + \hbar\omega_f a^\dagger a + \hbar g(a\sigma_1^+ + a^\dagger\sigma_1^- + a\sigma_2^+ + a^\dagger\sigma_2^-). \tag{51}$$

The index on the atomic operators indicates on which atom the operator operates. The Hamiltonian above assumes that the atoms are at equivalent positions relative to the field mode (for example just next to each other) since we have assumed identical coupling constants between the two atoms and the field-mode.

For simplicity we shall only look at the states in the first excitation manifold. We shall also assume that $\omega_a = \omega_f = \omega_0$. The Hilbert space is spanned by the bare states $|1,0,0\rangle$, $|0,1,0\rangle$ and $|0,0,1\rangle$, where $|a_1,a_2,b\rangle$ denote the excitation number in atoms 1 and 2, and in the field mode, respectively. Anticipating the result we shall work in the orthonormal base

$$(|1,0,0\rangle + |0,1,0\rangle)/\sqrt{2}, \qquad |0,0,1\rangle, \qquad (|1,0,0\rangle - |0,1,0\rangle)/\sqrt{2}.$$

The Hamiltonian matrix in this basis set is

$$H = \hbar \begin{pmatrix} \omega_0 & g\sqrt{2} & 0 \\ g\sqrt{2} & \omega_0 & 0 \\ 0 & 0 & \omega_0 \end{pmatrix} \tag{52}$$

We see that the state $(|1,0,0\rangle - |0,1,0\rangle)/\sqrt{2}$ does not couple to the other two states. Comparing this matrix with equation 5 we can deduce that the eigen-energies are

$$\omega_- = \omega_0 - \sqrt{2}g \tag{53}$$
$$\omega_+ = \omega_0 + \sqrt{2}g \tag{54}$$
$$\omega_3 = \omega_0. \tag{55}$$

The difference between the case when we only had one atom and one excitation is that the Rabi-frequency becomes a factor of $\sqrt{2}$ larger. This is due to the cooperative effect between the two atoms. However, the atoms only radiate when they have the same atomic phase (when they evolve in-phase, *i.e.* in the state $(|1,0,0\rangle + |0,1,0\rangle)/\sqrt{2}$). If the atoms are out of phase (*i.e.* in state $(|1,0,0\rangle - |0,1,0\rangle)/\sqrt{2}$) their emission interfere negatively, and the result is that they do not couple to the light-field at all. This state is a dark state, or a radiation trapped state. It can neither be excited or de-excited by the cavity photons.

4.2 Superradiance

As seen in the subsection above, atoms evolving in-phase couple more strongly to a light-field than a solitary atom. In Section 2 we learned that whereas the Rabi-frequency scaled as $\sqrt{n+1}$, with the number n of photons in the field mode, the emission rate into a quasi-mode scaled as $(n+1)$.

At low excitation the Rabi-frequency and the emission rate scales essentially the same way with the number of atoms n_a as the number of photons n. The emission rate (per atom) into a quasi-mode will hence scale proportionally to n_a. This is only true if the atoms are located close to each other. The atoms must essentially be closer that half a wavelength or so in order for them to be able to cooperate optimally, as we shall see in the next example.

Now let us consider the spontaneous emission decay of two atoms spaced some distance apart. We shall assume that the atomic dipole moments are parallel and not orthogonal, otherwise the two atoms will couple to modes of orthogonal polarisation and they will not cooperate. First assume that dipoles both point in the z-direction and are located on the x-axis, a distance Δx apart, see Figure 20. We shall use the results from subsection 2.2, and we note that if we ignore the possibility of photo-emission from one dipole followed by the photo-absorption by the second (or the same) dipole, we can easily calculate the decay rate. This approximation is called the pole-approximation and it neglects higher order corrections to the decay rate and to the radiative frequencies. For small systems the approximation is excellent (Knoester 1992; Björk *et al.* 1995). The inclusion of two atoms in the Weisskopf-Wigner theory really only modifies one parameter. When computing the sum over all coupling coefficients squared in Equation 19, there are two atomic wavefunctions to keep track of from Equation 3, so a factor $|\Psi(\mathbf{r}_{a1}) + \Psi(\mathbf{r}_{a2})|^2$ arises. For plane wave field modes this factor is easy to compute since the mode amplitude is fixed. The factor in the case depicted in Figure 20 becomes

$$| \exp(ik_x x_1) \pm \exp(ik_x x_2)|^2 = \begin{cases} 4\cos^2(\pi\Delta x \sin\theta/\lambda) \\ 4\sin^2(\pi\Delta x \sin\theta/\lambda) \end{cases}, \tag{56}$$

where Δx is the physical separation between the dipoles, and the plus and minus equations refers to the symmetric and anti-symmetric initial states given by $(|1,0,0\rangle \pm |0,1,0\rangle)/\sqrt{2}$, Hence the decay rate as a function of distance can be expressed

$$\begin{aligned} \Gamma &= \frac{1}{2}\frac{3\Gamma_0}{8\pi} \int_0^\infty d\omega\, f(\omega) \int_0^\pi d\theta\, 4 \left\{ \begin{array}{c} \cos^2[(k\Delta x \sin\theta)/2] \\ \sin^2[(k_x\Delta x \sin\theta)/2] \end{array} \right\} \sin^3\theta \int_0^{2\pi} d\varphi \\ &= \frac{3\Gamma_0}{2} \left(\frac{2}{3} \pm \frac{\cos(k\Delta x)}{k^2\Delta x^2} \pm \frac{(k^2\Delta x^2 - 1)\sin(k\Delta x)}{k^3\Delta x^3} \right) \end{aligned} \tag{57}$$

The pre-factor $1/2$ in the equation above comes from the normalisation factor $1/\sqrt{2}$ of the initial state wavefunction. We see that for the anti-symmetric initial condition, the decay rate is identical with that of a single atom in front of a perfect mirror with $2z_0 \rightarrow \Delta x$: see Equation 42, Figure 6 and Figure 20(a). That is because the perfect mirror creates the mirror image of the atomic dipole a distance $2z_0$ from the real dipole. This corresponds to the anti-symmetric solution.

Likewise, for two dipoles separated a distance Δz along their dipole moment axes, the expression for the decay rate becomes

$$\begin{aligned} \Gamma &= \frac{1}{2}\frac{3\Gamma_0}{8\pi} \int_0^\infty d\omega\, f(\omega) \int_0^\pi d\theta\, 4 \left\{ \begin{array}{c} \cos^2[(k\Delta z \cos\theta)/2] \\ \sin^2[(k_x\Delta z \cos\theta)/2] \end{array} \right\} \sin^3\theta \int_0^{2\pi} d\varphi \\ &= 3\Gamma_0 \left(\frac{1}{3} \mp \frac{\cos(kz_0)}{k^2 z_0^2} \pm \frac{\sin(kz_0)}{k^3 z_0^3} \right) \end{aligned} \tag{58}$$

An illustration of this setup and the ensuing decay rates is plotted in Figure 20.

The example above teaches us two lessons about collective radiative decay phenomena. The first is that multi-atom decay quickly becomes quite complex. There are many possible initial and intermediate state wavefunctions. Some are superradiant, and in general these are characterised by a high degree of symmetry. Others are sub-radiant, or even dark (Dicke 1954). In general detailed knowledge about the initial wavefunction is needed in order to correctly predict the decay rate. The second lesson is that the

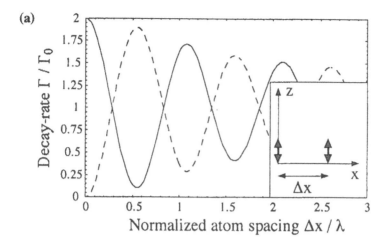

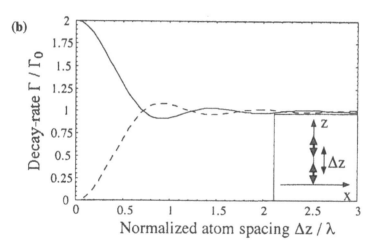

Figure 20. *The geometrical configuration, and the ensuing spontaneous decay rate for two parallel dipoles as a function of the normalised separation distance. The solid line is for the symmetric initial state, the dashed line is for the anti-symmetric initial state. In (a) the atoms are displaced normal to their dipole moment, in (b) they are displaced along their dipole moment axis.*

atoms must be located within a wavelength or so of each other in order to feature net cooperativity. When they are far apart they cooperate (the emission interfere positively) along certain directions, but the radiation is quenched along other directions (the emission interferes negatively). Integrated over all solid angles the net cooperative effects decrease rapidly with increasing distance. As will be shown below, this is also true for a collection of atoms along a line or in a plane. With a fixed atomic line (or surface) density, the net cooperativity will saturate when the system size approaches a wavelength (or wavelength squared).

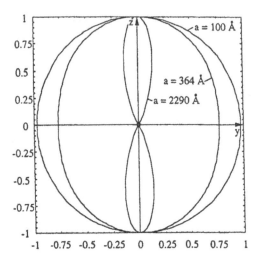

Figure 21. *Emission patterns for excitonic superradiant decay in a quantum well. The superradiance coherence radius a is the parameter. The values 100, 364 and 2290Å correspond to $a \approx a_{\text{Bohr}}$, $a = 1/k \equiv \lambda/2\pi$ and $a = \lambda$ in a GaAs quantum well.*

Let us now look at the superradiative decay of atoms in a thin film of thickness h, or of excitons in a quantum well. We shall assume that the atoms (excitons) are excited in a circular section of radius a. If the atomic density is denoted ρ, the number of atoms within the excitation area is $N = \pi a^2 h \rho$. For excitons in a thin quantum well (where h is of the order of the bulk semiconductor exciton Bohr-radius a_B), the oscillator strength per unit area f_A plays the same role as the number of excited atoms per unit area $h\rho$. Using essentially the same technique as described in the preceding paragraph, we can calculate both the emission pattern and the ensuing superradiative decay rate as a function of the excitation radius a (Björk *et al.* 1994). In Figure 21 the superradiance radiation pattern for GaAs quantum well Wannier excitons is shown for three different values of the excitation radius a. It is seen that as long as the radius is smaller than half a wavelength, the emission pattern of the collectively excited system is essentially the same as for a single excited atom (or a localised exciton). However, when the excitation radius equals or exceeds a wavelength the emission pattern becomes highly directional. For such a delocalised excitation, Heisenberg's uncertainty product allows conservation of transverse momentum, and provided the atomic excitation has no net transverse momentum, the radiation is emitted along a narrow lobe in the quantum well normal direction. In Figure 22 the superradiant decay time has been calculated for the same GaAs system. We see that initially, the decay time decreases as a^{-2} with increasing excitation radius a, but that the decay time saturates when the excitation radius exceeds half a wavelength. The saturated decay-time value is

$$\frac{1}{\tau_{\text{Atom}}} = \frac{3\rho\lambda^2 h}{8\pi\tau_0}\left[1 + \left(\frac{\sin(2\pi h/\lambda)}{2\pi h/\lambda}\right)^2\right], \tag{59}$$

for an atomic system, where τ_{Atom} is the decay time of a single excited atom. For

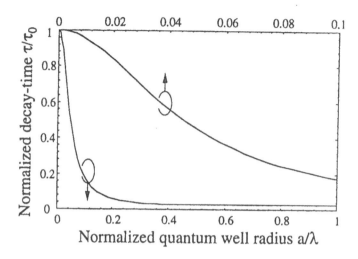

Figure 22. *The normalised excitonic superradiant decay time (inverse of decay rate) as a function of the exciton coherence radius a. In GaAs, τ_0 corresponds to about 1 ns and the minimum superradiant lifetime is predicted to be around 25 ps.*

atomic films much thinner than a wavelength, the square bracket in the expression above simplifies to two, and the expression reads

$$\frac{1}{\tau_{Atom}} = \frac{3\rho\lambda^2 h}{4\pi\tau_0}. \tag{60}$$

For quantum well Wannier excitons the corresponding relation reads

$$\frac{1}{\tau_{Wannier}} = \frac{3\pi q^2 h f_A}{4nmc_0}\left[1 + \left(\frac{\sin(2\pi h/\lambda)}{2\pi h/\lambda}\right)^2\right], \tag{61}$$

where f_A is the exciton oscillator strength per unit area, n is the quantum well refractive index, m is the electron mass and q is the elementary charge. Again, for thin slabs the expression simplifies as the square bracket equals two.

We see that even in a large collection of excited particles (or quasi-particles), net co-operativity extends only over distances of half or so wavelengths (in the direction transverse to the net momentum). For larger systems only the radiation pattern changes, the decay rate remains fixed. This has been ascribed to the impedance mismatch between the excited atomic system (consisting in a classical picture of a collection of small dipole antennas) and free space. The author's own preferred picture is that the radiative decay time saturates due to the poor match between the highly directional emission pattern and the isotropic density of states in free space.

4.3 Microcavity enhanced superradiance

A way to increase the superradiant decay rate further in a planar system is to enclose the atomic film (or quantum well) between two planar mirrors (Citrin 1994; Björk *et al.*

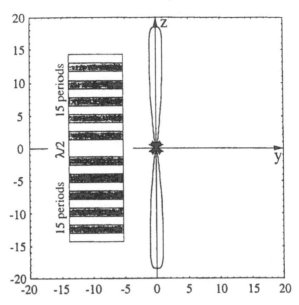

Figure 23. *The effective density of states at the resonant energy in the center of a GaAs/AlGaAs dielectric planar cavity. The assumed structure is shown inset.*

1994). The planar cavity formed between the mirrors will modify the density of states inside the cavity. Again, the modification is largest for cavities of wavelength size. In larger cavities averaging over the many allowed modes will wash out any cavity effects. In Figure 23 the effective density of states in the centre of a planar, half wavelength long cavity with dielectric Bragg mirrors is shown. It is seen that the density of states is highly enhanced in the forward direction. For moderate to high (say above 80%) mirror reflectivities the total number of states in this lobe is independent of the reflectivity, and hence, all of these states can be coerced into a lobe narrower than the lobe of the bare atomic film (or quantum well exciton) superradiative emission pattern. Since the decay rate is proportional to the integrated overlap between the density of states and the radiated emission as a function of angle (22), the superradiant decay rate will be enhanced by a factor equal to the ratio of density of states with and without the planar microcavity. The factor is roughly $(1 + R)\lambda/(2(1 - R)L_{\text{eff}})$, where R is the mirror reflectivity and L_{eff} is the effective planar cavity length. From this relation it seems like the decay rate can be increased without bound as $R \to 1$. However, this is not true, since, as the mirror reflectivity approaches unity one must include the radiation reaction (radiation re-absorption) properly. Remember that so far the decay rates we have calculated are based on a perturbation expansion where absorption processes have been neglected.

As seen from Figure 21 the superradiant emission from an extended excitation is emitted in a single quasi-mode. By matching the cavity to the excitation radius one can realise a situation where the extended excitation couples to a single cavity radiation mode. Hence, we are back to a situation similar to that described in Section 2.1 . If the cavity is perfect, *i.e.* does not let the radiation decay through the mirrors or otherwise, the emitted radiation will be re-absorbed by the atoms (excitons) and system will Rabi-

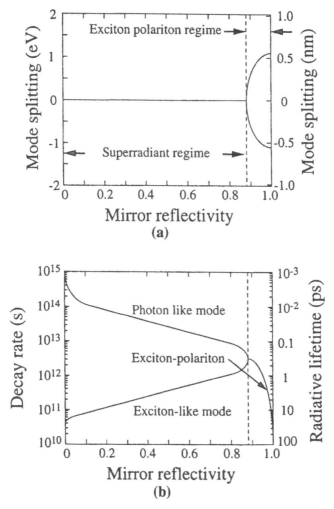

Figure 24. *(a) The calculated radiative energy difference of the two eigen-modes in a half wavelength long planar dielectric cavity with a single GaAs quantum well in the centre. (b) The calculated radiative decay rate of the two eigen-modes in a half wavelength long planar dielectric cavity with a single GaAs quantum well in the centre.*

oscillate. In Figure 24 the proper eigen-modes of the coupled atom-cavity system has been solved as a function of the mirror reflectivity. The two coupled modes give rise to two distinct system eigen-modes, characterised by their complex eigen-frequencies. The real part of the eigen-frequencies correspond to the transition frequency whereas the imaginary part corresponds to the decay rate. As long as the mirror reflectivity is low, only the decay rates will be modified. Since the coupling between the cavity and the atom is small, the eigenmodes have the same frequency. The two modes correspond to an exciton-like mode with (comparatively) small decay rate (although faster than the radiative decay rate of a bare atomic film or a quantum well exciton), and a photon-like mode with large decay rate. As predicted above, the decay rate of the exciton-like mode initially increases with increased mirror reflectivity, whereas the decay rate of the photon-like mode (quite naturally) decreases with increased reflectivity. However, at some reflectivity the two decay rates become equal (at $R \approx 0.87$ for a single 200Å thick GaAs quantum well). Increasing the reflectivity further will result in a decrease of the decay rate for both modes, but the frequency of the two modes will be split, *i.e.* under strong coupling, the two system eigenmodes have different real parts so that the frequency will be different for the two modes. The normal modes in this regime are called microcavity polaritons or dressed excitons. The modes now are neither excitonic nor photonic but are an equal mixture of a photon and an exciton. In the $N = 1$ manifold, denoting the excited photon mode (with the exciton in its ground state) $|0, 1\rangle$ and one excited exciton (with the photon in its ground state) $|1, 0\rangle$, the normal modes can be written $|\pm\rangle = (|0, 1\rangle \pm |1, 0\rangle)/\sqrt{2}$ (the expression is only exact if $R = 1$). One should note that an initially excited exciton with the field mode in its ground state is an equal superposition of the $|+\rangle$ and $|-\rangle$ modes, $|1, 0\rangle = (|+\rangle - |-\rangle)/\sqrt{2}$. Since the dressed states $|+\rangle$ and $|-\rangle$ periodically change their mutual phase due to their different eigen-frequencies, the energy will be periodically transferred between the bare field mode and the bare exciton mode. This is completely analogous to the single atom Rabi-flopping.

4.4   Dressed excitons

In the previous subsection we showed that the coupling between a single field mode and an exciton mode is completely analogous to the coupling between a single field mode and a single atom (in the $N = 1$ manifold). In the atom field case we talk about 'dressed atoms', indicating that the coupling modifies the bare atom state. In analogy, we can call the system eigen-modes in the coupled exciton field case 'dressed excitons'. These have recently been seen in GaAs microcavity systems (Weisbuch *et al.* 1992; Cao *et al.* 1995; Houdré, *et al.* 1994b). However, since the excitons per definition reside in a crystal matrix which influences the excitons' behavior, the situation in the exciton case is a bit more complicated. For one thing, the dressing modifies the dispersion relation of the excitons. In Figure 25 the dispersion (energy versus quantum well in-plane wavevector $k_{\parallel}$) is plotted for uncoupled and coupled excitons and planar microcavity field modes. In both cases the wavevector in the quantum well (and planar cavity) normal direction(s) is quantised. In the dispersion relation only the fundamental modes are plotted. It has also been assumed that the bare state modes (dotted) are resonant at $k_{\parallel} = 0$. It is seen that the normal mode eigen-frequencies split (at $k_{\parallel} = 0$) an amount given by the coupling factor. However, at larger in-plane wavevectors, the higher frequency dressed exciton mode becomes distinctly photon-like while the lower energy mode becomes exciton-like. In the given example, coupled modes with a $k_{\parallel}$ greater than $5 \times 10^5 \text{m}^{-1}$ can hardly be called dressed excitons. In this regime the modes have a distinct respective photon and exciton flavour, and a more proper description is to consider them as perturbed photon and exciton modes, respectively.

In Figure 26 the calculated absorption spectrum in the forward direction ($k_{\parallel}{=}0$) is plotted for a single GaAs quantum well embedded in a planar dielectric Bragg microcavity (Pau *et al.* 1995). The number of layers in the Bragg mirrors (*i.e.* the reflectivity)

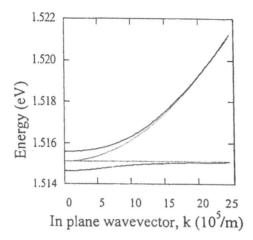

Figure 25. *The dispersion relation (energy v.s. quantum well in-plane wavevector $k_{\parallel}$) for the quantum well exciton and the cavity field. The dotted lines correspond to the dispersion relations for the uncoupled modes and the solid lines represent the dispersion of the coupled modes.*

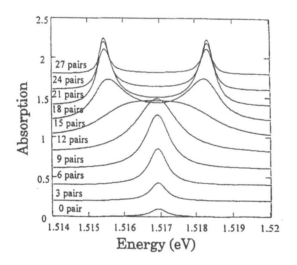

Figure 26. *The calculated absorption spectrum for a quantum well excitonic mode coupled to a planar dielectric cavity mode. The parameter is the number of Bragg mirror pairs, i.e. the mirror reflectivity. Compare this figure to Figure 24.*

is the parameter. It is seen that in the example, when the number of Bragg mirror pairs is around 15, the absorption peak is substantially wider than the absorption peak width of the bare quantum well (no mirror pairs). In this regime we have microcavity enhanced superradiance and consequently 'superabsorption' as explained above. If the

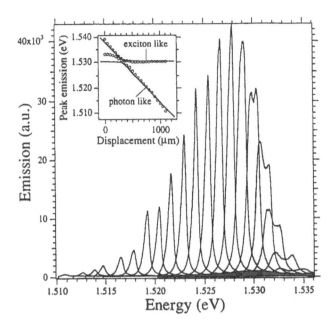

Figure 27. *Measured emission spectra for a GaAs quantum well planar dielectric mirror microcavity. The cavity was slightly tapered so probing the sample at different positions yield an exciton-cavity detuning. In the inset figure the peak energies are plotted versus the linear displacement. The solid lines correspond to a linear tapering, e.g. a linear dependence of detuning with displacement.*

mirror reflectivity is increased further, the absorption peak is split into two symmetric peaks, corresponding to the two dressed exciton modes. The information in this picture is equivalent to that in Figure 24 , except that the photon-like mode is 'omitted'. (The photon like mode has no absorption. It would show up in a transmittance or reflectance spectrum.) In Figure 27 a set of measured emission spectra is plotted. A planar, tapered cavity was used, so the cavity resonant wavelength (or energy) varies continuously with the displacement along the tapering direction. Again, from the inset figure (which displays the emission spectra peak positions as a function of the excitation spot displacement or detuning) it is seen that the two eigen-modes anti-cross instead of cross as would have been the case if the coupling was negligible (Houdré, *et al.* 1994a).

In Figure 28 the emitted intensity in the forward direction is plotted versus time, as recorded by a high temporal resolution streak camera. The black dots are the measurement results, the dashed line is there to guide the eye. The sample was excited with a sub-picosecond laser pulse, tuned to resonance. Peak *a* is the pump-pulse, scattered from the front microcavity mirror. It only serves as a means of accurately setting the zero of time. Peaks *b* and *c* are (damped) Rabi oscillation peaks. The damping is mostly due to scattering by thermal phonons even though the sample was held at a temperature of 4K. The solid line is a fit of the measured data. The solid line in Figure 29 is the Fourier transform of the fitting function in Figure 28. The dashed line is the spectrum recorded by a spectrometer simultaneously with the streak camera

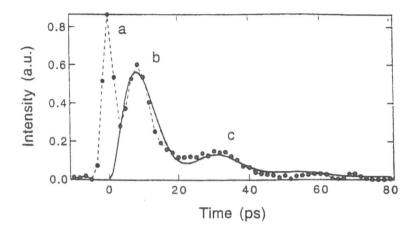

Figure 28. *Emitted intensity per unit time in the forward direction from a planar mirror microcavity excited at a five degree angle from the cavity normal. The dark dots are the measured intensities, the solid line is a theoretical fit. The dotted peak a correspond to scattered pump light. Peaks b and c correspond to Rabi oscillations.*

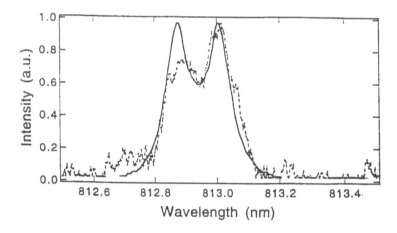

Figure 29. *Spectrum recorded simultaneously with the temporal data in the previous figure. The dotted line is the measured data, the solid line is the spectrum corresponding directly to the theoretical fit in the previous figure.*

recording. The asymmetry of the spectrum may be due to a slight detuning between the cavity and the excitonic modes, leading to an asymmetric spectrum (the higher energy mode being more photon-like and having smaller absorption/emission). As seen, the theoretical predictions are well reproduced in the experiments.

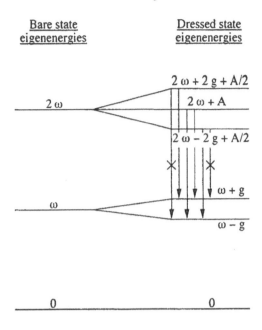

Figure 30. *Eigen-energies of the eigen-modes in the $N = 1$ and $N = 2$ manifold for coupled non-linear oscillators.*

4.5 Two-level atoms and excitons, differences and similarities

As mentioned above, the atom/field and the exciton/field systems are identical in the first excitation manifold. However, as we shall show below, there exist differences at higher excitation. When the excitation is small, and the relevant quantity is the Mott density which essentially is the excitation density required to create one exciton per Bohr area throughout the excitation spot, the exciton is essentially a bosonic particle. That is, as long as the excitons are few and far between they do not interact. Hence, in contrast to a single two-level atom, we can excite the exciton mode with two energy quanta. To outline the difference between the single dressed atom and the dressed exciton, we shall study the second excitation manifold of the latter. We have already solved the arbitrary excitation manifold for the single atom; recall that in each manifold there are two eigen-modes with a frequency splitting of $\Delta\omega = 2g\sqrt{N}$, where N is the manifold number. In the case of the exciton, the allowed bare state Hilbert space is greater than in the atom case; for example the $N = 2$ bare state space is spanned by the three states $|2,0\rangle$, $|1,1\rangle$, and $|0,2\rangle$. Hence, when the modes become coupled one expects three (possibly degenerate) dressed states. In Figure 30 the $N = 0$, $N = 1$ and $N = 2$ excitation manifolds are shown (when $A = 0$, see below). The dressed exciton eigen-frequencies in the second excitation manifold are $2\omega_0 + 2g$, $2\omega_0$ and $2\omega_0 - 2g$. In principle six different transitions can take place, with four different energies $\omega_0 + 3g$, $\omega_0 + g$, $\omega_0 - g$ and $\omega_0 - 3g$. However, when calculating the dipole transition element one easily finds that only the $\omega_0 \pm g$ transitions are allowed. We note that these are identically the same transition energies as from the first order manifold to the system ground state.

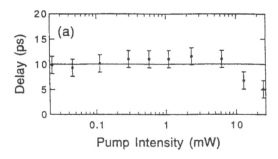

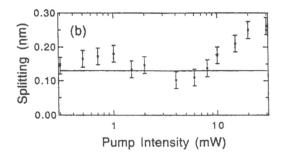

Figure 31. *The time delay between peak b and c in the emission intensity v.s. the pump intensity (a). The delay directly gives the Rabi period. In (b) the corresponding distance (in wavelength units) of the emission spectra peaks is plotted versus the pump intensity.*

It turns out that this is a systematic trait of the coupled exciton to field-mode system (or more generally, of two coupled bosonic modes). Between any two manifolds only two transition energies correspond to allowed transitions. Therefore the spectra always feature two peaks, no more and no less. This is not true for atoms where higher excitation manifold transitions feature multi-peaked spectra (Agarwal and Puri 1989). (It should be remembered, though, that for a collection of atoms, the collective atom excitation which is most conveniently described by a Bloch-vector is also essentially a bosonic mode at low excitation.) The exciton spectra pictured in Figure 27 and Figure 29 were all taken at low excitation (but probably in a manifold cascade starting at $N = 10^4$ or so) and they all feature two peaks. In Figure 31 (a) the measured time delay between the excitation pulse and the first Rabi-oscillation peak is plotted versus excitation pulse power. In (b) the peak-to-peak spectral splitting between the two peaks is plotted. We see, as expected, that the time delay and the energy splitting remains constant over several orders of magnitude in excitation energy. However, at sufficiently high excitation there is a deviation, the energy splitting becomes larger with increasing excitation. This signals the breakdown of the non-interacting exciton picture, and hence also of the description of an exciton as bosonic mode linearly coupled to a (bosonic) field-mode.

4.6   Non-linear effects

At high excitation, where the mean 'inter-excitonic distance' is small, the Pauli exchange interaction and the Coulomb interaction must be included in the Hamiltonian.
The latter of these forces is the dominant one, and it can simply be incorporated in
the Hamiltonian by addition of a single term. The total Hamiltonian when Coulomb
interaction has been taken into account reads

$$H = \hbar\omega_f a^\dagger a + \hbar\omega_e b^\dagger b + \hbar g(ab^\dagger + a^\dagger b) + Ab^\dagger b^\dagger bb, \tag{62}$$

where b is the (bosonic) exciton annihilation operator and the last term on the right
hand side of equation 62 is the Coulomb interaction term. If A is positive, the total
energy increases when there is more than one excitation in the exciton mode. If A
is negative, on the other hand, the Coulomb interaction is attractive. For excitons
the latter tends to be the case. The helium-like bi-excitons have lower energy than the
hydrogen-like excitons. In order to form a bi-exciton, the two constituent excitons must
have opposite spin, however, due to the Pauli-principle.

Inclusion of the non-linearity in the Hamiltonian will slightly change the dressed
state picture. First of all, the $N = 2$ excitation manifold levels are shifted by A
and $A/2$, respectively, as seen in Figure 30. Second of all, and more important, the
previously forbidden transitions become allowed. Hence the spectrum will no longer
feature simply two peaks, but (in principle) six. However, depending on the value
of the non-linear coefficient A, some of the peaks are degenerate. In Figure 32 the
calculated spectra are shown for $A = 0$, $A/g = -0.13$ and $A/g = -1.33$. We see
the emergence of new peaks as the non-linearity is increased, one also notes that the
spectrum becomes asymmetric. In Figure 33 the corresponding measured spectra are
shown. In (a) the pump polarisation is circular, resulting in the creation of excitons
with a fixed spin direction. In (b) the polarisation was switched to linear polarisation
(an equal superposition of left- and right-hand circularly polarised light). In the latter
case energetically more favourable bi-excitons can form. At smaller excitation power
the spectra only featured two peaks independent of the polarisation. We see that the
qualitative fit between the experiment and the theory is quite good, in spite of the
extremely simple Hamiltonian.

The excitonic non-linearity may possibly have some applications. It is well known
that optical non-linear effects in crystals and liquids are quite small. For substantial non-linearity large powers are needed. However, electrons interact quite strongly.
Therefore it may be favourable to convert the photons into microcavity excitons that
subsequently can interact (non-linearly) and relax to the ground state under photo-
emission. This real transition is different from the virtual (off-resonant) transitions
employed in most non-linear crystals used.

4.7   Dephasing

So far we have treated few mode coherent interactions (with the exception of radiative
decay). However, in a crystal matrix phonon scattering is important, even at low
temperatures. In this section we shall see that dephasing will usually act as a damping

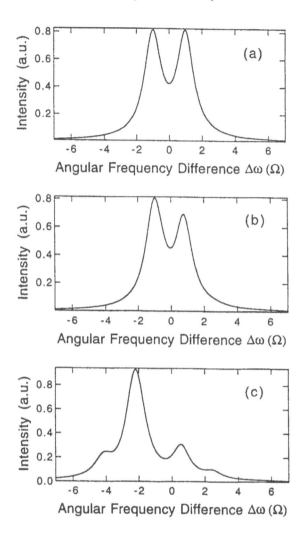

Figure 32. *Calculated spectra for the coupled non-linear oscillators. In (a) the oscillators are linear, $A/\Omega = 0$, in (b) $A/\Omega = -0.13$ and in (c) $A/\Omega = -1.33$.*

force, and that due to dephasing, coherent enhancements, such as superradiance will be quenched.

Our first example will be extremely simple. We shall look at the spontaneous decay of a collection of three identical atoms, with parallel dipole moments, located much closer than a wavelength apart. Assume, for simplicity that they are excited with a single energy quanta in a superposition-state, so that each atom has the same probability to be in the upper state. To make the model even simpler we shall assume that the atom's mutual dipole-moment phase is either zero or π. Denoting the state of an excited atom either $| \uparrow \rangle$ or $| \downarrow \rangle$ (depending on its phase) and the atom in its ground state $|0\rangle$,

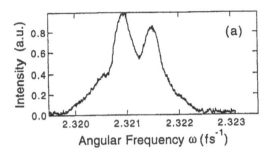

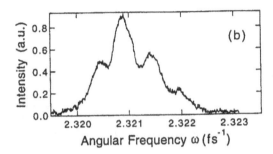

Figure 33. *Measured spectra for linear (a) and circular (b) polarisation.*

there are eight possible atomic states:

$$| \uparrow, \uparrow, \uparrow \rangle \equiv (| \uparrow, 0, 0 \rangle + | 0, \uparrow, 0 \rangle + | 0, 0, \uparrow \rangle)/\sqrt{3}$$

and similar constructions for $| \uparrow, \uparrow, \downarrow \rangle$, $| \uparrow, \downarrow, \uparrow \rangle$, $| \downarrow, \uparrow, \uparrow \rangle$, $| \uparrow, \downarrow, \downarrow \rangle$, $| \downarrow, \uparrow, \downarrow \rangle$, $| \downarrow, \downarrow, \uparrow \rangle$, and $| \downarrow, \downarrow, \downarrow \rangle$. In Table 1 we show the corresponding decay rates are tabulated.

State	Decay rate Γ/Γ_0	Excitation probability	Net decay rate
$| \uparrow, \uparrow, \uparrow \rangle$	3	1/8	$3\Gamma_0/8$
$| \uparrow, \uparrow, \downarrow \rangle$	1/3	1/8	$\Gamma_0/24$
$| \uparrow, \downarrow, \uparrow \rangle$	1/3	1/8	$\Gamma_0/24$
$| \downarrow, \uparrow, \uparrow \rangle$	1/3	1/8	$\Gamma_0/24$
$| \uparrow, \downarrow, \downarrow \rangle$	1/3	1/8	$\Gamma_0/24$
$| \downarrow, \uparrow, \downarrow \rangle$	1/3	1/8	$\Gamma_0/24$
$| \downarrow, \downarrow, \uparrow \rangle$	1/3	1/8	$\Gamma_0/24$
$| \downarrow, \downarrow, \downarrow \rangle$	3	1/8	$3\Gamma_0/8$

Table 1. *The decay rates for the different possible atomic states.*

It is seen that the first and last states are superradiant while the other six states are subradiant. If the three atoms were initially prepared in the $| \uparrow, \uparrow, \uparrow \rangle$ state, and subsequently were subjected to scattering by interaction with a phonon bath, the main effect of the phonon bath would be to completely randomise the phases of the three atoms. In our simplified model that would mean that all the states would be equally probable since they have the same net excitation per atom, only the atom's mutual phases differs. The state after interaction with the phonon bath would be a superposition state of the eight states listed in Table 1. From the table it is easy to see that the ensuing decay rate (sum of the last column = Γ_0) is identical to that of a single excited atom. Although shown for a simplified and small system, the same holds true in general, phonon scattering rapidly decoheres any superposition state into a statistical mixture state. The latter state has (on the average) the same decay rate as a single constituent particle (an atom or a localised exciton). Hence phonon scattering should be avoided if collective effects are to be seen. In our case most experiments were performed at 4K in order to minimise the phonon state occupation probability. Even so, the phonon scattering rate is on the order of 10ps, so only during the first few tens of a picosecond can cooperative effects be seen, as demonstrated in Figure 28. In order to 'beat out' the phonon scattering, the devices should be designed for maximum cooperativity, and hence the fastest possible temporal evolution.

To get a more quantitative picture of dephasing let us look at a system where two non-orthogonal field-modes couple to the same collection of atoms. Such a model suits the planar cavity particularly well since the planar cavity does not quantise the modes in the lateral direction. Hence if the system is excited by shining, for example, a Gaussian beam of light at an angle a few degrees from the cavity normal, the light emitted in the normal direction still couples weakly, via the excitation of the atoms (or excitons) to the pump-pulse. If no scattering took place, the only effect would be a slow and incomplete transfer of energy from the excitation pulse (which has been assumed to be resonant with the cavity, but has a finite in-plane wavevector $k_{||}$) to the cavity mode resonant in the cavity normal direction which has $k_{||} = 0$. The reason the transfer takes place is because we have assumed a finite system size, the atoms are located no more than a wavelength apart. Hence Heisenberg's uncertainty principle dictates that the excited atoms must radiate into modes with a range of k-vectors. In our model only two modes exist and they become weakly coupled.

In Figure 34 the slow transfer of energy can be seen. The excitation mode undergoes Rabi-oscillation, periodically exchanging energy with the atoms. However, since the excited atoms are confined in space, they couple (weakly) to modes with different momentum. Therefore, the mode resonant in the cavity normal direction slowly builds up energy. Of course this mode also exchanges energy periodically with the atoms. If there is no dissipation in the system the two modes evolve coherently at all times. Specifically, it is possible to beat the light in the two modes against each other in a interferometric measurement at any time and the visibility of the interference pattern will always be unity.

In Figure 35 the evolution of the same system with dissipation and scattering is depicted. We see that the Rabi-oscillation of the excitation mode is highly damped. The cavity normal mode, on the other hand, builds up energy much faster in this case. This is mainly due to scattering. The scattering time has been assumed to be two time units, and it is seen that after two time units the atomic state has been so thoroughly phase-randomised (or momentum-randomised) that it starts to couple

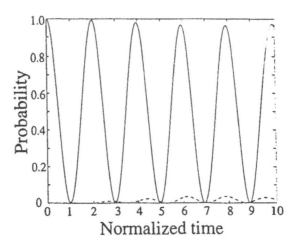

Figure 34. *Excitation probability as a function of time of the pump pulse (solid line) with a finite mean in-plane wavevector, and the cavity mode with no in-plane wavevector (dashed).*

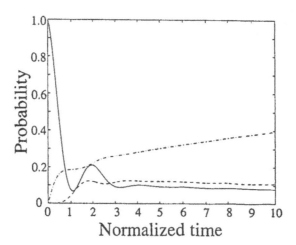

Figure 35. *Excitation probability as a function of time of the pump pulse (solid line), and the cavity mode with no in-plane wavevector (dashed). Since both modes have assumed to have dissipation, there is a finite (and growing) probability of finding the system in the ground state (with zero energy). This probability is drawn dash-dotted.*

equally well to the two field modes. Hence the two field modes have almost the same excitation probability. (The reason the excitation probability for the excitation mode [solid] is smaller than the cavity normal mode at large times is that the decay rate for the former has been assumed to be larger than the latter.) We see that both modes

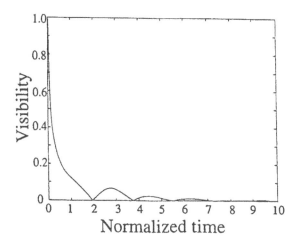

Figure 36. *Calculated visibility between the initial excitation light and the light in the cavity normal mode. (Strictly speaking, the visibility is the envelope function of the plotted graph.) It is seen that the visibility quickly goes to zero. This means that most of the energy in the cavity normal mode comes from phonon scattering of the excited atoms which leads to phase-randomisation, not the coherently coupled light from the excitation mode.*

undergo damped and attenuated Rabi-oscillation, but that the phonon scattering very quickly transfers excitation into the cavity normal mode. We also see that in spite of the fact that the decay-times of the two modes have been assumed to be two and ten time units, the real system decay is much slower because the phonons scatter the atomic state in and out of radiatively dark states. In fact, at any time greater than the scattering time constant, a substantial amount of energy resides in atomic dark states. Since only radiative decay has been assumed to exist in our model, these states effectively slow the system decay down. Finally in Figure 36, the calculated visibility between the initial excitation mode and the cavity normal mode is shown. To measure this visibility in reality, a time-delayed portion of the excitation light would be made to interfere with the light emitted from the cavity in the cavity normal direction. The fringe visibility as a function of the time-delay would be a measure of the visibility. What the curve shows is that most of the light in the cavity normal mode comes from (incoherent) atomic state scattering, only a small portion comes from the coherently coupled light. Nonetheless, the cavity normal mode undergoes (damped) Rabi-oscillation. It would have done so even in the total absence of coherent coupling—for example in an extended system where the transverse momentum is conserved. The reason is that the cavity normal mode starts in a well defined state (the ground state), so the initial light transferred by atom-phonon scattering will lead to a relatively well defined subsequent evolution. If the scattering and the decay were turned off at, say, time $t = 2$, the subsequent Rabi-oscillation would be rather easily observed, although complete energy transfer would not be possible.

We see that phonon scattering seriously and quite dramatically alters the temporal evolution of systems where collective effects are important. From an optoelectronic device point of view, dephasing destroys the coherence needed for the collective evolution. The initial superposition state will quickly evolve into a statistical mixture state that will have the same evolution as a single constituent particle. This is why so much effort has gone into the study of spontaneous emission alteration of a single atom (dipole) in a microcavity. Although in reality there are many orders of magnitude of excited atoms or free electrons in most microcavities, the rapid dephasing by phonon scattering will force the system to behave (on the average) like a cavity with a single dipole (of course the intensity of the emitted light will scale proportionally to the number of dipoles, but the temporal behaviour will be identical).

5   Conclusions

In this chapter we have tried to demonstrate the physics, technology and the possible benefits of spontaneous emission engineering. It was shown that in small systems (*i.e.* cavities) where an excited emitter interacts with only a few cavity modes, the spontaneous emission process can be quite different from the spontaneous emission decay in free space. It was also shown that there is an intimate relationship between the spontaneous emission rate of any mode and the stimulated emission rate of that mode. This is the underlying idea for threshold power reduction in microlasers, which at present is a very dynamic field.

While it is conceptually simple to modify both spontaneous emission patterns as well as spontaneous emission rates, technology sets limits on what can and what cannot be done at present in the laboratories. While it is relatively simple to modify the relative proportion of spontaneous emission emitted into different modes, it is hard to modify the absolute (total) rates in dielectric cavities.

It was also demonstrated that the radiative decay of an ensemble of collectively and coherently excited atoms is quite different from the decay of a single atom. The ensemble can decay either at superradiant rates or at subradiant rates depending on the initial conditions of the atomic state. We showed that there is an intimate connection between superradiance and spontaneous emission engineering. The cavity mirrors create virtual atoms and provided that the cavity is small enough, the real and virtual atoms can cooperate to feature subradiant decay (spontaneous emission inhibition) or superradiant decay (spontaneous emission enhancement).

It was finally shown that dephasing destroys any collective effects. The initial atomic ensemble is soon scattered into a mixture of all possible states. Some of these are superradiant while others are subradiant. The net (average) decay rate is the same as the single atom decay rate. Since phonon scattering is a very rapid process in semiconductors at room temperature, this explains why so much of the work in this field goes into single atom spontaneous emission engineering.

Acknowledgments

I would like to thank my colleagues at E L Ginzton Laboratory, Stanford University with whom he developed much of the understanding of cavity QED in semiconductor microlasers and cooperative phenomena of excitonic systems in particular. I would especially like to thank Professor Yoshihisa Yamamoto, Dr. Joseph Jacobson, Mr. Stanley Pau, Ms. Hui Cao and Mr. Gleb Klimovitch.

References

Agarwal G S, and Puri R R, 1989, Collapse and revival phenomenon in the evolution of a resonant field in a Kerr-like medium, *Phys Rev A*, **39**, 2969.

Baba T, Hamano T, Koyama F and Iga K, 1991, Spontaneous emission factor of a microcavity DBR surface emitting laser, *IEEE J Quant Electron*, **27**, 1347.

Baba T, Hamano T, Koyama F and Iga K, 1992, Spontaneous emission factor of a microcavity DBR surface emitting laser (II)—Effects of electron quantum confinement, *IEEE J Quantum Electron*, **27**, 1310.

Björk G, Yamamoto Y, Machida S and Igeta K, 1991, Modification of spontaneous emission rate in planar dielectric microcavity structures, *Phys Rev A*, **44**, 669.

Björk G and Yamamoto Y, 1991, Analysis of semiconductor microcavity lasers using rate equations, *IEEE J Quantum Electron*, **27**, 2386.

Björk G, 1994, On the spontaneous lifetime change in an ideal planar microcavity — Transition from a mode continuum to quantised modes, *IEEE J Quantum Electron*, **30**, 2314.

Björk G, Pau S, Jacobson J M and Yamamoto, Y, 1994, Wannier exciton superradiance in a quantum well microcavity, *Phys Rev B*, **50**, 17336.

Björk G, Pau S, Jacobson J M, Cao H and Yamamoto Y, 1995, Excitonic superradiance to exciton-polariton crossover and the pole approximation, to be published in Phys.Rev. B.

Brorson S D, Yokoyama H and Ippen E P, 1990, Spontaneous emission rate alteration in optical waveguide structures, *IEEE J Quantum Electron*, **26**, 1492.

Cao H, Jacobson J M, Björk G, Pau S and Yamamoto Y, 1995, Observation of dressed-exciton oscillating emission over a wide wavelength range in a semiconductor microcavity, *Appl Phys Lett*, **66**, 1107.

Citrin D S, 1994, Controlled exciton spontaneous emission in optical-microcavity-embedded quantum wells, *IEEE J Quantum Electron*, **30**, 997.

Chin M C, Chu D Y and Ho S-T, 1994, Approximate solution of the whispering gallery modes and estimation of the spontaneous emission coupling factor for microdisk lasers, *Opt Commun*, **109**, 467.

Chu D Y and Ho S-T, 1993, Spontaneous emission from excitons in cylindric dielectric waveguides and the spontaneous-emission factor of microcavity ring lasers, *J Opt Soc Am B*, **10**, 381.

Chu D Y, Chin M K, Bi W G, Hou H Q, Tu C W and Ho S-T, 1994, Double-disk structure for output coupling in microdisk lasers, *Appl Phys Lett*, **65**, 3167.

Collot L, Lefevre-Seguin V, Brune M, Raimond J M and Haroche S, 1993, Very high-Q whispering-gallery mode resonances observed on fused silica microspheres, *Europhys Lett*, **23**, 327.

Cutrer D M and Lau K Y, 1995, Ultralow power interconnect with zero-biased, ultralow threshold laser — How low a threshold is low enough?, *IEEE Photon Technol Lett*, **7**, 4.

De Neve H, Blondelle J, Baets R, Demeester P, Van Daele P and Borghs G, 1995, High efficiency planar microcavity LED's: comparison of design and experiment, *IEEE Photon Technol Lett*, **7**, 287.

De Martini F, Innocenti G, Jacobovitz G R and Mataloni P, 1987, *Phys Rev Lett*, **59**, 2955.

De Martini F and Jacobovitz G R, 1988, Anomalous spontaneous-stimulated-decay phase transition and zero-threshold laser action in a microscopic cavity, *Phys Rev Lett*, **60**, 1711.

Dicke R H, 1954, Coherence in spontaneous radiation processes, *Phys Rev*, **93**, 99.

Dodabalapur A, Rothberg L J, Miller T M and Kwock E W, 1994, Microcavity effects in organic semiconductors, *Appl Phys Lett*, **64**, 1994.

Drexhage K H, 1974, Interaction of light with monomolecular dye layers, in *Progress in Optics*, **12**, ed. Wolf E, (North Holland, New York).

Gigase Y B, Harder C S, Kesler M P, Meier H P and Van Zeghbroeck B, 1991, Threshold reduction through photon recycling in semiconductor lasers, *Appl Phys Lett*, **57** 1310.

Glauber R J and Lewenstein M, 1991, Quantum optics of dielectric media, *Phys. Rev. A*, **43**, 467.

Gourley P L, 1994, Microstructured semiconductor lasers for high-speed information processing, *Nature*, **371**, 571.

Houdré R, Weisbuch C, Stanley R P, Oesterle U, Pellandini P and Ilegems M, 1994, Measurement of cavity-polariton dispersion curve from angle-resolved photoluminescence experiments, *Phys Rev Lett*, **69**, 2043.

Houdré R, Stanley R P, Oesterle U, Ilegems M and Weisbuch C, 1994, Room-temperature cavity polaritons in a semiconductor microcavity, *Phys Rev B*, **49**, 16761.

Huffaker D L, Shin J and Deppe D G, 1994, Low threshold half-wave vertical-cavity lasers, *Electron Lett*, **30**, 1946.

Hulet R G, Hilfer E S and Kleppner D, 1985, Inhibited spontaneous emission by a Rydberg atom, *Phys Rev Lett*, **55**, 2137.

Hunt N E, Schubert E F, Logan R A and Zydzik G J, 1992, Enhanced spectral power density and reduced linewidth at 1.3μm in an InGaAsP quantum well resonant-cavity light-emitting diode, *Appl Phys Lett*, **61**, 2287.

Iwata A and Hayashi I, 1993, Optical interconnections as a new LSI technology, *IEICE Trans Electron*, **E-76-C**, 90.

Jahnke F, Henneberger K, Schafer W and Koch S W, 1993, Transient nonequilibrium and many-body effects in semiconductor microcavity lasers, *J Opt Soc Am B*, **10**, 2394.

Jaynes E T and Cummings F W, 1963, *Proc IEEE*, **51**, 89.

Karlsson A, Schatz R and Björk G, 1994, On the modulation bandwidth of semiconductor microcavity lasers, *IEEE Phot Technol Lett*, **6**, 1312.

Knoester J, 1992, Optical dynamics in crystal slabs: Crossover from superradiant excitons to bulk polaritons, *Phys Rev Lett*, **68**, 654.

Lei C, Deppe D G, Huang Z and Lin C C, 1993, Emission characteristics from dipoles with fixed positions in Fabry-Perot cavities, *IEEE Quantum Electron*, **29**, 1383.

Levi A F J, Slusher R E, McCall S L, Tanbun-Ek T, Coblentz D L and Pearton S J, 1992, Room temperature operation of microdisc lasers with submilliamp threshold current, *Electron Lett*, **28**, 1010.

Lin K-H and Hsieh W-F, 1991, Transient response of a thresholdless microdroplet dye laser, **16**, 1608.

Lin C C, Deppe D G and Lei C, 1994, Role of waveguide light emission in planar microcavities, *IEEE J Quantum Electron*, **30**, 2304.

Matinaga F M, Karlsson A, Machida S, Yamamoto Y, Suzuki T, Kadota Y and Ikeda M, 1993, Low threshold operation of hemispherical microcavity single-quantum-well lasers at 4 K, *Appl Phys Lett*, **62**, 443.

Michler P, Lohner A, Ruhle W W and Reiner G, 1995, Transient pulse response of In$_{0.2}$Ga$_{0.8}$As/GaAs microcavity lasers, *Appl Phys Lett*, **66**, 1599.

Milonni P W, 1994, *The quantum vacuum*, (Academic Press, New York).

Mohideen U, Slusher R, Jahnke F and Koch S W, 1994, Semiconductor microlaser linewidths, *Phys Rev Lett*, **73**, 1785.

Numai T, Kawakami T, Yoshikawa T, Sugimoto M, Sugimoto Y, Yokoyama H, Kasahara K and Asakawa K, 1993, Record low threshold current in microcavity surface-emitting laser, *Jpn J Appl Phys*, **32**, L1533-4.

Pau S, Björk G, Jacobson J, Cao H and Yamamoto Y, 1995, Microcavity exciton polariton splitting in the linear regime, *Phys Rev B*, **51** 14437.

Purcell E M, 1946, *Phys Rev*, **69**, 681.

Raizen M G, Thomason R J, Brecha R J, Kimble H J and Carmichael H J, 1989, *Phys Rev Lett*, **63**, 240.

Rice P R and Carmichael H J, 1994, Photon statistics of a cavity-QED laser: A comment on the laser-phase-transition analogy, *Phys Rev A*, **50**, 4318.

Schnitzer I, Yablonovitch E, Caneau C, Gmitter T J and Scherer A, 1993, 30% external quantum efficiency from surface textured, thin-film light-emitting diodes, *Appl Phys Lett*, **63**, 2174.

Shtengel G, Temkin H, Brusenbach P, Uchida T, Kim M, Parsons C, Quinn W E and Swirhun S E, 1993, High-speed vertical-cavity surface emitting laser, *IEEE Phot Technol Lett*, **5**, 1359.

Suzuki M, Yokoyama H, Brorson S D and Ippen E P, 1991, Observation of spontaneous emission lifetime change of dye-containing Langmuir-Blodgett films in optical microcavities, *Appl Phys Lett*, **58**, 998.

Ujihara K, 1991, Spontaneous emission and the concept of effective area in a very short cavity with plane-parallel dielectric mirrors, *Jpn J ApplPhys*, **30**, L901.

Weisbuch C, Nishioka M, Ishikawa A and Arakawa Y, 1992, Observation of the coupled exciton-photon mode splitting in a semiconductor quantum microcavity, *Phys Rev Lett*, **69**, 3314.

Yamamoto Y and Björk G, 1991, Lasers without inversion in microcavities, *Jap Journ Appl Phys* **30**, L2039.

Yamamoto Y, Machida S and Björk G, 1992, Micro-cavity semiconductor lasers with controlled spontaneous emission, *Opt and Quant Electron*, **24**, S215.

Yamanishi M, 1992, High-speed semiconductor light emitters based on quantum-confined field effect: developed devices and inclusion of quantum microcavities, *Jap J Appl Phys A*, **31**, 2764.

Yokoyama H and Brorson S D, 1989, Rate equation analysis of microcavity lasers, *J Appl Phys*, **66**, 4801.

Solid-State Laser Materials

Günter Huber

Universität Hamburg, Germany

1 Introduction

The progress in the area of diode pumped lasers has contributed very much to the renaissance in the field of solid-state lasers. With diode laser pumping it is possible to obtain higher efficiencies and to build rigid all solid-state devices with simpler and more compact design. Besides Nd^{3+} various efficient diode pumped rare earth lasers have been operated with Er^{3+}, Tm^{3+}, Ho^{3+}(Fan and Byer 1988, Fan et al. 1988, Esterowitz 1990), and Yb^{3+} (Lacovara et al. 1991, Payne et al. 1992).

The successful operation of Cr^{3+} (Walling et al. 1980, Struve et al. 1983, Payne et al. 1988, Payne et al. 1989, Scheps 1992) and Ti^{3+} as tunable room temperature lasers (Moulton 1982, Albers et al. 1986) has stimulated further research in transition metal ions. Interesting new results have been obtained with the ion Cr^{4+} (Petricevic et al. 1988, Zverev and Shestakov 1989, Jia et al. 1991, Kück et al. 1991) and very recently also with the divalent ion Cr^{2+} (Page et al. 1995).

Compact solid-state lasers in the visible spectral region are of potential interest, especially for display and high-density optical data storage applications. Recently optical efficiencies of more than 20% with respect to the pump power were obtained in Nd:YAG (Oka and Kubota 1992) and $Nd:YVO_4$ (Kitaoka et al. 1993) lasers by internal frequency doubling with a potassium titanyl phosphate (KTP) crystal.

An alternative approach presents generation of visible laser radiation by up-conversion schemes, which incorporate energy transfer processes or two step pump processes as ground state and excited state absorption.

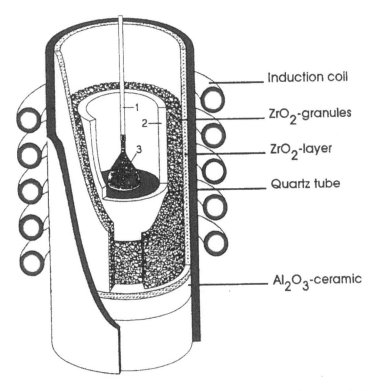

Figure 1. *Czochralski growth technique showing (1) pulling shaft and seed; (2) crucible, (3) growing laser crystal boule.*

The most important laser crystals are oxides and fluorides (Chai 1993). Laser crystals, laser transitions in crystals, and their properties are listed in detail in Kaminskii (1990). This chapter reviews the basic properties of some new laser crystals and describes new developments in the field of rare earth and transition metal solid-state lasers.

2 Crystal growth

Synthetic laser crystals can be grown from the melt, solution or gas phase. The most important technology for growing laser crystals is the Czochralski technique (Figure 1). With this method high purity and perfect crystals can be grown from congruent melts. The chemicals are mixed and melted in a crucible at temperatures, which can be above 2000°C. The crystal is grown under rotation and simultaneous pulling from the melt. Typical growth rates are 1–4mm/hour. Both, the rotation rate and pulling speed influence the shape of the interface between crystal and melt. Usually, a convex interface is used because the crystal growth is generally very stable under these conditions. If stable, however, a flat interface is preferred. In this case the crystal grows without a core and this results in very homogeneous crystalline boules. Prominent examples of flat interface growth are garnets like $Gd_3Ga_5O_{12}$ (GGG) and $Y_3(Sc,Ga)_2Ga_3O_{12}$. For $Nd:Y_3Al_5O_{12}$ (Nd:YAG) flat interface growth is more difficult, but also possible.

3 Rare earth doped laser crystals

Electric dipole transitions within the 4f shell are parity forbidden and admixtures of wavefunctions with opposite parity are necessary to create weak transition probabilities. Due to the screening of the outer filled $5s^2$ and $5p^6$ orbitals, electron phonon coupling is very weak. When doping occurs at centric sites, parity remains a good quantum number and all 4f–4f transitions remain electric dipole forbidden. The emission cross sections are then very small and not useful for laser applications. At acentric sites one observes electric dipole zero-phonon-transitions with very weak vibrational sidebands. Therefore, the operation of rare earth lasers is restricted to zero-phonon-lines with relatively small tunability ranges.

3.1 Nd lasers

Since diode-laser-pumped solid-state lasers are gaining importance, research on new Nd-doped materials with improved spectral properties for diode pumping is of increased interest again (Fan and Byer 1988). For example, $Nd:YVO_4$ is more efficient than Nd:YAG due to larger absorption and emission cross sections (Fields *et al.* 1987). Besides larger cross sections, higher concentrations of the active ion can improve the efficiency of diode-pumped lasers with short crystals. Because absorption and gain coefficients are increased, it is also easier to accomplish mode matching between pump and laser beam.

In many crystals the neodymium concentration is limited, because nonradiative decay of the upper laser level due to Nd^{3+}-Nd^{3+} cross relaxation occurs at high doping levels. In so-called stoichiometric neodymium laser materials, the Nd^{3+} ion is not a dopant but a part of the chemical compound. In these materials cross relaxation rates can be small due to a large distance between the lattice sites occupied by Nd^{3+} ions. Stoichiometric laser crystals like $NdAl_3(BO_3)_4$ were investigated in detail many years ago (Danielmeyer 1975, Huber 1980). Today, $(Nd,Y)Al_3(BO_3)_4$ (NYAB) is commercially available as a diode-laser-pumped self-frequency doubling solid-state laser. Large single crystals are rather difficult to produce, because $(Nd,Y)Al_3(BO_3)_4$ can be grown only from the flux. $(Nd,La)Sc_3(BO_3)_4$ has a structure similar to $(Nd,Y)Al_3(BO_3)_4$ and was first grown from the melt by Kutovoi *et al.* (1991).

For small, efficient diode-laser pumped solid-state laser systems (Fields *et al.* 1987, Kintz and Baer 1990, Ishimori *et al.* 1992, Sipes 1985) it is also desirable to decrease the temperature sensitivity which is caused by a narrow band absorption and the temperature drift of the diode-laser pump and to increase the tolerance in the wavelength selection (Fan and Byer 1988, Streifer *et al.* 1988). Therefore, it is necessary that the laser material features high absorption coefficients and large absorption linewidths. Disordered crystals, for instance, offer broad absorption lines but have the disadvantage of low cross sections due to the inhomogeneous broadening of the transitions. Therefore, the advantage of low temperature sensitivity of the pump process and the correspond-

crystal	concen-tration 10^{20}cm^{-3}	life-time μs	α cm^{-1}	σ_{abs} at 808nm 10^{-20}cm^2	FWHM (abs) nm	σ_{em} 10^{-20}cm^2	FWHM (em) nm
Nd:YAG	1.5	240	10	7.0	0.8	33	0.5
Nd:GVO E$\|$z	1.5	90	78	5.1	1.5	76	1.3
Nd(10%):LSB E$\|$x	5.1	118	36	7.1	3	13	4
Nd(10%):LSB E$\|$y	5.1	118	29	5.8	3	9	4
Nd(10%):LSB E$\|$z	5.1	118	10	1.9	5	5	4

Table 1. *Spectroscopic data of Nd-doped laser materials: YAG ($Y_3Al_5O_{12}$), GVO ($GdVO_4$), and LSB ($LaSc_3(BO_3)_4$)*

ing increased tolerance in pump-wavelength selection is correlated in such crystals with decreased inhomogeneous emission cross sections.

In Nd:GdVO$_4$, both broad homogeneous absorption lines and homogeneous emission lines feature high peak cross sections (Jensen *et al.* 1994). The accidental degeneration of the metastable $^4F_{3/2}$ level in GdVO$_4$ enhances the emission cross sections because the spectra condense into fewer emission lines. Nd:YVO$_4$, which is a similar compound, has a $^4F_{3/2}$ splitting of 18 cm^{-1} (Karayianis *et al.* 1974). Some relevant data for different laser crystals are listed in Table 1.

A comparison of the Nd:YAG and Nd:GdVO$_4$absorption spectra is shown in Figure 2. Under diode pumping the efficiency of Nd:GdVO$_4$ compares favourably with Nd:YAG as shown in Figure 3. The slope efficiencies of Nd:YAG and Nd:GdVO$_4$ are 53% and 57%, respectively.

The spectral properties of Nd^{3+}:LaSc$_3$(BO$_3$)$_4$ (Meyn *et al.* 1994) are also interesting for diode laser pumping. The maximum absorption and emission cross sections are observed for x-polarisation. They are similar to those of Nd:YAG, but the bands are three to five times broader.

The broadening mechanism is mainly homogeneous at 300K; at low temperatures inhomogeneous broadening is also observed. The absorption cross section σ_{abs} of Nd^{3+} in LaSc$_3$(BO$_3$)$_4$ at 808nm is as high as the corresponding absorption cross section of Nd^{3+} in YAG. However, the absorption coefficient α of Nd(10%):LaSc$_3$(BO$_3$)$_4$ is more than three times higher than the absorption coefficient of standard Nd(1.1%):YAG because of the higher concentration of Nd ions in LSB. The relatively low emission cross section (40% of Nd:YAG) is not a severe disadvantage, because Nd lasers are operating far above laser threshold in most configurations and a slightly higher threshold does not decrease the overall efficiency too much.

The optical slope efficiency of the Nd(10%):LaSc$_3$(BO$_3$)$_4$ laser is $\eta = 64\%$ (Meyn *et al.* 1994), which is close to the theoretical limit $\eta_{theor} = \lambda_{pump}/\lambda_{laser} = 76\%$. Since this slope efficiency is as high as the best values for Nd:YAG and Nd:YVO$_4$, the advantages of Nd^{3+}:LaSc$_3$(BO$_3$)$_4$ like broad bands, high absorption and gain coefficients due to high Nd^{3+} concentrations can be fully utilised.

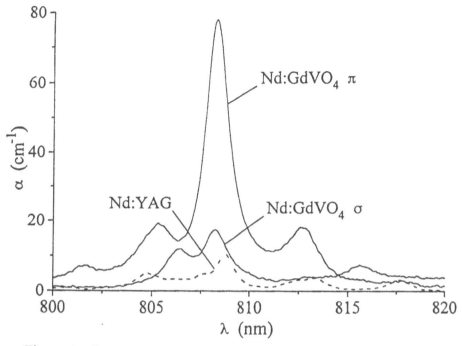

Figure 2. *Absorption spectra of Nd:YAG and Nd:GVO (Jensen et al 1994).*

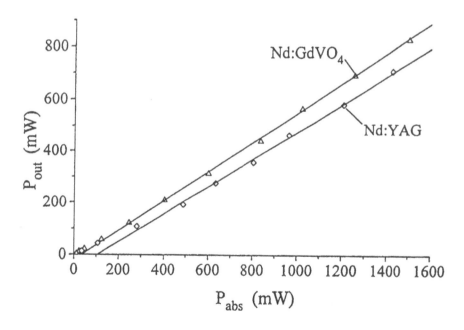

Figure 3. *Input vs. output power of diode pumped Nd:YAG and Nd:GVO (Jensen et al 1994).*

Active Ion	Crystal	λ (μm)	Reference
Tm^{3+}, Ho^{3+}	$Y_3Al_5O_{12}$(YAG)	2.08	Fan *et al.* 1988
Tm^{3+}	$Y_3Al_5O_{12}$ (YAG)	2.02	Esterowitz 1990
Tm^{3+}, Ho^{3+}	YLiF$_4$ (YLF)	2.31	Esterowitz 1990
Yb^{3+}	$Y_3Al_5O_{12}$ (YAG)	1.03	Lacovara *et al.* 1991
Yb^{3+}, Er^{3+}	$SrY_4(SiO_4)_3O$	1.55	Sourian *et al.* 1994
Yb^{3+}, Er^{3+}	$Y_3Al_5O_{12}$ (YAG)	1.64	Schweizer *et al.* 1995
Er^{3+}	$Gd_3Ga_5O_{12}$ (GGG)	2.8	Dinermann and Moulton 1992,Jensen *et al.* 1995
Er^{3+}	YLiF$_4$ (YLF)	2.8	Esterowitz 1990, Jensen *et al.* 1995

Table 2. *Examples of laser diode pumped* CW *rare earth lasers.*

3.2   Yb, Tm, Ho, and Er lasers

A number of other rare earth laser materials have been operated under diode pumping. Table 2 lists a selection of interesting crystals.

Tm^{3+}:YAG and Tm^{3+},Ho^{3+}:YAG are interesting laser crystals for eyesafe measurement techniques and medical applications, because biological tissue has strong absorption coefficients near 2 μm wavelength.

Figure 4 shows the energy level diagram and pumping scheme of diode pumped Ho and Tm lasers. The Tm^{3+} $^3F_4$ manifold is pumped by a AlGaAs diode laser. This is followed by either radiative and non-radiative relaxation to the $^3H_4$ state or a cross-relaxation process between adjacent ions, $Tm(^3F_4\text{-}^3H_4)$–$Tm(^3H_6\text{-}^3H_4)$, which converts one $^3F_4$ excited state into two $^3H_4$ states. For high Tm^{3+} doping densities this cross-relaxation process can be very efficient and can lead to an overall pump quantum efficiency of nearly 2. There is fast spatial energy migration among the Tm^{3+} ions, subsequent energy transfer to the Ho $^5I_7$ energy level, and finally laser action on the Ho $^5I_7$–$^5I_8$ transition. Without Ho-doping, the system can lase at the $^3H_4$-$^3H_6$ Tm-transition. Typical Tm-concentrations in the crystals are high (4 to 10%) in order to achieve efficient down conversion.

For 1.5 μm Er lasers the Er concentration must be kept low to avoid detrimental up-conversion mechanisms. Therefore, the absorption coefficients are too small for efficient laser operation. Co-doping with Yb^{3+} ions increases the absorption around 970nm considerably. The advantage of the increased absorption is greater than the additional Yb–Er upconversion losses. The energy is mainly absorbed by the Yb ions ($^2F_{7/2} \rightarrow {}^2F_{5/2}$ transition). The energy is transferred from the upper $^2F_{5/2}$ Yb level to the $^4I_{11/2}$ Er level by dipole–dipole interaction (see Figure 5). The $^4I_{11/2}$ Er level is depopulated by phonon relaxation and fills the upper laser level $^4I_{13/2}$.

With the help of the transfer mechanism in Figure 5, diode pumped CW operation is possible. However, the observed slope efficiencies have been quite small—well below 10% (Schweizer *et al.* 1995). The reason for the low efficiency is not really clear yet. There is however hope that optimised concentrations for a high transfer efficiency as

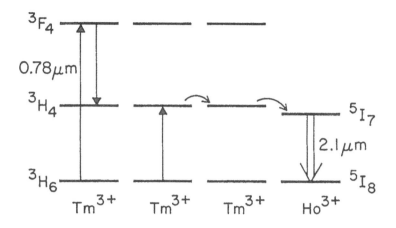

Figure 4. *Energy level scheme, illustrating pumping the $^3F_4$ Tm manifold, Tm–Tm cross relaxation, Tm–Tm energy migration, Tm–Ho energy transfer, and finally laser action on the Ho $^5I_7$–$^5I_8$ transition.*

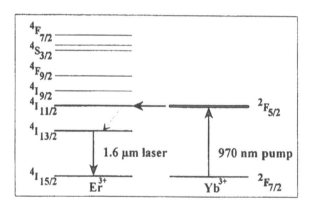

Figure 5. *Er^{3+} and Yb^{3+} energy levels with energy transfer pump mechanism.*

well as a favourable host for low re-absorption losses of the groundstate $^4F_{13/2}$ laser transition could yield much higher efficiencies.

Due to the extremely strong water absorption, 2.8 μm Er^{3+} lasers have very interesting applications in the medical field. The 2.8 μm Er^{3+} transition is known as a CW laser transition (Huber *et al.* 1988) despite the unfavourable lifetime ratio of the upper (1ms) and lower (4ms) laser levels. A fast up-conversion process from the lower laser level $^4I_{13/2}$ is responsible for recycling of the energy and creating a steady-state inversion between the upper ($^4I_{11/2}$) and the lower ($^4I_{13/2}$) laser levels. In order to create a strong Er–Er interaction and hence efficient up-conversion a relatively high Er concentration is necessary (10% to 50%). For Er:YLF the results for different doping levels are shown in Figure 6. The slope efficiency of this laser strongly depends as expected on the Er^{3+} concentration. For Er:YLF with a doping level of 15% the highest yet reported diode pumped slope efficiency is 35% at 2.8μm (Jensen *et al.* 1995). This result coincides

Günter Huber

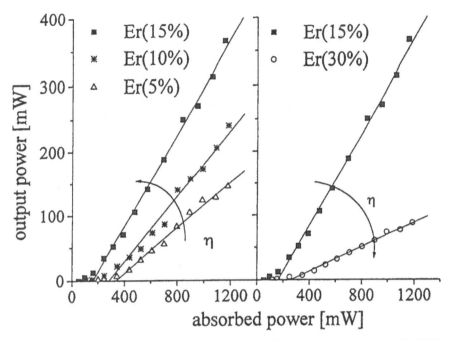

Figure 6. *Concentration dependent diode pumped laser experiments with Er:YLF at 2.8μm wavelength (Jensen et al 1995).*

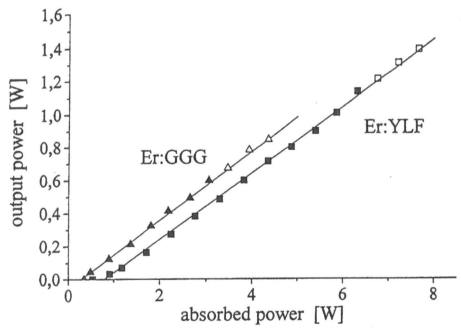

Figure 7. *High power 2.8μm Er lasers (full symbol:* cw; *open symbol: duty cycle 50%) (Jensen et al 1995).*

with the minimum pump threshold of 56mW. The slope efficiency corresponds to the quantum defect between pump and laser photons.

Advantages of the host material YLF are obvious at higher pump powers (Figure 7), when pumping with a 10 W pigtail InGaAs diode. The Er(10%):YLF crystal produces a pure CW laser output power of more than 1.1W at 2.8033μm at a threshold of 540mW and 20% slope efficiency. The Er(20%):GGG crystal has a lower threshold of 380mW and 21% slope efficiency, but thermal problems limit the laser performance at higher pump powers. The negative dn/dT value for YLF seems to by very favourable in this application.

4 Transition metal ion lasers

Transition metal ions mostly exhibit broadband emission because of the coupling between the electronic levels of the 3d electrons with the lattice vibrations. In addition, this electron-phonon coupling can yield temperature dependent nonradiative and radiative decay rates. The main interest in transition metal ions is based on their broad tunability.

4.1 Titanium

Today Ti^{3+}:Al_2O_3 (Moulton 1982, Albers *et al.* 1986) is already a widely used laser material which can be efficiently pumped by frequency doubled Nd lasers and CW Argon lasers in the blue-green spectral region. In spite of the relatively short lifetime, (3.2μs) flashlamp-pumping is also possible with reasonable efficiencies of the order of one percent.

Due to the absence of excited state absorption in the spectral region of the pump and fluorescence band, Ti^{3+}:Al_2O_3 is a very efficient and broadly tunable laser material. The tuning range covers almost the entire fluorescence band from 680 nm to 1100 nm. Due to the broadband behaviour of 3d–3d transitions it is always a problem to avoid excited state absorption (ESA) of pump and/or laser radiation. In Ti^{3+}:Al_2O_3 there is no ESA at either the pump wavelength or the laser wavelength; whereas for instance in Ti:YAlO$_3$ ESA strongly reduces the pump efficiency for the pump wavelengths in the blue-green spectral range (see Figure 8). For Ti:Al$_2$O$_3$ one can clearly see in Figure 9 the broad gain near 750nm for stimulated emission. In the pump region no parasitic ESA is present in case of Ti:Al$_2$O$_3$. The excited state absorption is probably connected with Ti^{3+} complex centres and might be a problem in most of the Ti^{3+} doped materials. Ti^{3+}:Al_2O_3 has this excited state absorption too, but the transition is located well above the pump band and does not influence the pump process (Figure 9).

4.2 Chromium

In contrast to Ti-sapphire, Cr^{3+} lasers can also be pumped by diode lasers via the $^4T_2$ absorption band in the red spectral region. Diode pumped operation of Cr^{3+} lasers has been demonstrated in Cr:LiCaAlF$_6$ (LICAF) and Cr:LiSrAlF$_6$ (LISAF) by Scheps

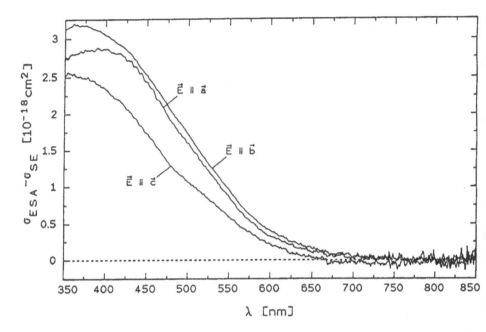

Figure 8. *ESA cross-sections σ_{ESA} of Ti:YAlO₃ for different polarisations (Danger et al. 1993).*

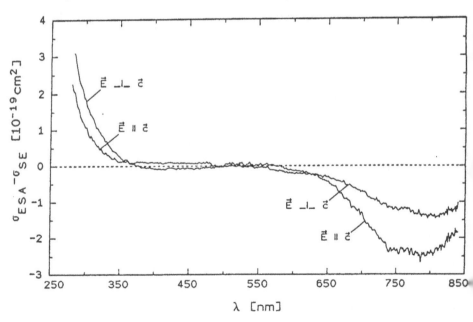

Figure 9. *ESA cross-sections σ_{ESA} and stimulated emission cross-sections σ_{SE} of Ti:Al₂O₃ for different polarisations (Danger et al 1993).*

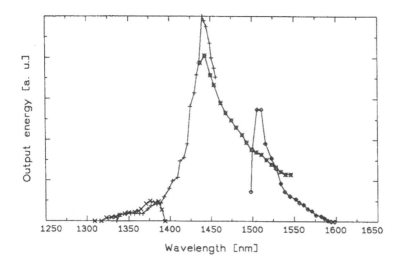

Figure 10. *Tuning range of a Cr^{4+}:YAG laser crystal when pumped with a pulsed Nd:YAG laser (Kück et al 1994).*

(1992). Both colquiriite minerals LICAF and LISAF are very efficient hosts for the Cr^{3+} ion (Payne *et al.* 1988, Payne *et al.* 1989) with slope efficiencies exceeding 50%.

Cr^{3+} lasers can also be influenced by ESA processes arising from both ^2E and $^4T_2$ excited states. In Cr^{3+}:Gd$_3$Sc$_2$Ga$_3$O$_{12}$ (GSGG) (Struve *et al.* 1983) the tunability of the laser (0.74–0.84μm) and the slope efficiency (28%) are limited by ESA at the wings of the fluorescence curve and by ESA of the pump photons, respectively. Uniaxial crystals like LICAF might have polarised ESA bands which do not influence special pump and laser polarisations. Due to the advantage of diode pumping, a revival of Cr^{3+} lasers is very likely.

The tetrahedrally coordinated Cr^{4+} ion is an attractive laser ion in the near-infrared spectral range 1.1–1.7μm. Until now room-temperature laser action has been achieved in Cr^{4+}-doped Mg$_2$SiO$_4$ (forsterite) (Petricevic *et al.* 1988), Y$_3$Al$_5$O$_{12}$ (YAG) (Zverev and Shestakov 1989, Jia *et al.* 1991, Kück *et al.* 1991), Y$_2$SiO$_5$ (YSO) (Koetke *et al.* 1993), and the mixed garnet system Cr^{4+}:Y$_3$Sc$_x$Al$_{5-x}$O$_{12}$ (YSAG) (Kück *et al.* 1994).

An absorption band near 1μm makes the system ideal for pumping with a Nd:YAG laser. The tuning range is broad and ranges in the case of Cr^{4+}:YAG for 1.32–1.6μm. Figure 10 shows the tuning range of Cr^{4+}:YAG under pulsed Nd:YAG excitation (Kück *et al.* 1994). Slope efficiencies of Cr^{4+} lasers are typically 10% for CW operation and 20% for pulsed operation. The problem is that the quantum efficiency at room temperature is, for example, only 33% for Cr^{4+}:YAG (Kück *et al.* 1995); this is due to electron-phonon coupling and corresponding non-radiative 3d–3d transitions. Other hosts have even lower quantum efficiencies.

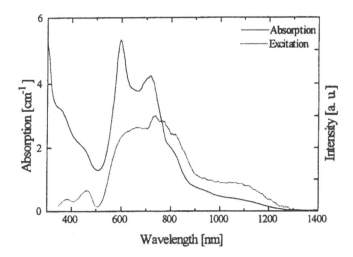

Figure 11. *Unpolarised absorption and excitation spectra of Cr:LiAlO₂ at room temperature (Kück et al 1995).*

Another problem is that not only tetrahedral Cr^{4+}, but also Cr^{3+} and/or octahedral Cr^{4+} is incorporated in these lattices. In contrast, $LiAlO_2$ exhibits tetrahedrally coordinated lattice sites only (Kück *et al.* 1995) and is therefore ideal for the investigation of the spectroscopic properties of transition-metal ions in fourfold coordination. Additionally, Cr^{4+} doped $LiAlO_2$ exhibits a room-temperature lifetime of $29\mu s$, which is the longest lifetime ever observed for the Cr^{4+} emission.

The unpolarised absorption and excitation spectra of the Cr^{4+}:$LiAlO_2$ crystal are shown in Figure 11. The absorption spectrum looks similar to that of Cr^{4+} doped forsterite. Three absorption bands are observed around 400, 700, and 1100nm due to the three spin-allowed transitions between the $^3A_2(^3F)$ ground state and the $^3T_1(^3P)$, $^3T_1(^3F)$, and $^3T_2(^3F)$ excited states, respectively.

The room temperature emission extends from 1.1 to $1.7\mu m$ with a maximum at 1290nm and a bandwidth of 241nm; the spectrum is shown in Figure 12. Both interesting wavelengths for fibres for optical communication at 1.3 and $1.55\mu m$ are covered. The emission belongs to the transition between the lowest level of the crystal field split $^3T_2(^3F)$ excited state to the ground state $^3A_2(^3F)$.

The room temperature lifetime and the lifetime at 12K of the excited state of Cr^{4+} in $LiAlO_2$ ion is 29 and $95\mu s$, respectively. In comparison to YAG, the corresponding lifetimes are 4 and $30\mu s$, respectively. The long lifetime at low temperature as well as the high ratio of room and low temperature lifetimes gives hope for a higher quantum efficiency of the Cr^{4+} emission in this material compared to YAG.

Recently, new lasers based on the divalent ion Cr^{2+} have been introduced (Page *et al.* 1995). Cr^{2+} doped II-VI compounds as ZnS and ZnSe show ^5E–$^5T_2$ luminescence bands in the range 1800–3000nm. When pumping with a pulsed Co^{2+}:MgF_2 laser near $1.88\mu m$, the slope efficiency of Cr^{2+}:ZnSe is 23%, (tuning range 2300–2500nm). Losses of laser crystals were high (10% per pass); this might be improved in the future.

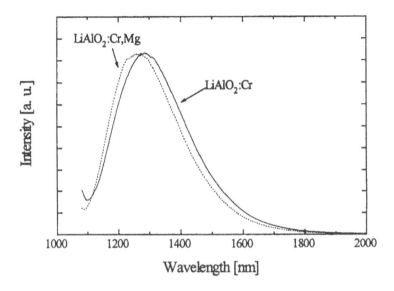

Figure 12. *Room temperature emission spectra of Cr:LiAlO₂ and Cr,Mg:LiAlO₂ (Kück et al 1995).*

5 Lasers in the visible spectral region

Lasers in the visible spectral region are of interest for many applications as display and high density optical data storage applications. Intracavity frequency doubling of CW Nd:YAG lasers for 1064–532nm is an efficient method of generation of visible laser radiation. Since semiconductor diode lasers became an important pump source for solid-state lasers, research in this field was revitalised again.

Another way of efficient generation of visible laser radiation is frequency doubling of near-infrared lasers, including diode lasers, with an external resonant cavity, in which case a single frequency laser is required and active control of the resonance frequency with high accuracy is necessary. A third possibility is the realisation of up-conversion lasers. Pumping in the near-infrared spectral region can lead to coherent radiation of photons in the red, green, and blue spectral ranges. This approach requires however very high pump intensities. Potential ions for such applications at room temperature are Pr^{3+}, Tm^{3+}, and Er^{3+}.

5.1 Intra-cavity second harmonic generation (SHG)

There is an optimum output coupling of the nonlinear crystal in intracavity doubled CW lasers like the optimum output coupling in a conventional laser. However, the key to efficient intracavity doubling is *low cavity losses*. If the cavity losses are small, already small nonlinearities (or short SHG crystals) can efficiently extract energy into the frequency doubled laser output. For instance, Nd:LSB (see section 3.1) is a low-loss material with a broad and strong absorption band at 808nm, making diode laser

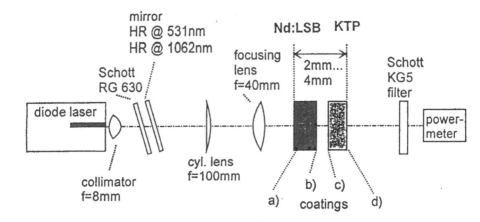

Figure 13. *Experimental setup of a diode pumped Nd:LSB laser. For explanation see text.*

pumping very efficient.

An experimental laser setup (Meyn and Huber 1994) is shown in Figure 13. A 3W diode laser array (SDL 2482) operating at 808nm is used as a pump source. The diode laser radiation is focused with an $f = 100$mm cylindrical lens and an $f = 40$mm spherical lens into a spot size of approximately 50μm $\times$ 200μm. Because of this elliptical pump beam profile the laser beam is transverse multimode. The laser crystal is 1.2 mm long and doped with 10% Nd. The absorption of the pump radiation is then higher than 95%.

The crystal is cut perpendicular to one principal axis of the indicatrix, which is between the two optical axes; thus the twofold monoclinic symmetry axis lies on the surface. The laser radiation is polarised perpendicular to the monoclinic axis. The crystal is polished plane parallel and coated (a) R = 99.97% at 1062nm and R < 1% at 808nm and (b) R < 0.1% at 1062nm and R = 40% (accidental) at 531nm. Surface (a) acts as one of the resonator mirrors. The KTP crystal is cut for type II phase matching at 1062nm. Coating (c) is anti-reflective for 1062 and 531nm, and the output coupler coating (d) is R > 99.99% at 1062nm and R < 2% at 531 nm. Different KTP crystals of lengths 0.5, 1, and 2 mm were used. The resonator length was slightly larger than the sum of the crystal lengths.

Figure 14 shows the input-output curves for different lengths of the nonlinear KTP crystal. The total conversion of diode pump radiation into green laser radiation is 25% at 2W incident pump power.

The Nd:LSB laser has many longitudinal modes. Therefore, chaotic intensity fluctuations that are due to sum-frequency generation can occur. The intensity noise could be reduced either by single frequency operation or by the introduction of a quarter-wave plate for the separation of the laser beam into two independent polarisation modes. However with the setup of Figure 13, stable emission without chaotic intensity fluctu-

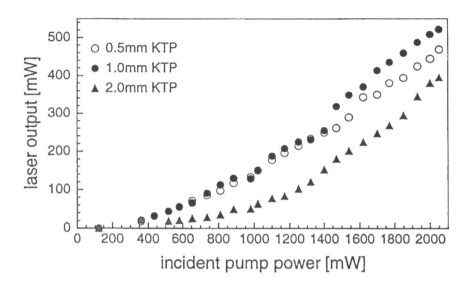

Figure 14. *Output power measured in the output coupler direction of the intracavity frequency doubled Nd:LSB laser (Meyn and Huber 1994).*

ations was observed (Meyn and Huber, 1994) if the KTP crystal was 1 mm long or shorter.

It was demonstrated (Baer 1986) that the chaotic fluctuations increase with higher nonlinear coupling. If the beam diameter remains almost constant in the cavity, the nonlinear coupling of a phase-matched nonlinear crystal with a given nonlinear coefficient is dependent only on its length. Therefore, one possible solution of the green problem is the use of a short nonlinear crystal with sufficiently low nonlinear coupling. This was also experimentally verified recently with a 2 mm long KTP crystal and Nd:YVO$_4$ in a 2.5 mm long cavity (MacKinnon and Sinclair 1994). By use of a low-loss laser and nonlinear crystals it is possible to obtain a good conversion efficiency, even with a KTP crystal length of 1 mm or less.

5.2   Room temperature cw up-conversion lasers

Lasers which emit at higher frequencies than the pump light usually are called up-conversion lasers. In these lasers the active ion is excited by internal up-conversion of near infrared or red light via multistep photon excitation or cooperative energy transfer and emits anti-Stokes visible light. In initial up-conversion laser research, cryogenic temperatures were required and the observed efficiencies were low (Herbert *et al.* 1990).

The advent and rapid improvement of high power laser diodes in the red and near infrared spectral ranges have caused new interest in the development of up-conversion lasers. The output wavelength of laser diodes can be tuned to match the absorption

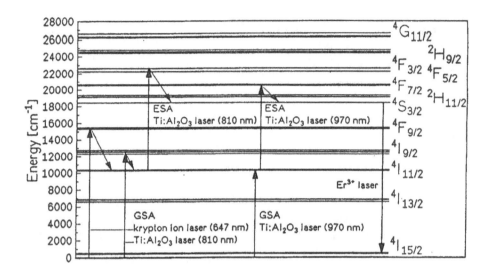

Figure 15. *ESA pump mechanisms for two step-pumped Er upconversion lasers.*

lines of the active laser ion, resulting in a substantial fraction of ions excited into higher energy levels, thus enhancing the up-conversion process.

Visible up-conversion lasing at room temperature has already been demonstrated in Tm-doped crystals (Nguyen *et al.* 1990, Thrash and Johnson 1992) and in various rare earth doped fluorozirconate fibres (Whitley *et al.* 1991). Er^{3+} is a very interesting ion for CW up-conversion to the green spectral region (Heine *et al.* 1994).

In recent experiments two step single ion pumping schemes, evaluated from ESA measurements, were used as shown in Figure 15 . Real interionic up-conversion processes would require Er^{3+} concentrations much higher than those used in the experiment (1% in most cases) of Heine *et al.* (1994).

There are two different ways of up-converting the infrared pump power to the upper laser level $^4S_{3/2}$ by a two step pumping process. The first possibility is dual wavelength pumping, using the 970nm absorption for populating the $^4I_{11/2}$ and 810nm for efficient population of the upper laser level. The second way is to use 810nm pump radiation alone, which is a more simple arrangement. However, the disadvantage of the latter pump scheme is the very weak ground state absorption at 810nm which is 15 times smaller than the 970nm absorption. The low Er concentration was also used to minimise reabsorption losses which are $\sim 10\%$ at the laser wavelength with an absorption cross section of $\sigma_{abs} = 0.2 \times 10^{-20}$ cm^2 and an emission cross section of $\sigma_e = 2 \times 10^{-20}$ cm^2.

The resulting input-output curve under Ti-sapphire pumping with 810nm radiation is shown in Figure 16. It should be mentioned, that for the results in Figure 16 the authors (Heine *et al.* 1994) have used a pump enhancement due to a feedback of the Ti-sapphire pump laser into the Er:YLF laser cavity. Figure 16 shows the pump power,

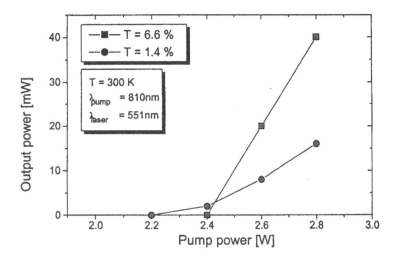

Figure 16. *Input-output curves of the CW Er up-conversion laser for two different output couplings (1.4 % and 6.6 %) at 551nm when pumped with a Ti-sapphire laser (Heine et al. 1994).*

measured without feedback. The feedback gave a pump intensity enhancement of about a factor of 4. These recent data lead to the conclusion, that diode pumped up-conversion lasers are very difficult to realise. One possibility might be pumping with a MOPA laser with resonant pump enhancement in the laser cavity of the Er^{3+} crystal.

5.3 Pr^{3+} lasers

Pr^{3+} is an interesting ion for visible laser applications because its large number of energy levels (Figure 17) offers many potential laser transitions. $Pr:CaWO_4$ was the first praseodymium laser realised in 1962 (Yariv *et al.* 1962). Since this time pulsed laser action of this ion has been obtained in many other hosts, an overview is given by Kaminskii (1990). Recently, CW lasers with Pr^{3+} were realised in $Pr:LaCl_3$ at low temperatures (Koch et al 1990) and in fibres (Percival *et al.* 1989, Allain *et al.* 1991, Durteste *et al.* 1991, Smart *et al.* 1991).

In a recent paper (Sandrock *et al.* 1994) room-temperature CW lasing of Pr:YLF, $Pr:GdLiF_4$ (Pr:GLF), and $Pr:KYF_4$ (Pr:KYF) was reported. An argon-ion laser was used as a pump source. The pump beam was focused with a lens of $f = 50$mm focal length onto the crystal placed in a linear, nearly concentric resonator. The average pump-beam waist was about 20μm in radius. The mirrors had a radius of curvature equal to 50mm. The input coupler was highly reflecting at the laser wavelength. Neither the transmissions of the used output couplers nor the crystal lengths have been optimised.

CW lasing has been observed in an 8.9mm long Pr:YLF crystal with a Pr concentration of 9.1×10^{19}cm^{-3}; several transitions are summarised in Table 3. The lifetime of the upper laser levels, which are the thermally coupled $^3P_1$ and $^3P_0$ states, was $\tau = 36\mu$s.

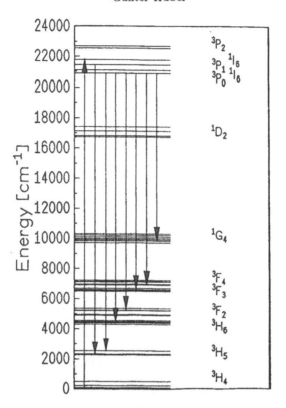

Figure 17. *Energy levels and laser transitions of Pr:YLiF₄ (Sandrock et al 1994).*

Transition	λ nm	Polari- sation	P_{thr} mW	n_{thr} cm^{-3}	P_{out} mW	η %
$^3P_1 \rightarrow {}^3H_5$	522	E ∥ c	163	1.2×10^{18}	144	14.5
$^3P_0 \rightarrow {}^3H_5$	545	E ∥ c	-	19	-	-
$^3P_0 \rightarrow {}^3H_6$	607	E ⊥ c	110	8.3×10^{17}	7	1.2
$^3P_0 \rightarrow {}^3F_2$	639.5	E ⊥ c	8	$6,0 \times 10^{16}$	266	25.9
$^3P_0 \rightarrow {}^3F_3$	697	E ∥ c	105	7.8×10^{17}	71	10.3
$^3P_0 \rightarrow {}^3F_4$	720	E ∥ c	98	7.3×10^{17}	40	7,2
$^3P_0 \rightarrow {}^1G_4$	907.4	E ∥ c	280	2.1×10^{18}	23	7.3

Table 3. *Results of the laser experiments on Pr:YLF obtained with different sets of mirrors.*

As examples, Figure 18 shows the input-output characteristics at three wavelengths. Very recently, Kerr-lens mode-locking was reported By Sutherland *et al.* (1995)

It is remarkable that lasing has been obtained at $\lambda = 639.5$ nm even with the transmission of the output mirror greater than 75% due to a high emission cross section, which could be estimated from the laser thresholds at different output coupling to be of

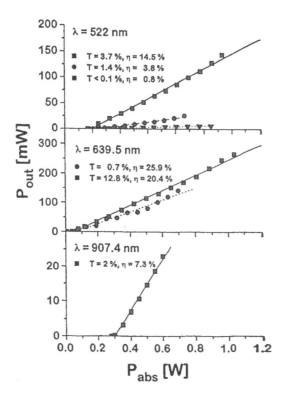

Figure 18. *Output vs. input power for the cw Pr:YLiF₄ laser at different laser wavelengths pumped at 457.9nm (Sandrock et al 1994).*

the order of 10^{-19} cm². The results show that Pr^{3+} is a very interesting ion for efficient CW laser emission in the visible spectral range. One of the main goals in the near future is the investigation of up-conversion pumping schemes for the visible laser transitions in Pr^{3+} doped crystals similar to schemes used in fibre lasers.

6 Conclusion

Among the class of vibronic transition metal lasers, in addition to the widely used Ti^{3+} systems, chromium lasers have regained interest due to the possibility of diode laser pumping in Cr^{3+} lasers. Tunable laser systems based on Cr^{4+} and Cr^{2+} also offer very interesting prospects. It seems that here, new generations of infrared lasers are emerging. Diode pumped rare earth lasers (Nd^{3+}, Tm^{3+}, Ho^{3+}, Er^{3+}, Yb^{3+}) have very good efficiencies for applications in the wavelength range between 1μm and 3μm. The ions Tm^{3+} and Er^{3+} (in some cases also Pr^{3+}) provide upconversion schemes which lead to visible CW laser operation with near infrared pumping. These systems are still under fundamental investigation and, from the commercial point of view, so far offer no alternative to frequency doubled lasers. However the simple scheme of upconversion lasers make them very attractive candidates for further research.

References

Albers P, Stark E, and Huber G, 1986. Continuous-wave laser operation and quantum efficiency of titanium-doped sapphire, *J Opt Soc Am,* **B 3** 134.

Allain J Y, Monerie M, Poignant H, 1991. Tunable CW lasing around 610, 635, 695, 715, 885 and 910 nm in praseodymium-doped fluorozirconate fibre, *Electron Lett,* **27** 189.

Baer T, 1986. Large-amplitude fluctuations due to longitudinal mode coupling in diode-pumped intracavity-doubled Nd:YAG lasers, *J Opt Soc Am,* **B3** 1175.

Chai B H T, 1993, Novel fluoride laser crystal materials, in *CLEO '93,OSA Techn. Dig. Series,* **11** 18.

Danger T, Petermann K, Huber G, 1993. Polarised and time-resolved measurements of excited-state absorption and stimulated emission in Ti:YAlO$_3$ and Ti:Al$_2$O$_3$,*Appl Phys,* A **57** 309.

Danielmeyer H G, 1975. Stoichiometric laser materials, *Festkörperprobleme,* **XV** 253.

Dinermann B Y, Moulton P F, 1992. CW laser operation of Er:YAG, Er:GGG, and Er:YSGG, in *OSA Proc on Advanced Solid-State Lasers,*eds, L.L. Chase and A.A. Pinto, **13** 152.

Durteste Y, Monerie M, Allain J Y, Poignant H, 1991. Amplification and lasing at 1.3 μm in praseodymium-doped fluorozirconate fibres,*Electron Lett,* **27** 626.

Esterowitz L, 1990. Diode-pumped holmium, thulium, and erbium lasers between 2 and 3 μm operating CW at room temperature,*Optical Engineering,* **29** 676.

Fan T Y and Byer R L, 1988. Diode-laser pumped solid-state lasers, *IEEE J. Quantum Electron,* **QE-24** 895.

Fan T Y, Huber G, Byer R L and Mitzscherlich P, 1988. Spectroscopy and diode laser-pumped operation of Tm;Ho:YAG,*IEEE J Quantum Electron* **QE-24** 924.

Fields R A, Birnbaum M, and Fincher C L, 1987. Highly efficient Nd:YVO$_4$ diode-laser end-pumped laser,*Appl Phys Lett,* **51** 1885.

Heine F, Heumann E, Danger T, Schweizer T, Huber G, Chai B H T, 1994. Green up-conversion continuous wave Er^{3+}:LiYF$_4$ laser at room temperature, *Appl Phys Lett,* **65** 683.

Herbert T, Wannemacher R, Lenth W, and Macfarlane R M, 1990. Blue and green CW up-conversion lasing in Er:YLiF$_4$,*Appl Phys Lett,* **57** 1727.

Huber G, 1980. Miniature neodymium lasers, in *Current Topics in Materials Science,* ed E Kaldis,(North Holland,Amsterdam) **4** 1.

Huber G, Duczynski E W, and Petermann K, 1988. Laser pumping of Ho-, Tm-, Er-doped garnet lasers at room temperature, *IEEE J Quantum Electron,* **QE-24** 920.

Ishimori A, Yamamoto T, Uchiumi T, Yagi S, Shigihara K, 1992. Pumping configuration without focusing lenses for a small-sized diode-pumped Nd:YAG slab laser, *Opt Lett,* **17** 40.

Jensen T, Diening A, Huber G, Chai B H T, 1995. A diode pumped 1.1W CW Er:YLF laser at 2.8 μm, in *CLEO'95, Baltimore,* paper CPO29

Jensen T, Ostroumov V G, Meyn J, Huber G, Zagumennyi A I, Shcherbakov I A, 1994. Spectroscopic characterisation and laser performance of diode-laser-pumped Nd:GdVO$_4$, *Appl Phys,* B **58** 373.

Jia W, Tissue B M, Lu, Hoffmann K R, and Yen W M, 1991. Near-infrared luminescence in Cr,Ca-doped yttrium aluminium garnet, in: *OSA Proc on Advanced Solid-State Lasers,* eds George Dubé,and Loyd Chase, **10** 87.

Kaminskii A A, 1990, Laser crystals. Their physics and properties, (Springer,Berlin) *Opt. Sci,*14

Karayianis N, Morrison C A, Wortman D E, 1974. Analysis of the ground term energy levels for triply ionised neodymium in yttrium orthovanadate, *J Chem Phys,* **62** 4125.

Kintz G J, Baer T, 1990. Single-frequency operation in solid-state laser materials with short absorption depths, *IEEE J Quantum Electron*, **QE-26** 1457.

Kitaoka Y, Ohmore S, Yamamoto K, Kato M, and Sasaki T, 1993. Stable and efficient green light generation by intracavity frequency doubling of Nd:YVO$_4$ lasers, *Appl Phys Lett*, **63** 299.

Koch M E, Kueny A W, Case W E, 1990. Photon avalanche upconversion laser at 644nm, *Appl Phys Lett*, **56** 1083.

Koetke J, Kück S, Petermann K, Huber G, Cerullo G, Danailov M, Magni V, Quian L F, Svelto O, 1993. Quasi-continuous wave laser operation of Cr^{4+}-doped Y$_2$SiO$_5$ at room temperature,*Opt Comm*, **101** 195.

Kück S, Hartung S, Petermann K, Huber G, 1995. Spectroscopic properties of Cr^{4+}-doped LiAlO$_2$, *Appl Phys*, **B 61** 33.

Kück S, Petermann K, and Huber G, 1991. Spectroscopic investigation of the Cr^{4+}-center in YAG, in *OSA Proc on Advanced Solid-State Lasers*, eds George Dubé and Loyd Chase, **10** 92.

Kück S, Petermann K, Pohlmann U, Huber G, 1995. Near-infrared emission of Cr^{4+}-doped garnets: Lifetimes, quantum efficiencies, and emission cross sections, *Phys Rev B*, **51** 17323.

Kück S, Petermann K, Pohlmann U, Schönhoff U, Huber G, 1994. Tunable room- temperature laser action of Cr^{4+}-doped Y$_3$Sc$_x$Al$_{5-x}$O$_{12}$, *Appl Phys*, **B58** 153.

Kutovoi S A, Laptev V V, and Yu Matsnev S., 1991. Lanthanum scandiumborate as a new highly efficient active medium of solid-state lasers, *Sov J Quantum Electron*, **21** 131.

Lacovara P, Choi H K, Wang C A, Aggarwal R L and Fan T Y, 1991. Room-temperature diode-pumped Yb:YAG laser, *Optics Letters*, **16** 1089.

MacKinnon N and Sinclair B D, 1994. A laser diode array pumped, Nd:YVO$_4$/KTP, frequency-doubled, composite-material microchip laser,*Opt Comm*, **105** 183.

Meyn J-P, Huber G, 1994. Intracavity frequency doubling of a continuous-wave, diode-laser-pumped neodymium lanthanum scandium borate laser, *Optics Lett*, **19** 1436.

Meyn J-P, Jensen T, Huber G, 1994. Spectroscopic properties and efficient diode-pumped laser operation of neodymium-doped lanthanum scandium borate, *IEEE J Quantum Electronics*, **30** 913.

Moulton P, 1982. Ti-doped sapphire: a tunable solid-state laser, *Opt News*, **8** 9.

Nguyen D C, Faulkner G E, Weber M E, and Dulick M, 1990. Blue upconversion thulium laser,*SPIE Solid State Lasers*, **1223** 54.

Oka M and Kubota S, 1992. Second-harmonic generation green laser for higher-density optical disks, *Jpn J Appl Phys*, **31** 513.

Page R H, DeLoach L D, Wilke G D, Payne SA, Krupke W F, 1995. A new class of tunable mid-IR lasers based on Cr^{2+}-doped II-VI compounds,*CLEO '95, Baltimore*, Paper CWH5.

Payne S A, Chase L L, Newkirk H W, Smith L K, and Krupke W F, 1988. LiCaAlF$_6$:Cr^{3+}: a promising new solid state laser material, *IEEE J Quantum Electron*, **24** 2443.

Payne S A, Krupke W F, Smith L K, DeLoach L D, and Kway W L, 1992. Laser properties of Yb in fluoro-apatite and comparison with other Yb-doped gain media, in *Conference on Lasers and Electro-Optics*, (OSA Technical Digest Series) **12** 540.

Payne S A, S.A. Chase S A, Smith L K, Kway W L, and Newkirk H W, 1989. Laser performance of LiSrAlF$_6$:Cr^{3+}, *J Appl Phys*, **66** 1051.

Percival R M, Phillips M W, Hanna D C, Tropper A C, 1989.Characterisation of spontaneous and stimulated emission from praseodymium (Pr^{3+}) ions doped into a silica-based monomode optical fiber, *IEEE J Quantum Electron*, **QE 25** 2119.

Petricevic V, Gayen S K, and Alfano R R, 1988. Laser action in chromium-activated forsterite for near-infrared excitation: is Cr^{4+} the lasing ion?, *Appl Phys Lett*, **53** 2590.

Sandrock T, Danger T, Heumann E, Huber G, Chai B H T, 1994. Efficient continuous wave-laser emission of Pr^{3+}-doped fluorides at room temperature, *Appl Phys*, B **58** 149.

Scheps R, 1992. Cr-doped solid state lasers pumped by visible laser diodes, *Optical Materials*, **1** 1.

Schweizer T, Jensen T, Heumann E, Huber G, 1995. Spectroscopic properties and diode pumped 1.6 μm laser performance in Yb-codoped $Er:Y_3Al_5O_{12}$ and $Er:Y_2SiO_5$, *Opt Comm*, **118** 557.

Sipes D L, 1985. Highly efficient neodymium: yttrium aluminium garnet laser end pumped by a semiconductor laser array, *Appl Phys Lett*, **47** 74.

Smart R G, Carter N, Tropper A C, Hanna D C, Davey S T, Carter S F, Szebesta D, 1991. Cw room temperature operation of praseodymium-doped fluorozirconate glass fibre lasers in the blue-green, green and red spectral regions, *Opt Comm*, **86** 337.

Sourian J C, Romero R, Borel C, Wyon C, Li C, Moncorgé R, 1994. Room-temperature diode-pumped continuous-wave $SrY_4 (SiO_4)_3O$: Yb^{3+}, Er^{3+} crystal laser at 1554nm, *Appl Phys Lett*, **64** 1189.

Streifer W, Scrifres D R, Harnagel G L, Welch D F, Berger J, Sakramoto M, 1988. Advances in diode laser pumps, *IEEE J Quantum Electron*, **QE-24** 883.

Struve B, Huber G, Laptev V V, Shcherbakov I A, and Zharikov E V, 1983. Tunable room-temperature CW-laser action in Cr^{3+}:GdScGa-garnet, *Appl Phys* B **30** 117.

Sutherland J M, Ruan S, French P M W, Taylor J R, and Chai B T H, 1995, Kerr-lens mode-locked Pr:YLF laser in the visible, in *Advanced Solid State Lasers*, eds Chai B T H, and Payne S A, OSA Proceedings Series, **24** 322

Thrash R J and Johnson L F, 1992. Tm^{3+} room temperature upconversion laser, in *Compact Blue-Green Lasers, Santa Fe, New Mexico*,(Technical Digest Series) **6,** paper ThB3.

Walling J C, Peterson O G, Jenssen H P, Morris R C, and O'Dell E W, 1980. Tunable alexandrite lasers, *IEEE J Quantum Electron* **QE-16** 1302.

Whitley T J, Millar C A, Wyatt R, Brierley M C, and Szebesta D, 1991. Upconversion pumped green lasing in erbium doped fluorozirconate fibre, *Electron Lett* **27** 1785.

Yariv A, Porto S P S, Nassau K, 1962. Optical maser emission from trivalent praseodymium in calcium tungstate, *J Appl Phys* **33** 2519.

Zverev G M and Shestakov A V, 1989. Tunable near-infrared oxide crystal lasers,in: *Tunable Solid State Lasers*, eds Shand M L and Jenssen H P, OSA Proceedings Series, **5** 66.

Diode-Pumped Solid-State Lasers

T Y Fan

Lincoln Laboratory
Massachusetts Institute of Technology, USA

1 Introduction

The key driver in solid-state laser technology over the past few years has been the development of high-power, efficient, and reliable semiconductor diode lasers that can be used as optical pump sources. Solid-state lasers, meaning the class of solid dielectric gain media as opposed to semiconductor gain media, have traditionally been optically pumped by either pulsed or continuous-wave lamps. The use of semiconductor diode lasers as pumps has enabled the development of solid-state lasers that are more efficient, more compact, more reliable, more stable, and in some cases less expensive than their lamp-pumped counterparts. In this chapter, I give an overview of the basic attributes of diode-pumped lasers and how they compare with lamp-pumped solid-state laser and diode lasers. Then I will discuss in further depth two areas in which diode-pumped lasers have enabled new regimes of operation for solid-state lasers. These are in the areas of quasi-three-level lasers and narrow-linewidth lasers.

Historically, diode-pumped solid-state lasers are not new; the first suggestion that semiconductor light sources could be used to optically excite solid-state lasers was in 1963 (Newman 1963); at that time it was suggested that GaAs LED's could be used to excite $Nd:CaWO_4$ lasers. The first diode-pumped laser was demonstrated in 1964 (Keyes and Quist 1964) who used a set of five GaAs diode lasers to pump a U^{3+}-doped CaF_2 laser. A photograph of that first laser is shown in Figure 1. The major difficulty in this experiment was that the entire laser, the pump diodes and the $U^{3+}:CaF_2$ laser rod, were placed in a liquid Helium dewar and held at cryogenic temperatures. The

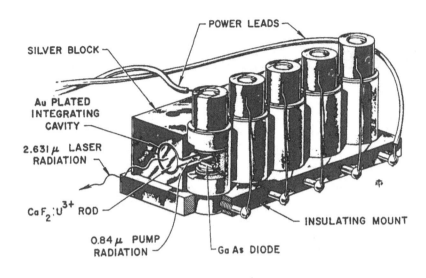

Figure 1. *The first diode-pumped solid-state laser.*

primary issue was that diode lasers did not operate at room temperature at that time. In
addition, the U^{3+}:CaF_2 laser operated better at low temperatures as the lower laser level
is near the ground state and consequently laser threshold is reduced at low temperatures.
Throughout the rest of the 1960's and the 1970's there was a low level of effort in
diode-pumped lasers, using both LED's and laser diodes as pump sources. However,
these experiments were always constrained by the low power and low reliability of
the semiconductor diode light sources themselves. The key in diode-pumped solid-state
laser technology was the development of high-power, reliable semiconductor diode lasers,
particularly AlGaAs diode lasers operating in the 0.8μm wavelength region (Streifer
et al. 1988, Endriz *et al.* 1992). These AlGaAs diodes enabled diode pumping of the
most widely developed solid-state gain media, Nd^{3+}-doped laser materials. Overviews
of the early work in diode-pumped lasers are available (Byer 1988, Fan and Byer 1988)
as is an overview of more recent developments (Hughes and Barr 1992).

Diode-pumping has several advantages relative to lamp-pumped solid-state lasers,
which include higher overall efficiency, higher pump density, lower thermal loading
of the gain medium, and reduced amplitude and frequency fluctuations. There are
disadvantages as well which include higher cost, particularly in devices with high energy
per pulse, and lack of short-wavelength pump photons, which places some limitations on
the laser gain media that can be pumped with diodes. Another important comparison
is the use of diode lasers to pump gain media versus using the diode lasers directly.
Here, the diode-pumped solid-state laser has advantages in the areas of high-peak-power
capability, frequency stability, and high power in a good laser beam. The advantages
of using a diode laser directly are simplicity and efficiency.

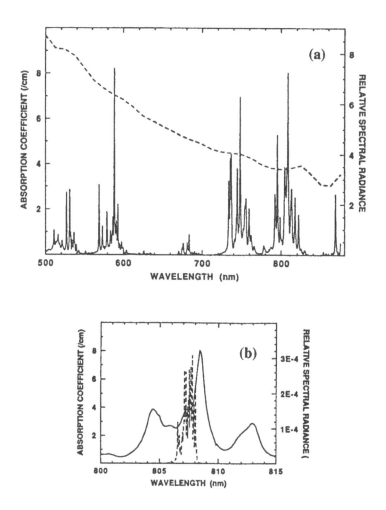

Figure 2. *Absorption spectrum of Nd:YAG overlaid by the output of (a) a lamp and (b) a diode laser.*

1.1 Comparison of diode pumping and lamp pumping

Let me address the comparisons in more detail. Diode-pumped lasers can be more efficient than lamp-pumped lasers because of the narrowband spectral output of the diode lasers. This can be seen in Figure 2 which shows the spectral outputs of a lamp and a diode laser overlaid on the absorption spectrum of the most important solid-state laser material Nd:YAG which is typically operated at $1.06\mu m$. The Nd:YAG absorption spectrum has significant regions where there is essentially no absorption, so large portions of the lamp output pass through the Nd:YAG without being absorbed. In contrast, diode lasers can be tuned directly onto an absorption feature, meaning that much more efficient utilisation of the pump radiation is achieved. The electrical-to-optical efficiencies in lamps can be high, as high as 60% to the spectral region with wavelengths less than $1\mu m$, while AlGaAs diode lasers operating near $0.8\mu m$ can have efficiencies as high as 50%.

The fact that there is less heating of the Nd:YAG with diode pumping can also be seen from this figure. The difference in energy between a pump photon and a laser output photon goes into heating the Nd:YAG, and it can be seen in Figure 2 that the average pump photon has a higher energy in lamp pumping than in diode pumping. Diode-pumping leads to lower thermal loading than lamp pumping in Nd^{3+} lasers (Chen *et al.* 1990), but this does not necessarily have to be true for all gain media; whether it holds true depends on the details of the pump spectrum and laser wavelength. However, for most diode-pumped systems a reduction in thermal loading is observed relative to their lamp-pumped counterparts.

Higher pumping densities can be obtained using diode lasers as pump sources rather than lamps as a result of diode lasers' directionality and high radiance (power per unit area per unit solid angle, also called *brightness*) and narrowband spectral output (also called *spectral brightness*), as opposed to the omnidirectional output of a lamp and its broadband output. The high directionality and radiance of a diode laser means that the output can be focused to a small cross sectional area, and the narrow bandwidth means that the radiation can be efficiently absorbed in a short distance. It is easy to show that low-power (20mW) single-stripe diode lasers can be used to obtain over an order of magnitude higher pump densities in Nd:YAG than high-power (10kW) CW arc lamps. This has enabled the demonstration of efficient quasi-three-level lasers (lasers with the lower level near the ground state) at room temperature, which had not been previously demonstrated in lamp-pumped devices. These lasers require high pump densities in order to reach laser threshold, as will be discussed in a later section. The high pump intensities have also enabled the development of upconversion lasers and fibre lasers and amplifiers. Upconversion lasers are those devices in which the output laser wavelength is shorter than the pump wavelength. Population of the upper laser levels occurs by multistep excitation, either through excited-state absorption or by energy-transfer upconversion. In either case, high pump intensities are required for efficient excitation to the upper laser levels, because for two-step excitation processes, the efficiency of the excitation scales with the pump density. Fibre lasers take advantage of the ability to focus the pump radiation into a small cross-sectional area thereby enabling efficient coupling of the pump radiation into a fibre. These devices are covered elsewhere in this text (Hanna 1995).

Diode-pumped lasers have reduced amplitude and frequency fluctuations relative to lamp-pumped lasers because of the reduced heating in the gain medium and less technical noise. A large fraction of the frequency fluctuations in lasers are caused by variations in optical path length due to vibrations and temperature changes. Since the thermal loading is less in diode-pumped lasers, pump fluctuations cause smaller changes in refractive index and therefore smaller changes in optical path length. Also, many lamp-pumped lasers are cooled with flowing gas or liquid while diode-pumped lasers are often conduction cooled. This means that the diode-pumped lasers are typically in a less acoustically noisy environment which gives them an advantage in optical path length stability. In lamp-pumped solid-state lasers, the narrowest 3dB linewidth that had been observed was 120kHz; in typical single-frequency diode-pumped lasers, these linewidths are approximately 5kHz.

One of the drawbacks of diode pumping is that the shortest-wavelength commercially available diode lasers today are at about 630 nm. This means that some solid-state laser transitions, such as many in Pr^{3+}, cannot be pumped directly by diodes (although upconversion Pr^{3+} lasers have been demonstrated) but can be pumped directly by lamps. This is not a major drawback as many of the lasers that cannot be pumped by diodes are not of significant economic importance. The other major disadvantage of diode pumping relative to lamps is the large cost, particularly for high energy per pulse solid-state lasers. Diode lasers are peak power limited, in other words, even if diode lasers are pulsed, their peak output power cannot really be increased. Thus, for high energy per pulse systems, large numbers of diode lasers are required. While progress has been made in reducing the cost of diode lasers, for high energy systems, diode lasers are much more costly than lamps. For CW operation or for low-energy lasers (mJ level currently), the economics of diode lasers become more favourable because of the fewer diodes required, and what is observed commercially is that the major inroads for diode-pumped lasers are for CW and low energy per pulse lasers.

1.2   Diode lasers versus diode-pumped lasers

Diode lasers are certainly simpler, more efficient, and less costly than diode-pumped lasers, and diode lasers are generally preferred if they are adequate for a particular application. Diode-pumped lasers have advantages in those applications when high average power with good beam quality is needed, when high peak power is needed, or when frequency stability or narrow linewidth is needed. Solid-state lasers can essentially act as radiance-enhancing beam combiners for diode lasers; multiple, incoherent diode lasers are used to optically pump the solid-state laser and generate a single output beam. The peak-power advantage of a solid-state laser is related to the upper-state (metastable level) lifetimes: in typical rare-earth-doped solid-state lasers the upper-state lifetimes are tens of microseconds to a few milliseconds whereas in diode lasers, the electron-hole recombination times are typically on the order of nanoseconds. Solid-state lasers can be pumped by diode lasers for a period of the order of the upper-state lifetime, and then this stored energy can be released in a short Q-switched pulse lasting of the order of a nanosecond. This brings a peak-power enhancement of greater than 10^4 relative to the peak diode power. With the relatively short recombination time in diode lasers, little energy storage is possible.

Solid-state lasers have better frequency stability and narrower linewidth than diode lasers. The *Schawlow-Townes linewidth* of lasers $\Delta\nu_{ST}$, the component of laser linewidth caused by spontaneous emission, is given by

$$\Delta\nu_{ST} = \frac{h\nu}{2\tau_c P} \qquad (1)$$

where $h\nu$ is the laser photon energy, P is the output power, and τ_c is the cavity lifetime which is given by

$$\tau_c = \frac{2L}{c\delta_c} \quad . \qquad (2)$$

L is the length of the cavity, c is the speed of light, and δ_c are the round-trip cavity losses (including the loss for output coupling). The cavity lifetime is proportional to the cavity length and decreases with increasing cavity loss (lower Q). Solid-state laser materials have low loss ($<10^{-3}$ cm^{-1}) relative to diode lasers (>1 cm^{-1}) and the laser cavities for solid-state lasers are typically longer than those for diode lasers. A single-frequency solid-state laser operating with 1mW output power can have Schawlow-Townes linewidths of less than 1Hz whereas a simple single-stripe diode laser has a linewidth of greater than 1MHz. Frequency changes of solid-state lasers with respect temperature are also less than for diode lasers. Monolithic Nd:YAG lasers tune at a rate of approximately 3GHz/K whereas a typical AlGaAs diode laser tunes at a rate of about 30GHz/K, and therefore the long-term frequency stability is better than for simple diode lasers.

2   Quasi-three-level lasers

One area of solid-state lasers that has dramatically benefited from diode pumping is the performance of quasi-three-level lasers; these operate on laser transitions that have lower laser levels close enough to the ground state that in thermal equilibrium there is a significant population density in the lower laser level. This is distinguished from the ideal four-level laser model, in which there is no population in the lower laser level, and from the ideal three-level model, in which the lower laser level is the ground state. The fact that there is population in the lower laser level means that higher pump densities are needed than in a four-level laser such as Nd:YAG. The ability to attain high pump densities has enabled efficient room-temperature operation of such rare-earth laser transitions as the $^4F_{3/2} - ^4 I_{9/2}$ transition in Nd^{3+} near 0.94μm, the $^2F_{5/2} - ^2 F_{7/2}$ transition in Yb^{3+} near 1.03μm, the $^3F_4 - ^3 H_6$ transition in Tm^{3+} near 2.0μm, the $^5I_7 - ^5 I_8$ transition in Ho^{3+} near 2.1μm, and the $^4F_{13/2} - ^4 I_{15/2}$ transition in Er^{3+} near 1.5μm. While all these laser transitions had been demonstrated previously with lamp pumping, the efficiencies tended to be low, and in many cases the lasers needed to be cooled well below room temperature to operate. In order to understand these quasi-three-level lasers it is first useful to understand the energy-level structure and the basic spectroscopy of rare-earth materials. Much of the discussion in this section was originally published in an earlier article (Fan 1993b).

2.1   Energy levels and spectroscopy in trivalent rare earths

It is useful to understand the energy-level diagrams for rare earth ions since they are the most common type of laser centre in solid-state lasers. Rare-earth atoms have outer electron shells with a configuration of $4f^N 6s^2$. When rare-earths are doped into solids, the two $6s$ electrons and one of the $4f$ electrons go into bonding. The positions of the energy levels are determined by three factors: the *Coulomb interaction* between $4f$ electrons, *spin-orbit coupling*, and the *crystal field* of the host material. The *Coulomb interaction* splits the $4f$ configuration into terms that are separated by typically 10,000cm^{-1} while the *spin-orbit* interaction splits the terms into manifolds that are separated by typically 3,000cm^{-1}. The interaction with the *crystal field* is the

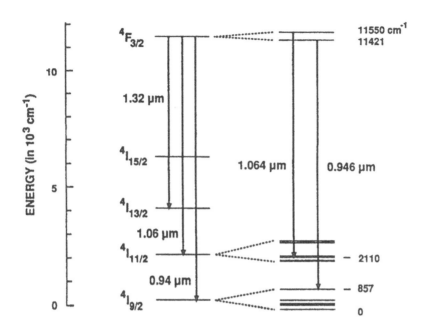

Figure 3. *Energy-level diagram for Nd:YAG showing the manifold positions and the crystal-field splittings.*

weakest effect and only splits the energy levels within a manifold by typically 200cm^{-1}. As an example, Figure 3 shows the energy-level diagram of Nd:YAG. The ^{4}I term is split from the ^{4}F term by the Coulomb interaction by about 10,000cm^{-1}. The ^{4}I term is split into four manifolds by spin-orbit coupling, and the manifolds are separated by about 2,000cm^{-1}. Finally, the crystal field splits a manifold into levels separated by 100–200cm^{-1}.

The small crystal-field splitting is a consequence of the fact that the $4f$ electrons are well shielded from external interactions because their radial extent is small compared with the other electron shells. One implication of the *weak* crystal field is that the positions of the manifolds only vary slightly as rare-earth ions are doped from host to host since neither the Coulomb interaction nor spin-orbit coupling are changed by the crystal field. The other implication is that the $4f$ energy levels can be treated as if the rare-earth ions are isolated impurities; even at relatively high concentrations of rare earth ions (1×10^{22}cm^{-3}), the $4f$ energy levels do not form a band structure since, effectively, the $4f$-shell electrons do not interact with the nearest-neighbour rare-earth ion. This means that the spectra of doped rare-earth ions appear more like line spectra than band spectra.

This energy-level diagram is more complicated than that assumed for the ideal three- or four-level laser model, and it would be undesirable to need to have separate rate equations to describe the populations of each of these many levels. However, there are simplifying assumptions based on the idea of *quasi-thermal equilibrium* within a manifold; in other words, the population distribution among crystal-field splittings

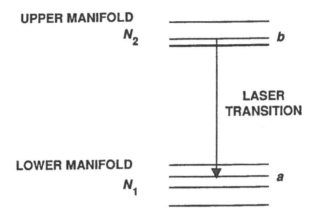

Figure 4. *Idealised upper and lower manifolds for a quasi-three-level laser. The upper-laser level is denoted by b and the lower by a. The population densities of the two manifolds are N_2 and N_1 for the upper and lower manifolds respectively.*

within a manifold can be given by the Boltzmann distribution as if the material is in thermal equilibrium. This assumption is valid to the extent that the *thermalisation time*, which is the time for relaxation within a manifold, is short compared with the *manifold fluorescence lifetime*, which is the inverse of the sum of the radiative and nonradiative transition rates out of the manifold.

Consider the energy level diagram in Figure 4 in which there is an upper manifold and a lower manifold that are each split by the crystal field. Quasi-thermal equilibrium implies that the ratios of the populations in each crystal-field splitting within a manifold are constant for all times. The fraction of the population within the level a of the lower manifold is given by

$$f_a = \frac{\exp(-E_{\ell a}/kT)}{Z_\ell} \tag{3}$$

where

$$Z_\ell = \sum_{i=1}^{m} \exp(-E_{\ell i}/kT) \quad . \tag{4}$$

f_a is the fractional population within the level a, $E_{\ell a}$ is the energy of the level a, kT is the thermal energy, and Z_ℓ is the partition function of the lower manifold. $E_{\ell i}$ are the energies of the levels in the lower manifold which has m levels. The fractional population in level b of the upper manifold f_b can be similarly found by calculating the partition function over the levels of the upper manifold. Degeneracy has been ignored in this calculation because for many cases in rare-earth ions, the degeneracies of the levels are the same.

The condition for laser gain is that there be a population inversion, *i.e.* that the population density in the upper level be greater than that for the lower level. Thus for gain on the transition between levels b and a

$$\Delta N = f_b N_2 - f_a N_1 > 0 \tag{5}$$

which implies

$$f_b N_2 > f_a N_1 \tag{6}$$

where N_2 and N_1 are the total population densities of the upper and lower manifolds, respectively. If we assume that the dynamics of the excited states are such that the ions are either in the upper or lower laser manifolds (true in most quasi-three-level lasers), then

$$N_1 + N_2 = N_t \tag{7}$$

where N_t is the *total* dopant concentration; then the fraction of the population that needs to be in the upper manifold to attain population inversion β_{min} is given by

$$\beta_{min} = \frac{f}{1+f} \tag{8}$$

where f is equal to f_a/f_b and $\beta = N_2/N_t$. Table 1 lists some values of f for quasi-three-level laser transitions in YAG at 300 K.

Ion	Transition	Wavelength (μm)	f
Nd^{3+}	$^4F_{3/2} - ^4I_{9/2}$	0.946	0.12
Ho^{3+}	$^5I_7 - ^5I_8$	2.097	0.17
Yb^{3+}	$^2F_{5/2} - ^2F_{7/2}$	1.03	0.066

Table 1. *Boltzmann occupation factors for quasi-three-level transitions in YAG at 300 K*

Clearly, for these transitions only a relatively small fraction of the population needs to be in the upper manifold in order to attain population inversion, in contrast to a three-level laser like ruby in which a large fraction of the population must be excited to the upper laser level for inversion.

The above discussion assumes a relatively simple case in which there are nonoverlapping transitions at the laser wavelength. Figure 5 shows the fluorescence spectra of Nd:YAG in the 940nm wavelength region and the fluorescence spectrum of Ho:YAG near 2.1μm. In Nd:YAG, the transitions do not overlap in wavelength, and the initial and final energy levels for the transitions are clearly defined. In this case, it is easy to calculate f_a and f_b. It is also easy to measure both the *spectroscopic cross sections* for the transition σ and the *effective cross section* σ_{eff}. The spectroscopic cross sections are those that are defined between individual crystal-field levels, and from reciprocity the spectroscopic absorption and emission cross sections are equal within degeneracy factors, whereas the effective cross sections are used when only the manifold population is known. In absorption, the *effective absorption cross section* $\sigma_{a,eff}$ is related to the spectroscopic cross section by

$$\sigma_{a,eff} = \sigma f_a \quad . \tag{9}$$

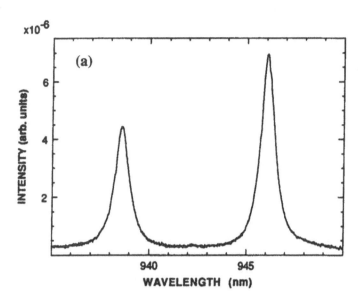

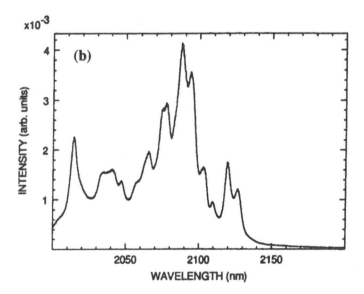

Figure 5. *Fluorescence spectra of (a) Nd:YAG near 940 nm and (b) Ho:YAG near 2.1µm. The Nd:YAG features can be assigned to specific transitions between crystal-field splittings. For the Ho:YAG transitions near 2.090 and 2.097µm, multiple overlapping transitions can be identified that contribute to gain.*

In other words the absorption coefficient is given in equilibrium by $\sigma f_a N_t$ which equals $\sigma_{a,\text{eff}} N_t$. Analogously, the *effective emission cross section* $\sigma_{e,\text{eff}}$ can be defined as $\sigma_{e,\text{eff}} = \sigma f_b$. In a situation with many overlapping transitions at the laser wavelength (such as Ho:YAG at 2.090μm), the story is more complicated, and simple spectroscopic cross sections and the Boltzmann occupation factors f_a and f_b cannot be used to describe the laser transition. It is more useful to use the effective emission and absorption cross sections. While f_a and f_b cannot be used to describe the transition, it can be shown that

$$f = \frac{\sigma_{a,\text{eff}}}{\sigma_{e,\text{eff}}}. \tag{10}$$

2.2 Laser rate equations

Lasers are often described by idealised three- and four-level laser models. Here, we contrast these models with a quasi-three-level laser model and show why diode-laser pumping has enabled efficient quasi-three-level lasers at room temperature. The interpretation in this section was developed previously (Fan 1993b).

The energy-level diagrams and transition rates for three-level, four-level, and quasi-three-level laser models are shown in Figure 6. The key difference between these idealised three and four-level laser models and the quasi-three-level models is that the crystal-field splittings are taken into account in the latter case. For these three models, the rate equations describing the change in population inversion density with time can be written as

$$\frac{d\Delta N}{dt} = 2W_{13}N_1 - \frac{N_t + \Delta N}{\tau_2} - 4W_{21}\Delta N \qquad \text{(three level)} \quad (11)$$

$$\frac{d\Delta N}{dt} = W_{14}N_1 - \frac{\Delta N}{\tau_3} - 2W_{31}\Delta N \qquad \text{(four level)} \quad (12)$$

and

$$\frac{d\Delta N}{dt} = f_b(f_b + f_a)W_{14}N_1 - \frac{f_a N_t + \Delta N}{\tau}$$
$$- 2(f_b + f_a)W_{31}\Delta N \qquad \text{(quasi three level)} \quad (13)$$

under the assumption of spatially uniform pumping and extraction. Here, ΔN is defined as

$$\Delta N = \begin{cases} N_2 - N_1 & \text{(three level)} \\ N_3 - N_2 = N_3 & \text{(four level)} \\ f_b N_2 - f_a N_1 & \text{(quasi three level)} \end{cases} \tag{14}$$

N_1, N_2 and N_3 are population densities in manifolds 1, 2, and 3, and N_t is the total dopant concentration, and consequently ΔN is the population difference between the upper and lower laser levels. The rate constants W_{xy} are given by

$$W_{xy} = \frac{\sigma I_{xy}}{h\nu_{xy}}. \tag{15}$$

The subscript xy denotes a transition between manifold x and manifold y, σ is the spectroscopic cross section of the transition, and I_{xy} is the intensity of the light

THREE LEVEL

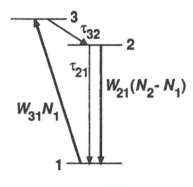

FOUR LEVEL

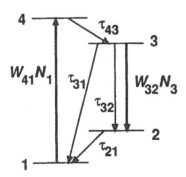

QUASI-THREE LEVEL

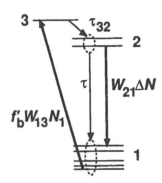

Figure 6. *Energy-level diagrams for idealised laser models. From left to right: three level laser, four-level laser, and quasi-three-level laser. Heavy lines indicate absorption and stimulated emission while the lighter lines indicate relaxation processes.*

at photon energy $h\nu_{xy}$. f'_b is the *Boltzmann occupation factor* for the initial state of the pumping transition in the ground-state manifold. The *spontaneous emission rates* $1/\tau_x$ are the total spontaneous relaxation rates out of the upper manifold. It has been assumed that relaxation from the pump level to the upper laser level is infinitely fast.

The first term of the right-hand sides of equations (11) (12) and (13) represents *pumping*, the second term is *spontaneous emission*, and the third term is *stimulated emission*. In equations (11) (12) (13), the factors of 2 in the stimulated emission terms arise because the laser is assumed to have a *standing-wave cavity*; thus the population sees the circulating intensity twice per round-trip. There is another factor of two in the pumping term and stimulated emission terms describing the three-level laser because a single absorbed pump photon or a single stimulated photon causes the population inversion to change by two. It is possible to see by inspection of equations (11) (12) (13) that a three-level laser is intrinsically *less* efficient than a four-level laser. The stimulated-emission rate is given in all cases by $2W_{xy}\Delta N$; the spontaneous-emission rate in the four-level laser case is proportional to ΔN while in the three-level laser case it is proportional to $N_t + \Delta N$ which is a much larger quantity. The quasi-three-level case scales between the four-level and three-level lasers because the spontaneous-emission rate is proportional to $f_a N_t + \Delta N$; as f_a approaches zero, the efficiency approaches that of a four-level laser.

The other rate equation required to describe these lasers is the change in *intracavity circulating intensity* with time. This rate equation is the same for each of the models and is given by

$$\frac{dI_{ba}}{dt} = cW_{ba}\Delta N h\nu_{ba} - \frac{I_{ba}}{\tau_c}. \tag{16}$$

The subscripts *ba* denote the laser transition (*i.e.* I_{ba} is the intensity on the laser transition and $h\nu_{ba}$ is the photon energy at the laser wavelength). The first term on the right-hand side of equation (16) is the growth in intracavity circulating intensity due to stimulated emission, and the second term is the decrease in circulating intensity due to cavity losses. It is assumed here that the gain medium occupies the entire cavity length; this assumption has no impact on the calculated efficiency or threshold.

2.3 Laser threshold

In the limit of steady-state pumping in which the time derivatives in equations (11) (12) (13) and (16) are zero, the *threshold population inversion density* ΔN_{th} can be found from equation (16) in the limit where I_{xy} goes to 0. ΔN_{th} is the same for all three models and is given by

$$\Delta N_{\text{th}} = \frac{\delta_c}{2\sigma L}. \tag{17}$$

This is the typical solution derived for the threshold population inversion density if the condition for threshold is that the round-trip gain be equal to the round-trip loss.

The *total threshold pump power* absorbed by the gain element is given by the product of the pumping density at threshold, the pump photon energy, and the volume of the gain element. The pumping density at threshold can be found by substituting equation (17) into equations (11) (12) and (13) and then solving these equations for $W_{1z}N_1$ where

z is either 3 or 4. Thus the absorbed powers at threshold are given by

$$P_{th} = \frac{1}{4\sigma\tau}(2\sigma N_t L + \delta_c)h\nu_p A \qquad \text{(three level)} \qquad (18)$$

$$P_{th} = \frac{1}{2\sigma\tau}(\delta_c)h\nu_p A \qquad \text{(four level)} \qquad (19)$$

and

$$P_{th} = \frac{1}{2(f_a + f_b)\sigma\tau}(2f_a\sigma N_t L + \delta_c)h\nu_p A \qquad \text{(quasi three level)} \qquad (20)$$

A is the cross sectional area of the gain element. From equations (18)–(20), we can understand the scaling laws of the threshold power and understand the reason that quasi-three-level lasers have become more important given the improvements in diode laser pump sources.

In the three-level laser, there are two terms that contribute to threshold power. The first term is the power required to reach $\Delta N = 0$ while the second term is the additional amount of power required to overcome the round-trip cavity losses. For CW three-level lasers, this second term is usually negligible compared with the first term, and consequently, the threshold pump power is proportional to the volume of the gain element. In the four-level laser, the only term that contributes to threshold is the need to overcome the cavity loss; the threshold power is proportional to the cross-sectional area of the gain element. Finally, in the quasi-three-level laser, there are two terms that contribute to the threshold power; these terms represent the same contributions as in the three-level laser. The difference relative to the three-level laser is that even for low cavity loss, the power needed to reach population inversion is not necessarily dominant since this power is proportional to f_a; in the limit f_a that approaches zero, this term vanishes. As before, the contribution to threshold power, dictated by the need to invert the population, scales proportionally with the volume of the gain element while the contribution to threshold power required to overcome intracavity loss scales with gain element cross-sectional area.

The interpretation that can be drawn from these threshold power scaling laws is that the gain element should be made small in order to minimise absorbed threshold power. In the case of a four-level laser it is sufficient to make the gain element cross-section small, but in the case of quasi-three-level and three-level lasers, both the cross-sectional area and gain element length need to be minimised. With diode pumping small gain elements can be used and operation well above threshold can be achieved in quasi-three-level lasers.

The essential engineering trade offs for quasi-three-level lasers can be found in equation (20). In the first term of equation (20) which is the power required to reach the condition $\Delta N = 0$, there are four multiplicative factors; these are $f_a\sigma N_t L$. N_t cannot be made small because as N_t is reduced to zero, none of the pump radiation is absorbed. σ is actually irrelevant because it is cancelled by a σ outside the bracket. That leaves f_a and L. Lamps are limited in the power per unit length that they can supply. In other words, to get higher power from a lamp, a lamp is made longer. Thus, L is not a consideration in achieving threshold in a lamp-pumped laser; the required threshold power scales proportionally with the maximum available lamp power. Consequently, because of the low pump densities achievable with lamps, the engineering solution is to make f_a small by decreasing the temperature. In diode-pumped lasers, L can be made small by pumping along the cavity axis and tuning the diode pump to an absorption peak in end-pumped lasers, and therefore f_a can be larger thus allowing operation of many quasi-three-level lasers at room temperature. In side-pumped devices, the cross-sectional area can be made small by tuning the diode onto a pump feature, which does not reduce L but instead reduces A. The term $2f_a\sigma N_t L$ in equation (20) has a simple interpretation; it is equal to the round-trip absorption at the laser wavelength for the gain element in thermal equilibrium. Thus the round-trip absorption at the laser wavelength due to lower state population is often referred to as the *reabsorption loss*.

2.4 Spatial dependence and bleaching

The above models in equations (11)–(13) and in (16) assume uniform pumping and extraction. However, for more accurate modelling of laser performance, it is important to include spatial dependence for both pumping and extraction. Models have been developed for CW end-pumped quasi-three-level lasers that incorporate *Gaussian-intensity transverse spatial dependence* (Fan and Byer 1986, Risk 1988), and these models have been shown to give good agreement with experiment. The models predict that near threshold, the differential quantum efficiency can be significantly less than 1, but that far above threshold, this differential quantum efficiency approaches 1. Thus, quasi-three-level lasers can be nearly as efficient as four-level lasers when operated far above threshold.

Another aspect of spatial dependence in modelling calculations is *bleaching*, which necessarily occurs in three-level and quasi-three-level lasers. Bleaching is the reduction in ground-state population, and therefore the absorption coefficient for pump radiation, as a result of population in the upper laser level. The CW models discussed above address the absorbed pump power, but in calculations of the overall efficiency of lasers, the fraction of the pump power absorbed needs to be known. The reduction in absorption coefficient at the pump wavelength in a three-level laser such as ruby is significant since half of the population must be in the upper manifold at threshold. In contrast, in most of the quasi-three-level lasers, the fraction of population in the upper manifolds at threshold is more modest; therefore, in CW operation, the reduction in absorption coefficient at the pump wavelength may cause only a small error in an overall efficiency calculation. On the other hand, in energy-storage calculations as in Q-switching or amplifiers for pulsed lasers, bleaching must be considered in calculating the overall efficiency and determining the optimum design (Krupke and Chase 1990, Fan 1992). An example of how bleaching impacts the performance of Q-switched quasi-three-level lasers is shown schematically in Figure 7, which shows the output energy of CW-pumped, repetitively Q-switched lasers as a function of period between pulses. For an ideal four-level laser, in which there is an assumption of no ground-state bleaching, there are two asymptotes that define the laser performance in the high and low repetition-rate regimes. These asymptotes are given by

$$E_{\text{out}} = \begin{array}{ll} P_{\text{CW}}\tau_f & \text{(low repetition rate)} \\ P_{\text{CW}}t_p & \text{(high repetition rate)} \end{array} \qquad (21)$$

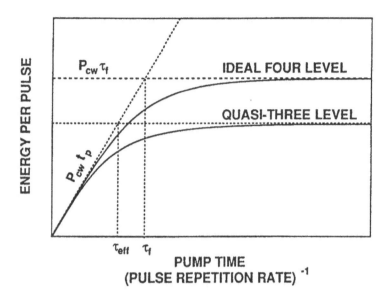

Figure 7. *Projected performance of* CW*-pumped, repetitively Q-switched four-level and quasi-three-level lasers.*

where P_{CW} is the CW output power of the laser, t_p is the period between pulses, and τ_f is the fluorescence lifetime of the upper laser level. These asymptotes intersect at a period between pulses equal to the fluorescence lifetime in an ideal four-level laser model (Siegman 1986). The intersection of these asymptotes is often called the *effective storage time* and in an ideal four-level laser it should be equal to the fluorescence lifetime. Experimental observations of effective lifetimes less than τ_f have often been interpreted as a sign of parasitic processes, such as undesired upconversion. This interpretation may be partially correct, but in quasi-three-level lasers, this effective lifetime is necessarily less than the fluorescence lifetime because of increased bleaching of the ground-state manifold with longer time between pulses (Fan and Lacovara 1992, Nabors 1994). This causes the output energy in the low-repetition-rate regime to be below $P_{CW}\tau_f$ because the fraction of the incident pump power that is absorbed is lower than in the high-repetition-rate regime, and consequently, the effective storage time is a function of laser design, even in the absence of parasitic effects.

2.5   A quasi-three-level laser example: Yb:YAG

Room-temperature, quasi-three-level laser operation has been demonstrated now on several transitions in rare-earth ions with diode laser pumping including the $^2F_{5/2}-^2F_{7/2}$ transition in Yb^{3+} near $1\mu m$ (Lacovara *et al.* 1991). This laser is an ideal quasi-three-level laser, and there is only one excited-state manifold because of its simple energy level diagram, which is a result of the outer $4f$-shell being an f-electron hole. Consequently, there is no possibility of parasitic effects such as excited-state absorption, upconversion, or concentration quenching in principle. Yb^{3+} laser systems, particularly Yb:YAG, also offer advantages relative to Nd:YAG for diode laser pumping. These advantages include a wider absorption band for reduced need of thermal management of the diodes, longer upper-state lifetime for improved energy storage relative to Nd:YAG, and lower heat generation in the laser gain element per unit pump power for reduced thermo-optic distortion. Another aspect of Yb:YAG lasers is the output wavelength of $1.03\mu m$; frequency doubling the output gives 515 nm which is the wavelength of green Ar^+ lasers.

The energy level diagram of Yb:YAG and the room-temperature absorption and fluorescence spectra are shown in Figure 8. The main gain line at $1.03\mu m$ is on a transition between the lowest-lying crystal-field splitting of the upper manifold and a level at $612cm^{-1}$ in the ground-state manifold. There are two absorption lines, one each at 0.968 and $0.943\mu m$, with essentially the same absorption coefficient. The feature at $0.943\mu m$ is preferred for diode pumping because of its larger absorption bandwidth. The effective gain cross section at $1.03\mu m$ has been measured to be $2.3 \times 10^{-20}cm^2$ and the radiative lifetime of the $^2F_{5/2}$ manifold has been measured to be 0.95 ms (Sumida and Fan 1994). Concentration quenching of the lifetime has been observed at doping concentrations of 10 at.%, despite the fact that intrinsic quenching should be absent, and has been attributed to rare-earth impurities such as Tm^{3+}. Relatively low thermal loading in Yb^{3+} laser materials is expected because of the small difference between the pump photon energy and the laser photon energy; the difference in the photon energies goes into heating. Based purely on this difference, known as the *quantum defect*, the heat dissipated in the gain element should be only 0.087 of the pump power, if the laser were to be pumped at $0.94\mu m$ and the laser were to operate at $1.03\mu m$. Calorimetric measurements have shown that this fractional heating is <0.11 of the pump power (Fan 1993a). The *radiative quantum efficiency*, which is the fraction of the absorbed photons that result in a fluorescence photon, can be inferred to be >0.95 from this measurement.

We have performed CW laser experiments at room temperature with both $Ti:Al_2O_3$ pumping and InGaAs diode laser pumping with high slope efficiency (Fan 1993b). The laser results are shown in Figure 9. With $Ti:Al_2O_3$ pumping a power slope efficiency of 72% is obtained by fitting the four highest power points to a straight line. This power slope efficiency is higher than any $0.81\mu m$-pumped, $1.06\mu m$ Nd^{3+} laser demonstrated to date to our knowledge. This shows that quasi-three-level lasers can indeed be efficient. This power slope efficiency corresponds to a differential quantum efficiency of 0.79. The InGaAs-diode-pumped laser has a power slope efficiency of 60% corresponding to a differential quantum efficiency of 0.66. This power slope efficiency is comparable to that obtained in diode-pumped Nd:YAG lasers. The differential quantum efficiencies are in agreement with the values projected by the quasi-three-level laser models. The slightly lower differential quantum efficiency for the diode-pumped laser is primarily due to the fact that it is not being operated as far above threshold. In addition, the spatial overlap is usually not quite as good with a diode laser as it is with a $Ti:Al_2O_3$ laser, which leads to the somewhat higher threshold.

Others have also observed high slope efficiencies in Yb:YAG lasers; in waveguide lasers grown by liquid phase epitaxy 80% slope efficiencies with respect to absorbed

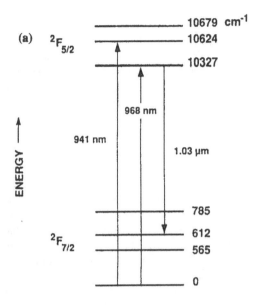

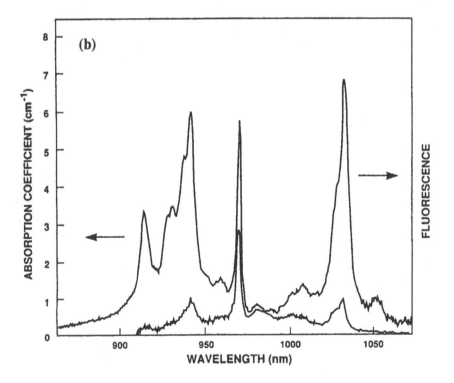

Figure 8. (a) Energy-level diagram and (b) absorption and fluorescence spectra of 6.5% doped Yb:YAG at 300 K.

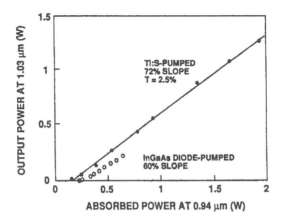

Figure 9. *Results from room-temperature Yb:YAG lasers pumped by Ti:Al₂O₃ lasers or InGaAs diode lasers. (Output is power limited).*

diode power have been observed (Chartier *et al.* 1994). Threshold in such devices is lower than in bulk lasers because of confinement of both the pump beam and laser mode. In another experiment, high CW powers have been obtained >10 W (Giesen *et al.* 1994, Brauch *et al.* 1995). This experiment used multiple diode lasers to pump the Yb:YAG, and the high power is obtained by virtue of the low thermal loading in the gain element. Progress is also being made in the areas of high-power (1 W CW) single-frequency Yb:YAG lasers and high-energy normal mode Yb:YAG lasers (Sumida and Fan 1994).

3   Laser linewidth

Diode pumping has enabled narrow linewidth solid-state lasers to approach the *Schawlow-Townes linewidth limit*. Figure 10 shows progress in solid-state laser linewidths over time. With the advent of diode-pumping, linewidths have decreased by over five orders of magnitude. Most measurements of laser linewidth have been performed by heterodyning two solid-state lasers on a photodetector and then measuring the resultant beat frequency signal on a RF spectrum analyser (which measures the power spectrum of the signal out of the photodetector). The laser linewidth is determined by measuring the 3dB full width (half power) of the central peak of the signal on the spectrum analyser and then dividing the full width by two to get the full width of the laser line (the Schawlow-Townes lineshape is *Lorentzian* and the heterodyne linewidth obtained by beating two identical lasers with Lorentzian lineshapes is expected to be Lorentzian with double the width). However, for many applications, such as those using coherent detection, the laser linewidth measured in this fashion is meaningless unless the lineshape is known as well. In most lasers, the lineshape is not actually Lorentzian owing to the presence of other types of noise besides spontaneous

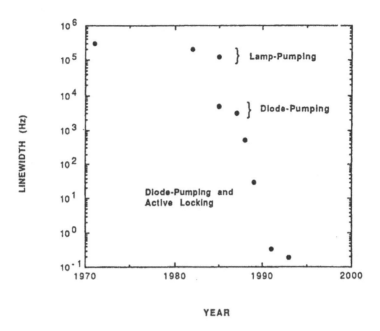

Figure 10. *Solid-state laser linewidths as a function of year. Note the large decrease in linewidth with diode pumping.*

emission, which also cause frequency variation. These other noise sources include mirror vibration and temperature variations, which both can cause changes in the optical path length of a laser cavity and therefore the output frequency.

There has been much work on ways of characterising frequency stability and linewidth of oscillators (Rutman 1978, Walls and Allan 1986). This section will illustrate the inadequacy of heterodyne beat frequency measurements as typically measured on a spectrum analyser, discuss other ways of characterising frequency stability and linewidth, and finally discuss ways of obtaining performance beyond the limits set by Schawlow-Townes using examples from the literature.

3.1   Limitations of heterodyne measurements using a spectrum analyser

Figure 11 shows the spectrum analyser output for two different lasers: one is the spectrum obtained by heterodyning two external-cavity diode lasers, and the other is a spectrum from two Nd:YAG microchip lasers. In each case the laser linewidth has been measured to be approximately 5kHz, but qualitatively the spectra are very different. The central peak of the solid-state laser spectrum drops down to the noise floor of the spectrum analyser at a frequency offset of just 30kHz from the centre of the peak whereas the spectrum for the diode laser drops more gradually to the noise floor as the offset frequency increases. Additionally the solid-state laser spectrum shows peaks that are offset by around 600kHz and are about 30dB weaker than the central peak.

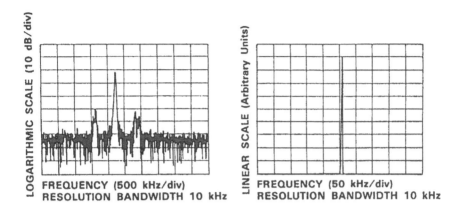

(a) **HETERODYNED Nd:YAG MICROCHIP LASER SPECTRUM**

(b) **Heterodyne Spectrum of Two Free-running GaInAs External Cavity Diode Lasers**

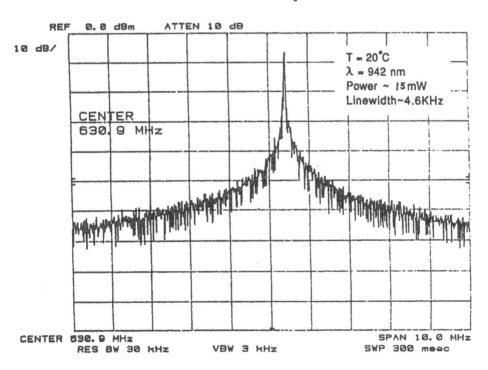

Figure 11. *Heterodyne beat spectrum measured with two identical lasers. (a) Nd:YAG microchip lasers and (b) external cavity diode lasers. The 3dB linewidths are similar but the lineshapes are very different.*

The spectra in Figure 11 illustrate two of the difficulties in many of the measurements of the linewidth. First, the 3dB linewidth does not tell the entire story; there is a lineshape as well. In this example, the diode laser spectrum fits a Lorentzian whereas the solid-state laser spectrum does not fit a Lorentzian. Second, in RF spectrum analyser measurements of this type it is difficult to distinguish between amplitude fluctuations and frequency fluctuations, as the sidelobes in the solid-state laser spectrum are due to relaxation oscillations, which are amplitude fluctuations. In fact, because of this difficulty, the RF spectrum (*i.e.* the power spectrum) of an oscillator is not recommended as a measure of frequency stability or linewidth (Walls and Allan 1986) although it may serve as a reasonable qualitative measure.

3.2   Improved measures of laser frequency stability

The output of a laser (or any other electronic oscillator) can be represented by

$$E(t) = [E_0 + \varepsilon(t)] \sin(2\pi\nu_0 t + \phi_n(t)) \tag{22}$$

where E_0 is the nominal output electric field, and ν_0 is the nominal frequency of the oscillator. $\varepsilon(t)$ and $\phi_n(t)$ are fluctuations in the amplitude of the electric field and the phase that are related to noise (*i.e.* amplitude noise and phase noise). In an ideal oscillator $\varepsilon(t)$ and $\phi_n(t)$ are both equal to 0, but this is nonphysical. Of course, phase and frequency are related so the instantaneous frequency can be rewritten as

$$\nu(t) = \nu_0 + \frac{d\phi_n(t)}{2\pi dt} = \nu_0 + \nu_n(t) \tag{23}$$

where $\nu_n(t)$ is the frequency noise. It is the phase or frequency noise that is fundamentally related to linewidth or frequency stability, and consequently in order to get a better measure of frequency stability it is the phase or frequency noise that needs to be characterised.

The measure that is used to characterise phase fluctuations is the spectral power density of phase noise, represented by $S_\phi(f_m)$ and having units of rad^2/Hz. Similarly the measure that is used to characterise frequency fluctuations is the spectral power density of frequency noise, represented by $S_\nu(f_m)$ and having units of Hz^2/Hz. Here, f_m is called the *offset frequency*, which can be thought of as the frequency at which the phase (frequency) is being modulated by the noise. Because of the relation between phase and frequency the spectral power density of phase noise can be written as

$$S_\phi(f_m) = S_\nu(f_m)/f_m^2 \tag{24}$$

An analogy can be made between the power spectral density of an electrical signal and the spectral power density of phase (frequency) fluctuations. Consider an electrical signal $V(t)$. The power spectrum of $V(t)$ is given in units of power per unit bandwidth or equivalently V^2/Hz. The way this is obtained in practice is to measure the electrical signal with a spectrum analyser, which gives information on the frequency components in $V(t)$ and their magnitude. Equivalently if we can measure $\phi_n(t)$ or $\nu_n(t)$ then a power spectral density of phase and frequency noise can be determined, which gives information on the size of the phase (frequency) fluctuations and what frequency components (*i.e.* offset frequencies) are in these fluctuations. Here, I discuss only frequency

(a)

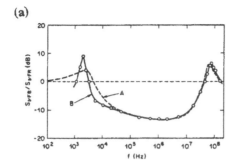

(b)

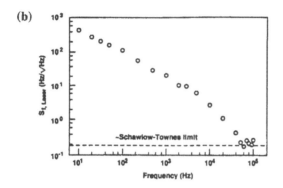

Figure 12. *Frequency noise spectra from two free running lasers; (a) a diode laser (Ohtsu and Tabuchi 1988)(©1988 IEEE), and (b) a diode-pumped solid-state laser (Day et al. 1992)(©1992 IEEE).*

domain measures of frequency stability, such as phase and frequency noise; another important measure of frequency stability is the *Allan variance*, which is a time domain measurement that I do not discuss here.

The spectral power density of frequency fluctuations for a Lorentzian is white; in other words, $S_\nu(f_m)$ is a constant with

$$S_\nu(f_m) = \frac{\Delta\nu_{ST}}{\pi} \tag{25}$$

where $\Delta\nu_{ST}$ is the Lorentzian linewidth (Owens and Weiss 1974). (In this article, I will freely switch between $S_\nu(f_m)$ with units of Hz^2/Hz and its square root with units of Hz/Hz$^{1/2}$; both are used in the literature, and their usage is not consistent between the cited authors).

Figure 12 shows two frequency noise spectra from representative lasers, a diode laser (Ohtsu and Tabuchi 1988), and a diode-pumped solid-state laser (Day *et al.* 1992). At high offset frequencies, the noise is white in both of these lasers and is near the level given by the Schawlow-Townes limit, but at lower offset frequencies the noise rises well above the Schawlow-Townes limit. This behavior is due to noise sources, in addition to spontaneous emission, that occur at low frequencies. Two examples of noise sources that are expected to contribute to frequency noise at low offset but not at higher offsets are frequency fluctuations due to cavity-length changes associated with mirror vibration and temperature fluctuations.

The Schawlow-Townes linewidth is sometimes termed the *fundamental linewidth* of a laser because the noise source which causes the finite linewidth, spontaneous emission, must occur in a laser. Mirror vibrations and temperature fluctuations can also be viewed, at some level, as being necessary. Mirror vibrations have two components, one is due to the acoustic environment, but the other is thermal fluctuations due to kT noise (Zayhowski 1990), which is fundamental. In optically pumped solid-state lasers, fluctuations in pump power cause temperature fluctuations which in turn change the refractive index and therefore the optical path length of the cavity. A fraction of these pump fluctuations is due to shot (quantum) noise and therefore can be viewed as being fundamental as well.

The kT-induced mirror vibrations lead to a lineshape that is *Gaussian*, which has a tail that falls off much more rapidly than a Lorentzian (Zayhowski 1990). In heterodyne beat frequency measurements of free-running microchip solid-state lasers, the Gaussian portion of the lineshape dominates the 3dB linewidth measurement but the tails still fit a Lorentzian. The effect of these mirror vibrations is highest at low offset frequencies (the magnitude of the mirror vibration decreases with vibrational frequency). The frequency fluctuations caused by pump fluctuations are also largest for low offset frequencies (Day *et al.* 1992). Consequently, the laser lineshape is not a pure Lorentzian in solid-state lasers because of these two types of $1/f$ noise, but at high enough offset frequencies, the frequency noise has the characteristic of a Lorentzian. Diode lasers tend to have lineshapes that are closer to Lorentzian because the ratio of spontaneous emission noise to other noise sources is larger for a diode laser than for a solid-state laser. The large improvement in solid-state laser linewidth seen in Figure 10 has been achieved largely by reducing the frequency fluctuations at low offset frequency. In lamp-pumped solid-state lasers, the amount of technical noise, such as vibrations induced by flowing liquid coolant and pump-induced refractive index fluctuations, is much greater than in diode-pumped lasers.

3.3   Impact of frequency noise on measurements

Now, consider the effects of frequency noise on a measurement. Figure 13 shows a Michelson interferometer with an optical path difference between the legs of $\Delta\ell$. Such an interferometer can be used to measure changes in $\Delta\ell$ due to mirror vibration or the change in refractive index between legs. Alternatively it can be used to measure changes in ν, and when operated in this mode, the interferometer acts as a *frequency discriminator*, *i.e.* as a device that provides an output that is a measure of the optical frequency. The output of the interferometer P_{out} has a characteristic such that

$$P_{\text{out}} \sim 1 + \cos\left(\frac{2\pi\nu\Delta\ell}{c}\right) \qquad (26)$$

At a proper bias point, $2\pi\nu\Delta\ell/c$ equal to $m\pi/2$ where m is odd, the output of the interferometer has a linear slope with respect to $\nu\Delta\ell$; changes in output are then directly proportional to changes in $\nu\Delta\ell$. The sensitivity of measurement of $\Delta\ell$ will be

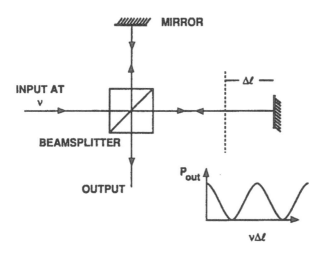

Figure 13. *Michelson interferometer with path length difference $\Delta\ell$ and a laser frequency ν.*

limited by the frequency noise since for fixed $\Delta\ell$ there is still a varying output from the interferometer because the frequency is not constant. The limitations in the measurement of $\Delta\ell$ imposed by frequency noise will be much greater if the changes in $\Delta\ell$ are at low characteristic frequencies relative to high because low-frequency changes will be masked much more by the $1/f$ frequency noise at low offset frequencies. Measurements of the change in $\Delta\ell$ at high characteristic frequencies can be used to show why 3dB linewidth measurements are inadequate for characterising frequency stability. If the 3dB linewidth is dominated by the noise at low offset frequencies but the characteristic frequency of $\Delta\ell$ fluctuations is sufficiently high that the frequency noise is Schawlow-Townes limited, then an improvement in 3dB linewidth by eliminating $1/f$ noise will have no impact on the measurement of $\Delta\ell$.

For fixed $\Delta\ell$, ν can be measured. The frequency fluctuations are $\nu_n(t)$, and this frequency discriminator can be used to measure $S_\nu(f)$. Fluctuations in ν lead to proportional fluctuations in P_{out} which lead to fluctuations in an electrical signal from a photodetector that monitors P_{out}. The electrical signal can then be measured on an RF spectrum analyser to obtain its power spectrum and by knowing the calibration factor between the size of fluctuation in electrical signal caused by a fluctuation in ν, the spectral power density of frequency noise can be obtained.

One of the limitations of a Michelson interferometer as a discriminator is that its sensitivity is not the same for all offset frequencies. For an offset frequency of $c/2\Delta\ell$ there is a null in the response of the discriminator, and the responsivity is significantly reduced for offset frequencies greater than $c/4\Delta\ell$. Consequently, for measuring large

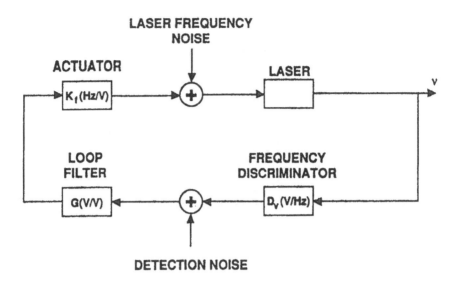

Figure 14. *Block diagram of a system for producing reduced frequency noise.*

offset frequencies, $\Delta\ell$ cannot be too large. On the other hand, for high sensitivity to low offset frequencies, $\Delta\ell$ should be made as large as possible (conversely for highest sensitivity to $\Delta\ell$ changes, ν should be made large), so the optimum value for $\Delta\ell$ will be dependent on the expected frequency-noise characteristics and the offset-frequency range over which a measurement is to be made.

3.4   Beyond the Schawlow-Townes limit

The Schawlow-Townes linewidth limit is often termed the *fundamental linewidth* in lasers. However, it is possible to have laser linewidths that are narrower (*i.e.* frequency noise characteristics that are better) than the Schawlow-Townes limit. The basic idea is that if the frequency fluctuations can be measured in real time, then it may be possible to correct them using a feedback loop. A block diagram of such a system is shown in Figure 14. The laser is assumed to be perfect with some additive frequency noise. The frequency fluctuations are detected using a discriminator which provides an error signal for a feedback loop. The feedback loop provides a signal to an actuator that adjusts the frequency of the laser to correct for the frequency noise. Examples of actuators (*i.e.* tuners) are piezoelectric transducers which can be used to change the cavity length and intracavity electro-optic crystals which can change the optical path length of the cavity.

An example of an experiment in which a diode laser was line narrowed is shown in Figure 15 (Ohtsu and Tabuchi 1988) which illustrates the diode laser frequency noise spectrum before feedback correction. The result with the feedback loop is illustrated in Figure 16 which shows the ratio of the spectral power density of frequency noise with feedback to that without feedback. The results show that the spectral power density of frequency noise is reduced by 10dB but only over a limited bandwidth. The limited bandwidth is because there are many elements of the feedback loop with finite

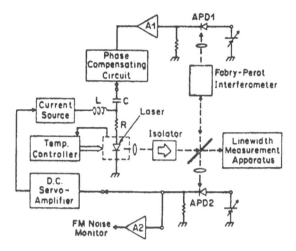

Figure 15. *Schematic of an experiment to reduce frequency noise in semiconductor lasers. The Fabry-Perot etalon is the frequency discriminator and is used by biasing the diode laser frequency so that it is on the side of a transmission peak of the etalon. The feedback loop uses the drive current to the diode laser as the actuator that adjusts the diode laser frequency (Ohtsu and Tabuchi 1988)(©1988 IEEE).*

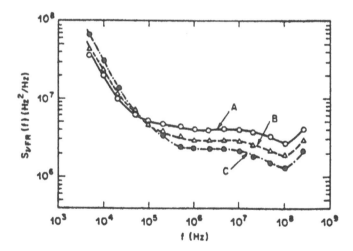

Figure 16. *Ratio of the power spectral density of frequency noise with feedback to that without feedback for the experiment in Figure 15 (Ohtsu and Tabuchi 1988)(©1988 IEEE).*

bandwidth such as the discriminator, the electronics, and the actuator, which in this case is electrical current. The frequency noise due to Schawlow-Townes noise is white. Consequently it is not possible to correct over all offset frequencies, but at least over limited frequency ranges, improvements in frequency noise can be achieved.

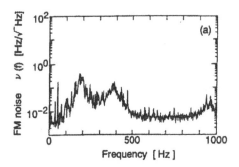

Figure 17. *Frequency noise spectrum of a Nd:YAG laser locked to a Fabry-Perot reference cavity using Pound-Drever locking. The laser output power is 3.8mW and the spectral density of frequency noise imposed by Schawlow-Townes is 290mHz/Hz$^{1/2}$ (Uehara and Ueda 1993).*

Solid-state lasers have also been demonstrated with frequency noise spectral power densities better than the Schawlow-Townes limit (Uehara and Ueda 1993, 1994). Diode-pumped single-frequency Nd:YAG ring lasers were locked to a Fabry-Perot reference cavity using the Pound-Drever locking technique. This locking technique acts as a frequency discriminator for the laser frequency by yielding an error signal proportional to the difference between the laser frequency and a resonance of the reference cavity. A feedback loop is then used to apply a correction signal to piezoelectric transducers attached to the lasers. The frequency noise spectrum is shown in Figure 17; the Schawlow-Townes limit for the spectral density of frequency noise is 290mHz/Hz$^{1/2}$ given the characteristics of the laser. For offset frequencies between 10Hz and 1kHz, the spectral density of frequency noise is less than the Schawlow-Townes limit. The poorer performance in the 100–500Hz range relative to the rest of the range was attributed to acoustic effects. Above 30kHz offset frequency, the spectral density of frequency noise is expected to be at the Schawlow-Townes limit as the feedback loop has its unity gain point at 30kHz, and acoustic and thermal effects are expected to be negligible at these offset frequencies.

What limits the improvements in frequency noise by use of discriminators and feedback? As previously mentioned, there is only a limited bandwidth over which improvement can be achieved because of finite bandwidth actuators and loop electronics. In addition, there are propagation time delays in these systems which will limit the upper frequency that can be corrected. The signal-to-noise ratio on the error signal places a bound on the magnitude of the correction; if the frequency fluctuations cannot be measured because of the presence of noise on the output of the discriminator, then they cannot be corrected. A fundamental source of noise in the detection process is *shot noise*, which occurs because photons are discrete. Uehara and Ueda derived an expression that relates the best frequency noise performance that can be achieved in the presence of shot noise. This shot-noise-limited performance is achieved only in the limit of infinite feedback loop gain and in the limit where shot noise dominates other noise

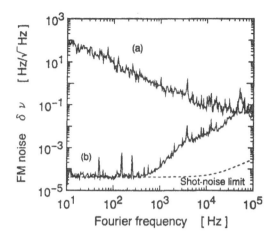

Figure 18. *Shot-noise limited frequency-noise performance achieved by using a discriminator to detect frequency fluctuations and a feedback loop for correction. The Schawlow-Townes limited frequency noise is 40 mHz/Hz$^{1/2}$ (Uehara and Ueda 1994).*

sources such as noise in the electronics. In the experiment of Figure 17, the shot-noise limit was 381μHz/Hz$^{1/2}$, well below the observed noise level. This indicates that sources of noise, other than shot noise, are limiting the performance and/or the feedback loop gain is insufficiently high to reach the shot-noise limit. By using a novel photodetector arrangement and boosting the gain of the feedback loop, shot-noise-limited performance was achieved as shown in Figure 18 (Uehara and Ueda 1994). One application for these low frequency-noise sources is in gravity-wave detection (Lewis 1995), which uses interferometers to detect path length changes. The signals are expected to be in the frequency range of a few Hz to a few kHz. Consequently good performance is needed in this offset frequency range.

Acknowledgements

This work was supported by the Department of the Air Force (U.S.A.).

References

Brauch U, Giesen A, Karszewski M, Stewen Chr, Voss A, 1995, Multiwatt diode-pumped Yb:YAG thin disk laser continuously tunable between 1018–1053nm, *Opt Lett* **20** 713.

Byer R L, 1988, Diode-pumped solid-state lasers, *Science* **239** 742.

Chartier I, Wyon C, Pelenc D, Ferrand B, Shepherd D P, and Hanna D C, 1994, High slope efficiency and low threshold in a diode-pumped epitaxially grown Yb:YAG waveguide laser, in *New Materials for Advanced Solid State Lasers*, eds Chai B H T, Payne S A, Fan T Y, Cassanho A, and Allik T H (Materials Research Society, Pittsburg, 1994) p179.

Chen T S, Anderson V L, and Kahan O, 1990, Measurements of heating and energy storage in diode-pumped Nd:YAG, *IEEE J Quantum Electron* **26** 6.

Day T, Gustafson E K, and Byer R L, 1992, Sub-hertz relative frequency stabilization of two diode laser pumped Nd:YAG lasers locked to a Fabry-Perot interferometer, *IEEE J Quantum Electron* **28** 1106.

Endriz J G, Vakili M, Browder G S, DeVito M, Haden J M, Harnagel G L, Plano W E, Sakamoto M, Welch D F, Willing S, Worland D P, and Yao H C, 1992, High power diode laser arrays, *IEEE J Quantum Electron* **28** 952.

Fan T Y and Byer R L, 1986, Modelling and CW operation of a quasi-three-level 946 nm Nd:YAG laser, *IEEE J Quantum Electron* **23** 605.

Fan T Y and Byer R L, 1988, Diode laser pumped solid-state lasers, *IEEE J Quantum Electron* **24** 895.

Fan T Y and Lacovara P, 1992, Modelling of energy storage Yb:YAG lasers and amplifiers, in *OSA Proceedings on Advanced Solid-State Lasers*, eds Chase L L and Pinto A A (Optical Society of America, Washington DC) **13** p190.

Fan T Y, 1992, Optimizing the efficiency and stored energy in quasi-three-level lasers, *IEEE J Quantum Electron* **28** 2692.

Fan T Y, 1993a, Heat generation in Nd:YAG and Yb:YAG, *IEEE J Qu Elec* **29** 1457.

Fan T Y, 1993b, Quasi-three-level lasers, in *Solid State Lasers: New Developments and Applications*, eds Inguscio M and Wallenstein R (Plenum Press, New York) p189.

Giesen A, Hugel H, Voss A, Wittig K, Brauch U, and Opower H, 1994, Scalable concept for diode-pumped high-power solid-state lasers, *Appl Phys B* **58** 365.

Hanna D C, 1995, this volume.

Hughes D W and Barr J R M, 1992, Laser diode-pumped solid-state lasers, *J Phys D: Appl Phys* **25** 563.

Keyes R J and Quist T M, 1964, Injection luminescent pumping of $CaF_2:U^{3+}$ with GaAs diode lasers, *Appl Phys Lett* **4** 50.

Krupke W F and Chase L L, 1990, Ground-state depleted solid-state lasers: principles, characteristics and scaling, *Opt Quantum Electron* **22** S1–S22.

Lacovara P, Choi H K, Wang C A, Aggarwal R L, and Fan T Y, 1991, Room temperature diode-pumped Yb:YAG laser, *Opt Lett* **16** 1089.

Lewis M A, 1995, Sleuthing out gravitational waves, *IEEE Spectrum* **32** (5) 57.

Nabors C D, 1994, Q-switched operation of quasi-three-level lasers, *IEEE J Quantum Electron* **30** 2896.

Newman R, 1963, Excitation of the Nd^{3+} fluorescence in $CaWO_4$ by recombination radiation in GaAs, *J Appl Phys* **34** 437.

Ohtsu M and Tabuchi N, 1988, Electrical feedback and its network analysis for linewidth reduction of a semiconductor laser, *J Lightwave Technol* **6** 357.

Owens D K and Weiss R, 1974, Measurement of the phase fluctuations in a He-Ne Zeeman laser, *Rev Sci Instruments* **45** 1060.

Risk W P, 1988, Modelling of longitudinally pumped solid-state lasers exhibiting reabsorption losses, *J Opt Soc Am B* **5** 1412.

Rutman J, 1978, Characterization of phase and frequency instabilities in precision frequency sources: fifteen years of progress, *Proc IEEE* **66** 1048.

Siegman A E, 1986, *Lasers* 1028 (University Science Books, Mill Valley, CA).

Streifer W, Scifres D R, Harnagel G L, Welch D F, Berger J, and Sakamoto M, 1988, Advances in diode laser pumps, *IEEE J Quantum Electron* **24** 883.

Sumida D S and Fan T Y, 1994, Impact of radiation trapping on fluorescence lifetime and stimulated emission cross section measurements in solid-state laser media, *Opt Lett* **19** 1343.

Uehara N and Ueda K, 1993, 193 mHz beat linewidth of frequency-stabilized laser-diode-pumped Nd:YAG ring lasers, *Opt Lett* **18** 505.

Uehara N and Ueda K, 1994, Ultrahigh-frequency stabilization of a diode-pumped Nd:YAG laser with a high-power-acceptance photodetector, *Opt Lett* **19** 728.

Walls F L and Allan D W, 1986, Measurements of frequency stability, *Proc IEEE* **74** 162.

Zayhowski J J, 1990, Microchip lasers, *Lincoln Lab. J* **3** 427.

.

Fibre Lasers

D C Hanna

University of Southampton, England

1 Introduction

The history of the fibre laser is an interesting illustration of how a good idea can, at first, fail to prosper if it arrives before its time. The first fibre laser, demonstrated by Snitzer (1961) consisted of a neodymium (Nd^{3+})-doped glass fibre, side-pumped by a pulsed flash-lamp. Absorption of pump light was therefore very inefficient. However, even with these early fibres, one of their striking attributes, the ability to provide very high gain amplification, was convincingly displayed. In fact a gain of 47dB was reported in Nd-doped fibre (Ross and Snitzer 1972). The demonstration of another important attribute of fibre lasers, their low threshold pump power requirement, had to await the availability of a suitable laser for end-pumping, *i.e.* launching the pump light into the active core from the end of the fibre (Stone and Burrus 1973). The next step, which finally launched fibre lasers and amplifiers into prominence, was the demonstration of lasing and amplification in monomode silica fibres, pumped by low power semiconductor lasers (Poole *et al.* 1985, Mears *et al.* 1985). By far the most important of these fibre devices was the erbium-doped fibre amplifier (EDFA, Mears *et al.* 1986), providing gain at the wavelength of the third telecommunication window (1.5μm). Such is the importance of the EDFA for telecommunications that the volume of research on this device has far exceeded that on other fibre lasers and amplifiers. The book by Desurvire (1994) testifies to the volume of work on EDFAs.

Nevertheless it is now becoming more widely appreciated that fibre lasers and amplifiers may have much to offer in other areas of application. Thus, while fibre lasers have generally been thought of as low power devices, in the milliwatt region, they are now seen as attractive, simple sources of high brightness at the power level of several watts. Fibre lasers have also been regarded, on account of their length, as unpromising candidates for single-frequency operation. However, recent results from fibre lasers incorporating an optically-written grating in the fibre core, have shown that distributed feedback operation (DFB) is achievable, with the robust single-frequency oscillation that is characteristic of such devices.

Further important developments in fibre lasers are based on the use of glasses with a low phonon energy, such as ZBLAN. This greatly expands the range of amplifying transition compared to silica since the reduced rate of multiphonon decay means that many more levels have lifetimes long enough to serve as upper laser levels. Particularly striking results from the exploitation of this capacity have been seen in the area of infrared-pumping of visible upconversion lasers.

This chapter will therefore concentrate on these developments in fibre lasers, with potential applications outside the domain of optical telecommunications, which has so far formed the driving motivation for so much fibre laser work. First however in Section 2, a brief review will be made of the basic features of fibre lasers. Section 3 will review developments based on ZBLAN fibres, with particular emphasis on upconversion lasers. Section 4 will briefly review the use of fibre gratings to control the spectral properties of fibre lasers. Finally Section 5 will discuss developments in high power fibre lasers and amplifiers, including amplification of short pulses. A brief discussion of power and energy limitations of fibre lasers will be included.

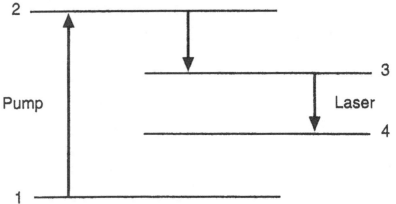

Figure 1. *Energy level scheme of four-level laser. For a three-level laser, the lower laser level (4 in the figure) is the same as the ground level 1.*

2   Basic principles of fibre lasers

The small cross-section area of the active core of the fibre means that a high pump intensity, and hence a high inversion density can be achieved for a modest pump power. This in turn implies that a high gain can be produced from a small pump power, hence a low pump power requirement to reach threshold. For a 4-level laser, *i.e.* where the lower level of the laser transition is empty (see Figure 1), the gain at frequency ν for a pump power P absorbed in a core of area, A_{core} is (Hanna 1993)

$$\text{gain} = \frac{\sigma_{em}(\nu)\tau P\phi_p}{h\nu_p A_{core}} \qquad (1)$$

where ν_p is the pump frequency, $\sigma_{em}(\nu)$ is the stimulated emission cross-section at frequency ν, τ is the lifetime of the upper level, and ϕ_p is the pumping quantum efficiency (the fraction of pump photons that produce excitation of the upper level).

Equation 1 indicates the inverse dependence of gain on the core area. Expressing the decay lifetime τ in terms of the radiative lifetime τ_{rad}, one has

$$\tau = \phi_{rad}\tau_{rad} \qquad (2)$$

where ϕ_{rad} is the radiative quantum efficiency, *i.e.* the fraction of ions in the upper laser level that decay via the laser transition under consideration. From (1) and (2) it is seen that the gain is proportional to $\phi = \phi_{rad}\phi_p$, the product of the pump and radiative quantum efficiencies. On the other hand it can be shown that the slope efficiency is independent of ϕ_{rad} and is in fact proportional to ϕ_p. Thus, even for a low radiative quantum efficiency, while this reduces gain and hence increases threshold, nevertheless a high slope efficiency can be obtained if $\phi_p \sim 1$. This is true of any laser but it has particular relevance in fibre lasers where low thresholds can be obtained even for low quantum efficiency. So, transitions which look unpromising in bulk media can still have low enough thresholds in fibre form to be diode-pumped and show a high slope-efficiency.

By expressing τ_{rad} in terms of the integrated emission cross section, we can express the gain in a simple and useful form which clearly indicates the relevant parameters (Hanna 1993). First we define numerical aperture of the fibre in terms of the core and cladding refractive indices by $NA = (n_{core}^2 - n_{clad}^2)^{1/2}$. To simplify we assume the emission lineshape to have a Lorentzian form and we assume the V parameter of the fibre $(V = 2\pi a(NA)/\lambda)$ just meets the condition for monomode propagation of the signal, $(V = 2.4)$, corresponding to the condition that higher order modes are cut-off and only the LP_{01} mode can propagate. The gain for a four-level laser transition can then be expressed, in dB per milliwatt of absorbed pump power as

$$\text{gain} = \frac{(NA)^2\phi}{10^4\Delta\nu\,h\nu_p} \qquad \text{(dB/mW)} \qquad (3)$$

where $\Delta\nu$ is the full width half maximum of the laser transition. With typical numbers substituted in Equation (3), e.g. ~ 0.2, $\phi \sim 1$, $\Delta\nu \sim 3THz$, gains of several dB/mW are predicted, as borne out in practice.

For a three-level laser, where the lower level of the laser transition is also the ground level, the gain expression can be approximated as

$$\text{gain} \simeq \frac{(I_p/I_{sat} - 1)}{(I_p/I_{sat} + 1)} \frac{(NA)^2\phi}{10^4\Delta\nu\,h\nu_p} \qquad \text{(dB/mW)} \qquad (4)$$

differing from (3) by the first term, indicating that net gain is achieved only if the pump intensity exceeds the saturation intensity of the pump transition I_{sat}, *i.e.* the pump intensity that excites sufficient ions to reduce the absorption coefficient to half its unpumped value. For a 3-level system this corresponds to the groundstate population being halved, *i.e.* equal populations in ground and upper levels. I_{sat} is given by

$$I_{sat} = \frac{h\nu_p}{\sigma_a \tau} \tag{5}$$

where σ_a is the pump absorption cross section. Typical values for I_{sat}, with typical value of $A_{core} \sim 10^{-11} \text{m}^2$, correspond to pump powers in the milliwatt region. Thus pump saturation is very easily produced and in fact the expression (4), when $I_p \gg I_{sat}$, as is often the case, reduces to the same as the 4-level expression in (3), reflecting the fact that strong pumping essentially empties the lower laser level.

Given the ease with which the pump absorption can be saturated it follows that typical pump powers can bleach the absorption over lengths very much exceeding the small-signal extinction length. In fact as a rough guide one can say that if the pump intensity exceeds the saturation intensity by a factor of x, then the pump will penetrate through the fibre to a distance x times the extinction length.

Two further comments should be added, by way of indicating how conditions in a fibre can be very different from those familiar in bulk lasers. First, since gains can be extremely high for low pump powers, this can easily lead to strong amplification of spontaneous er.ission (ASE). Thus photons emitted spontaneously at one end of the fibre can I amplified in a single pass through the fibre, to an intensity level which leads to strong depopulation, via stimulated emission, of the population inversion. This process limits the maximum gain that can be achieved; typically this becomes significant in the region of 30-40dB of single-pass gain. If strong feedback occurs from just one end of the fibre then the maximum gain is halved, to 15-20dB/pass. One consequence of ASE is to limit the range of tuning of a fibre laser into the low-gain wings of a transition. If sufficient gain is achieved for oscillation in wings, then the gain at the centre could be very much higher and subject to strong ASE. This will clamp the maximum gain available in the wings and will also reduce efficiency of lasing in the wings. If very broad tuning is desired, then every effort should be made to keep the resonator loss as low as possible for oscillation wavelengths in the wings of the transition.

ASE also imposes a limitation on the maximum energy that can be stored in an active fibre medium. One can re-express (1), to give the gain per unit stored energy as

$$\frac{\text{gain in dB}}{\text{energy}} = \frac{4.34}{A_{core}F_{sat}} = \frac{4.34}{E_{sat}} \tag{6}$$

where F_{sat} is the saturation fluence, given by $h\nu_{em}/\sigma_{em}$, and $E_{sat} = A_{core}F_{sat}$ is the saturation energy. Typical values for Nd^{3+} give a value of $\sim 3\text{dB}/\mu\text{J}$ for the above ratio. Thus, the stored energy is limited to $\sim 10\mu\text{J}$, since this would give a 30dB gain, with consequent ASE. For the Er^{3+}, $1.5\mu\text{m}$ transition having a much smaller σ_{em} (corresponding to the 30 times larger radiative lifetime), the figures are 30 times greater, *i.e.* a few hundred microjoules of energy could be stored.

Such considerations are of relevance to schemes using fibres for amplification of pulses and Q-switching. Of course, under pulsed conditions it is also necessary to bear in mind the damage limitations that would restrict the peak power that can be propagated within the small cross-sectional area of the core.

3 Low-phonon energy glasses and upconversion

A cursory examination of the energy levels of the triply ionised rare-earths (RE^{3+}), as shown in the 'so-called' Dicke diagram (Dicke and Crosswhite 1963), suggests a wealth of potential transitions if each indicated energy level is considered as a potential upper laser level. However, in practice, nonradiative decay from one level to another can greatly restrict the number of candidates for upper laser levels. The decay process of concern is a radiationless decay via the emission of one or more phonons ('multiphonon-decay'). The probability of such a decay process occurring is a very strong function of the number of phonons that need to be emitted in the process. Thus if there is a sufficiently large energy gap (*i.e.* requiring a large enough number of phonons to bridge the gap) between some level and the nearest level below it, then the multiphonon decay rate from this upper level can be negligible in comparison to typical radiative decay rates for the rare earths. As one goes from three phonons to five phonons being required to bridge the energy gap, so the nonradiative decay rate goes from being typically much faster to much slower than radiative decay rates. This behaviour is indicated in the following expression for the multiphonon decay rate.

$$A = C[n(T) + 1]^p \exp(-\alpha \Delta E) \tag{7}$$

where ΔE is the energy gap, $p = \Delta E / \hbar \omega$ is the number of phonons of energy $\hbar \omega$ that bridge the gap. Here $\alpha = \log(\epsilon/\hbar\omega)$, ϵ and C are host-dependent constants and

$$n(T) = \frac{1}{\exp(\hbar\omega/kT) - 1}$$

The relevant phonon energy which determines the nonradiative decay rate is the maximum energy in the phonon energy spectrum of the host material. For fused silica this is $\sim 1150 \text{cm}^{-1}$. This results in rapid nonradiative decay for energy gaps of less than 4500cm^{-1}. It is worth noting that estimates of energy gaps made from the simplified Dicke diagram should be treated with caution since the Stark splittings are not indicated there. Since the highest Stark levels in a given manifold can be several hundred cm^{-1} separated from the lowest level, the energy gap to the next manifold can be much less than apparent in the Dicke diagram.

Thus RE-doped silica offers a relatively limited number of suitable candidates for (metastable) upper laser levels, the most effective being the all-important $^4I_{13/2}$ level in Er^{3+}, as well as the $^2F_{5/2}$ level in Yb^{3+} and the $^4F_{3/2}$ level in Nd^{3+}. The need for a glass host of lower phonon energy to allow exploitation of other levels, and capable of being fabricated into a fibre, has been met by ZBLAN glass (France *et al.* 1992). The glass consists of a mixture of heavy metal fluorides (ZBLAN is an acronym for Zirconium, Barium, Lanthanum, Aluminium and Sodium (Na)), and has a maximum phonon energy of 590cm^{-1}. Originally the fibre was developed as a possible route to ultralow-loss fibre, relying on the fact that the longer wavelength of its infrared transmission edge (a consequence of the lower phonon energy) would allow the loss minimum to be at a longer wavelength than for silica and hence with a lower Rayleigh-scattering loss (λ^{-4} dependence). In practice these very low losses have been difficult to achieve due to rare-earth impurities contaminating the Lanthanum constituent. Nevertheless ZBLAN fibre has found use, with deliberate addition of RE impurity, most prominently at first (Ohishi

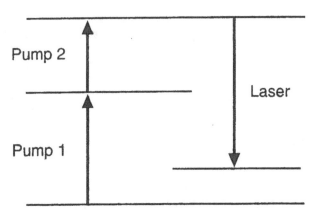

Figure 2. *Energy level scheme for upconversion laser. Two (as shown) or more pump photons are absorbed sequentially, followed by laser emission of a more energetic photon.*

1991) with regard to the Pr^{3+} transition at $1.3\mu m$ ($^1G_4$–$^3H_5$), offering the prospect of a fibre amplifier to offer for the second telecom window what the EDFA offers for the third window. However, the $^1G_4$ level lifetime is still significantly shortened by non-radiative decay, even in ZBLAN, thus making this $1.3\mu m$ amplifier much less effective (requiring much higher pump power) than the EDFA. The search for suitable glasses of even lower phonon energies continues. Apart from this possible $1.3\mu m$ application in telecom, these low phonon energy glasses also offer the prospect of long wavelength infrared lasers. Since the low phonon energy provides both infrared transmission and metastability of the upper laser level, which in the case of a long wavelength infrared transition necessarily implies a level close below. Er-doped ZBLAN has for example produced a $3.4\mu m$ laser ($^4F_{9/12}$– $^4I_{9/21}$, Tobben 1992) and a $3.9\mu m$ laser in Ho:ZBLAN (Schneider 1995) has been demonstrated.

The other area in which doped ZBLAN fibres have come into prominence is in up-conversion lasers. The basic idea is shown schematically in Figure 2. Two (or more) photons are absorbed sequentially, exciting the ion to a high level, from which emission occurs in a single, more energetic photon. A motivation for interest in such a scheme is the possibility of using a low power, cheap, and reliable, infrared diode laser as pump source to generate visible radiation. This would meet many of the requirements for compact visible laser sources, and is seen as one possible route to that goal, to be compared with alternative approaches such as frequency doubling of infrared diode lasers, or direct generation of visible light from wide-band-gap semiconductor lasers.

The notion of achieving upconversion lasing in fibres will occur to anyone looking at, for example, Er-doped silica when pumped in the infrared. Seemingly copious amounts of green fluorescence are generated by upconversion pumping of the $^4S_{3/2}$ level, followed by green emission to the ground ($^4I_{15/2}$) level. Likewise Tm^{3+} in silica shows strong blue fluorescence via upconversion. However examination of the fluorescence power in Tm-doped silica (Hanna *et al.* 1989) revealed that despite the bright appearance, only a small excited state population could be produced in a silica host, due to rapid multi-phonon decay of the various levels involved, and that a host of low phonon energy such as ZBLAN would be much more suitable. The first upconversion fibre laser was in fact

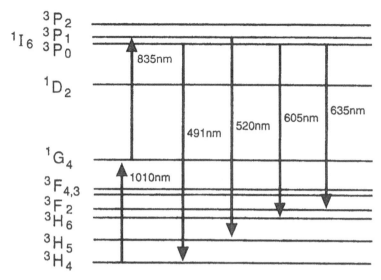

Figure 3. *Energy level scheme for upconversion lasing in Pr:ZBLAN. Pumping can be via absorption of two pump photons, as shown or via energy transfer from an excited Yb co-dopant to achieve the first step, followed by a further pump photon absorption.*

demonstrated in Tm:ZBLAN (Allain 1990a), operating at liquid nitrogen temperature and producing blue light at 451nm and 480nm. Since then extensive investigations were made (Oomen 1992) of the relevant spectroscopic parameters of Tm^{3+} in ZBLAN with a view to its upconversion potential. This Tm result was followed shortly after by a demonstration of green upconversion lasing in Ho:ZBLAN (Allain 1990b). This result gave a true indication of the capability of upconversion fibre lasing, since it occurred at room temperature, with high slope efficiency and had sufficient gain to lase from bare uncoated fibre ends. This level of performance and ease of operation was in marked contrast to the results hitherto obtained in bulk lasers, requiring cryogenic temperatures and being constrained by low gains and high pump power requirements. However, since it required a red pump laser, the Ho:ZBLAN system, while handsomely demonstrating the principle, did not meet the practical requirements of upconversion from the infrared. This was demonstrated with two results obtained in Pr:ZBLAN fibre, see Figure 3. Allain et al (1991) pumped the Pr system with a single pump wavelength, by way of Yb^{3+} as a co-dopant. A pump photon was absorbed by the Yb to the $^2F_{5/2}$ level, and the Yb excitation then transferred non-radiatively to a neighbouring Pr ion, which was thus promoted to the $^1G_4$ level. From the $^1G_4$ level a second pump photon took the Pr to the $^3P_0$ level, from which red lasing on the $^3P_0$–$^3F_2$ transition occurred. The scheme of Smart *et al.* (1991]) used two different pump photons, at 1.03μm for the first step $^3H_4$–$^1G_4$, and then 840nm for the second step $^1G_4$ to $^3P_0$. In this case lasing was observed on all the visible transitions from blue, (491nm),$^3P_0$–$^3H_4$: green (520nm), $^3P_0$–$^3H_5$: orange (610nm), $^3P_0$–$^3H_6$: red (635nm),$^3P_0$–$^3F_2$. This was the first demonstration of blue upconversion lasing at room temperature.

This upconversion scheme in Pr:ZBLAN has since been examined in some detail (Tropper *et al.* 1994) and in particular has been subjected to various experimental improvements. Better fibre has been used (the quality of Pr:ZBLAN fibre improved

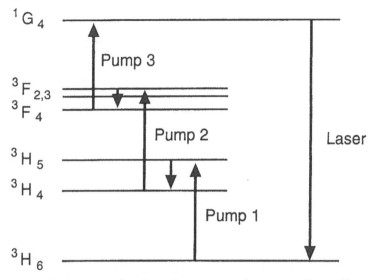

Figure 4. *Energy level scheme for three-photon pumped upconversion in Tm:ZBLAN. All three photons can have the same energy. A single blue photon (480nm) is emitted.*

rapidly as a result of efforts aimed at the $1.3\mu m$ amplifier), pump power has been retroflected to improve its absorption (Zhao *et al.* 1995), leading to 22mW of blue output with an overall efficiency of 7.5% In particular the Yb:Pr scheme has worked in the blue (initially only working in the red in the experiments of Allain 1991) and has been pumped by a single diode laser (Xie and Gosnell 1995) with a slope efficiency of 3%. These figures show promise for a practical blue source.

Two other upconversion schemes should also be mentioned. The green light so familiar from EDFAs can be made to lase efficiently in Er: ZBLAN (Whitley *et al.* 1991, Allain *et al.* 1992). Pumped by diodes at 800nm or 980nm, thresholds as low as 30mW have been observed (Massicott *et al.* 1993, Piehler 1994) and output power up to 11.7mW for 150mW of pump. While these figures are promising, the competition from other green sources is considerable, since efficient frequency doubling can be achieved rather effectively from diode-pumped miniature lasers and microchip lasers. Blue light generation is less well developed from the latter sources, there being less convenient laser transitions involved in place of the 1064nm Neodymium transition. So, blue upconversion lasing still generates interest and beside the Pr transition referred to above, there is a very efficient upconversion transition in Tm:ZBLAN. This scheme (Grubb *et al.* 1992) involves absorption of three photons, a single wavelength ($\sim$1120–1150nm) sufficing for all three steps (Figure 4). Lasing, at 480nm is on the $^1G_4$–$^3H_6$ transition, down to a Stark level of ground manifold (quasi-3-level operation). Despite the need for three pump photons, and the quasi-3-level nature of the scheme, this laser has operated very efficiently, 30%-40% efficient, and has shown low enough threshold for pumping by diode laser. In fact, a blue output of >100mW has been observed with diode pumping (Waarts *et al.* 1995).

These upconversion fibre lasers have reached the point where efficient diode-pumped room temperature operation is demonstrated. As sources they still need further refinement and indeed have some serious limitations. For example, a convenient means of modulation, such as exists for semiconductor lasers, is absent. Single-frequency selection is also not straightforward. Finally it should be added that the long term stability of ZBLAN fibre, particularly for conditions of intense blue lasing, is not certain and indeed there is evidence of unwanted light-induced darkening under some conditions (Barber *et al.* 1995).

4 Gratings for fibre lasers

Feedback has been introduced into fibre lasers in a variety of ways. Simple, uncoated, cleaved ends can suffice when high-gain is available. More commonly mirrors have been butted to the ends of the fibre, although this is probably not practical for anything other than the laboratory environment. Coatings have also been directly applied to the ends of the fibre. High reflectivity without the use of coatings has been achieved by means of fibre loop reflectors. All of these approaches have drawbacks and deficiencies. Recently however, the matter of providing feedback for fibre lasers has been transformed by the availability of fibre gratings, *i.e.* Bragg gratings fabricated in the core of the fibre by side illumination with UV light.

The basic principles have been described in review articles, (Russell 1993, Morey *et al.* 1994). Ultraviolet light of an appropriately short wavelength (249nm from KrF lasers, 244nm from frequency doubled Ar lasers, and 193nm from ArF lasers have all been used) provides side-illumination of the fibre in such a way as to produce an interference fringe pattern of UV light in the fibre core. Techniques for producing the interference patterns include the use of a two beam interference arrangement (as used for producing holographic gratings) or, more simply, by use of a phasemask. The UV light produces an index increase in the fibre core, the greater the intensity the greater the index increase, so the spatially modulated UV intensity within the interference pattern produces a spatially modulated refractive index variation. This will act as a Bragg reflector for light of appropriate wavelength, matched to the period of the grating. In the case of the two-beam interference set-up, the period can be selected by appropriate choice of angle between the interfering beams. This freedom is not available with the phase-mask approach, although the techniques using a phase-mask are easier to apply.

As reflectors for fibre lasers, these gratings offer many attractive features. First, since they form an integral part of the fibre, so the laser resonator can be both compact and robust. They provide very effective line-narrowing, to a frequency bandwidth of the order of (or rather less than) $c/2nL$ where L is the grating length and n the effective refractive index. If the grating length is comparable to the overall length of the fibre laser then single-frequency operation is rather easily achieved. The extreme version of this is where the grating occupies the full length of the active fibre. Such distributed feedback (DFB) fibre lasers have been demonstrated (Kringlebotn *et al.* 1994, Asseh *et al.* 1995, Sejka *et al.* 1995), with the excellent single-frequency robustness that is characteristic of such lasers. Very narrow linewidth operation is achieved, an advantage that these fibre lasers have over semiconductor lasers whose small dimension and relatively high loss result in a rather large Schawlow-Townes limited bandwidth. Unlike a bulk grating, which, when used in a laser can provide extensive tuning by tilt of the incidence angle, a fibre grating offers a relatively limited tuning range. Temperature tuning via the temperature-dependence of length and refractive index gives tuning rates of the order of 1–2GHz/°C. Tuning can also be achieved by stretch (the percentage tuning corresponding essentially to the percentage stretch, hence limited typically to less than 1%) and, less conveniently, but much more extensively, by compression (Ball *et al.* 1994).

This ability to tune the wavelength of peak reflection for the grating reflector is one example of how these reflectors offer more versatility than conventional (*e.g.* multi-layer dielectric coatings) mirrors. Other benefits over conventional mirrors include their excellent frequency discriminations when used as dichroics, allowing efficient transmission of a pump wavelength very close to the resonated lasing wavelength. A novel feature of these grating reflectors is their ability to provide chirped reflectivity *i.e.* the period being varied spatially along the grating, so that a pulse incident on the grating can have different spectral components reflected from different regions of the grating and hence be dispersed in time. This has been used in dispersion compensation and chirped pulse amplification as will be described in the next section.

Thus fibre gratings offer a major step in versatility of fibre systems in general and fibre lasers in particular. There is still much to be done in grating development. For example, the origin of the photorefractive behaviour is by no means fully understood. At first the requisite photosensitivity was displayed only in silica fibre containing GeO_2 as a dopant. Other glass compositions are now being developed for their photosensitivity and indeed grating fabrication has been demonstrated (with lasing via feedback from these gratings) in ZBLAN fibre [Taunay 1994].

5   High power, cladding pumped fibre lasers

In solid-state lasers any heat that is generated in the pumping process, by, for example, nonradiative decay processes, has to be removed to the surface of the material by thermal conduction. For heat to flow by thermal conduction, a temperature gradient is necessary, leading to stress as a result of the inhomogeneous thermal expansion. The limit to mean power output in many solid-state lasers is set by the fracture limit associated with the thermally induced stress. However, fibre lasers have significant advantages in the matter of heat removal. They can be made long, for example by reducing the dopant concentration. The low-loss and absence of diffraction spread permits this freedom over choice of length. Thus heat dissipation can take place over a long length, and since the fracture limit, *i.e.* the maximum tolerable heat absorbed per unit length is of the order of 100 W/m, (Alcock *et al.* 1986, Hanna *et al.* 1989), independent of diameter, it can be seen that in principle very high power operation of fibre lasers is possible without fracture problems. In fact the problems of highest priority in scaling fibre lasers to high power, is that of getting enough pump power into the active core.

If the core dimensions are such that the fibre is monomode, then for end-launching of pump light directly into the core one needs a pump source which itself is essentially diffraction-limited. High power diffraction-limited pump sources have been used, for example a CW Nd:YAG laser used to pump a Tm-doped silica fibre, leading to the

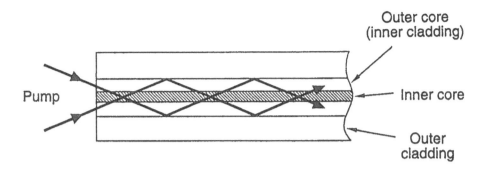

Figure 5. *Schematic diagram of double clad fibre. End pumping into the inner cladding, with guidance of the pump at the interface between inner and outer-cladding, ultimately leads to all of the pump being absorbed into the active (monomode) core.*

first demonstration of a fibre laser at the 1W power level (Hanna *et al.* 1990). Direct pumping with diode lasers is a more attractive proposition however, since this provides the route to a cheap and compact system. On the other hand, high power diode lasers typically have a rather poor beam quality, unsuitable for direct launch into the end of the fibre core. The solution to this problem is provided by the elegant technique, referred to as cladding-pumping using a so-called double-clad fibre, first demonstrated by Snitzer *et al.* (1988).

The principle is shown in Figure 5. The core, which may be monomode, lies within a lower index inner-cladding, which in turn lies within an outer-cladding of yet lower index. Pump light can be end-launched into the inner-cladding, with a much less stringent beam quality requirement compared with launching into the core. This pump light, guided in the inner-cladding is then progressively absorbed into the core as it propagates in the inner-cladding. A longer length of fibre is needed to achieve this absorption, by a factor of the order of the ratio of inner-cladding area to core area. Provided propagation losses of the pump and of the lasing mode in the core are not excessively increased by the extra length of fibre, then efficient pumping and efficient monomode lasing can result. Thus the cladding-pumping scheme can provide a very simple means of enhancing the brightness of a (diode) pump source by efficiently converting it to a monomode laser output.

The first results, obtained with a Nd-doped fibre (Snitzer *et al.* 1988), gave convincing confirmation of the effectiveness of the idea, with more recent results showing $\sim$ 5W of monomode output at $\sim 1\mu$m obtained for 15W of multimode diode pump power (Po *et al.* 1993). The principle works particularly well for a four-level laser where reabsorption loss due to population in the lower laser level is not an issue. Thus one can increase the inner-cladding area, necessarily increasing the fibre length to achieve absorption, until propagation loss become significant. With a three-level or quasi-three-level laser a more demanding requirement must be met, *i.e.* that of ensuring a sufficient intensity at the launch to invert the population sufficiently to overcome the reabsorption loss. A greater demand is therefore placed on the focusing of the pump source and hence on its beam being configured so that it can be launched into the minimum area of inner-cladding. As examples of three-level lasers, both Er and Yb:Er double-clad fibres have been successfully operated, achieving power levels in the region of 1 Watt in recent work (Minelly *et al.* 1995). Developments in the effective shaping of beams from diode bars, to a well-confined circular spot (Clarkson *et al.* 1994, Clarkson and Hanna 1995), hold promise for launching substantial diode laser power (>10W) into double-clad fibres of small inner-cladding diameter.

Thus, to some extent the problem of launching high power into the fibre core is solved via cladding-pumping. It is natural then to address the question of the maximum power that can be handled, before some form of optically induced damage sets in. Here there is some uncertainty over what the ultimate limits might be. As a benchmark indicating what power level can be handled, the result of Zollner *et al.* (1995) can be quoted. In their experiment, using a double-clad Nd-doped fibre a total diode laser power of 38W was launched into an inner cladding of 400μm diameter. The output, from the (slightly multimode) 10μm diameter core, was 9.2W, in a beam with beam quality factor M^2 ~2. In principle it should be possible to achieve monomode operation, by judicious choice of numerical aperture, up to say 20μm diameter. So, assuming the same intensity as for the 10μm core, then monomode power output in the 40W range should be possible. Larger powers can be obtained by going to a larger core diameter, but the M^2 would then increase in proportion to the increased core diameter. The higher power would be achieved at a constant brightness. Scaling up the core diameter in this way means a greater heat absorption per unit length, and the fracture limit of ~ 100W/m (thermal load) will ultimately intervene. Lower doping and hence longer fibre can in principle allow further power scaling free from fracture. Other limiting factors will then be nonlinear processes, such as stimulated Raman scattering, and of course the propagation loss for pump and laser. Zenteno (1993) has given a discussion of various limiting factors. Considerable further work will be needed to establish what power levels are in practice achievable and maintainable in a safe and reliable way. However it is clear already that fibre lasers are not confined to the low powers that was characteristic in their early days when pumped by low power diode lasers.

A few further comments are worth making on double-clad fibre lasers. First, while the motivation of using the double-clad fibre has been a quest for high power, it should also be noted that the double-clad geometry offers a relaxation in the positioning toler-ance for efficient launch of the pump light. By enabling more efficient pump launch, the overall efficiency can be increased. A modest ratio of inner-cladding to core diameter is sufficient. This was demonstrated for a double-clad Yb-doped silica fibre laser (Pask *et al.* 1994). Yb-doped silica is well suited to efficient operation (Pask *et al.* 1995), and with the double-clad fibre, having a relatively small diameter ratio of ~ 3 for inner-cladding to core, a slope efficiency with respect to *incident* power of 80% was achieved, very close to the quantum limit and the highest efficiency reported for a fibre laser. It is also worth noting that this double-clad geometry was compatible with fabrication of gratings in the core via side illumination with KrF laser light, and the gratings were used to force the Yb laser to oscillate at a specific wavelength of choice.

In addition to providing high power lasing, cladding-pumped fibre can also produce high gain. This combination of high power and high gain is a new feature, also beginning to appear in bulk lasers, giving new opportunities in the architecture of light sources.

A good example of this is a recent result (Minelly *et al.* 1995) in which a mode-locked output from an erbium-doped silica fibre laser, producing 200fs pulses at a 50MHz repetition rate with 3pJ energy per pulse, has been amplified in an Er-doped fibre preamplifier, followed by a cladding-pumped Yb:Er fibre power amplifier to give 380fs pulses with an energy of 20nJ, and an average power of 1W. To keep the generated peak power in the Yb:Er fibre amplifier to acceptable levels, the technique of chirped pulse amplification was used. This involved reflecting the initial (unamplified) mode-locked pulses from a chirped fibre grating so as to produce a greatly lengthened (and hence lower power) chirped pulse. This pulse, even after amplification had a low enough power to keep self-phase-modulation effects to an acceptable level. A second chirped grating served to recompress the final, energetic, pulses. This result should help to further dispel the notion that fibre lasers are characteristically low power devices, since the power levels achieved in this essentially all-fibre system are comparable to those achieved in systems that make use of Ar laser pumping.

References

Alcock I P, Ferguson A I, Hanna D C, and Tropper A C, 1986, *Optics Comm* **58** 405

Allain J Y, Monerie M, and Poignant H, 1991, *Electron Lett* **27** 1156

Allain J Y, Monerie M, and Poignant H, 1990a, *Electron Lett* **26** 166

Allain J Y, Monerie M, and Poignant H, 1990b, *Electron Lett* **26** 261

Allain J Y, Monerie M, and Poignant H, 1992, *Electron Lett* **28** 111

Asseh A, Storoy H, Kringlebotn J T, Margulis W, Sanhlren B, Sandgren S, Stubb R, and
    Edwall G, 1995, *Electron Lett* **31** 969

Ball G A, and Morey W W, 1994, *Optics Letts* **19** 23

Barber P R, Paschotta R, Tropper A C, and Hanna D C, 1995, *Opt Lett* **20** 2195

Clarkson W A, Neilson A B, and Hanna D C, 1994, *CLEO-94 Anaheim* paper CThL2, 360

Clarkson W A, and Hanna D C, 1995, Two mirror beam-shaping technique for high-power
    diode bars, *Accepted by Electron Lett*

Desurvire E, 1994, *Erbium-doped fiber amplifiers* (Wiley Interscience, New York)

Dicke G H, and Crosswhite H M, 1963, *Appl Opt* **2** 675

France P W, Drexhage M G, Parker J M, Moore M W, Carter S F, and Wright J V, 1990,
    Fluoride Glass Optical Fibres (Blackie/CRC Press Inc)

Grubb S G, Bennett K W, Cannon R S, and Humer W F, 1992, *Electron Lett* **28** 1243

Hanna D C, McCarthy M J, and Suni P J, 1989b, Fibre Lasers and Light Sources,
    SPIE Proc **1171** 160

Hanna D C, Percival R M, Perry I R, Smart R G, Townsend J E, and Tropper A C, 1990,
    Opt Comm **78** 187

Hanna D C, Perry I R, and Lincoln J R, 1990, *Opt Comm* **80** 52

Hanna D C, 1993, Fibre Lasers in *Solid State Lasers: New Developments and Applications*
    Ed. M Inguscio, (Plenum, New York)

Kringlebotn J T, Archambault J L, Reekie L, and Payne D N, 1994, *Opt Lett* **19** 2101

Massicott J F, Brierley M C, Wyatt R, Davey S T, and Szebesta D, 1993,
    Electron Lett **29** 2119

Mears R J, Reekie L, Poole S B, and Payne D N, 1985, *Electron Lett* **21** 738

Mears R J, Reekie L, Poole S B, and Payne D N, 1986, *Electron Lett* **22** 159

Minelly J D, Barnes W L, Laming R I, Morkel P R, Townsend J E, Grubb S G, and Payne
    D N, 1993, *IEEE Photonics Tech Lett* **5** 301

Minelly J D, Galvanauskas A, Fermann M E, Harter D, Caplen J E, Chen Z J, and Payne D
N, 1995, *Opt Lett* **20** 1798

Morey W M, Ball G A, and Meltz G, 1994, *Optics and Photonics News* **5** 8

Ohishi Y, Kanamori T, Kitagawa T, Takahashi S, Snitzer E, and Sigel G, 1991,
Optics Letts **16** 1747

Oomen E W J L, 1992, *J of Luminescence* **50** 317

Pask H M, Archambault J L, Hanna D C, Reekie L, Russell P St J, Townsend J E, and
Tropper A C, 1994, *Electron Lett* **30** 863

Pask H M, Carman R J, Hanna D C, Tropper A C, Mackechnie C J, Barber P R and Dawes
J M, 1995, *IEEE J of Selected Topics in Quantum Electron* **1** 2

Piehler D, and Craven D, 1994, *Electron Lett* **30** 1759

Po H, Cao J D, Laliberte, Minns R A, Robinson R F, Rockney B H, Tricca R R, and Zhang
Y H, 1993, *Electron Lett* **29** 1500

Poole S B, Payne D N, and Fermann M E, 1985, *Electron Lett* **21** 737

Ross B, and Snitzner E, 1972, *IEEE J Quant Electron* **6** 361

Russell P S J, Archambault J L, and Reekie L, 1993, *Physics World, October* p.41

Sejka M, Varming P, Habner J, and Kristensen M, 1995, *Electron Lett* **31** 1445

Smart RG, Hanna D C., Tropper A C, Davey S T , Carter S F, and Szebesta D, 1991,
Electron Lett **27** 1307

Schneider J, 1995, *Electron Lett* **31** 1250

Snitzner E, 1961, *Phys Rev Lett* **7** 444

Snitzer E, Po H, Hakini F, Tumminelli R, and McCollum B C, 1988, *in Optical Fiber Sensors
Vol.2, OSA Technical Digest Series* paper PD5

Stone J, and Burrus C, 1973, *Appl Phys Lett* **23** 388

Taunay T, Niay P, Bernage P, Xie W, Poignant H, Boj S, Delavaque E, and Monerie M, 1994
Opt Lett **19** 1269

Tobben H, 1992, *Electron Lett* **28** 1361

Tropper A C, Carter J N, Lauder R D T, Hanna D C, Davey S T, and Szebesta D, 1994,
J Opt Soc Am B **11** 886

Waarts R G, Mehuys D G, and Welch D F, 1995, *CLEO-95* paper CPD-24

Whitley T J, Millar C A, Wyatt R, Brierley M C, and Szebesta D, 1991,
Electron Lett **27** 1785

Xie P, and Gosnell T R, 1995, *Digest Advanced Solid-State Lasers* Memphis, paper WD-1

Zellmer H, Willamewski U, WellingH, Unger S, Perchel V, Müller H-R, Kirchkof J, and Albers
P, 1995, *Opt Lett* **20** 578

Zenteno L, 1993, *J Lightwave Technol* **9** 1435

Zhao Y, Fleming S, and Poole S, 1995, *Opt Comm* **114** 285

Optical Parametric Oscillators: continuous wave operation

Miles J Padgett

St Andrews University, Scotland.

1 Introduction; a brief history of OPOs

Investigations into optical parametric oscillators (OPOs) began thirty years ago when Giordmaine and Miller (1965) demonstrated a pulsed source of coherent radiation based on optical parametric oscillation in $LiNbO_3$. Three years later this was followed by the first report by Smith *et al.* (1968) of continuous wave (CW) output from an optical parametric oscillator based on $Ba_2Na_2Nb_5O_{15}$. Since these early demonstrations, a steady stream of new materials has achieved ever increasing efficiencies. A significant breakthrough occurred in 1984 when Donaldson and Tang (1984) developed an OPO based on crystalline urea, pumped in the ultraviolet (UV) giving a visible output. Soon this was superseded by Fan *et al.* (1988) who demonstrated the use of β-barium borate (BBO) as the non-linear crystal. Having better long term damage properties than urea, the discoveries of BBO and subsequently of LiB_3O_5 (LBO) by Chen *et al.* (1989) have acted as a catalyst for future work and formed the basis for several pulsed commercial OPO systems.

A critical design parameter in OPOs is the pump power required to reach threshold. This is proportional to the product of the losses at signal, idler and pump wavelengths. For pulsed operation, it is sufficient for the OPO cavity to be resonant at just one of these wavelengths, for example at either the signal or the idler wavelength. A device of this kind is called a singly resonant oscillator (SRO). Generally, the threshold power for an SRO is too high for convenient CW pump sources, although recently a device based on KTiOPO$_4$ (KTP) has been operated in the infrared in such a configuration (Yang *et al.* 1993). One solution is to make the OPO resonant at both signal and idler fields. Such a device is termed a doubly resonant oscillator (DRO) and thresholds below 10mW have been reported (Nabors *et al.* 1989). However, when configured as a DRO, the OPO is over-constrained by the requirement to satisfy energy conservation, phase matching and support resonances of both the signal and idler fields simultaneously within the a single cavity. This often leads to frequency instabilities in the output of a DRO as sub-nanometer changes in cavity length or MHz changes in pump frequency cause mode hops in the signal and idler fields (Falk 1971, Eckardt *et al.* 1991). Consequently, CW OPO's have had the reputation of being difficult systems to control.

2   Resonance of signal and idler fields within OPOs

This section considers the resonance requirements for the signal and idler fields within OPOs. Specifically, it examines how these resonance conditions influence the stability requirements for maintaining single mode-pair outputs within a CW OPO

2.1   Requirements for resonance of the signal and idler fields

In optical parametric down-conversion, an input pump wave at frequency ν_p is converted into two outputs, the signal and idler, at frequencies ν_s and ν_i, respectively. These frequencies must obey the energy conservation relation, i.e.

$$\nu_p = \nu_s + \nu_i \tag{1}$$

The ratio of ν_s and ν_i is determined by the phase-matching within the crystal which is essentially the conservation of momentum. For perfect phase-matching this condition is as follows:

$$\frac{2\pi}{c}(n_p\nu_p - n_s\nu_s - n_i\nu_i) = 0 \tag{2}$$

where n_p, n_s and n_i are the refractive indices of the pump, signal, and idler fields respectively, and c is the velocity of light.

To form an optical parametric oscillator, resonance is provided by feedback using cavity mirrors for either, or both, the signal and idler frequencies. These cavity resonances have a significant effect on the frequency tuning of the OPO outputs (Eckardt *et al.* 1991). A device with feedback at only one of the signal or idler frequencies is referred to as a singly-resonant OPO (SRO), in which the exact frequency of the resonant field is that of the cavity mode closest to the optimum phase-matching condition. The frequency of the non-resonant field is merely that dictated by Equation 1, making control of the OPO frequencies comparatively straight-forward.

There have been two reports of CW SROs. The first used a multi-Watt, frequency-doubled Nd:YAG laser to pump a near-degenerate type II phase-matched KTP OPO (Yang *et al.* 1993 and Yang *et al.* 1994). The second used a pump-enhancement scheme in which a multi-Watt argon ion laser was resonated strongly within a singly-resonant OPO cavity which used LBO as the nonlinear material (Robertson *et al.* 1994). However, at present, the high pump power thresholds (> 1 W), and the subsequent high intensities in the nonlinear materials ($\approx$ MW/cm^2), required for singly-resonant operation, remain a limitation on CW implementations.

2.2 Simultaneous resonance of the signal and idler fields

To reduce the pump power required to reach threshold, thereby opening up the prospect of using virtually any single-frequency CW laser as the pump source, CW OPOs are configured routinely as doubly-resonant OPOs (DROs), in which both the signal and idler are brought to resonance simultaneously. Continuous-wave thresholds at the mW-level are reliably obtained from such devices. However, to operate a DRO, four conditions must be satisfied simultaneously; energy-conservation, phase-matching, and cavity resonances for the signal and idler frequencies. In general, a DRO is over-constrained, and this introduces complications in the tuning of these devices (Smith 1993, Falk 1971 and Padgett *et al.* 1994).

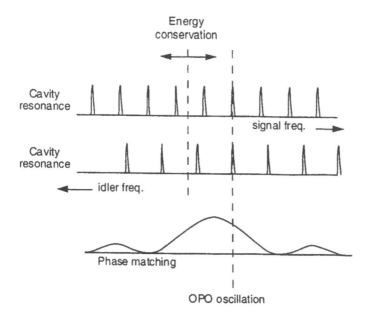

Figure 1. *Simultaneous requirements for energy conservation, cavity resonance and phase matching in a* CW OPO.

Figure 1 shows the interrelation between the various constraints. (Note that the axes for the signal and idler frequencies are reversed.) Any vertical line drawn through both the signal and idler frequency axes satisfies the requirement for energy conservation, the phase-match condition is adequately satisfied for a range of signal and idler frequencies and the selectivity is imposed by the requirement for simultaneous resonance of the signal and idler fields. We denote the signal (idler) cavity length by $L_s (L_i)$, and the length of the OPO crystal by l. The resonance condition for the signal and idler are

$$\nu_s = \frac{m_s c}{2(L_s + (n_s - 1)l)} \qquad \nu_i = \frac{m_i c}{2(L_i + (n_i - 1)l)} \tag{3}$$

where m_s (m_i) is the longitudinal mode number of the signal (idler), n_s (n_i) is the refractive index of the signal (idler) within the nonlinear material, and c is the velocity of light. To calculate, to good approximation, numerical values for the tuning rates, the dispersion terms can be ignored. A change in the length of the signal (idler) cavity of ΔL_s (ΔL_i) causes the resonant frequency of the signal (idler) field to change by $\Delta \nu_s$ $(\Delta \nu_i)2$ such that,

$$\Delta \nu_s = \frac{-\Delta L_s \nu_s}{L_s + (n_s - 1)l} \qquad \Delta \nu_i = \frac{-\Delta L_i \nu_i}{L_i + (n_i - 1)l} \tag{4}$$

This equation can be expressed in terms of the free spectral range, FSR_s (FSR_i) of the signal (idler) cavity,

$$\Delta \nu_s \approx \frac{-2\Delta L_s \nu_s (FSR)_s}{c} \qquad \Delta \nu_i \approx \frac{-2\Delta L_i \nu_i (FSR)_i}{c} \tag{5}$$

The signal and idler oscillation frequencies will be determined as a compromise between the optimum phase-matching condition and the requirement that the frequencies must satisfy the conservation of energy and lie close to a resonant mode of the cavity. Pairs of signal and idler modes that satisfy energy conservation are termed mode-pairs. Usually, within the phase-matching bandwidth, there are several mode-pairs for which the signal and idler lie close enough to the cavity modes for the OPO to oscillate. Comparatively small changes in either the cavity mode frequencies or the pump frequency cause the OPO output to switch from one mode-pair to another. Depending on the mis-match in the $FSRs$ of the signal and idler fields, the new mode-pair is either adjacent to the original (a mode-hop), or many mode-pairs removed (a cluster-hop). Mode-pairs are said to be in the same cluster if the sum of m_s and m_i remains the same.

2.3   Mode hopping in DRO OPOs

The mis-match in the $FSRs$ dictates the level of cavity length/pump frequency detuning required to cause a hop to an adjacent mode-pair. Assuming that the original mode-pair is exactly on resonance, then by tuning the cavity mode frequencies by a total amount equal to the mis-match in the $FSRs$, namely $\Delta(FSR)$, a mode-hop will occur. Therefore, the condition for a mode-hop is given by

$$\Delta(FSR) \approx \frac{2(\Delta L_s \nu_s (FSR)_s + \Delta L_i \nu_i (FSR)_i)}{c} \tag{6}$$

For a single-cavity OPO, in which the signal and idler fields are resonant in the same cavity, the change in cavity length required to achieve a mode-hop is

$$\Delta L_{hop} \approx \frac{\Delta(FSR)}{2(FSR)} \lambda_p \tag{7}$$

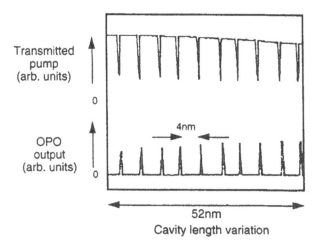

Figure 2. *Intensity of transmitted pump and* OPO *output as a function of cavity length in a Type-II phase matched, doubly resonant* OPO.

where λ_p is the free-space pump wavelength. This expression can be written in terms of the refractive indices of the signal and idler frequencies within the nonlinear crystal:

$$\Delta L_{\text{hop}} \approx \frac{\lambda_p}{2} \frac{|n_s - n_i| l}{(L + (\bar{n} - 1)l)} \tag{8}$$

where $\bar{n}$ is the average refractive index for the signal and idler fields. Figure 2 shows the mode hoping behaviour of a type-II phase-matched OPO. Note the regular intervals at which OPO output is obtained.

The second parameter that affects the tuning behaviour of the OPO is the pump frequency. The change in pump frequency required to cause a mode-hop is given by

$$\Delta\nu_{(\text{p-hop})} = \Delta(FSR) \tag{9}$$

2.4 Pump power threshold in DRO OPOs

The pump power required to reach threshold is dependent on the degree to which the resonance conditions for the signal and idler modes are satisfied and can be understood in terms of Figure 3, in which the linear scales for the signal and idler are again reversed.

Approximate expressions can be obtained for the maximum fluctuations in the cavity length or pump frequency that can be tolerated if the output of the OPO is to be maintained to a single mode-pair. From Figure 3, it can be deduced that the detuning, $\delta\nu$, while maintaining operation on a single mode-pair, is given by,

$$\delta\nu = \Delta\nu_s + \Delta\nu_i \leq \frac{(FSR)_s}{2F_s} + \frac{(FSR)_i}{2F_i} \tag{10}$$

where $\Delta\nu_s$ ($\Delta\nu_i$) is the half-width at half-maximum of the signal (idler) resonance, and F_s (F_i) is the cavity finesse for the signal (idler) frequency.

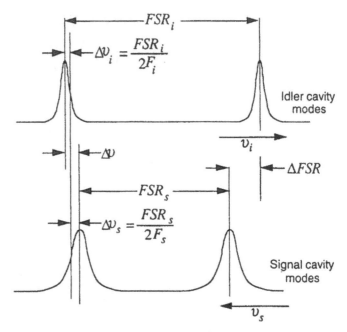

Figure 3. *The requirement for simultaneous resonance of the signal and idler fields in a doubly resonant* OPO.

We compare the above expression to Equation 15 in Eckardt *et al.* (1991) which derives the standard expression of the threshold pump power, P_{th}, for a DRO; i.e.

$$P_{th} \propto \left(\frac{1}{\text{sinc}^2(\Delta kl/2)}\right)\left(\frac{\pi^2}{F_i F_s}\right)\left(1 + \left(\frac{2(\Delta\nu_s + \Delta\nu_i)F_s F_i}{F_i(FSR)_s + F_s(FSR)_i}\right)^2\right) \tag{11}$$

The first term relates to the increase in threshold that arises from imperfect phase matching. In practise the phase-matching bandwidth is many times larger than the mode spacing (as a rule of thumb, the phase matched bandwidth is 2–3 clusters wide independent of phase-match geometry or detuning from degeneracy). From Equation 17, this gives the approximate phase-matching bandwidth of a single-cavity DRO to be

$$\Delta\nu_{\text{phase-match}} \approx \frac{2(FSR)^2}{\Delta(FSR)} \tag{12}$$

The second term related simply to the reduction in threshold that can be expected from making the OPO doubly resonant.

2.5 Stability requirements in DRO OPOs

The third term in Equation 11 is related to our Equation 10, in which $\Delta\nu$ corresponds to the detuning required to double the threshold of operation for the particular mode-pair. So that comparison can be drawn between various configurations, we take the point where the threshold doubles as an indication of the range over which the OPO operates on a single mode-pair (Colville *et al.* 1994). By substituting Equation 5 for the change in signal and idler frequencies, Equation 10 becomes

$$\left|\frac{4}{c}(\Delta L_s \nu_s (FSR)_s + \Delta L_i \nu_i (FSR)_i)\right| \leq \frac{(FSR)_s}{F_s} + \frac{(FSR)_i}{F_i} \tag{13}$$

The cavity length stability required to maintain the output to a single mode-pair is as follows:

$$\Delta L_{\text{stab}} \approx \pm\frac{\lambda_p}{4}\left(\frac{1}{F_s} + \frac{1}{F_i}\right) \tag{14}$$

Similarly, the pump frequency stability requirement to maintain the output to a single mode-pair within a single-cavity DRO is as follows:

$$\Delta\nu_{\text{p-stab}} \approx \pm\frac{(FSR)}{2}\left(\frac{1}{F_s} + \frac{1}{F_i}\right) \tag{15}$$

The requirements for cavity length and pump frequency stability are summarised in the Figure 4. The double-resonance condition can be maintained by control of either the cavity length or the pump frequency. Combining Equations 14 and 15 gives the following relationship between the perturbation of pump frequency or cavity length, and the required change of the other, to maintain the simultaneous resonance of the signal and idler fields.

$$\Delta\nu_p \approx -2\Delta L\left(\frac{(FSR)}{\lambda_p}\right) \tag{16}$$

From Equation 16 it follows that, even within a single-cavity OPO, smooth frequency tuning can be obtained by changing the cavity length by ΔL while controlling simultaneously the pump frequency to maintain the double-resonance condition. Coarse frequency tuning can be obtained by changing the phase-matching relation.

Although interrelated, it should be emphasised that a change in cavity length or pump frequency affects the output of the OPO somewhat differently. Within any given phase-matching geometry, the precise pump frequency determines the exact frequencies of the signal and idler fields. Therefore, assuming that a particular mode-pair is on exact simultaneous resonance, small perturbations in the cavity length result in a detuning of the cavity mode resonances away from the generated frequencies which, although not changing these frequencies, does lead to a decrease in conversion efficiency. In contrast, perturbations in the pump frequency can result in both a direct change in the generated frequencies and a reduced conversion efficiency.

For type I phase-matching, near frequency-degeneracy, the difference in the FSRs of the signal and idler fields is small. In this case the cavity length and the pump frequency stability requirements (Equations 7–9) to prevent a mode-hop become more stringent than the stability requirements to hold a single mode-pair above threshold (Equations 14 and 15). As a result, under detuning of either the cavity length or the

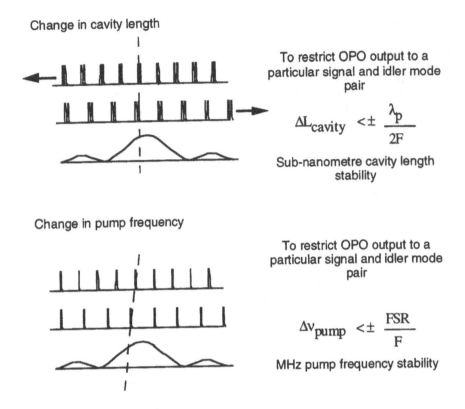

Figure 4. *The stability requirements for cavity length and pump frequency to maintain simultaneous resonance of the signal and idler modes.*

pump frequency, the output of a type I near-degenerate OPO will mode-hop between different signal and idler mode-pairs while remaining above threshold. Under these conditions, the stability requirements needed to maintain the output of the OPO to a single mode-pair are dictated by Equations 8 and 9. Figure 5 shows the experimentally observed rapid mode-hop behaviour of a type-I near-degenerate phase-matched OPO (Henderson *et al.* 1995) .

2.6 Cluster hopping in DRO OPOs

In addition to the mode-hop and stability criteria, the detuning required to produce a cluster-hop is also of interest. This is particularly the case if the OPO is operating significantly above threshold with a large Δ(FSR) or high cavity finesse for which simultaneous resonance of a signal and idler mode pair in the central cluster is not guaranteed.

For pump powers significantly above threshold, output on more than one cluster may be obtained. For cavity lengths lying between those corresponding to a mode-pair

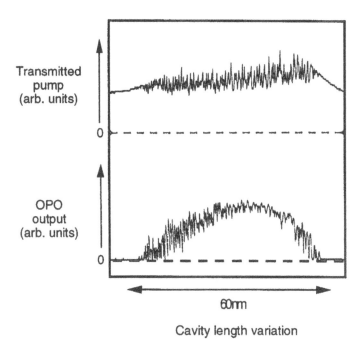

Figure 5. *Intensity of the transmitted pump and* OPO *outputs as a function of cavity length in a type-I phase matched, doubly resonant* OPO.

in the central cluster, the pump power may be sufficient for a mode-pair in an adjacent cluster to reach threshold. As previously, a well defined period can be observed which corresponds to the mode-hop interval. At intermediate cavity lengths oscillation is observed on adjacent clusters, many mode-pairs removed. Figure 6 shows the multi-cluster operation of a type-II phase-matched OPO pumped well above threshold.

The number of modes, N, between the centre of one cluster and the centre of the next is given simply by

$$N = \frac{(FSR)}{\Delta(FSR)} \tag{17}$$

Therefore, from Equation 7, we can obtain the cavity length $\Delta L_{\text{cluster}}$ or pump frequency $\Delta \nu_{\text{p-cluster}}$ detuning required to move from the centre of one cluster to the centre of the next,

$$\Delta L_{\text{cluster}} = \frac{\lambda_p}{2} \tag{18}$$

and

$$\Delta \nu_{\text{p-cluster}} = (FSR) \tag{19}$$

The position of the mode clusters with respect to optimum phase-matching depends on the specific cavity configuration. The cavity detuning required to move between cluster centres is given by Equation 18 and is equal to the interval between pump resonances in the OPO cavity. This is clearly shown in Figure 7, where the spacing

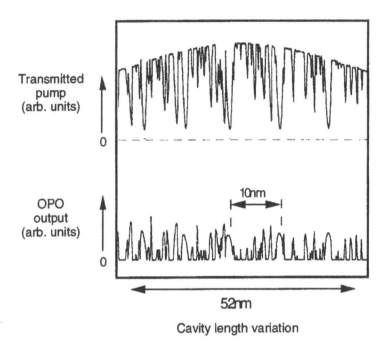

Figure 6. *Intensity of the transmitted pump and* OPO *outputs as a function of cavity length in a type-II phase-matched , doubly resonant* OPO *pumped well above threshold.*

between cluster centres (maximum OPO output power) is equal to the spacing between pump resonances (maximum transmitted pump power). Note that in the centre of the scan both clusters have similar, but higher thresholds, which corresponds to the clusters being located symmetrically either side of optimum phase-matching condition.

The previous equations have been derived ignoring the influence of dispersion. If the dispersion term is large, then it can alter significantly the calculation of $\Delta(FSR)$ which in turn influences the calculated cavity length or pump frequency detuning required to cause a hop to the adjacent mode-pair. As an alternative to the approximate analytical expressions, a computer model based on Equation 4 has been written to calculate the threshold pump power for all possible signal and idler mode-pairs under different combinations of the cavity length, the pump frequency, and the phase-matching temperature. However, in most cases, the numerically derived values differ from those values predicted by the above equation by less than 10%.

One solution to the demanding requirements on the stability of the cavity length is to use a monolithic design where the cavity is formed from the coated end-faces (or total internal reflection) of the nonlinear crystal itself. The inherently stable cavity length combined with a frequency stable diode-pumped, solid-state laser enables the output of the OPO to be restricted to a single signal and idler mode pair (Nabors *et al.* 1989). However, this requires the nonlinear material to be suitable for precision grinding, polishing and /or dielectric coating. In addition, the fixed cavity geometry and lack of

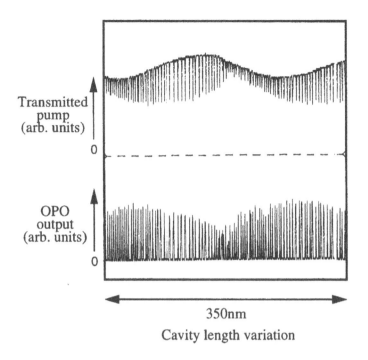

350nm
Cavity length variation

Figure 7. *As Figure 6 to different scale*

intra-cavity components restricts the monolithic OPO to simple configurations.

Recent work by Stevenson *et al.* (1995) has shown the use of index-matching fluid to join the nonlinear crystal to the cavity mirrors. This gives a pseudo-monolithic construction without requirement for cating the nonlinear crystal

2.7 Stability requirements in SRO OPOs

For an SRO, the frequency of the resonant wave corresponds to the cavity mode closest to optimum phase-matching. As has been demonstrated by a number of reported SRO configurations, no additional frequency selection is required to ensure single mode output(Robertson *et al.* 1994,Yang *et al.* 1993). In a true SRO, the required stability for cavity length to maintain single mode output is identical to that of a conventional laser, namely,

$$\Delta L_{\text{stab}} \approx \pm \frac{\lambda_{\text{res}}}{4}, \tag{20}$$

where λ_{res} is the wavelength of the resonant wave. The stability required for the pump laser is more difficult to calculate since it depends on the detailed behaviour of the phase-matching. Assuming that the phase-matching is such that the ratio between the frequencies of the signal and idler fields remains approximately constant for a small change in pump frequency, then the stability required of the pump frequency is

$$\Delta\nu_{\text{stab}} \approx \pm(FSR)\frac{\lambda_{\text{res}}}{\lambda_{\text{nonres}}}, \tag{21}$$

where λ_{nonres} is the wavelength of the non-resonant wave.

If pump enhancement is introduced then the stability requirements become more stringent. The stability requirements for the cavity length are those to maintain the pump on resonance, namely

$$\Delta L_{\text{stab}} \approx \pm\frac{\lambda_p}{4F_p}, \tag{22}$$

where F_p is the finesse of the cavity at the pump field. The requirements for the pump frequency stability are

$$\Delta\nu_{\text{stab}} \approx \pm\frac{(FSR)}{2F_p}. \tag{23}$$

In practise, for a pump enhanced SRO, the cavity length is servo-locked to the pump frequency, or the pump frequency servo-locked to the cavity, to hold the resonance condition. It is then sufficient for the non-locked parameter to satisfy the standard SRO stability conditions in order to maintain single mode output. Leaving aside the problems of increased threshold that are associated with SROs, it can be seen that single mode operation of such configurations is easier to obtain than for the corresponding DRO.

3   Continuous frequency tuning from CW OPOs

This section reviews and compares options for obtaining smooth frequency tuning from continuous wave optical parametric oscillators OPOs. Specifically we compare

- dual-cavity OPOs in which the signal and idler fields are resonant in separate cavities,

- single-cavity OPOs incorporating methods for controlling independently the resonance frequencies of the signal and idler fields,

- single-cavity OPOs pumped using tunable sources and single-cavity,

- singly-resonant OPOs with and without pump enhancement.

For each of the configurations we consider the requirements on cavity stability, pump frequency stability, pump power threshold, mirror/crystal coating requirements and methods for frequency control.

3.1   Threshold pump power for various cavity configurations

A key parameter when designing a CW OPO is the pump power required to reach threshold. The simplest configurations to analyse are those based on ring resonators where the travelling wave nature of the electric fields means the pump, signal and idler fields only interact once per round trip and consequently that the mirror phase shifts have no influence on the conversion efficiency. In the standing wave case, there are two interactions per round trip with a relative phase shift between them which depends on the phase shifts associated with the mirror reflections and other dispersive intra-cavity components. This determines the position of nodes and anti-nodes of the electric fields within the nonlinear crystal and hence the conversion efficiency of the parametric process. However, a small relative dephasing of the three waves caused by the mirror reflections or other intracavity dispersion can be compensated by a slight detuning of the phase-matching within the nonlinear material resulting in only a comparatively small increase in threshold (Debuisschert *et al.* 1993).

The threshold pump powers for various configurations of OPO are summarised as shown in Table 1 (assuming confocal focussing and no mirror dephasing) where

$$K = \frac{n_p^2 \epsilon_0 c^4}{2\pi^2 |d_{\text{eff}}|^2 (1 - \delta^2)^2 \nu_p^3} \tag{24}$$

and the degeneracy factor

$$\delta = \frac{2\nu_s - \nu_p}{\nu_p} \tag{25}$$

and all other symbols have their usual meaning.

Configuration	Threshold	Typical values
DRO with single pass pump	$K\pi^2/(F_sF_i)$	10's mW
DRO with pump enhancement	$K\pi^2/(4F_sF_iE)$	1's mW
SRO with single pass pump	$K\pi/F$	1's W
SRO with pump enhancement	$K\pi/(2FE)$	100's mW

Table 1. *The threshold pump powers for various configurations of* OPO. *E is the enhancement of the pump field due to the pump resonance.*

3.2 Single cavity OPOs with independent control of the resonant frequencies of the signal and idler fields

Until recently all CW OPOs have been configured as DROs with the signal and idler fields resonant in the same cavity. As discussed above, the requirement for simultaneous resonance for the signal and idler fields means that a doubly-resonant, single-cavity OPO is over constrained. When using a fixed frequency pump source and a single cavity to resonate both the signal and idler fields, continuous tuning over an appreciable range of the OPO is not possible. This is because to maintain energy conservation, $(\omega_i + \omega_s = \omega_p)$, the signal and idler frequencies are required to tune in opposite directions. A single cavity cannot be used to tune continuously the OPO while maintaining both the signal and idler fields on resonance. An additional intracavity component is required to control independently the resonance frequencies of the signal and idler modes. Within a type-II phase matched OPO the orthogonal polarisation states of the signal and idler fields

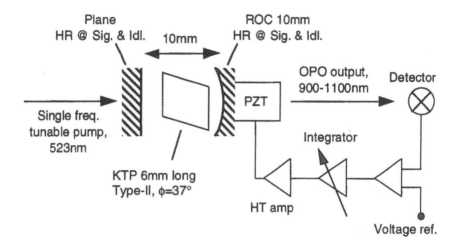

Figure 8. *Schematic layout of a single cavity* OPO *pumped with a tunable pump source.*

allows the use of a birefringent element to control independently the resonant frequencies of the two fields. One approach is to use the electro-optic coefficient of the nonlinear crystal itself to tune the resonant frequencies of the orthogonally polarised signal and idler fields. Using KTP as the nonlinear crystal, and a single-mode frequency-stabilised, krypton-ion laser at 530.9nm, simultaneous control of the cavity length and the electro-optic birefringence gave continuous and smooth tuning of the OPO output frequency of ~400MHz (Lee and Wong 1992). In practise, the tuning range is limited by the electro-optic tuning range of the crystal.

3.3   Single cavity OPOs pumped with a tunable pump source

When using a tunable source to pump a CW OPO such as a Ti:sapphire laser (Colville *et al.* 1994), the OPO signal and idler frequencies can be controlled by adjusting the frequency of the pump. The OPO can be configured as a single-cavity DRO with a servo-control on the cavity length to maintain the simultaneous resonance of the signal and idler fields. As the pump frequency is varied, the signal and idler frequencies both tune in the same sense with a tuning rate which depends on the degeneracy factor (Polzik *et al.* 1993). We have reported such a device using a frequency doubled Nd:YLF laser to pump a single-cavity, doubly-resonant, KTP OPO (Henderson *et al.* 1995) . Control of the crystal angle gives coarse tuning about frequency degeneracy from 900–1100nm. A simple cavity length servo maintains the double resonance condition and tuning of the pump frequency over ~10GHz gives tuning of the signal and idler frequencies of ~5GHz each. See Figure 8.

Single-cavity OPOs are easier to align than their dual-cavity (see below) equivalents and have no requirements for cavity dividing optical components.

As mentioned above, stable, single-frequency operation of the device was achieved using a 'side of fringe' locking technique. The total signal and idler power output

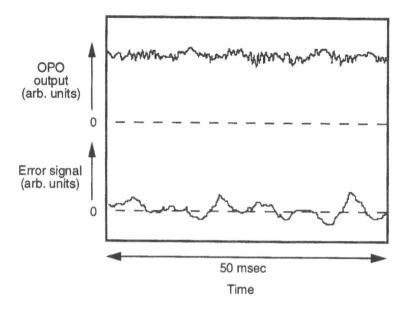

Figure 9. *The* OPO *output and the error signal as a function of time.*

was monitored on a standard photodiode, the output from which was compared to a voltage reference. The resulting difference voltage formed the error signal, which after integration was used to control the length of the OPO cavity to keep a constant output power, and hence maintain the double-resonance condition. Furthermore, the cavity length servo kept the OPO locked to the same signal and idler mode pair as the pump frequency was changed continuously. Thus, the smooth tuning of the pump laser was followed by the smooth tuning of the OPO. The particular mode pair to which the OPO was locked was selected by setting the initial OPO cavity length prior to activating the locking electronics. The OPO frequency could thus be tuned discretely through the mode pairs of each cluster. When locked, the output power of the OPO exhibited a modulation of around 20% at kHz frequencies, as a result of the limited bandwidth of the locking electronics. The OPO remained locked to the same mode pair for 20 minutes without mode-hopping. The spectral output was monitored on a confocal interferometer (1.5GHz free spectral range), while the laser output spectrum was observed using a plane-plane interferometer (7.5GHz free spectral range). In accordance with energy conservation and phase matching, when the laser was tuned over its maximum 9GHz range, the OPO signal and idler frequencies tuned continuously over approximately equal 4.5GHz ranges. Figure 9 shows the OPO output power while the system was locked, as the frequency of the pump laser was tuned.

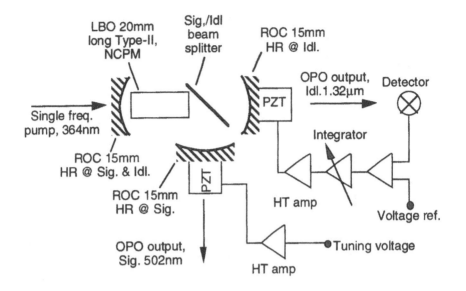

Figure 10. *Schematic layout of a dual cavity* OPO.

3.4   Signal and idler fields resonant in separate cavities

In a dual-cavity OPO, as shown in Figure 10, the signal and idler fields are resonated in separate cavities which share a common section at the nonlinear crystal (Wong 1992). The additional degree of freedom means that the OPO is no longer over-constrained and resonant frequencies of the signal and idler cavities can be tuned independently. The length of one of the cavities is servo controlled to maintain the double resonance condition and the OPO outputs can be tuned by varying the length of the other. Assuming the frequency of the pump remains fixed, the frequencies of the signal and idler fields tune by the same frequency interval in opposite directions. A critical component in this configuration is the cavity divider, which needs to have a low insertion loss for both signal and idler waves if the OPO is to maintain a low threshold pump power. When using a type-II NCPM geometry a polarisation beam splitter allows the low loss separation of the signal and idler fields. Previously we reported a dual-cavity oscillator using LBO as the nonlinear material (Colville *et al.* 1994) . The cavity divider was an enhanced Brewster plate polariser giving greater than 99.7% efficiency for the reflection and transmission of the signal and idler waves respectively. In principle, simultaneous control of each cavity and the phase matching condition allows smooth and continuous tuning over an arbitrarily large range. In practice, factors such as parasitic pump resonances have to be overcome if the tuning range is to be extended beyond about a few hundred MHz.

Figure 11 shows the experimentally observed tuning range obtained by scanning one of the cavities and relying on the servo control to adjust the length of the other to maintain the double resonance condition. The tuning range was limited by parasitic pump resonance.

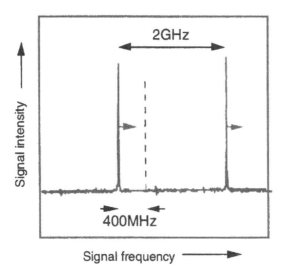

Figure 11. *The single mode pair output of a dual cavity* OPO, *showing the limit of smooth tuning.*

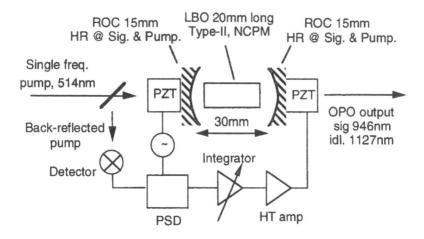

Figure 12. *Schematic layout of a singly resonant* OPO *with pump enhancement.*

3.5   Singly resonant OPOs with/without pump enhancement

The high sensitivity of CW OPOs to perturbations of pump frequency and cavity length arises for the requirement within a DRO for simultaneous resonance of the signal and idler fields. This is problem is removed by configuring the OPO as an SRO. In an SRO the operating frequency is simply set by the cavity mode of the resonant wave which lies closest to line centre with the frequency of the non-resonant wave being determined

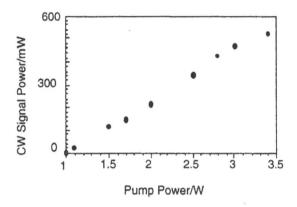

Figure 13. *Output power of the singly resonant* OPO *as a function of pump power*

by the conservation of energy. Scanning the cavity length changes the frequency of
the resonant wave without any complications arising from constraints on the frequency
of the non-resonant wave. Although a true CW SRO based on KTP has been reported
(Yang *et al.* 1993), for most nonlinear materials the increased threshold is too high for
existing pump sources. Recently, we reported a pump enhancement scheme where the
pump field is resonantly enhanced within the single OPO cavity (Robertson *et al.* 1994).

Using LBO as the nonlinear material, the argon ion pump source at 514.4nm was
resonantly enhanced to increase the pump field by thirty times to give SRO operation
with a threshold of 1 watt.

3.6   Comparison of techniques for smooth and continuous tuning of CW OPOs

OPO Configuration	Advantage	Disadvantage
Dual Cavity, fixed freq.pump	Fixed freq.pump	Coating requirement and complex alignment
Single Cavity with tunable pump	Simple cavity design	Needs tunable pump
SRO with pump enhancement	Ease of alignment and good capture range for lock	Limited tuning range in simple configuration

Table 2. *Comparison of techniques for smooth and continuous tuning of* CW OPOs.

4 Acknowledgements

This chapter comprises an overview of ideas and experimental results originating in the continuous-wave OPO group at St Andrews University and is based on the work of F G Colville, M H Dunn, A J Henderson, M J Padgett, G Robertson, T R Stevenson and J Zhang.

5 References

Chen C, Wu Y, You G, Li R, and Lin S, 1989, *J Opt Soc Am* **B6** 616.
Colville F G, Padgett M J, and Dunn M H, 1994a, in *Proc of Advanced Solid-State Lasers (Utah 1994), OSA Tech Proc.*
Colville F G, Padgett M J, and Dunn M H, 1994b, *Appl Phys Lett* **64** 1490.
Colville F G, Ebrahimzadeh M, Sibbett W, and Dunn M H, 1994, *Appl Phys Lett* **64** 1765.
Debuisschert T, Sizmann A, Giacobino E, and Fabre C, 1993, *J Opt Soc Am* **B10** 1668.
Donaldson W R, and Tang C L, 1984, *Appl Phys Lett* **44** 25.
Eckhardt R C, Nabors C D, Kozlovsky W J, and Byer R L, 1991, *J Opt Soc Am* **B8** 646.
Falk J, 1971, *J Quantum Electron* **QE7** 230.
Fan Y X, Eckhardt R C, Byer R L, Nolting J, and Wallenstein R, *Appl Phys Lett* **53** 2014.
Giordmaine J A, and Miller R C, 1965, *Phys Rev Lett* **14** 973.
Henderson A J, Padgett M J, Zhang J, Sibbett W, Dunn M H, 1995, *Optics Lett* **20** 1029.
Henderson A J, Padgett M J, Zhang J, Sibbett W, Dunn M H, 1995, *Optics Comm* **119** 256.
Lee D, and Wong N C, 1992, *Opt Lett* **17** 13.
Nabors C D, Eckhardt R C, Kozlovsky W J, and Byer R L, 1989, *Opt Lett* **14** 1134.
Padgett M J, Colville F G, and Dunn M H, 1994, *IEEE Quant Elec* **30** 2979.
Polzik E S, Mabuchi H, and Kimble H J, 1993, *CLEO, OSA Technical Digest Series (Washington D.C. 1994)* **11** 434
Robertson G, Padgett M J, and Dunn M H, 1994, *Opt Lett* **19** 1735
Smith R G, 1973, *IEEE J Quant Electron* **QE9** 530
Smith R G, Geusic J E, Levinstein H J, Rubin J J, Singh S, and Van Uitert V G, 1968, *Appl Phys Lett* **12** 308.
Stevenson T R, Colville F G, Dunn M H, and Padgett M J, *Opt Lett* **20** 722.
Wong N C, *Phys Rev A* **45** 3176.
Yang S T, Eckhardt R C, and Byer R L, 1993, *Opt Lett* **18** 971.
Yang S T, Eckhardt R C, and Byer R L, 1994, *Opt Lett* **19** 475.

Pulsed Parametric Oscillators

Majid Ebrahimzadeh

University of St Andrews, Scotland

1 Introduction

Since the invention of the laser, a major area of research activity in laser physics and photonics has been centred on the development of optical sources capable of providing coherent radiation in a continuously tunable manner. Traditional methods for the generation of such radiation are based on the stimulated emission between quantised energy levels in conventional atomic or molecular laser gain media with extended fluorescence bandwidths. In media where the quantised levels are sufficiently broadened to form continuous bands, tunable emission can result from transitions between such bands. This is the basic operation principle in tunable dye, colour-centre, or vibronic lasers such as the now well-established Ti:sapphire laser. While conventional tunable lasers of this type have proved highly effective in providing coherent radiation in different spectral regions, the tuning range available to many of these systems is often limited to, at best, a few hundred nanometers. Moreover, with the exception of colour-centres and some newly emerging vibronic lasers, the wavelength coverage of most traditional tunable sources is confined mainly to the visible spectrum.

The potential of nonlinear optical mixing techniques as an alternative means of providing laser radiation at new wavelengths was realised more than three decades ago. Following the demonstration of second harmonic generation of light in quartz, and hence the existence of substantial nonlinearities, many experiments in nonlinear optics were performed. These experiments were made possible due to the significant increase in the spectral intensity provided by the laser radiation. Soon after, the importance of the optical parametric process as a powerful tool for the generation of widely tunable

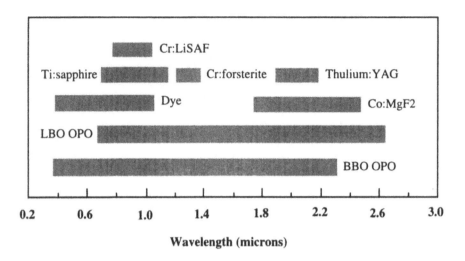

Figure 1. *The wavelength coverage of a number of well-established tunable sources and that of two* OPO *devices recently demonstrated with* BBO *and* LBO.

radiation was recognised. The first experimental demonstration of a tunable optical parametric oscillator (OPO) was reported by Giordmaine and Miller (1965). The main attraction of OPOs derives from their exceptionally broad tuning range (typically several hundred nanometers) and the ability to cover those wavelength regions not directly accessible to conventional tunable lasers. The OPO is also an all-solid-state device which can be made highly compact, is long-lived, efficient, and remarkably easy to use. In Figure 1 the tuning range of a number of widely established tunable sources is compared with that of two OPO devices based on the new nonlinear crystals of β-BaB$_2$O$_4$ (BBO) and LiB$_3$O$_5$ (LBO). The BBO OPO is continuously tunable from 350nm in the ultraviolet to 2.4μm in the infrared, while the LBO device can cover a wavelength range from 650nm to 2.6μm, both using a single nonlinear crystal. As a comparison, the coverage of the somewhat limited tuning range available to dye lasers shown in Figure 1 requires a total of nearly twenty different dyes, each with an effective tuning bandwidth of only 20–30nm.

2   Basic concept

In its simplest form, the OPO consists of an anisotropic nonlinear crystal enclosed within an optical resonator formed by a pair of mirrors, as in a conventional laser oscillator (Figure 2). When the crystal is irradiated with an intense optical *pump* field at frequency ω_3, the nonlinear polarisation gives rise to two new fields, the *signal* at ω_1 and the *idler* at ω_2, such that

$$\omega_3 = \omega_2 + \omega_1 \tag{1}$$

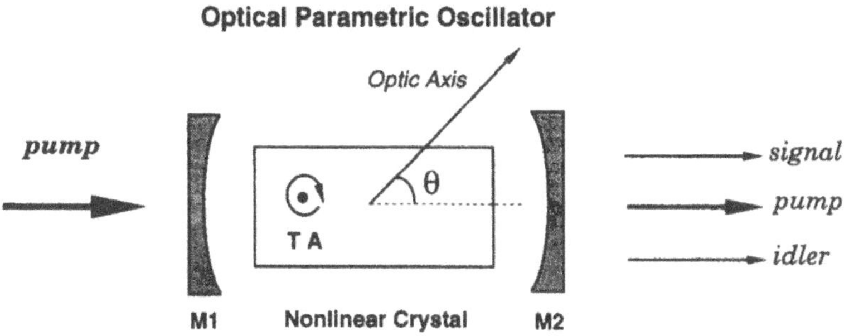

Optical Parametric Oscillator

Figure 2. *Schematic of an optical parametric oscillator* (OPO). *The device generates tunable signal and idler output for a fixed-frequency pump by varying the phase-match angle θ through rotation of the crystal about the tuning axis* (TA). *Tuning may also be achieved by varying the temperature or by applying an electric field to the crystal at a fixed phase-match angle.*

This simple relation, which can also be regarded as an energy conservation condition, implies that the total photon energy is preserved during the parametric generation process. The frequencies ω_1 and ω_2 cannot be uniquely determined by this condition alone, as there exist a continuous range of (ω_1, ω_2) pairs that can satisfy Equation 1 for a given pump frequency ω_3. Out of the infinite number of possible combinations (ω_1, ω_2) allowed by Equation 1, only those satisfying the condition

$$\Delta k = k_3 - k_2 - k_1 = 0 \qquad (2)$$

can be generated as macroscopic fields. This relation, which is known as the *phase-match* condition, may also be viewed as a momentum conservation condition. The generated signal and idler fields at ω_1 and ω_2 will undergo amplification as they propagate through the crystal if the phase-match condition (Equation 2) among the optical fields is maintained throughout the interaction length. In practice, this is often achieved by exploiting the birefringence of the anisotropic crystal to compensate for dispersion (see Section 4). Under this condition, the three optical fields interact constructively as they propagate through the medium and the generated waves at ω_1 and ω_2 experience sustained growth at the expense of the pump. Unlike conventional laser sources where optical gain is a consequence of population inversion, in parametric devices, gain is provided through the interaction of optical fields arising from the nonlinear polarisation exhibited by the medium.

The phase-match condition (Equation 2), which can be expressed in an alternate form as

$$n_3\omega_3 - n_2\omega_2 - n_1\omega_1 = 0 \qquad (3)$$

is also the basic tuning mechanism for the OPO. For a given pump frequency ω_3 and propagation direction k_3, tuning can be accomplished by altering the phase-match condition (Equation 2), so that Equation 1 is satisfied for a new pair of frequencies (ω_1', ω_2'), say. This is, in practice, achieved by altering the refractive indices of the medium along the direction of propagation through rotation of the crystal (*angle-tuning*), changing its temperature (*temperature-tuning*), or by applying an electric field to the material (*electro-optic tuning*). In this way, continuously tunable radiation over extended spectral regions can be conveniently generated from a fixed-frequency input pump field using a single gain medium.

3   Theory of parametric interactions

When a transparent dielectric material is subjected to electromagnetic radiation, the force associated with the incident electric field produces a displacement of electrons in the medium with respect to the nuclei. In the case where the electric field associated with the radiation is small, the induced polarisation of the medium (the electric dipole moment per unit volume) is linearly proportional to the applied field, according to

$$\mathbf{P} = \varepsilon_0 \chi^{(1)} \mathbf{E} \tag{4}$$

where $\mathbf{P}$ and $\mathbf{E}$ are the polarisation and electric field vectors, respectively, $\chi^{(1)}$ is the *linear susceptibility* of the medium, and ε_0 is the permittivity of free space. This is the regime of *linear* optics. In general, $\chi^{(1)}$ is dependent on the frequency, ω, of the optical radiation.

The linear dependence of the induced polarisation on the electric field strength is, in fact, only an approximation and is valid only for small fields. In the case where the optical electric field strength is comparable to the intra-atomic electric field, as with the laser radiation, the polarisation response of the medium becomes nonlinear in the applied field. In this case, the relation between $\mathbf{P}$ and $\mathbf{E}$ can be expressed as a power series expansion. In a simplified scalar form this can be written

$$P = \varepsilon_0 \left[\chi^{(1)} E + \chi^{(2)} E^2 + \chi^{(3)} E^3 + \ldots \right] \tag{5}$$

where $\chi^{(2)}$, $\chi^{(3)}$,... are the nonlinear susceptibilities of the medium. This is the regime of *nonlinear* optics. In general, $\mathbf{P}$ and $\mathbf{E}$ are not parallel and are connected through tensor relations. The susceptibilities $\chi^{(1)}$, $\chi^{(2)}$, $\chi^{(3)}$,... are tensors of the second, third, fourth, and higher ranks, respectively. The magnitude of the tensor components rapidly decreases with increasing rank of nonlinearity ($\chi^{(1)} : \chi^{(2)} : \chi^{(3)} \sim 1 : 10^{-8} : 10^{-16}$). It is thus not surprising that the exploitation of most nonlinear optical effects was not possible prior to the invention of the laser.

Almost all the linear and nonlinear optical effects of practical importance are described by the first three terms in Equation 5. The linear term involving $\chi^{(1)}$ gives rise to the index of refraction, absorption, dispersion, and birefringence of the medium. Most of the interesting nonlinear optical effects, however, arise from the terms of electric polarisation which are quadratic or cubic in the electric field. The quadratic polarisation gives rise to the phenomena of second harmonic generation, sum- and difference-frequency mixing, and parametric generation, while the cubic term is responsible for third harmonic generation, the optical Kerr effect, self-focusing, self-phase-modulation, and phase conjugation. In the discussion of parametric generation, we are concerned only with the quadratic nonlinear term involving $\chi^{(2)}$. This quantity is usually referred to as the *second-order nonlinear susceptibility* and is non-zero only in non-centrosymmetric media. In media which possess inversion symmetry in their crystalline structure, $\chi^{(2)}$ reduces to zero. In this Section, we discuss the underlying physical principles responsible for second-order nonlinear optical processes and derive the fundamental equations governing parametric generation in anisotropic, nonlinear media. We shall adopt MKS units throughout, as most of the later literature on the subject use this system of units.

3.1 Wave propagation in nonlinear media

In order to derive the equations governing parametric interactions, we consider the propagation of electromagnetic waves within an anisotropic nonlinear medium. We start by writing down Maxwell's wave equation for a non-absorbing, non-conducting dielectric medium containing no free charge, that is

$$\nabla^2 E = \mu \varepsilon_0 \frac{\partial^2 E}{\partial t^2} + \mu \frac{\partial^2 P}{\partial t^2} \tag{6}$$

The formalism may now be developed further by separating the total polarisation of the medium, P, into two components; a linear component given, in scalar notation, by

$$P_L = \varepsilon_0 \chi^{(1)} E \tag{7}$$

and a nonlinear component arising from the higher-order susceptibilities in Equation 5, namely

$$P_{NL} = \varepsilon_0 \left[\chi^{(2)} E^2 + \chi^{(3)} E^3 + \ldots \right] \tag{8}$$

As we are interested in the nonlinear polarisation term, we shall be seeking solutions to Maxwell's equation (6), with P_{NL} as the source term. Therefore, let us re-write the wave equation in terms of P_{NL} by substituting for P in Equation 6, using Equations 7 and 8. Assuming that $\chi^{(1)}$ is time-independent (*i.e.* no dispersion), we obtain

$$\nabla^2 E = \mu \varepsilon \frac{\partial^2 E}{\partial t^2} + \mu \frac{\partial^2 P_{NL}}{\partial t^2} \tag{9}$$

where $\varepsilon = \varepsilon_0 (1 + \chi^{(1)})$. This is the wave equation describing the propagation of electromagnetic fields in nonlinear media. Since we are interested in second-order nonlinear effects, we can ignore higher-order terms than $\chi^{(2)}$ in Equation 8, and write P_{NL} in the reduced form as

$$P^{(2)} \sim \varepsilon_0 \chi^{(2)} E^2 \tag{10}$$

Further, to simplify the treatment, we will assume that the electromagnetic waves are propagating collinearly in the z-direction. Substituting for P_{NL} in Equation 9 using the last result, we obtain the wave equation for second-order nonlinear processes, namely

$$\frac{\partial^2 E}{\partial z^2} = \mu \varepsilon \frac{\partial^2 E}{\partial t^2} + \mu \frac{\partial^2}{\partial t^2} \left(\varepsilon_0 \chi^{(2)} E^2 \right) \tag{11}$$

3.2   Coupled wave equations

In the case of second-order nonlinear processes including parametric generation, we are usually concerned with the interaction of three harmonic waves at frequencies ω_1, ω_2 and ω_3, with $\omega_3 = \omega_2 + \omega_1$. To simplify the analysis, we assume that the three fields are infinite uniform plane waves and ignore the effects of double-refraction and focusing. Although this is rarely the case in practice, where anisotropic media and focused Gaussian beams are involved, the simplified analysis demonstrates the essential features of the parametric generation process. We will, therefore, restrict consideration to a total field $\mathbf{E}$ consisting of three infinite uniform plane waves at frequencies ω_1, ω_2 and ω_3. The total instantaneous field is, therefore, of the form

$$\mathbf{E}(z,t) = \mathbf{E}_1(z,t) + \mathbf{E}_2(z,t) + \mathbf{E}_3(z,t), \tag{12}$$

with

$$\mathbf{E}_m(z,t) = \frac{1}{2}\left[\mathbf{E}_m(z)\,e^{i(k_m z - \omega_m t)} + c.c.\right] \tag{13}$$

In writing Equation 13, we have used the complex function description of the optical field in which the real electric field $\mathbf{E}_m(z,t)$ (which is the quantity involved in the wave equation) is expressed in terms of the complex amplitude $\mathbf{E}_m(z)$, which itself is given by

$$\mathbf{E}_m(z) = |\mathbf{E}_m(z)|\,e^{i\phi_m} \qquad m=1,2,3 \tag{14}$$

with $|\mathbf{E}_m(z)|$ as the real field amplitude. Also, since the waves at the three different frequencies are interchanging energy through $\chi^{(2)}$ as they propagate through the medium, the respective amplitudes will, in general, vary with position. Hence, the complex field amplitude in Equation 13 is expressed as functions of z.

We proceed by substituting Equation 12, using Equation 13, into the wave equation (11) and then separating the resulting equation into three components at the three different frequencies, which separately must satisfy the equation. Assuming that the complex field amplitudes $\mathbf{E}_m(z)$ are all time-independent and the field amplitudes vary only slowly over distances compared to a wavelength, the solution of the wave equation (11) can be shown to be, in tensor notation

$$\frac{d\mathbf{E}_1(z)}{dz} \sim -\frac{i\omega_1}{2}\left(\frac{\mu}{\varepsilon_1}\right)^{1/2}\chi^{(2)}_{(\omega_1=\omega_3-\omega_2)}{:}\mathbf{E}_3(z)\,\mathbf{E}_2^*(z)\,e^{i\Delta k z} \tag{15}$$

We can also derive similar equations for each of the other field components at ω_2 and ω_3. For convenience, we write down these equations in an explicit form, as

$$\frac{d\mathbf{E}_1(z)}{dz} \sim -\frac{i\omega_1}{2}\left(\frac{\mu}{\varepsilon_1}\right)^{1/2}\chi^{(2)}_{(\omega_1=\omega_3-\omega_2)}{:}\mathbf{E}_3(z)\mathbf{E}_2^*(z)\,e^{i\Delta k z} \tag{16}$$

$$\frac{d\mathbf{E}_2(z)}{dz} \sim -\frac{i\omega_2}{2}\left(\frac{\mu}{\varepsilon_2}\right)^{1/2}\chi^{(2)}_{(\omega_2=\omega_3-\omega_1)}{:}\mathbf{E}_3(z)\,\mathbf{E}_1^*(z)\,e^{i\Delta k z} \tag{17}$$

$$\frac{d\mathbf{E}_3(z)}{dz} \sim -\frac{i\omega_3}{2}\left(\frac{\mu}{\varepsilon_3}\right)^{1/2}\chi^{(2)}_{(\omega_3=\omega_2+\omega_1)}{:}\mathbf{E}_2(z)\,\mathbf{E}_1(z)\,e^{-i\Delta k z} \tag{18}$$

where $\Delta k = k_3 - k_2 - k_1$ is the phase-mismatch between the three optical fields. These equations, which are commonly referred to as *coupled wave equations*, are the starting points in the analysis of a wide range of nonlinear optical effects involving the second-order susceptibility $\chi^{(2)}$, including parametric generation. They are the fundamental equations describing the propagation of optical fields in three-wave mixing processes, where the frequencies of the fields satisfy Equation 1. It is important to note from Equations 16 to 18 that the electric field amplitudes are coupled to one another through the second-order nonlinear susceptibility $\chi^{(2)}$. This coupling is the mechanism responsible for the interaction of optical fields in nonlinear media and is the basis of many of the observed nonlinear optical phenomena including second harmonic generation, frequency mixing, and parametric generation.

Physically, the three-wave mixing process can be viewed in terms of the distortion in the electron charge clouds in the medium, driven by the optical fields at ω_3, ω_2 and ω_1. These distortions lead to an oscillating polarisation at any combination of the frequencies, which in turn radiate optical waves at the corresponding frequencies. The direction of power flow between the interacting fields depends on their relative phase $\phi = \phi_3 - \phi_2 - \phi_1$ as well as the initial amplitude of the optical fields at the input to the medium. It is generally argued that the relative phase ϕ adjusts itself to an optimum value that ensures power flow from the strong field to the weaker fields. The optimum relative phase depends on the strength of the nonlinear interaction as well as the magnitude of Δk. In the parametric generation process, the optimum relative phase is given by $\phi = \phi_3 - \phi_2 - \phi_1 = -\pi/2$ in the case where $\Delta k = 0$ and the pump field at ω_1 is strongly depleted. In the case where the pump is not significantly depleted, the more general solution of ϕ has been given by Smith (1970). Under the condition of optimum relative phase, the generated fields at ω_1 and ω_2 experience maximum growth, while the power in the strong input pump field at ω_3 is gradually drained with propagation through the medium.

3.3 Parametric amplification and gain

The coupled wave equations (16–18) can be used to obtain the equations governing parametric amplification and gain in a nonlinear medium. The exact solutions to these equations are involved, but in a limiting case that demonstrates the essential features of parametric amplification, it is assumed that the pump wave maintains constant amplitude throughout the interaction region. This is equivalent to assuming that the power drained from the pump at ω_3, by the generated signal and idler fields, is small compared to the input pump power (*i.e.* small pump depletion). Under this approximation, Equation 16 and the conjugate of Equation 17 can be solved as a pair of coupled linear differential equations. It can then be shown, in the limit of small gain, that the generated waves at ω_1 and ω_2 undergo amplification according to (Harris 1969)

$$G_\ell(\omega_1, \omega_2) = \frac{I_\ell(\omega_1, \omega_2)}{I_0(\omega_1, \omega_2)} - 1 \cong \frac{\omega_1 \omega_2 \mu_0 \left(\chi^{(2)}\right)^2 \ell^2}{2n_1 n_2 n_3 c} I_0(\omega_3) \left[\frac{\sin^2(\Delta k \ell/2)}{(\Delta k \ell/2)^2}\right] \quad (19)$$

where G_ℓ represents the net single-pass power gain at ω_1 and ω_2 after propagation through an interaction length ℓ. The quantity $I = \frac{1}{2}nc\varepsilon_0|E(z)|^2$ is the intensity (in W/m^2) and I_0 refers to the intensity at the input to the nonlinear medium. It is seen from Equation 19 that gain at ω_1 and ω_2 is proportional to the pumping intensity at ω_3, the square of the nonlinear susceptibility and interaction length, as well as the frequencies and refractive indices of the coupled fields. Moreover, the parametric gain exhibits an oscillatory dependence on the quantity $\Delta k \ell / 2$, in the form of a sinc^2 function. Gain is at its maximum when $\Delta k \ell = 0$ and drops to zero at $\Delta k \ell / 2 \cong \pm \pi$. Thus, in analogy with a conventional laser, Equation 19 describes the gain lineshape of the parametric amplifier with a line centre at $\Delta k = 0$ and a full-width half-maximum (FWHM) gain bandwidth defined by $|\Delta k \ell| \cong 2\pi$.

4 Phase-matching

From Equation 19, it is clear that for the attainment of maximum gain in a parametric amplifier, we require that $\Delta k \ell = 0$. Since the interaction length ℓ has to be finite, this condition implies that Δk must be set to zero. However, in practice, this is often not possible because material dispersion inhibits the maintenance of phase unison among the coupled fields over the interaction length. This leads to an increase in Δk, with the result that the interacting fields gradually become out of step and lose their relative optimum phase with propagation through the medium. Material dispersion sets an upper limit to the maximum interaction length useful in the parametric generation process. This is called the *coherence length*, ℓ_c, which is defined as the interaction length over which parametric gain is reduced by half of its peak value at $\Delta k = 0$, due to dispersion. It is given by $\ell_c = \pi / \Delta k$, and for parametric generation it can be shown to be $\ell_c = 1/[2(n_3/\lambda_3 - n_2/\lambda_2 - n_1/\lambda_1)]$, with λ as the free-space wavelength. As an example, if we consider parametric generation in a crystal of BBO pumped with a frequency-tripled Nd:YAG laser at $\lambda_3 = 355$nm, with generated waves at $\lambda_2 = 532$nm and $\lambda_1 = 1064$nm, we obtain $\ell_c \sim 5\mu$m! This means that the maximum useful crystal length is limited to only $\sim 5\mu$m, resulting in trivial gain. Therefore, we may conclude that in the presence of dispersion, the magnitude of nonlinear gain will be too negligible for any meaningful growth of the generated waves.

The practical technique for satisfying the phase-match condition, $\Delta k = 0$, is to exploit the birefringence of optically anisotropic media to offset dispersion. In general, anisotropic nonlinear media in which second-order nonlinear processes can occur, exhibit the phenomenon of birefringence. In such media, the index of refraction (and hence the magnitude of the k-vector) of a wave travelling through the medium depends not only on its frequency, but also on its direction of propagation as well as its state of polarisation. Therefore, by judicious choice of polarisation states and crystal orientation, it is possible to maintain phase-matching at every point along the direction of propagation for the particular combination of frequencies (ω_3, ω_2, ω_1). Under this condition, ℓ_c becomes infinite and the maximum useful interaction length becomes limited only by the available crystal length.

The particular choice of polarisation directions for phase-matching depends on the optical characteristics of the crystal and the sign of its birefringence. The full description of crystal categories and types of phase-matching is beyond the scope of this treatment and can be found elsewhere (for example, Dmitriev *et al.* 1991). For the present discussion it is sufficient to note that, with regard to their optical properties, birefringent crystals may be classified into either *uniaxial* or *biaxial* categories. The optical characteristics of such crystals are often described in terms of a three-dimensional index surface called the *optical indicatrix*. The principal axes of the surface are defined by three principal refractive indices n_x, n_y, and n_z. In uniaxial crystals $n_x = n_y \neq n_z$, so that the optical indicatrix is a spheroid (or an ellipsoid of revolution about the z-axis). In this case, it is customary to define the principal axes of the indicatrix in terms of the *ordinary* and *extraordinary* indices n_o and n_e, instead of n_x (or n_y) and n_z. In biaxial crystals with n_x, n_y, n_z all different the optical indicatrix is a general ellipsoid with three unequal axes. In uniaxial crystals, there exists a unique direction along which all allowed polarisation states experience the same refractive index. In biaxial crystals, two such directions may be identified. These directions define the *optic* axes of the crystal. In uniaxial crystals, the optic axis coincides with the principal axis, z, of the indicatrix, while in biaxial crystals the two optic axes generally lie in different directions from the principal axes of the indicatrix. In both crystal classes, it is in principle possible to achieve phase-matching by appropriate choice of polarisations and propagation direction. In general, there are two distinct types of phase-matching, referred to as *type I* (or parallel) and *type II* (or orthogonal) phase-matching. In type I phase-matching, the polarisation vectors of the generated waves at ω_1 and ω_2 are parallel (and orthogonal to that of the pump at ω_3), whereas in the type II process the polarisation vectors at ω_1 and ω_2 are perpendicular (with the pump polarisation parallel to one of the generated waves). Moreover, when the propagation direction is along one of the principal axes of the indicatrix, the phase-matching is termed *noncritical*, while for any other direction it is referred to as *critical* phase-matching.

Whether a material can be phase-matched, in either the type I or type II scheme, depends on the values of its refractive indices, the strength of its birefringence and the particular combination of frequencies involved. The general procedures for determining the locus of phase-matching directions for three-wave interactions involving ω_3, ω_2, and ω_1 have been outlined by Dmitriev *et al.* (1991) for uniaxial and biaxial crystals. While in uniaxial crystals the computation of phase-matching loci is relatively straightforward, in biaxial crystals it is often more convenient to restrict consideration to a plane formed by the principal axes of the indicatrix, because of the mathematical complexities involved. The determination of angular locus over which phase-matching can occur is useful for processes such as harmonic generation and sum or difference frequency mixing. For parametric generation, it is more important to obtain the phase-matching curves and tuning behaviour of the signal and idler at ω_1 and ω_2 as a function of the phase-match angle, for a fixed pump frequency ω_3. The described phase matching method is known as *birefringent phase-matching* and is the most commonly employed technique in parametric devices. Other methods of phase-matching include quasi-phase-matching in periodic structures, total internal reflection, and non-collinear interaction.

4.1 Effects of phase-mismatch

In the preceding discussion, we established how the use of birefringent phase-matching can overcome dispersion, thus leading to zero phase-mismatch ($\Delta k = 0$), infinite coherence length, and maximum gain in the parametric amplification process. In practice, however, the attainment of perfect phase-matching is not possible because ofthe finite frequency spectrum and spatial divergence of the light beams involved in the three-wave interaction. As a result, all the frequency and angular components of the interacting waves cannot be brought into phase-matching simultaneously and Δk increases. This departure from the ideal phase-match condition means that despite compensation for dispersion, the maximum useful interaction length is no longer infinite, but some finite value depending on the degree of phase-mismatch. The increase in the magnitude of Δk arising from the finite bandwidth of the coupled waves is often measured in terms of the quantity $\partial(\Delta k)/\partial \omega$, which can be calculated from the refractive index data for the material. This is then used to obtain the *spectral acceptance bandwidth* of the crystal by confining the mismatch to within the half-maximum points of the parametric gain curve, namely $|\Delta k \ell| \cong 2\pi$. The sensitivity of phase matching to the angular deviations can also be determined in a similar way by evaluating the quantity $\partial(\Delta k)/\partial \theta$. This, together with the condition $|\Delta k \ell| \cong 2\pi$, defines the *angular acceptance bandwidth* of the crystal in the θ-direction. The acceptance bandwidths of the crystal are a measure of maximum spectral and angular deviations that can be tolerated for a given crystal length, before parametric gain is reduced to one half of its peak value at $\Delta k=0$. Alternatively, they can be used to determine the maximum useful interaction length, given the spectral linewidth and divergence of the beams involved in the three-wave interaction.

To maximise gain, it is generally desirable to use phase-matching geometries and nonlinear materials that yield large acceptance bandwidths. In this context, noncritical phase-matching (NCPM) configurations are particularly useful because the small sensitivity of the refractive indices to angular variations in this geometry ensures large angular acceptance bandwidths. In parametric devices, the acceptance bandwidths are commonly calculated relative to the parameters of the pump beam. In order to maximise gain, it is important to use near-diffraction-limited pump beams with low divergence and narrow spectral bandwidth. In pulsed nanosecond and picosecond parametric devices, the linewidth of the pump pulses is sufficiently narrow to maintain $\Delta k \approx 0$ for practical crystal lengths of tens of millimetres. In femtosecond devices, on the other hand, the maximum useful interaction lengths are limited to only a few millimetres because of the large bandwidths associated with femtosecond pulses. It should also be noted that phase-matching can also be sensitive to the variations in the refractive indices through changes in the crystal temperature. This sensitivity can be similarly determined from the expression $\partial(\Delta k)/\partial T$, which defines the *temperature acceptance bandwidth* for the particular phase-matching geometry.

5   Walk-off effects

In addition to phase-mismatch, spatial and temporal walk-off play an important role in the operation of parametric devices. Returning to Equation 19, it is clear that the magnitude of gain in a parametric amplifier is also strongly dependent on the interaction length through ℓ^2. Walk-off effects can severely limit the maximum useful interaction length, even in the presence of perfect phase-matching, thus resulting in serious degradation in the nonlinear gain. Such effects are a consequence of the reduction in the spatial or temporal overlap of the interacting beams with propagation through the nonlinear medium.

Spatial walk-off arises from crystal double-refraction. In the treatment of three-wave nonlinear interactions above, we have assumed that the propagation direction is the same as the direction of the wave-normal or the k-vector. In birefringent anisotropic media, however, the direction of energy flow or the ray direction is generally different from the wave-normal direction. This means that in a collinearly phase-matched interaction, although the k-vector of the interacting waves are in the same direction, the corresponding rays walk off from one another with propagation through the medium. This results in a reduction in the interchange of energy between the pump and the generated fields with propagation through the medium. The nonlinear interaction can, therefore, become ineffective after the waves have travelled a finite distance through the medium. This distance, which sets an upper limit to the maximum useful crystal length in the presence of spatial walk-off, is known as the *aperture length* and is given by

$$\ell_a = \frac{\sqrt{\pi}\,\omega_0}{\rho} \tag{20}$$

where ω_0 is the input beam radius and ρ is the double-refraction angle. If, for example, we take BBO with $\rho \approx 5°$ and a beam waist radius of 100μm, we obtain $\ell_a \approx 2$mm. Therefore, regardless of phase-matching, the effective interaction length is limited to only 2mm due to spatial walk-off. It is clear from Equation 20 that the effects of beam walk-off can be minimised by using large beam waists and crystals with small double-refraction. However, since double-refraction is a consequence of birefringence, the latter approach is not always desirable as it can constrain phase-matching. Spatial walk-off effects can be eliminated by employing NCPM geometries where the beams propagate along one of the principal index axes of the crystal. Since in these directions the crystal does not exhibit double-refraction, the wave-normal and ray directions become coincident and ℓ_a becomes infinite. The strength of the nonlinear interaction is thus maximised and tightly focused beams and long interaction lengths can be used to achieve high nonlinear gains. Therefore, the use of NCPM geometry is not only advantageous due to large angular acceptance bandwidths, but also because it avoids the deleterious effects of spatial walk-off. Another technique for minimisation of beam walk-off is the use of cylindrical focusing in the plane of walk-off (Kuizenga 1972). As illustrated in Figure 3, this technique can substantially increase the interaction length beyond that available under spherical focusing, resulting in substantial increase in parametric gain. Other strategies to combat spatial walk-off include optimal mode-matching (Boyd and Kleinman 1968), two-crystal walk-off compensated geometries (Bosenberg *et al.* 1989b), and non-collinear interaction (Wachman *et al.* 1991).

As well as spatial walk-off, temporal walk-off effects also have a strong influence on three-wave interactions and can diminish nonlinear gain by limiting the useful interaction length. Such effects are of particular relevance in the ultrashort pulse regime where optical pulses with femtosecond and picosecond temporal durations are involved. When considering the propagation of such pulses in dispersive media, it is more appropriate to describe the interaction in terms of the group velocities because this is the speed at which the pulse envelope, and hence the pulse energy, travels through the medium. The phase velocity, on the other hand, defines the speed at which the different spectral components contained within the pulse propagate in the medium. Temporal walk-off arises from the mismatch in the group velocity of the interacting pulses, with the ef-

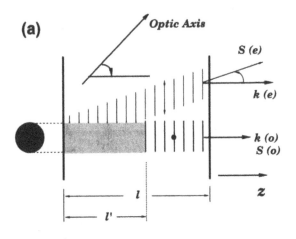

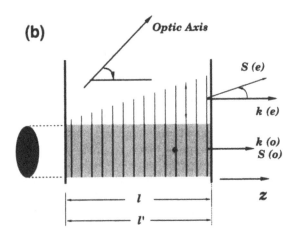

Figure 3. *Walk-off minimisation with elliptical focusing. In (a) the pump beam is focused to a circular waist, while in (b) the pump is focused to an elliptical waist in the plane containing the k-vector and the optic axis. The shaded area represents the region of spatial overlap between the ordinary (o) and extraordinary (e) beam, and k and S refer to the wave-normal and ray directions, respectively. In (a) the effective interaction length is limited by spatial walk-off, so that $\ell' < \ell$. In (b) the interaction length is increased to the full crystal length, so that $\ell' \approx \ell$. The nonlinear gain is thus substantially increased in going from (a) to (b), as long as the power density remains unchanged.*

fect that the nonlinear coupling and energy transfer between the pump and parametric pulses is reduced with propagation through the crystal. Therefore, despite matching of the phase velocities, the efficiency of nonlinear interaction can be severely degraded by the mismatch in group velocities. Temporal walk-off can be quantified in terms of the so-called *pulse-splitting length* which, for two collinearly propagating pulses, is defined as the distance after which they separate by a path equal to their pulse width, and is given by

$$\ell_{12}^g = \Delta\tau \left(\frac{1}{v_1^g} - \frac{1}{v_2^g} \right)^{-1} \tag{21}$$

where $\Delta\tau$ is the full-width at half-maximum pulse width and $v^g = \partial\omega/\partial k$ is the group velocity. From the equation it is clear that the effects of temporal walk-off are more pronounced with shorter pulse durations and larger mismatch in the group velocities. As an example, if we consider parametric generation of ultrashort pulses in KTiOPO$_4$ (KTP) under type II phase-matching, we obtain splitting lengths of typically 2mm for 100fs pump pulses, while for 1ps pulses longer splitting distances of up to 20mm are available (Nebel *et al.* 1995). On the other hand, if we take a material with a small group velocity mismatch, such as type I phase-matched LBO, the pulse splitting lengths increase to 5mm for 100fs pulses and up to 50mm for 1ps pulses (Ebrahimzadeh *et al.* 1995). For optical pulses longer than 1ps, the effects of group velocity walk-off can generally be ignored. We may, therefore, conclude that as well as spectral acceptance bandwidth requirements, temporal walk-off places an additional constraint on the maximum useful crystal length in ultrashort-pulse parametric devices. The effects of temporal walk-off can be minimised by using materials with small group velocity mismatch or by employing noncollinear geometries (Di Trapani *et al.* 1995). It should be noted that group velocity walk-off can also affect the temporal and spectral pulse characteristics in three-wave parametric interactions through pulse broadening and compression or by modifying the chirp content of the pulses. The exact influence of temporal walk-off in this case depends on material dispersion, input pump pulse characteristics, and the wavelengths involved in the interaction.

6 Material requirements

The choice of the nonlinear crystal is clearly the vital ingredient in the design of parametric devices. As discussed in Section 3, the basic requirement for a material to be suitable for second-order nonlinear mixing processes is that it should be non-centrosymmetric and anisotropic. In the discussion of phase-matching and walk-off effects, we also established that for efficient parametric generation it is desirable to use materials with large acceptance bandwidths, small double-refraction, and low group velocity walk-off in cases where ultrashort pulses are involved. We also highlighted the merits of NCPM geometries with regard to large angular acceptance bandwidths and zero spatial walk-off. In practice, however, one of the most important considerations in the choice of a suitable nonlinear crystal is its optical damage tolerance. If we return to Equation 19, we see that parametric gain is also linearly dependent on the input pump intensity, $I_o(\omega_p)$. The higher the input intensity, the larger the nonlinear gain and hence the stronger the parametric interaction. In general, attainment of operation threshold in parametric devices requires high pumping intensities of typically tens of MW/cm^2 (and often many times higher), because of the small magnitude of the second-order nonlinear polarisation, the origin of parametric gain. This means that the nonlinear crystal must be able to withstand power densities of at least 5–10 times these values for efficient conversion. If the material damage tolerance is too low, parametric generation will be suppressed by the onset of damage and threshold will not be reached, regardless of all other design criteria being satisfied.

Optical damage may be caused by several different mechanisms and can be of various types. It can be caused by thermal heating, induced absorption, self-focusing, stimulated Brillouin scattering, dielectric breakdown, or several other mechanisms. Accurate determination of the optical damage threshold for a given material is often difficult, as it depends on the wavelength of the radiation, pulse duration, crystal quality, and the exact experimental conditions. However, optical damage is generally more severe under pulse pumping than in continuous-wave (CW) operation, because of the high energy fluences present under pulsed conditions. In addition, in most materials damage tends to increase with the increase in the pulse duration from the femtosecond and picosecond to the nanosecond temporal regime. This is again, in the main part, due to the increase in the energy fluence associated with longer optical pulses. As a general trend, material damage also appears to increase with shorter pumping wavelengths and higher photon energies. Some of the new nonlinear crystals such as BBO and LBO exhibit damage thresholds as high as 10–20GW/cm^2 under pulse pumping (Dmitriev *et al.* 1991). In comparison, the corresponding damage tolerance of most classical materials such as LiNbO$_3$ can be as much as two orders of magnitude lower at less than 100MW/cm^2 under the same pumping conditions. For efficient parametric conversion, it is imperative to use materials with high optical damage thresholds, particularly when nanosecond pulses are involved. The optical quality of the crystal and its surface preparation and cleanliness are also vitally important for minimisation of damage. In particular, the material must be free from optical inhomogeneities and impurities, and exhibit low absorption and scattering loss at the wavelengths involved.

It is also clear from Equation 19 that parametric gain has a square dependence on the magnitude of the second-order nonlinear susceptibility through $(\chi^{(2)})^2$. Therefore, it is important that the nonlinearity of the medium is sufficiently high to allow efficient parametric conversion. In its most general form, the nonlinear susceptibility is a tensor with 27 components. However, the tensor is subject to certain symmetry conditions which reduce the number of independent elements to a maximum of 10 and also allow a (3×6) matrix representation of the tensor. In contracted notation, the elements of this matrix are represented by $\chi_{im}^{(2)}$, with i and m representing the rows and columns of the matrix. In the treatment of second-order nonlinear effects, however, it is also customary to refer to the elements of the piezoelectric d-tensor rather than the susceptibility tensor. This is possible because the two tensors are of the same form, with the respective components related through $\chi_{im}^{(2)} = 2d_{im}^{(2)}$. The components d_{im} are known as the *nonlinear coefficients* of the medium and in the definition of the second-order polarisation according to Equation 10, the units of d_{im} are metre/volt. The magnitude and sign of d_{im} and the forms of the piezoelectric d-matrices depend on crystal structure and symmetry. The efficiency of the nonlinear coupling in a given direction, however, is determined by the *effective* nonlinear coefficient, d_{eff}. The functional form of d_{eff} depends on the propagation direction, the polarisation of the respective waves, and those d_{im} coefficients contributing to the interaction. The form of d_{eff} can be determined by taking the projection of the generated polarisations along the directions of the respective optical electric fields. The expressions for d_{eff} in crystals of different symmetry classes have been tabulated by, for example, Dmitriev *et al.* (1991).

Regardless of walk-off effects and phase-matching conditions, the coefficients d_{im} determine the fundamental strength of nonlinearity in the medium, with d_{eff} determining the resultant nonlinear coupling in the particular interaction geometry. Thus, for efficient operation of parametric devices, it is vitally important to use materials with sufficiently large nonlinear coefficients and phase-matching and propagation geometries that yield maximum d_{eff}. It is important also to note that because of the direct correlation between the refractive index of a medium and its susceptibility, the higher nonlinearity of the medium the larger will be its refractive indices. Therefore, according to Equation 19, the parametric gain can be lowered, but this reduction in gain due to the increase in refractive indices is generally more than compensated for by the higher nonlinear susceptibility. This is often reflected in the figure-of-merit for the material which is defined as d_{eff}^2/n^3. In addition to a high optical damage threshold, large nonlinearity, and favourable phase-matching characteristics, other material requirements include a wide transmission range and good mechanical and chemical properties. In particular, the material must be mechanically hard, non-hygroscopic, and chemically stable. It must also be available in bulk form and large size and be able to take a good polish.

It is thus clear that practical operation of parametric devices requires nonlinear materials with favourable linear and nonlinear optical properties and laser pump sources of sufficient intensity and high spectral and spatial coherence. The somewhat stringent demands on the pump beam and nonlinear crystal quality have in the past rendered parametric devices as impractical sources of tunable radiation. Although the first operation of a parametric oscillator was reported more than three decades ago (Giordmaine and Miller, 1965), the development of practical parametric devices was for many years hampered by the absence of suitable laser pump sources and nonlinear materials with desirable optical and mechanical properties. In recent years, however, there has been a revival of interest in parametric devices for the generation of widely tunable radiation prompted, in the main part, by the emergence of a new generation of nonlinear materials. Progress in crystal growth has led to the development of new nonlinear crystals such as BBO, KTP, LBO, MgO:LiNbO$_3$, and KNbO$_3$ (KNB), which exhibit excellent optical and mechanical characteristics and damage thresholds far higher than many classical materials. The parallel progress in laser technology has resulted in the availability of new laser sources with improved output beam coherence and high intensity for use as pumps for parametric devices. These advances have led to the realisation of numerous new parametric devices based on novel design concepts and capable of providing tunable radiation from the ultraviolet to the infrared. The operation of parametric devices has now also been extended to the femtosecond regime, thus establishing these devices as truly practical sources of tunable radiation from the pulsed nanosecond to the ultrashort-pulse picosecond and femtosecond temporal domain.

7   Parametric oscillator devices

Of the numerous parametric devices developed to date, the optical parametric oscillator (OPO) represents the most prevalent configuration. As discussed previously, attainment of sufficient nonlinear gain to reach operation threshold in parametric devices requires high pumping intensities. However, it is possible to achieve a substantial increase in gain by resonating the generated fields in an optical cavity, as in a conventional laser oscillator. In this way, the pump power thresholds are reduced to levels accessible by many of the existing laser sources. The cavity mirrors may provide feedback at either of the generated signal or idler waves, in which case the device is known as a *singly-resonant oscillator* (SRO), or at both, in which case it is referred to as a *doubly-resonant oscillator* (DRO). In either form of the device, there is a threshold condition at which the single-pass parametric gain becomes sufficiently large to overcome the total round-trip cavity losses at the resonated wavelength(s). The parametric wave(s) thus experience a net round-trip gain and can build up to intensity levels on the order of the input pump, hence allowing useful coherent output to be extracted from an optical parametric oscillator.

The steady-state oscillation threshold of an optical parametric oscillator can be obtained from the solution of the coupled-wave equations (16)–(18) for the signal and idler fields, by balancing the single-pass parametric gain with round-trip cavity losses. The main result is that the SRO configuration in which only one of the generated waves is resonant yields the highest threshold. Specifically, if we consider an OPO resonator in which the total fractional round-trip power losses at the signal and idler are represented by α_s and α_i, respectively, then for small losses it can be shown that the threshold pump power with signal only resonant ($\alpha_i = 1$, α_s small) is $2/\alpha_i$ times higher than with both the signal and idler resonant (α_s, α_i small). This means that the oscillation threshold for a SRO ($\alpha_i = 1$, α_s small) is 100 times larger than the equivalent DRO with 2% loss at the idler ($\alpha_i = 0.02$, α_s small). The most important implication of this result is that the attainment of oscillation threshold in SROs is generally beyond the reach of CW pump lasers and practical operation of these devices requires high-peak-power pulsed pump sources. On the other hand, operation of DROs is readily attainable under both CW and pulsed excitation at modest power levels. The major drawback of DROs, however, is the inherent amplitude and frequency instabilities and tuning difficulties imposed by the double-resonance condition and the requirement for stringent control of the pump laser frequency and oscillator cavity length. For this reason, SROs are generally preferred in practice, albeit at the expense of higher oscillation thresholds. Other OPO resonator architectures include triply-resonant and pump-enhanced oscillators in which different degrees of feedback are also provided for the pump. This approach can yield further reductions in oscillation threshold from pure DRO and SRO configurations, but the requirement for simultaneous resonance of additional fields places further demands on the pump frequency and OPO cavity length stability. In the following discussion, we review some of the recent progress in optical parametric oscillator devices. We will confine the discussion to pulsed OPOs operating in the nanosecond, picosecond, and femtosecond temporal domain, as most of the recent practical developments have been in this area.

7.1 Nanosecond OPOs

Because of the high peak powers readily available from pulsed nanosecond laser sources, OPOs operating in this temporal regime have traditionally been the most extensively developed and utilised of all parametric devices. Indeed, the first successful demonstration of an OPO was a nanosecond oscillator based on $LiNbO_3$ and pumped by a frequency-doubled, Q-switched Nd:CaWO$_4$ laser at 0.529μm (Giordmaine and Miller, 1965). Following this demonstration, several devices based on other classical materials including $LiIO_3$, $CdSe$, Ag_3AsS_3, $Ba_2NaNb_5O_{15}$, and $AgGaS_2$ were developed, covering wavelength regions mainly in the infrared spectrum. The pump sources for these OPOs were predominantly Q-switched neodymium lasers and their harmonics, although pulsed ruby lasers and chemical lasers such as the HF laser were also used as pumps. However, for many years, material damage remained a major limitation to the operation of these devices, even at moderate power levels.

The basic operating principles in nanosecond OPOs are the same as in CW oscillators. However, because of the transient nature of pulse pumping, the steady-state threshold analysis of CW OPOs is not strictly applicable in this case and a modified model taking account of the dynamic behaviour of the OPO is necessary. By using a time-dependent gain analysis, Brosnan and Byer (1979) have derived the equations governing the threshold pump fluence in pulsed OPOs. The main conclusions of their work relate to the strong dependence of OPO threshold on the characteristic rise time of the oscillator, which is a measure of the time required for the parametric gain to build up from noise to oscillation threshold (Pearson *et al.* 1972). In order for the OPO to switch on, it is essential that the oscillator rise time is shorter than the pump pulse duration. For efficient operation, however, the rise time must be minimised so that the parametric waves are rapidly amplified above the threshold level in a time significantly shorter than the pump pulse interval. Therefore, the rise time represents an effective loss in pulsed OPOs, with a direct impact on oscillation threshold. Physically, the influence of rise time on the operation of pulsed OPOs derives from the instantaneous nature of nonlinear gain and the lack of gain storage in the parametric process, which means that no amplification can occur outside the temporal window of the pump pulse. Therefore, to exploit maximum gain, it is essential that the parametric waves are amplified in as short a time interval as possible in the presence of the pump pulse. In practice, this can be achieved by minimising the OPO cavity length to allow a maximum number of round-trips over the pump pulse interval. In nanosecond OPOs, cavity lengths of a few centimetres or shorter are often practicable, yielding as many as a hundred round-trips for pump pulses of a few nanoseconds duration. The OPO rise time can also be further reduced by using longer pump pulse durations (without sacrificing peak power) and minimising intracavity parasitic losses.

The emergence of new nonlinear materials over the past decade has led to the development of practical nanosecond OPOs capable of providing high-power and widely tunable radiation in new regions of the optical spectrum. The predominant pump sources have continued to be frequency-converted, flashlamp-pumped, Q-switched Nd:YAG or Nd:YLF lasers, but ultraviolet excimer lasers have also been shown to be highly effective as OPO pumps, particularly given their high-average-power capability. Of the numerous new devices demonstrated, the type I phase-matched BBO OPO pumped by the

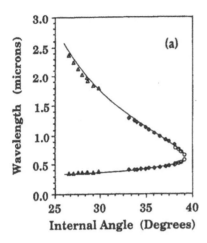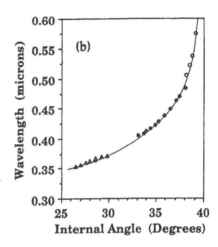

Figure 4. *(a) Tuning range of type I* BBO OPO *pumped by a XeCl excimer laser at 308nm. The* OPO *is tunable from 350nm in the near ultraviolet to 2.4μm in the near infrared by angular rotation of the crystal through ~ 18°. (b) The short-wavelength signal coverage of the* BBO OPO *from 350nm to near degeneracy at 600nm. The solid curve is the calculated tuning range and the experimental points are the output from the* OPO. *The different symbols represent the different mirror sets used.*

frequency-tripled Nd:YAG laser at 355nm (Cheng *et al.* 1988) or by the XeCl excimer laser at 308nm (Komine 1988, Ebrahimzadeh *et al.* 1990) has been established as the most versatile and practical source of tunable nanosecond pulses with a tunable range extending from the near-ultraviolet to the near-infrared. These devices can provide tunable nanosecond pulses over continuous range from typically 400nm to 2.5μm with a single crystal and only one or two sets of optimally-designed cavity mirrors to resonate the short-wavelength branch of the tuning range up to degeneracy (Figure 4). Output pulse energies from a few millijoules to hundreds of millijoules are available in pulses of 1–10ns duration, at 1–100Hz repetition-rate and extraction efficiencies in excess of 30%. Because of the high optical damage threshold of the crystal, energy scaling of BBO OPOs to the multijoule level is also attainable with the use of high-energy pulsed pump sources. The upper limit to the conversion efficiency and output pulse energy in BBO OPOs is generally set by damage to the OPO mirrors and crystal coatings, which can be caused either by the pump pulses or by the circulating signal intensities. However, these may be circumvented by using novel pumping and crystal configurations (Bosenberg *et al.* 1989a,1989b), longer pump pulse durations, and shorter cavity lengths. One of the major drawbacks of the type I BBO OPO, however, is the spectral linewidth of the output. Without any bandwidth control, the linewidth of the BBO OPO typically varies from 0.1nm to as much as 12nm near wavelength degeneracy, which can limit the use of the device in spectroscopic applications. However, by employing techniques such as grating feedback, intracavity etalon, and external injection-seeding it is possible to achieve linewidth reduction to below 0.1nm across the tuning range of theBBO OPO. The use of type II phase-matching and walk-off compensated geometries (Bosenberg *et al.* 1990) and manipulation of pump laser spatial profile and filtering of the output (Haub *et al.* 1995) have also been shown to yield significant linewidth reduction in BBO OPOs.

In addition to the BBO OPO, numerous nanosecond oscillators based on other nonlinear crystals including LBO, KNB, KTP and its new arsenate isomorphs such as $KTiOAsO_4$ (KTA) have also been developed using a variety of phase-matching geometries and pumping configurations. In particular, because of its temperature-tuning capability, versatile phase-matching characteristics, broad transparency ($160nm-2.6\mu m$), and high damage threshold, LBO has proved to be highly attractive for use in nanosecond OPOs operating in ultraviolet, visible, and the near-infrared. While the effective nonlinear coefficient of LBO ($\sim1pm/V$) is considerably lower than that of BBO ($\sim3pm/V$), its NCPM capability and small double-refraction permits the use of longer interaction lengths, thus allowing comparable efficiencies to be maintained. The crystal has been used for parametric generation under both type I and type II temperature-tuned NCPM with XeCl excimer lasers (Ebrahimzadeh *et al.* 1991) and frequency-tripled flashlamp-pumped Nd:YAG lasers (Hanson and Dick, 1991). The small spatial walk-off in LBO has even enabled the development of an all-solid-state, widely tunable OPO under type I critical phase-matching using a frequency-tripled, Q-switched, diode-pumped Nd:YAG laser (Cui *et al.* 1993).

Other nonlinear crystals such as KTP, KNB, and KTA have been of particular interest for wavelength generation in the difficult $3-5\mu m$ spectral range because of their longer infrared transmission cutoff beyond $3\mu m$. While these materials exhibit significantly lower optical damage thresholds than LBO, they possess considerably larger nonlinear coefficients. Of these, KTP has been by far the most extensively used material, in which nanosecond parametric oscillation out to $3.2\mu m$ has been demonstrated with the fundamental of the Nd:YAG laser at $1.064\mu m$ (Kato, 1991). Under type II NCPM, all-solid-state pumping of a nanosecond KTP OPO has also been reported with a diode-laser-pumped, Q-switched Nd:YAG laser (Marshall *et al.* 1991). However, an infrared cutoff wavelength of $\sim4.3\mu m$ as well as absorption bands near $2.8\mu m$ and $3.2\mu m$ has limited the scope of use of KTP over the whole of its transmission window. On the other hand, the arsenate isomorphs of KTP such as KTA exhibit increased transparency with little or no absorption up to $\sim 5.3\mu m$, while maintaining similarly attractive phase-matching and nonlinear optical properties to KTP. As in KTP, the most efficient interaction in KTA is type II, which is possible under critical phase-matching in any of the principal planes or under NCPM along a principal axis. The potential of this material for mid-infrared parametric generation was recently demonstrated with the use of a flashlamp-pumped Nd:YAG laser at $1.064\mu m$ as the pump source (Bosenberg *et al.* 1994). By using type II critical phase-matching in the xz-plane, continuously tunable output in the $1.5-5\mu m$ spectral range was generated at conversion efficiencies of up to 20%. The nonlinear crystal KNB has also been successfully used for nanosecond pulse generation in the $2-4\mu m$ spectral range. However, the relatively high optical damage threshold, superior optical and mechanical properties and availability in large size and good optical quality makes KTA a superior choice of material for practical near-infrared to mid-infrared nanosecond OPOs, at the present time. Other arsenate crystals such as

RbTiOAsO$_4$ (RTA) and CsTiOAsO$_4$ (CTA) also hold promise for parametric generation in the important 3–5μm spectral region.

7.2 Picosecond OPOs

Optical parametric oscillators also offer a highly effective means for the generation of optical pulses with picosecond temporal durations. However, unlike nanosecond devices, operation of ultrashort-pulse OPOs of this type is attainable only under *synchronous pumping* conditions. This is because the temporal window of the pump pulses in this case is too narrow to allow sufficient number of round-trips for the build up of the parametric wave(s) over the pump envelope, even for practical OPO cavity lengths as short as a few millimetres. To circumvent this difficulty, the OPO resonator length is matched to the length of the pump laser, so that the round-trip transit time in the OPO cavity is equal to the repetition period of the pump pulse train. In this way, the resonated parametric pulses experience amplification with consecutive coincidences with the input pump pulses as they make consecutive transits through the nonlinear crystal. In practice, the technique of synchronous pumping is restricted to relatively high pulse repetition rates (>50MHz). At lower repetition frequencies, the required cavity lengths for synchronous pumping become too long to be practical for most purposes. The SPOPO offers distinct advantages over conventional mode-locked lasers for many applications. In particular, the output pulses from the SPOPO exhibit lower timing jitter relative to the pump pulses than other synchronously-pumped lasers with gain storage, because of the instantaneous nature of parametric gain. This makes the SPOPO highly suitable for applications such as high-resolution time-domain spectroscopy.

In general, synchronously-pumped OPOs (SPOPOs) may be classified into either CW or pulsed devices. In CW SPOPOs, the input pump radiation comprises a continuous train of ultrashort pulses, whereas in pulsed SPOPOs the pump consists of trains of ultrashort pulses contained within a nanosecond or microsecond envelope. With regard to their operating characteristics, CW SPOPOs may be treated as steady-state devices in the same way as CW OPOs, but with the peak pump pulse intensity determining the nonlinear gain. As such, the steady-state analysis of CW OPOs is similarly applicable to CW SPOPOs. On the other hand, the operating dynamics of pulsed SPOPOs is analogous to nanosecond oscillators where a transient analysis taking account of rise time effects is necessary to adequately describe the SPOPO behaviour. In either case, however, additional effects pertinent to the operation of SPOPOs including temporal walk-off and group-velocity dispersion also have to be included in the model. In picosecond SPOPOs where pulse durations of more than 1ps are involved, such temporal effects can generally be neglected for practical crystal lengths of up to 20mm. However, they become increasingly important in femtosecond SPOPOs involving pulses of 100 fs or shorter, as discussed in Section 7.3. The plane-wave analysis of SPOPOs under steady-state and transient operation has been presented by Becker et al. (1974), and more recently by Cheung and Liu (1991) for CW oscillators with Gaussian beams. One of the main conclusions of their analyses is that in the absence of group velocity dispersion and temporal walk-off, the output pulses from a SPOPO are always shorter than the input pump pulses, with the parametric pulses broadening towards to the pump pulse length with increasing pump depletion. They also show that it is possible to obtain conversion efficiencies as high as ~70% in such devices with suitable choice of design parameters.

For most practical applications, the CW SPOPO is the desired device configuration because the output consists of a truly continuous train of ultrashort pulses with identical intensities and durations. In pulsed SPOPOs, on the other hand, both the intensity and amplitude of the output pulses can vary across the pulse envelope, so that the output does not constitute a truly repetitive pulse train. However, because of the significantly higher peak powers available from pulsed mode-locked than the equivalent CW mode-locked pump lasers, operation of SPOPOs is more readily attainable under pulsed conditions, particularly in SRO configurations of practical interest. For this reason, the majority of SPOPOs to date have been pulsed oscillators, pumped predominantly by the mode-locked and Q-switched (MLQS) Nd:YAG laser or its harmonics. Of these, the type I singly-resonant BBO SPOPO pumped by the second and third harmonic of the MLQS flashlamp-pumped Nd:YAG laser has been shown to be a versatile source of picosecond pulses across a broad spectral range from around 400nm to over 3μm (Bromley *et al.* 1989, Burdulis *et al.* 1990). Output pulses with durations of 10–100ps in Q-switched envelopes of a few millijoules at conversion efficiencies approaching 30% are currently available from these devices. In addition to BBO, the crystals KTP and LBO have also been used in pulsed SPOPOs pumped by the second harmonic of MLQS Nd:YAG and Nd:YLF lasers in the green. In particular, use of type I temperature-tuned NCPM in LBO has enabled the development of widely tunable picosecond SPOPOs in all-solid-state SRO configurations by using the frequency-doubled output of a MLQS diode-pumped Nd:YLF laser at 523nm (Ebrahimzadeh *et al.* 1993). In addition to SRO devices, a number of pulsed doubly-resonant SPOPOs based on BBO, KTP, LBO, and Ba$_2$NaNb$_5$O$_{15}$ have also been reported with the use of MLQS Nd:YAG pump lasers.

In the meantime, considerable advances have been made in the development of practical CW picosecond SPOPOs. Because of the relatively low peak powers available from CW mode-locked picosecond pump lasers, the early attempts in this area were confined to DRO devices, with the concomitant disadvantages of amplitude and spectral instabilities of the output. However, with the availability of high-power CW mode-locked laser sources and the emergence of novel mode-locking techniques, the operation of CW SPOPOs in SRO configurations has recently become a practical reality. The majority of the CW singly-resonant SPOPOs demonstrated to date have been based on the CW mode-locked neodymium laser and its second harmonic as the pump source. In most cases, the attainment of short pump pulses of sufficient intensity has necessitated the use of coupled-cavity techniques or external pulse compression. Particularly noteworthy has been the development of all-solid-state CW singly-resonant SPOPOs based on KTP (McCarthy and Hanna 1992) and LBO (Hall *et al.* 1993, Butterworth *et al.* 1993) and pumped by the frequency-doubled output of coupled-cavity mode-locked Nd:YLF lasers. These oscillators have been shown to provide truly continuous and highly stable output pulses with durations of 1–5ps at repetition rates as high as 130MHz. The LBO SPOPO can generate continuously tunable output from 650nm to above 2.5μm with a single crystal, while the KTP device is tunable over 900nm to 1.2μm. Average output powers of tens of milliwatts at efficiencies of up to 30% are routinely available from these devices. High-power operation of CW singly-resonant SPOPOs has also recently been reported with the use of flashlamp-pumped CW mode-locked neodymium lasers. By using the fundamental output of a Nd:YLF laser at 1.053μm, total average output powers of up to 2.8 W have been obtained from a KTP SPOPO for 14W of pump, with 800mW in the idler beam near 3.2μm (Grasser *et al.* 1993). This device generated signal pulses of 12ps duration for 40ps input pump pulses at 76MHz repetition rate. A similar mid-infrared oscillator was also demonstrated by Chung and Siegman (1993) who used the temporally compressed output pulses from a Nd:YAG laser at 1.064μm to pump a KTP SPOPO. Average output powers of up to 350mW in pulses of 2–3ps at 75MHz were generated for 4W of pump power.

More recently, the operation of CW singly-resonant picosecond SPOPOs has been achieved with the use of the self-mode-locked Ti:sapphire laser as the pump source. A particularly attractive feature of this approach is that the tunability of the Ti:sapphire pump source allows wavelength tuning in the SPOPO without resort to angle phase-matching. This minimises intracavity reflection losses caused by crystal rotation, thus maintaining maximum efficiency across the available tuning range. By using type II NCPM in KTP and a self-mode-locked Ti:sapphire pump laser, Nebel *et al.* (1993) generated tunable picosecond pulses in the 1–1.2μm and 2.3–2.9μm spectral range by tuning the Ti:sapphire laser over 720–853nm. Average output powers of up to 700mW in 1.2ps pulses at 82MHz repetition rate were produced for 1.6W of pump. The combination of tunable Ti:sapphire laser with temperature-tuned LBO under type I NCPM has also been shown to provide a highly versatile source of picosecond pulses (Ebrahimzadeh *et al.* 1995) with continuous tunability over 1–2.4μm (Figure 5). This SPOPO could generate output powers of up to 325mW for 1.2W of pump in pulses of $\sim$1–2ps durations at 81MHz repetition rate. The current trends in CW picosecond SPOPOs point towards the development of new devices for spectral regions beyond 3μm which are of interest for many spectroscopic applications. In this context, the arsenate crystals such as KTA, RTA, and CTA hold promise as suitable material candidates in combination with mode-locked Nd:YAG or Ti:sapphire pump lasers.

7.3   Femtosecond OPOs

Optical parametric oscillators operating in the femtosecond time domain represent a new class of tunable laser sources. The high peak powers available from ultrashort femtosecond pulses are particularly suited to the exploitation of the small nonlinear gain in parametric devices. However, because of the short temporal duration (typically less than 100fs) and large spectral content (typically greater than 10nm) associated with femtosecond pulses, additional effects such as group velocity dispersion, temporal walk-off, and spectral acceptance bandwidths play an important role in the operation of these devices. Group velocity dispersion often leads to pulse broadening in the SPOPO, while temporal walk-off can degrade nonlinear gain or even modify the temporal characteristics of the output pulses. The spectral acceptance bandwidth of the crystal can also reduce gain as well as set a lower limit to the minimum attainable pulse duration from the SPOPO. The influence of temporal walk-off and spectral acceptance bandwidth can be minimised by using short crystal lengths of typically 1–2mm, while the effects of group velocity dispersion can be overcome by the inclusion of dispersion compensation in the SPOPO cavity, as in conventional femtosecond lasers.

Although crystal lengths of 1–2mm are too short to provide sufficient gain in pi-

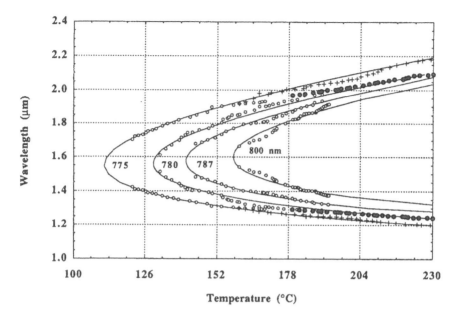

Figure 5. *Wavelength tuning in Ti:sapphire-pumped picosecond* LBO SPOPO *with type I temperature-tuned* NCPM. *The* SPOPO *is continuously tunable from 1.2µm to 2.2µm, with potential tuning over 1–2.4µm. The different curves correspond to the different Ti:sapphire pump wavelengths.*

cosecond and nanosecond oscillators, the considerably higher peak powers available with femtosecond pulses can adequately compensate for this shortfall in gain. However, the high in-crystal peak intensities (typically $>10\text{GW}/\text{cm}^2$) can also induce additional higher-order nonlinear effects such as self-phase-modulation in femtosecond SPOPOs. Such effects which are not generally present in picosecond and nanosecond oscillators often lead to spectral broadening and chirping of the output pulses and become more pronounced with longer interaction lengths. The spectral broadening due to self-phase-modulation can, however, be exploited for subsequent compression of output pulses. Interestingly, the high intensities in femtosecond SPOPOs can also result in non-phase-matched mixing processes over a coherence length of the crystal where, in addition to the signal and idler beams, several other wavelengths are also generated to a detectable level. The design of femtosecond SPOPOs involves a careful consideration of the additional temporal and spectral effects as well as other phase-matching and material parameters relevant to the operation of parametric devices.

The first demonstration of a femtosecond SPOPO was a CW oscillator based on a 1.4mm crystal of KTP and pumped by 170fs pulses from a colliding-pulse mode-locked dye laser at 620nm (Edelstein *et al.* 1989). To access the high peak intensities necessary for oscillation, the KTP crystal was pumped at the intracavity focus of the dye laser in a non-collinear phase-matching geometry. This singly-resonant device produced 220fs signal pulses at milliwatt average power levels in the near-infrared. The approach was subsequently extended to an externally-pumped oscillator where increased signal powers of up to 30mW were generated (Mak *et al.* 1992). In the meantime, Laenen *et al.* (1990) reported a pulsed femtosecond SPOPO based on non-collinearly phase-matched BBO and pumped by microsecond pulse trains from a frequency-doubled Nd:glass laser at 527nm. Output pulses with durations of $\sim$ 90fs over the range 700nm-1.8μm were obtained for 800fs input pump pulses.

Soon after, the emergence of the self-mode-locked Ti:sapphire laser provided a laser source capable of pumping CW femtosecond SPOPOs at significantly higher powers and many times above threshold. This enabled the development of a Ti:sapphire-pumped KTP-based femtosecond SPOPO capable of producing 175mW of average power in the near-infrared in 62fs pulses at 76MHz repetition rate (Fu *et al.* 1992). At the same time, Pelouch *et al.* (1992) demonstrated a similar device generating hundreds of milliwatts of average power in 135fs pulses in the near-infrared by using a 2.5W Ti:sapphire laser. This scheme was soon extended to KTA, CTA, and RTA (Powers *et al.* 1994) with the aim of providing high-repetition-rate femtosecond pulses in the near- to mid-infrared spectral regions. Average output powers of hundreds of milliwatts in sub-100fs pulses out to 3.6μm have been demonstrated from these devices. By using alternative collinear NCPM schemes in KTP in combination with Ti:sapphire pump tuning, Dudley *et al.* (1994) also reported a femtosecond near-infrared SPOPO generating 40fs output pulses, while Reid *et al.* (1995) demonstrated a similar device with a pump power threshold as low as 50mW. The spectral range of Ti:sapphire-pumped femtosecond SPOPOs has also been extended to the visible by using two distinct approaches. Ellingson and Tang (1993) generated tunable femtosecond pulses in the visible through intracavity second harmonic generation of SPOPO signal pulses. By exploiting the high signal intensities available internal to the SPOPO cavity and a 47μm BBO doubling crystal placed at an additional intracavity focus, output powers of 240mW in sub-100fs pulses were obtained over a wavelength range 580-657nm. Reid *et al.* (1995) later reported a similar device based on RTA and capable of delivering tunable 70fs pulses in the 620-660nm spectral range at 170mW average power. An alternative approach was demonstrated by Driscoll *et al.* (1994) who used a frequency-doubled Ti:sapphire laser at 400nm to directly pump a visible femtosecond SPOPO based on BBO. This device generated 30fs signal pulses with average powers of up to 100mW from 566 to 676nm.

An interesting feature of femtosecond SPOPOs which has also been observed in other ultrashort-pulse picosecond oscillators is the wavelength tuning available through cavity length mismatch. The wavelength tuning occurs because the cavity length detuning introduces a loss at the signal wavelength by reducing the synchronism between the pump and signal pulses. To maintain synchronism and optimise gain, the signal shifts to a more favourable wavelength with a group velocity that satisfies a constant round-trip time. This cavity length tuning which was first observed by Edelstein *et al.* (1989) is a useful mechanism for fine-tuning the output wavelength, often by as much as 50nm.

Finally, it is important to note that as well as OPOs, parametric generation in resonator-free, single-pass or travelling-wave configurations is also attainable, but large pumping intensities on the order of 1–100GW/cm^2 are often necessary to reach threshold. This is the basic principle of operation in optical parametric amplifier (OPA) and travelling-wave optical parametric generator (TOPG) devices (Danielius *et al.* 1993). The pump sources for OPAs and TOPGs generally comprise regenerative, multipass, and chirped pulse amplification schemes to provide pulses of sufficiently high energy (1μJ-

1mJ) and peak power (10MW–10GW) to allow large single-pass gains (typically greater than 10^{10}) to be achieved in a single or double-pass through the crystal. However, the large pumping intensities in this case place even more stringent demands on material damage, so that practical operation of these devices generally requires crystals with high optical damage thresholds such as BBO and LBO. At present, OPA and TOPG systems can provide output pulses with picosecond and femtosecond temporal durations over a wide spectral range from the near ultraviolet to the infrared and pulse energies from a few microjoules to hundreds of microjoules are available at repetition rates of up to 250kHz.

8 Conclusions

This discussion has aimed to provide an insight into the operation of OPOs, with particular emphasis on pulsed oscillators operating in the nanosecond, picosecond and femtosecond temporal domain. We have discussed the basic principles of parametric generation and have outlined the criteria important in the design of parametric devices such as the choice of nonlinear material and laser pump source, tuning and phase-matching considerations, optical damage, and walk-off effects. In general, it is often not possible to satisfy all the design requirements simultaneously, so that practical operation of parametric devices often necessitates a compromise among the various design parameters. The unprecedented level of research interest and the rapid advances in this area over recent years have established OPOs as truly practical tunable light sources more than thirty years after their potential was first recognised. Many of the devices demonstrated as laboratory prototypes have found their way to the commercial market place in a period of only a few years and continue to find numerous new applications from spectroscopy and environmental studies to optical frequency synthesis and quantum optics. The current efforts in this area are directed towards the development of devices for new spectral regions that have so far remained inaccessible, while future trends point to the realisation of novel ultracompact, all-solid-state OPOs based on integrated quasi-phase-matched structures and semiconductor nonlinear optics.

References

Becker M F, Kuizanga D J, Phillion D W, and Siegman A E, 1974, *J Appl Phys* **45** 3996.

Bosenberg W R, Cheng L K, and Tang C L, 1989a, *Appl Phys Lett* **54** 13.

Bosenberg W R, Pelouch W S, and Tang C L, 1989b, *Appl Phys Lett* **55** 1952.

Bosenberg W R and Tang C L, 1990, *Appl Phys Lett* **56** 1819.

Bosenberg W R, Cheng L K, and Bierlein J D, 1994, *Appl Phys Lett* **65** 2765.

Boyd G D and Kleinman D A, 1968, *J Appl Phys* **39** 3597.

Bromley L J, Guy A, and Hanna D C, 1989, *Opt Commun* **70** 350.

Brosnan S J and Byer R L, 1979, *IEEE J Quantum Electron* **15** 415.

Burdulis S, Grigonis R, Piskarskas A, Sinkevicius G, Sirutkaitis V, Fix A, Nolting J, and Wallenstein R, 1990, *Opt Commun* **74** 398.

Butterworth S D, McCarthy M J, and Hanna D C, 1993, *Opt Lett* **18** 1429.

Cheng L K, Bosenberg W R, and Tang C L, 1988, *Appl Phys Lett* **53** 175.

Cheung E C and Liu J M, 1991, *J Opt Soc Am B* **8** 1491.

Chung J and Siegman A E, 1993, *J Opt Soc Am B* **10** 2201.

Cui Y, Withers D E, Rae C F, Norrie C J, Tang Y, Sinclair B D, Sibbett W, and Dunn M H, 1993, *Opt Lett* **18** 122.

Danielius R, Piskarskas A, Stabinis A, Banfi G P, Di Trapani P, and Righini R, 1993, *J Opt Soc Am B* **10** 2222.

Di Trapani P, Andreoni A, Foggi P, Solcia C, Danielius R, and Piskarskas A, 1995, *Opt Commun* **119** 327.

Dmitriev V G, Gurzadyan G G, and Nikogosyan D N, 1991, *Handbook of Nonlinear Optical Crystals* (Springer-Verlag).

Driscoll T J, Gale G M, and Hache F, 1994, *Opt Commun* **110** 638.

Dudley J M, Reid D T, Ebrahimzadeh M, and Sibbett W, 1994 *Opt Commun* **104** 419.

Ebrahimzadeh M, Henderson A J, and Dunn M H, 1990, *IEEE J Quantum Electron* **26** 1241.

Ebrahimzadeh M, Robertson G, and Dunn M H, 1991, *Opt Lett* **16** 767.

Ebrahimzadeh M, Hall G J, and Ferguson A I, 1993, *Opt Lett* **18** 278.

Ebrahimzadeh M, French S, and Miller A, 1995, *J Opt Soc Am B* **12** 2180.

Edelstein D C, Wachman E S, and Tang C L, 1989, *Appl Phys Lett* **54** 1728.

Ellingson R J and Tang C L, 1993, *Opt Lett* **18** 438.

Fu Q, Mak G, and van Driel H M, 1992, *Opt Lett* **17** 1006.

Giordmaine J A and Miller R C, 1965, *Phys Rev Lett* **14** 973.

Grasser Ch, Wang D, Beigang R, and Wallenstein R, 1993, *J Opt Soc Am B* **10** 2218.

Hall G J, Ebrahimzadeh M, Robertson A, Malcolm G P A, and Ferguson A I, 1993, *J Opt Soc Am B* **10** 2168.

Hanson F and Dick D, 1991, *Opt Lett* **16** 205.

Harris S E, 1969, *Proc IEEE* **57** 2096.

Haub J G, Johnson M J, Powell A J, and Orr B J, 1995, *Opt Lett* **20** 1637.

Kato K, 1991, *IEEE J Quantum Electron* **27** 1137.

Komine H, 1988, *Opt Lett* **13** 643.

Kuizenga D J, 1972, *Appl Phys Lett* **21** 570.

Laenen R, Graener H, and Laubereau A, 1990, *Opt Lett* **15** 971.

McCarthy M J and Hanna D C, 1992, *Opt Lett* **17** 402.

Mak G, Fu Q, and van Driel H M, 1992, *Appl Phys Lett* **60** 542.

Marshall L R, Kasinski J, Hays A D, and Burnham R L, 1991, *Opt Lett* **16** 681.

Nebel A, Fallnich C, Beigang R, and Wallenstein R, 1993, *J Opt Soc Am B* **10** 2195.

Pearson J E, Ganiel U, and Yariv A, 1972, *IEEE J Quantum Electron* **8** 433.

Pelouch W S, Powers P E, and Tang C L, 1992, *Opt Lett* **17** 1070.

Powers P E, Tang C L, and Cheng L K, 1994, *Opt Lett* **19** 1439.

Reid D T, Ebrahimzadeh M, and Sibbett W, 1995, *Opt Lett* **20** 55.

Reid D T, Ebrahimzadeh M, and Sibbett W, 1995, *J Opt Soc Am B* **12** 1157.

Smith R G, 1970, *J Appl Phys* **41** 4121.

Wachman E S, Pelouch W S, and Tang C L, 1991, *J Appl Phys* **70** 1893.

Ultrashort Pulse Sources

Wilson Sibbett

University of St Andrews, Scotland

1   Introduction

Optical science is a vibrant activity within research and development which is leading to new technologies that can be expected to deliver much economic return and quality-of-life enhancements. Within this general field, the generation, characterisation and application of ultrashort light pulses represents one of the most active topic subgroups in current laser-related international research. The interest in ultrashort-pulse laser sources derives, at least in part, from the inter-disciplinarity of the applications-base that continues to expand in the picosecond and femtosecond regimes that are of particular relevance to physics, chemistry, biology and optoelectronics. An especially notable feature of modern vibronic gain media, such as titanium-doped sapphire for example, is the extended bandwidth which offers excellent tunability and the potential for yet shorter pulses. Within this latter context, the possibilities for the achievement of ultra-high, terawatt/petawatt ($10^{12}/10^{15}$W) peak power levels through optimised oscillator-amplifier system combinations open up many new and exciting novel aspects of atomic and plasma physics.

As the field of laser physics matures, it is clearly evident that the efficiency and practicality has become an increasingly important feature of the lasers that are being developed. For this reason, in considering the methodologies that can be applied to generate ultrashort laser pulses, some emphasis is given within this overview to schemes that will ultimately afford an optimum robustness, reliability and practicality. It follows, quite naturally in my view, that any compact and user-compatible ultrashort-pulse laser will involve semiconductor diode lasers, at least as the optical pumping source. Due consideration will therefore be given to 'all-solid-state' modelocked laser configurations.

By way of format, the basics of the technique of modelocking are introduced using a qualitative description and this forms a physically-based framework to which the so-called active, passive and hybrid schemes are referred. Because of their relative importance, particular attention has been directed towards the generation of frequency-tunable femtosecond pulses from broadband gain media. Both the coupled-cavity mode-locking (Kean *et al.* 1989) and self-modelocking (Spence *et al.* 1991) techniques in which the optical Kerr effect is exploited are highlighted and set in a context that is readily traceable to the basic concepts of the phase-locking of the resonator modes that are discussed at the outset. One of the pedagogical aims in this chapter is to establish the conditions under which the Kerr nonlinearity can be accessed even at low levels of pump power (less than 100mW) and to outline the limitations on ultrashort pulse evolution arising from intracavity dispersive effects in respect of representative diode-pumped vibronic lasers.

2   Basic Principles of Laser Modelocking

In its simplest form, a laser resonator comprises a gain medium that is incorporated within either a standing-wave (Fabry-Perot) or travelling-wave (ring) optical cavity as illustrated in Figure 1(a), (b) respectively.

On the assumption that a fundamental (Gaussian spatial profile) transverse mode (TEM$_{00}$) beam is produced by the laser, then a discrete sequence of longitudinal (or axial) modes can be associated with either type of resonator configuration. The frequency, ν_m, of the mth longitudinal mode that corresponds to a standing-wave cavity length L or to a travelling-wave cavity perimeter P can be expressed as:

$$\nu_m = \frac{mc}{2nL} \qquad \text{or} \qquad \nu_m = \frac{mc}{nP} \tag{1}$$

where c is the speed of light in vacuum and n is the average value of the refractive index. It follows, therefore, that the frequency separation, $\delta\nu$, between adjacent longitudinal

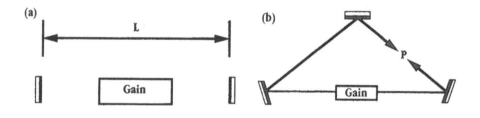

Figure 1. *Standing-wave (a) and travelling-wave (b) resonator configurations*

modes will be:

$$\delta\nu = \frac{c}{2nL}\left(\text{or}\,\frac{c}{nP}\right) \qquad \text{ie} \qquad \delta\omega = \frac{\pi c}{nL}\left(\text{or}\,\frac{2\pi c}{nP}\right) \tag{2}$$

The number of modes that oscillate is determined by the spectral bandwidth $\Delta\nu_l$, over which the gain exceeds the losses in the laser. When the laser is free-running (i.e. unmodelocked operation) the relative phases between the modes are randomly fluctuating and the output intensity has stochastic or noisy characteristics. The resultant field may then be expressed as a sum over $2N + 1$ modes

$$E(t) = \sum_{m=-N}^{N} \frac{1}{2}E_m \exp i[(\omega_0 + m\delta\omega)t + \phi_m] + \text{c.c.} \tag{3}$$

The corresponding total intensity, I, will be given by

$$I \sim [E(t)]^2 = \sum_{m=-N}^{N} \frac{1}{2}|E_m|^2 \tag{4}$$

This is just the sum of the individual mode intensities as is expected for phase-incoherent components.

By contrast, when the longitudinal modes, each assumed to have amplitude E_0, are forced to maintain a fixed phase relationship, then the phase ϕ_m, can be regarded as a linear function of m. Supposing that the intermode phase difference is $\delta\phi$ (i.e. $\phi_m - \phi_{m-1} = \delta\phi$) then:

$$\phi_m = \phi_0 + m\delta\phi \tag{5}$$

Substitution of this form of ϕ_m into equation 3 and its subsequent analytical summation produces the resultant field $E(t)$ given below;

$$E(t) = E_0 \frac{\sin\left[\dfrac{2N+1}{2}(\delta\omega t + \delta\phi)\right]}{\sin\left[\dfrac{\delta\omega t + \delta\phi}{2}\right]} \exp i(\omega_0 t + \phi_0) + \text{c.c.} \tag{6}$$

or

$$E(t) = A(t)e^{i(\omega_0 t + \phi_0)} + \text{c.c.} \tag{7}$$

The maximum values for the amplitude terms $A(t)$ appear with a periodicity, T, given by

$$T = \frac{2\pi}{\delta\omega} \equiv \frac{1}{\delta\nu} \equiv \frac{2nL}{c}\left(\text{or}\,\frac{nP}{c}\right) \tag{8}$$

T is the cavity round-trip period, also denoted by T_{RT}, and the reciprocal of the cavity-frequency, f_{cav}. It can also be deduced from equation (5) that the pulse duration (FWHM) $\Delta\tau$, is given by:

$$\Delta\tau = \frac{2\pi}{2N\delta\omega} \equiv \frac{1}{2N\delta\nu} \equiv \frac{1}{\Delta\nu_l} \tag{9}$$

(N.B. This pulse duration, $\Delta\tau$ is related to the inverse of the lasing bandwidth (FWHM) as distinct from the fluorescence or gain bandwidth of the lasing medium.) This phase-locking process thus leads to a transformation from a stochastic intensity output to a periodic sequence of discrete pulses as illustrated in Figure 2.

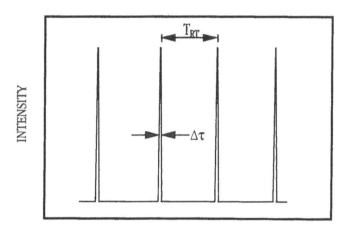

Figure 2. *Pulse sequence from a modelocked laser*

It follows from the relationship in Equation 9 that the shortest pulses will be produced under the conditions where the maximum number of longitudinal modes can be phase-locked. In practice, this can be an especially demanding requirement for broad bandwidth gain media and significant intensity-induced nonlinear effects and intracavity dispersion can give rise to serious dephasing influences (see Sections 3.3, 3.4). Also, it should be pointed out here that a more rigorous treatment of the modelocking process implies that $\Delta\tau\Delta\nu_l \geq K$ where the constant, K, is pulseshape-related. For example, $K = 0.441$ for Gaussian intensity profiles and $K = 0.315$ for sech-squared intensity profiles. Thus, whenever it can be established for a particular pulse shape that the duration-bandwidth product is close to the theoretically predicted value for K then the pulses are described as bandwidth-limited or Fourier-transform limited. Unless the laser is carefully controlled, the $\Delta\tau\Delta\nu_l$ product is generally greater than K but interestingly, some degree of asymmetry in pulse profiles can lead to duration-bandwidth products that are *less* than the expected K value for an identified pulseshape. Ultrashort pulses must therefore be carefully characterised in terms of their specific temporal and spectral features and the complementary data relating to amplitude and phase information from more recently developed diagnostic methods such as the frequency-resolved optical grating technique (Kane and Trebino 1993) are of vital importance.

From Equation 6 it can be seen that the peak pulse amplitude will be $(2N + 1)E_0$ with a corresponding intensity of $(2N + 1)^2 I_0$ where I_0 is the assumed equal intensity of the individual modes. Clearly, when a large number of longitudinal modes have been properly phase-locked together, then a very substantial enhancement in peak intensity can be realised. For instance, when the self-modelocking technique was initially reported for a titanium-sapphire laser (Spence *et al.* 1991) the bandwidth was ∼14nm at wavelengths around 870nm for bandwidth-limited pulses produced at a repetition period of 12ns. This corresponds to in excess of 66,000 longitudinal modes that had been phase locked together to produce the 60fs duration pulses. Impressively, the peak powers associated with these directly generated pulses was approximately 100kW!

3 Modelocking Techniques

In 1964 Di Domenico introduced the concept of modelocking. Since that time many clever approaches have been conceived for the modelocking of a wide range of laser types. Although excellent review papers have been written on the subject of picosecond and subpicosecond pulse generation (New 1983) the topic of femtosecond pulse generation has not yet been comprehensively reviewed. However, the 1994, Volume 58 special-feature issue of the journal, Applied Physics B contains several review-type articles that are relevant to the subject matter of this chapter.

By way of categorisation it is customary to describe modelocking procedures as active, passive or hybrid (a combination of active and passive). For active modelocking, a periodic modulation of the gain or loss is provided either at the cavity frequency, f_{cav} or at a multiple of f_{cav} by a driving signal that is derived from a source that is external to the laser. A good example of this is the radio-frequency (RF) electrical signal applied to the transducers on intracavity acousto-optic modulators that are frequently used with argon-ion (or Kr-ion) or Nd:YAG (YLF) lasers. Generally, both acousto-optic and electro-optic modulators find application in active modelocking through either amplitude modulation (AM modelocking) or phase modulation (FM modelocking). By contrast, passive modelocking is where the intracavity radiation induces a periodic modulation that is automatically synchronised to the cavity frequency.

3.1 Active Modelocking

Amplitude modulation of the longitudinal modes of a laser cavity can be conveniently applied in terms of either a periodic loss or gain at the cavity frequency, f_{cav}. With an acousto-optic modulator (see Figure 3, for an ion laser) a RF electrical signal is applied at a frequency f_{RF} to the transducer on a silica block or prism having an appropriate material thickness such that a standing acoustic wave can be established. The consequent diffractive loss arising from the associated phase grating has periodic minima at a modulation frequency $f_m = 2f_{RF}$. Thus, when f_m is chosen to equal the cavity frequency f_{cav}, this amplitude modulation leads to the transfer of power to sideband frequencies. This can be justified in a rather simple manner by considering the modal field representations in the following way.

Suppose the amplitude and frequency of the longitudinal mode that experiences the maximum available gain are E_0 and ω_0 respectively. Thus:

$$E(t) = E_0 \cos \omega_0 t \tag{10}$$

When this mode is modulated at a frequency, ω_m, and amplitude, E_m, then:

$$E'(t) = E(t) + E_m \cos \omega_m t \tag{11}$$

where $E'(t)$ is the modulated mode amplitude. This may be written as:

$$E'(t) = E(t)(1 + M \cos \omega_m t) \tag{12}$$

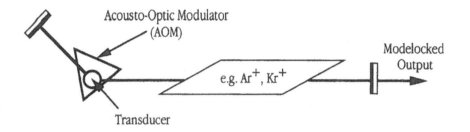

Figure 3. *Representative scheme for an acousto-optically modelocked ion-laser.*

where the modulation depth, M, is given by:

$$M = \frac{E_m}{E_0} \tag{13}$$

Combining Equations 10 and 12 gives:

$$E'(t) = E_0 \cos \omega_0 t + \frac{M E_0}{2}[\cos(\omega_0 - \omega_m)t + \cos(\omega_0 + \omega_m)t] \tag{14}$$

or

$$E'(t) = E_0 \cos \omega_0 t + \frac{M E_0}{2} \cos(\omega_0 + \delta\omega)t + \frac{M E_0}{2} \cos(\omega_0 + \delta\omega)t \tag{15}$$

Because the sideband frequencies are coincident with the frequencies of the neighbouring modes at $\omega_0 \pm \delta\omega$, it can be appreciated that power can be coherently coupled and exchanged from the initial mode centred at frequency ω_0 to the adjacent modes at frequencies $\omega_0 \pm \delta\omega$. Thus, sustained and properly controlled amplitude modulation enables a phase-locking of all those longitudinal modes that reach the lasing threshold. In the example given above in Section 2, this would imply a phase-locking of $2N$ 'sideband-frequency' modes with an equivalent total bandwidth of $2N\delta\omega$.

In practice, various sources of noise and phase drift in the modulator and in the laser cavity usually means that the phase-locking is compromised such that bandwidth-limited pulse durations are seldom achieved in acousto-optically modelocked lasers. Typical pulse durations for actively modelocked continuous-wave lasers range from 50ps (Nd:YLF) through to 70–100ps (Ar+, Kr+, Nd:YAG). For the pulse repetition rates which are generally around 80–100 MHz the average powers from the modelocked ion lasers are usually 1–2W and 10–20W for the solid-state counterparts. Interestingly, shorter, higher peak power pulses can be produced by such lasers (Nd:YLF or YAG in particular) using the coupled-cavity modelocking (see Section 3.3) and self-modelocking (see Section 3.4) techniques where the optical Kerr effect nonlinearity is exploited.

An alternative AM modelocking scheme involves modulation of the gain. This can be provided either by optically-pumping a slave laser by an already modelocked pump laser or in the case of a semiconductor diode laser it can be provided by a direct modulation of the injection drive current. In both cases the frequency of the gain

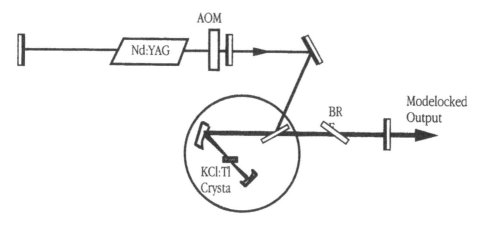

Figure 4. *Representative synchronously-modelocked laser*

modulation is generally matched to the cavity frequency or to a multiple of it. For the so-called synchronously modelocked laser a configuration of the type shown in Figure 4 is appropriate. Each Nd:YAG pump laser pulse produces transient gain that enables a pulse to be established in the slave colour-centre laser cavity. A combination of modulated periodic gain, gain saturation (especially in organic dye laser media), and perhaps additional nonlinear effects (see Section 3.2) leads to the generation of slave laser pulses that are substantially shorter than the pump pulses. For typical pump pulse durations around 100ps from actively modelocked ion or Nd:YAG pump lasers, synchronously modelocked dye laser pulses in the picosecond/subpicosecond regime are obtained (Johnson and Simpson 1983) and similarly for colour-centre lasers (Mollenauer 1985).

An attractive feature of a synchronously modelocked laser is that its output can be tuned across much of the inherent gain spectrum of the lasing medium. However, the provision of transient enhanced gain via an already modelocked pump source gives rise to serious spectral, temporal and timing-jitter deficiencies in such a laser. The origin of such instabilities can be traced to particular influences of the spontaneous emission process on the dynamics of the pulse formation under the conditions of synchronous pumping. Significantly, as the slave-laser cavity length is adjusted to obtain the optimally short pulses, these ultrashort pulses become very susceptible to large-scale perturbations that originate from the noise background (New 1990). It follows, therefore, that a substantially improved pulse generation status would be obtained by injecting into the slave-laser cavity a relatively weak coherent signal having an intensity that is sufficient to overwhelm the noise background while not compromising the pulse forming dynamics. This idea was first confirmed experimentally for a synchronously-pumped dye laser (Beaud *et al.* 1990) and more recently for colour-centre lasers (Mollmann and Gellermann 1994).

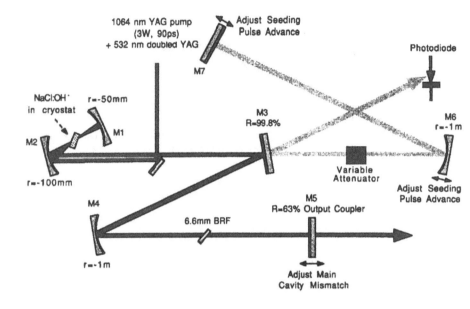

Figure 5. *Synchronously modelocked NaCl:OH- colour-centre laser with coherent pho-ton seeding*

From a pedagogical viewpoint, it is quite instructive to consider the effect of using an empty external cavity to feed back a highly attenuated ($10^{-5} - 10^{-12}$) replica of the pulse that circulates in the main cavity. The timing for this weak feedback coherent-seeding pulse is arranged such that it arrives at the gain medium slightly in advance of the main-cavity pulse so that the destabilising effects of the spontaneous emission can be suppressed. Thus, for the cavity arrangement of the type illustrated in Figure 5, a dramatic stability improvement in modelocking performance can be achieved as confirmed by the autocorrelation, spectral and RF power spectra data included in Figure 6. In common with observations relating to the injection seeding of single longitudinal mode ('spectrally pure') lasers, it is impressive that the performance of a synchronously-modelocked laser can be affected in such a major way by an extremely weak but coherent seed signal. With a semiconductor laser it is convenient to achieve periodic gain simply by modulating the injection current to the amplifying medium. This can be attempted in the millimetre-wave band at the inherent cavity-frequency of the chip (typically 120–180 GHz) but this is technically quite difficult. By utilising the alternative scheme of configuring integrated modulator/laser monolithic structures it has been shown that modulation frequencies up to 40 GHz can be achieved (Tucker *et al.* 1989). The use of external resonator configurations and RF modulation of injection currents has also proved to be relatively convenient. By applying high quality

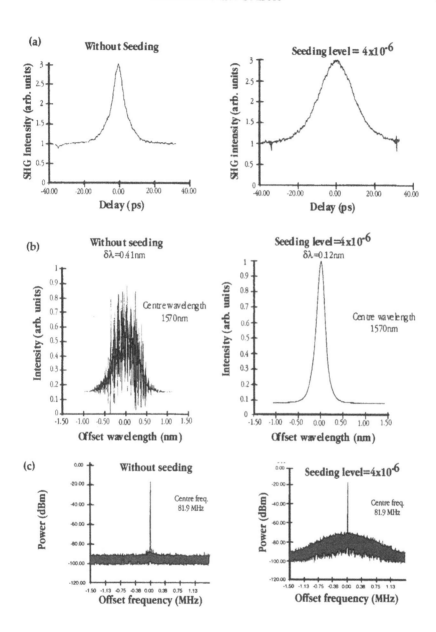

Figure 6. *(a) Autocorrelations, (b) optical and (c) RF-power spectra, for synchronously modelocked colour-centre laser with and without coherent photon seeding.*

anti-reflection coatings to the chip facets and/or by suitable orientation of the facets relative to the gain stripe, the cavity features of the diode device itself can be suppressed such that the chip behaves essentially as a discrete amplifying medium. An

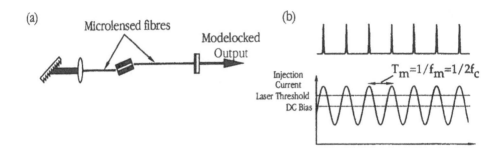

Figure 7. *Schematic for external-cavity, actively modelocked diode laser (a) with illustrated conditions for modulated injection current (b)*

external-cavity semiconductor laser arrangement such as that illustrated in Figure 7(a) for an InGaAsP diode laser can thus be configured. By supplementing a sub-laser-threshold dc injection current with an RF current then the laser can reach threshold periodically as indicated in Figure 7(b). If the gain medium is located at the centre of the external standing-wave resonator then the modulated injection current should have a frequency of $2mf_{cav}$ where m is the integer. This ensures that the luminescence that propagates towards, and returns from the cavity reflectors experiences gain during each transit of the diode chip. Also, this choice of set-up enables the counter-propagating intracavity flux to interfere within the gain medium and so pulse shortening through gain-saturation effects can be enhanced through the coherent interaction. Picosecond pulses have thus been reliably produced by this basic type of active modelocking scheme in AlGaAs and InGaAsP diode lasers.

When FM modelocking is employed, it is customary to locate a phase modulator (electro-optic or Pockels effect device) close to one of the cavity reflectors. The application of a RF drive signal induces a sinusoidal variation in the phase shift experienced by the intracavity radiation that passes through the modulator. As a consequence the optical carrier frequency is shifted everywhere except at the extrema of the phase-sinusoid. The cumulative frequency shifts progressively displace the frequency of the circulating radiation away from the peak region of the gain spectrum such that pulses are only generated at the two extrema of the phase shift during each modulation cycle. (The presence of some optical nonlinearity generally leads to some spectral and temporal distinctiveness between the two pulse sequences that are produced at these two extrema.) It is also perhaps worth noting that in the frequency-domain description, the sinusoidal phase modulation produces a sequence of sidebands (with a Bessel-function amplitude series) for each cavity longitudinal mode rather than just the two sidebands that arise with amplitude modulation. However, for the relatively small phase shifts that are generally encountered in practical arrangements, only a few sidebands have sufficient amplitudes to command serious consideration. Thus a coherent power coupling of the modes takes place in much the same fashion as applies to AM-modelocking.

FM-modelocking schemes have been applied successfully to a number of laser systems and notably doped-fibre lasers. Both integrated-optical modulators and bulk modulator devices have been employed and pulse durations in the picosecond regime are frequently obtained.

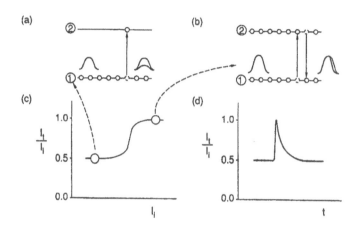

Figure 8. *Basic absorption saturation features of a two-level medium*

3.2 Passive Modelocking

For most of the 1970's and early 1980's reference to passively modelocked lasers generally implied that saturable absorber media, which usually take the form of dye solutions or multiple quantum well semiconductor structures, were involved. In the case of the frequently employed saturable absorber dye solutions the behaviour can be understood by assuming that it can be described in simplified terms to behave as a two-level energy manifold as illustrated in Figure 8. At low levels of propagating intensity the predominant lower state population gives rise to net absorption and so the low-light-level transmission is significantly less than 100% (see Figure 8a,c). For significantly higher intensities, promotion of molecules from energy state 1 to energy state 2 leads to a reduction in the absorption. Indeed, at a sufficiently high intensity the situation depicted in Figure 8b,d may be reached where the population densities in states 1 and 2 may become equal. In other words, the excitation of the dye molecules result from absorption from the propagating radiation. When the populations in states 1 and 2 become equal then the net absorption is zero and the absorber is said to be saturated, bleached or transparent. The saturable absorber can thus be regarded as an 'optical switch' which opens completely in this transparency condition. Recovery of the absorption will occur with a time that is characteristic of the molecules in their solvent host as illustrated in Figure 8d. Thus for a saturable absorber placed within a laser resonator the circulating intracavity flux will influence the transmissivity of this passive medium and a periodic amplitude (or intensity) modulation is thus induced at cavity frequency f_{cav}. As mentioned earlier, an amplitude modulation at the cavity frequency when applied to a particular longitudinal mode will lead to sideband generation at frequencies that are coincident with those of the nearest neighbour modes and phase-locking is thus achieved through an intensity-dependent and entirely passive process. The behaviour of saturable absorbers may also be readily described in the time domain. Given that the output of intracavity radiation of a free-running laser has a stochastic intensity characteristic then the evolution of discrete ultrashort pulses can be considered. In doing so, it is appropriate to mention at the outset that fast or slow recovery saturable absorbers may be employed. For instance, the saturable absorber dye media used with the early flashlamp-pumped Nd:glass laser had fast recovery times, $\tau_{recovery} \leq 10$psec. It is therefore reasonable to suppose that under the appropriate pumping and laser operating conditions that selection could be made of the most intense fluctuation or intensity spike in the circulating radiation. In other words, only this most intense feature would saturate the absorber to the point that it received sufficient preferential gain to evolve at the expense of less intense radiation features. By this means a single discrete pulse could develop and circulate within the resonator such that some transmission through the output coupler of the resonator would thereby constitute the passively modelocked output pulse train or sequence. The typical pulse durations of ~2-10ps are comparable to the recovery time of the saturable dye solution and durations in this regime were observed in the initial, build-up stage of the pulse train produced by lamp-pumped Nd:glass lasers (von der Linde 1972).

With slow saturable absorbers of the type that are used with passively modelocked dye lasers, the recovery time of ~1 nsec is not too different from the cavity periods of several nanoseconds. The evolution of discrete ultrashort laser pulses in this case relies upon a combination of the saturation of the absorption in the absorber medium and of the gain in the laser medium. The pulse evolution can be readily understood by following the illustrative sequence of Figure 9. Saturation of the absorption will most probably be accomplished by a family of fluctuations and a quite extended burst of radiation 'noise' having a duration $\sim \tau_{recovery}$ will be preferentially transmitted. After a substantial number of cavity periods the enhanced intensity of this noise burst may be such that the available population inversion in the active medium may be insufficient to provide gain for all of this selected signal. Consequently, the trailing portion of the burst may receive a deficit of amplification. This is the process of *gain saturation*. It can therefore be seen that the establishment of periodic transparency in the saturable absorber leads to a selection of a dominant noise burst while gain saturation is responsible for subsequent temporal shortening of the evolving pulse. Both the saturable absorption and gain influence the pulse-shaping processes that arise. This process continues during many round trips until a steady-state condition is achieved. This may be where the increasing bandwidth requirement of the shortening pulse reaches a limit, set by a tuning element for example, or where some additional nonlinearity-related effects may arise to compete with the pulse shortening mechanisms. In a flashlamp pumped dye laser the pulse durations may be ~ 1-5 psec, but in a CW laser, subpicosecond (~200fsec)

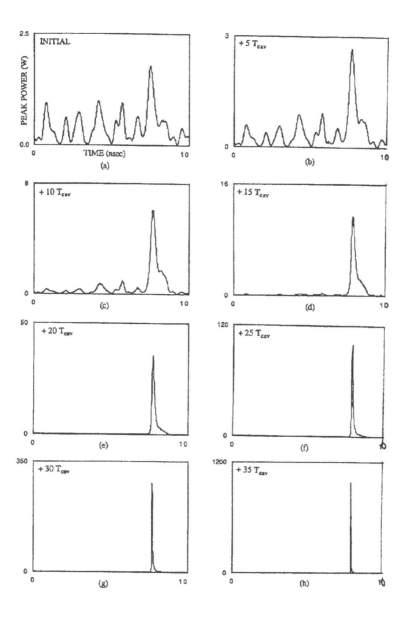

Figure 9. *Combined effects of saturated absorption and gain on pulse evolution*

pulses can be generated in the type of arrangement shown in Figure 10 (Fork *et al.* 1981). For this latter CW ring dye laser the propagation of intracavity flux is permitted in both clockwise and counter-clockwise directions and the pulse evolution process is initiated when the interacting (or colliding) flux has sufficient intensity to saturate the

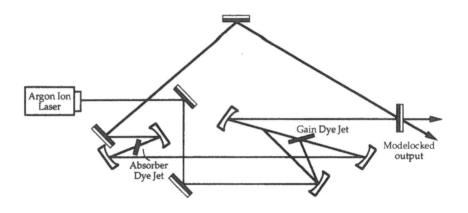

Figure 10. *Schematic of a basic, colliding-pulse modelocked* CW *dye laser*

absorption of the passive medium. It is probably worth noting here that the counter-propagating initial intensity fluctuation patterns, and indeed the subsequent evolving pulses, are unlikely to be identical because of distinct phase and intensity variations experienced at the reflectors and gain medium by the bi-directional intracavity radiation. Also, coherence or interference effects occurring in the saturable absorber will also lead to *amplitude* and *phase related* behaviour rather than simply *intensity-dependent* saturation effects in the passive medium. (Additional optical Kerr effects that influence the passive modelocking are dealt with in the next section.) Because of this configuration, the system is usually described as a *colliding-pulse-modelocked* or CPM dye laser (Fork *et al.* 1981).

The most common gain medium for CPM dye lasers is rhodamine 6G in ethylene glycol solvent and diethyloxadicarbocynanine iodide (DODCI) in ethylene glycol solvent for the saturable absorber medium. With this dye combination, pulse durations are typically $\sim$ 100–200fsec at the optimum operating wavelength around 620–630nm (Fork *et al.* 1981). Although the average output power from such a CPM dye laser is generally only a few milliwatts, the intracavity peak powers $\sim$ 10kW are sufficient to give rise to substantial intensity-dependent phase modulation effects. In particular, the self-phase modulation arising in the absorber-dye solvent and resulting from off-resonant absorber and gain saturation effects (Valdmanis *et al.* 1985) impresses significant amounts of frequency chirp on the circulating ultrashort pulses.

This represented a severe limitation to the pulse shortening process and it was elegantly circumvented by reconfiguring the resonator of CPM dye lasers to include provision for adjustable intracavity group velocity dispersion, GVD, (Fork *et al.* 1984). The scheme that was first implemented involved the incorporation of two pairs of Brewster-angled prisms within a ring resonator as illustrated in Figure 11. The relative separations of prisms P_1, P_2 and P_3, P_4 constitute a controllable amount of negative group velocity (or anomalous) dispersion whereas the transverse movement of prism P_3 pro-

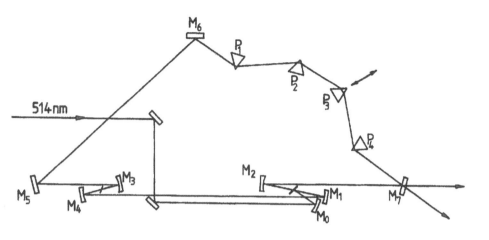

Figure 11. CPM *dye laser having adjustable intracavity group-velocity dispersion*

vides an adjustable path length and hence adjustable positive group velocity (or normal) dispersion (Fork *et al.* 1984). By this means the spectral broadening due to self-phase modulation can be exploited such that optimally short, bandwidth-limited, femtosecond pulses can be generated.

With GVD-compensation in a passively modelocked CW rhodamine 6G dye laser (DODCI as saturable absorber) pulses as short as 27fsec have been directly produced using a six-mirror ring resonator (Valdmanis *et al.* 1985) and 19fsec when the seven-mirror ring-resonator scheme of Figure 11 was employed (Finch *et al.* 1988). Typical output average powers from such CPM dye lasers is generally in the 1–5mW regime which corresponds to extracavity pulse peak power levels in the 100s W regime. These laser pulses have moderate peak powers and with further amplification using high repetition-rate ($\sim$ 5 kHz) copper vapour laser-pumped dye amplifiers (Knox 1988) effective post-amplifier extracavity temporal compression could be applied. Thus, with the generation of optical pulses as short as 6fsec around 630nm (Fork *et al.* 1987), new time-domain spectroscopic studies began to be pioneered in the truly femtosecond regime (Brito-Cruz *et al.* 1986). Although the dye laser pulses were essentially untunable, the generation of self-phase-modulated, white light probe spectra by the amplified pulses was of direct applicability.

The technique of GVD-compensated colliding-pulse modelocking has also been exploited in LiF:F_2^+ colour-centre laser arrangements so that additional spectral regions could be covered (Langford *et al.* 1989). With other colour-centre gain media (notably KCl:Tl, NaCl:OH) femtosecond pulses within the 1400–1700nm range can be generated more conveniently using the coupled-cavity modelocking schemes described below in Section 3.3.

3.3   Coupled-cavity modelocking

The concepts of active, passive and hybrid modelocking have also been extended to
laser configurations that comprise a main cavity which includes the gain medium and a
control or coupled cavity that contains an appropriate nonlinear optical element. This
was most impressively demonstrated by the colour-centre-based soliton laser that was
first reported by Mollenauer and Stolen (1984). An especially attractive feature of
this type of arrangement was that femtosecond pulses could be produced with essen
tially passive modelocking characteristics without any tuning restrictions imposed by
a saturable absorber compound. Instead, the optical Kerr-effect in an anomalously
dispersive monomode fibre was exploited as a means of producing optical solitons
Subsequent theoretical studies (Blow and Wood 1988) and related experimental demon
strations (Kean *et al.* 1989) confirmed that a more general coupled-cavity modelocking
could be established in which the generation of bright optical solitons was not a prereq
uisite. (This modelocking scheme involving a coupled nonlinear cavity is also referred
to as additive-pulse modelocking (Mark *et al.* 1989) and occasionally as interferential
modelocking (Morin and Piché 1989).

The coupled nonlinear cavity approach is exemplified by the schematic of Figure 12
(Although a nonlinear Fabry-Perot is assumed here, similar physical insights may be
applied to alternative configurations such as those involving Michelson and Sagnac in
terferometers.) The gain medium may be excited at a constant level or by a weakly
modulated signal or synchronously by a modelocked pump laser. It follows, therefore
that the radiation that circulates within the main (or master) cavity may take the form
of stochastic noise, noise bursts, or the pulses that evolve from synchronous pumping
An intensity component of this radiation propagates within the nonlinear coupled cav
ity and on its return it thereby interacts with the radiation in the main cavity. An
appreciation of this interaction can give important insights into the initial processes
that lead to pulse evolution and the subsequent shortening/shaping by which pulse
durations into the sub-100fs region can be reliably established.

It is instructive to consider the cavity radiations using a frequency-domain descrip
tion. At optical power levels where intensity-induced nonlinear phase shifts are pro
duced (particularly in the control cavity) the Fabry-Perot resonators have dynamically
varying resonances (or modes). As a consequence of this dynamic process a subset of
the many oscillating cavity modes can become phase-coupled such that primitive pulse
can evolve to have durations in the picosecond regime. (This mechanism has distinc
similarities with the modelocking that has been observed by French and co-workers
for linear control cavity configurations in which the position of one resonator mirror
is oscillated.) For pulse-shortening into a femtosecond regime it is necessary to have
access to significant spectral broadening. This can be conveniently ensured when a
nonlinear element such as monomode optical fibre or semiconductor amplifier has been
incorporated into the control cavity (Kean *et al.* 1989).

The spectrally extended character of the optical signal returning from the control
cavity can have a major influence upon the pulse that circulates within the main cavity
For example, a self-phase-modulated control pulse that experiences substantial group
velocity dispersion can mean that some of its spectral components actually precede
the propagation of the main-cavity pulse through the gain medium. The consequen

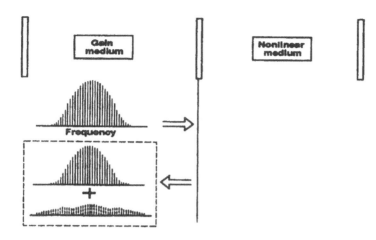

Figure 12. *Generalised scheme for coupled-cavity modelocking*

competitive demand for amplification can therefore precondition the gain-status of the laser medium that is subsequently experienced by the main-cavity pulse. This can imply the preferential gain may enable a pulse to grow with a central wavelength that is significantly shifted from the original noise burst or pulse in the main cavity (Zhu *et al.* 1991). Another key aspect of pulse refinement relates to the competition for gain that arises in the spectral wing regions of the circulating pulses. To highlight this point, particular reference should be made to the relatively weak but spectrally up/down-shifted regions of the control-cavity signal that returns to the gain medium in the main cavity (see Figure 12). For this low-intensity feedback radiation, it is not expected that it would greatly affect the intense central region of the main-cavity-pulse spectrum. By contrast, however, this re-entrant control signal amplitude will be sufficient to influence and even dominate the demand for gain at the wings of the circulating pulse spectrum. This constitutes a self-amplitude modulation and so if the configuration affords appropriate frequency and phase matching between the main and control cavities then an enhanced coupling of the longitudinal modes will result. It is for this reason that the interaction that occurs within the gain medium is highlighted in Figure 12. Thus the associated pulse shortening will lead to increased peak intensities and this process will ensure that a significantly enlarged laser bandwidth will be engaged in the steady-state condition. This explains, in rather simplified terms, how the pulse durations can shorten from the picosecond to the femtosecond regime.

Other experimental work (Mark *et al.* 1989) confirmed the importance of the coherent interaction between the pulses in the main cavity and those returning from control cavities. Indeed, these researchers developed a time-domain model for the pulse-shortening mechanism in terms of an additive-pulse modelocking scheme. Interestingly, independent research carried out by Ouellette and Piché on the passive modelocking of a CO_2 laser, involved the implementation of a Michelson-type, dual-branch cavity with a bulk nonlinear element (germanium) in one branch. The theoretical basis for this approach led to the term interferential modelocking (Morin and Piché 1989). Taken to-

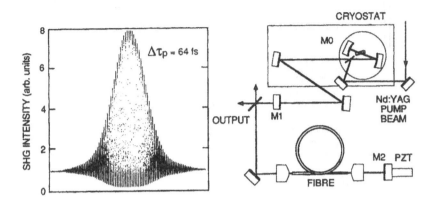

Figure 13. *Cavity configuration for the coupled-cavity modelocked KCl:Tl colour-centre laser. Also shown is a representative interferometric autocorrelation trace where the pulse duration was 64fs*

gether, these studies served to confirm that the original soliton laser of Mollenauer and Stolen (1984) and the passively modelocked CO_2 laser of Ouellette and Piché (1986) were particular examples of the more general scheme of coupled-cavity modelocking.

Although semiconductor amplifier (Kean *et al.* 1989) and passive waveguide (Grant *et al.* 1993) devices have been used successfully, the more typical nonlinear component in the control cavity is a length of monomode optical fibre. A representative experimental arrangement for a nonlinear Fabry-Perot type of coupled cavity is illustrated in Figure 13 for a KCl:Tl colour-centre laser. For this work, the nonlinear-cavity incorporated an erbium-doped erbium fibre which exhibits an enhanced nonlinearity compared to Er- free fibre (Zhu *et al.* 1989). The inset showing an interferometric autocorrelation trace of the output pulse sequence, confirms the excellent quality and inherent stability of this coupled-cavity modelocking procedure.

As an alternative coupled-cavity configuration, the nonlinear Michelson scheme has been extensively evaluated with colour-centre gain media (Grant and Sibbett 1981). A major practical benefit afforded by the nonlinear Michelson-cavity is that it is possible to retain the piezo-controlled relative phase adjustment between the main and control cavity branches while allowing a second piezo-mounted mirror incorporated within the main body of the laser cavity (Figure 14) to simultaneously control the length of both branches of the laser. By this means, the pulse repetition frequency of the coupled-cavity modelocked laser could be synchronised to that of a crystal oscillator. The implementation of this scheme in a KCl:Tl laser system enabled the long-term pulse timing jitter to be reduced to less than 200fs in the 50Hz-50kHz frequency range (Walker *et al.* 1992). This synchronisation technique can also be adapted very effectively for the amplitude stabilisation of a coupled-cavity modelocked NaCl:OH- colour-centre laser (Kennedy *et al.* 1993). The laser-active colour-centres in this crystal host have a relatively short

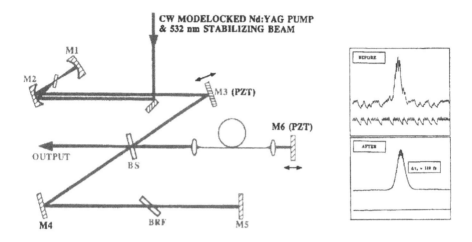

Figure 14. *Amplitude-stabilised, coupled-cavity modelocked NaCl:OH- colour-centre laser. Inset: autocorrelation traces before and after cavity stabilisation*

upper-state lifetime of 150ns. When the gain crystal is excited by a modelocked Nd:YAG pump laser then instabilities can arise whenever the coupled-cavity modelocking (having passive characteristics) begins to dominate the synchronous modelocking (having active characteristics). This is because of beating between the repetition frequencies of the colour-centre and the pump Nd:YAG lasers. The provision of intracavity piezo-control enables the coupled-cavity laser output to be synchronised to that of the pump laser (rather than to a reference oscillator) and so the beating effects which give rise to substantial parasitic amplitude modulation can be eliminated. It has thus been possible to demonstrate an amplitude-stabilised, coupled-cavity modelocked NaCl:OH- laser that is fully compatible to demanding source-specifications relating to time-resolved spectroscopy in the near-infrared (Kennedy *et al.* 1993 (See inset of Figure 14.))

Coupled-cavity modelocking has become well established for a range of other types of picosecond and femtosecond lasers. Good examples for the picosecond regime are those of Nd:YAG (YLF) lasers especially where diode laser pumping is employed (Malcolm *et al.* 1990). In addition to the femtosecond colour-centre lasers mentioned above, the coupled-cavity modelocking of titanium-doped sapphire lasers has also been demonstrated. Although pulse durations as short as 90fs have been reported (Spence *et al.* 1991) the higher operating powers and higher group velocity dispersion effects in the control fibres rendered these systems to be significantly less practical than their colour-centre counterparts. Fortunately, the self-modelocking technique that is described in the next section has been shown to be better suited to controlled ultrashort-pulse generation in the higher power vibronic lasers.

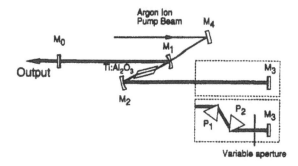

Figure 15. *Archetypal configuration of a self-modelocked titanium-sapphire laser*

3.4   Self-modelocking technique

On the basis of extensive experimental evaluations of coupled-cavity modelocking in a Ti:sapphire laser using both nonlinear Fabry-Perot and Michelson cavity arrangements, we reported (Spence *et al.* 1991) that the behaviour of our laser was not strongly dictated by the nonlinearities in the optical fibre contained in the control cavity. It was deduced that the nonlinearity inherent in the $Ti:Al_2O_3$ laser rod was influencing the modelocking in a very distinctive fashion. In fact, the inference was that in the case of a suitable gain medium, the inherent optical Kerr-effect would give rise to intensity-dependent self-phase modulation and self-focusing that would be sufficient to dominate the operational status of the laser - notably the establishment of a stable modelocking regime. However, for a single cavity configuration such as that illustrated in Figure 15 a key issue is the mechanism by which an initial pulse (or noise burst) can build up to an intensity level that is sufficient to induce appreciable and exploitable nonlinear effects.

For the original demonstration of self-modelocking in a $Ti:Al_2O_3$ laser (Spence *et al.* 1991) a fairly extended ($\sim$1.5–2.0m) resonator geometry was used which led to reduced beam diameters in the gain medium. For the pump mode volume produced in the gain medium by the argon-ion beam, it was thus possible for the $Ti:Al_2O_3$ laser to operate simultaneously in both a higher-order transverse mode and the fundamental transverse mode. Because of the Brewster-angled rod, the typical higher-order modes were TEM_{03}, TEM_{05} etc. In the initial setup, the resonator had an essentially concentric geometry and in common with an equivalent Fabry-Perot interferometer, it had the feature that the axial mode frequency separation was degenerate for all of the transverse modes (See Figure 16(a)). It should be recognised, however, that when the $Ti:Al_2O_3$ laser operates simultaneously in two transverse modes there is gain competition because a higher-order transverse mode is spatially non-degenerate with the fundamental counterpart (see illustration in Figure 16(b)). For this reason it can be appreciated that the gain competition that arises constitutes a type of self-amplitude modulation which was previously introduced to describe the physical basis of coupled-cavity modelocking (Section 3.3). As a consequence, therefore, this amplitude modulation at the cavity frequency, taken together with the initial mode beating in the homogeneously-broadened

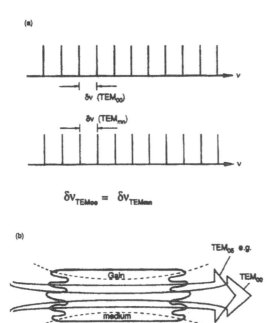

Figure 16. *(a)Degenerate mode frequencies associated with a fundamental (TEM$_{00}$) and higher-order (TEM$_{mn}$) transverse modes of a concentric resonator and (b) illustration of spatially distinct propagating fundamental and higher-order transverse modes*

Ti:Al$_2$O$_3$ medium, leads to a phase coupling of longitudinal modes and a primitive pulse can evolve. Provided that this primitive signal has sufficient intensity it can then experience the principal optically nonlinear effects of self-phase modulation and self-focusing that are highlighted later in this section.

It is this combination of *self*-amplitude modulation, *self*-phase modulation and *self*-focusing that led to the term *self*-modelocking. The technique is also frequently referred to as Kerr-Lens modelocking (Spinelli *et al.* 1991).

Although the original demonstration of self-modelocking involved a dominant TEM$_{00}$ mode accompanied by a weaker, 'enabling' TEM$_{0n}$ mode this merely represented one example of a passive mechanism by which the modelocking process could be initiated. It was recognised and stated from the outset that alternative passive or active intracavity amplitude (or intensity) modulation at the cavity frequency would produce a similar initiating effect. Indeed, the Author suggested that a very weak concentration saturable absorber solution would suffice to initiate what was undoubtedly a primarily self-modelocking process. The validity of this suggestion was later confirmed by several research groups who showed that a *cocktail* of weak dyes would permit a relatively wide tuning range (Sarakura and Ishida 1992). In our own laboratories, thin, polished samples ($<100\mu$m) of semiconductor-doped glasses (eg type RG 830 Schott Filter) have been used. Several active initiation schemes for the modelocking have also been demonstrated. These have included cavity-length modulation (Spinelli *et al.* 1991), coupled-cavities (Keller *et al.* 1991), regenerative acousto-optic modulation (Spence *et al.* 1991) and synchronous gain modulation (Spielmann *et al.* 1991). It probably bears repeating here that any one of these modulation options serves to assist in the evolution of an enabling pulse that ultimately leads to a modelocked operation in preference to a free-running CW regime.

The subsequent process by which picosecond and femtosecond pulses can evolve from the enabling pulse (or noise burst) mentioned above must now be considered. For such pulses propagating through a Ti:Al$_2$O$_3$ rod, the time-varying intensity-induced refractive index changes give rise to substantial self-phase modulation. This additional bandwidth when combined with appropriate control of the intracavity group velocity dispersion means that significant pulse shortening can take place. Moreover, the accompanying spatial variation of the induced nonlinear phase shift (*i.e.* the self-focusing effect) will also influence the kinetics of the pulse shaping and shortening process. For instance, the incidence of self-focusing will produce a distinct gain-guiding contribution in respect of the propagating laser beam through the pumped region of the Ti:Al$_2$O$_3$ rod. Additionally, even modest gain saturation and intensity-induced self-focusing effects will complement this gain-guiding influence. Consequently, it is the interplay of these intensity-induced nonlinear effects that gives rise to a stable self-modelocking status where femtosecond pulses can be produced across an uncompromised tuning range for a particular gain medium.

It should be pointed out that cavity configurations can be selected to enhance the self focusing. The basis for this is conveyed in Figure 17 where the gain medium is assumed to exhibit a significant self-focusing effect at higher intracavity optical power levels. From Figure 17(a) it can be seen that a hard aperture that is deliberately and strategically located at an induced beam waist region in the cavity can provide a useful self-amplitude modulation whereby the transmission increases at a higher intensity level (this is comparable with the behaviour of a saturable absorber). This type of beam/spot size modification has been successfully implemented in both femtosecond and picosecond versions of self-modelocked lasers. Alternatively, a soft-aperture counterpart can exist as illustrated in Figure 17(b) where the laser beam diameter can be modified within the pump mode volume. This gives rise to a combined self-focusing and gain-guiding influence.

As an example of a frequently-used cavity arrangement for a femtosecond vibronic laser, Figure 18 shows the principal features of a regeneratively-initiated, self-modelocked titanium-sapphire laser (Spence *et al.* 1991). With CW pumping from the argon-ion laser, initial longitudinal mode beating can provide an optical modulation which can be detected by a photodiode. This electronic output can be suitably filtered, pre-amplified, frequency-divided and amplified to provide a drive signal to an intracavity acousto-optic modulator as indicated. Because the laser radiation itself provides the input signal for this process, it is referred to as regenerative modulation (or modelocking). When a stable femtosecond-pulse regime has been established, it is usually possible to switch off

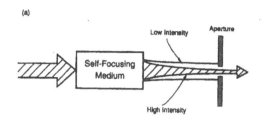

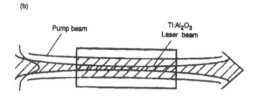

Figure 17. *Self-focusing effects involving (a) a hard aperture, and (b) a soft aperture with a gain guiding influence*

this acousto-optic modulation and the laser will continue to operate. However, for some cavity configurations, self-modelocking does not reinitiate itself automatically following some perturbation to the laser. For this reason, this regenerative modulation is generally retained on a continuous basis, rather than just for the initiation of modelocking. Therefore, although the modulator material can contribute some group velocity dispersion, this can be compensated by appropriate adjustment of the intracavity prisms P_1, P_2. A stable sequence of femtosecond pulses (durations generally less than 50fs; see inset of Figure 18) can be produced with average power levels in excess of 1W for argon-ion pump powers around 10W. The peak powers associated with the pulses generated directly by a self-modelocked Ti:Al$_2$O$_3$ laser are thus in the 500kW range. This is typically three orders of magnitude greater than that for the femtosecond pulses produced by a passively modelocked CW dye laser! Access to the full tuning range afforded by the broadband gain media is also available through this self-modelocking technique. For the Ti:Al$_2$O$_3$ crystal, this is $\sim$700–1070nm but two or three mirror sets for the laser cavity are generally required to give this spectral coverage.

From the viewpoint of generating pulses having durations in the sub-10fs range, then a broad gain bandwidth is essential. Moreover, particular attention has to be paid to third and higher-order dispersive effects in the laser to ensure that optimally short pulses can be produced (Lemoff and Barty 1993). To date, the shortest pulses having durations of 8fs, have been produced by self-modelocked Ti:Al$_2$O$_3$ lasers where the dispersion compensation has been achieved either using selected prism pairs (Zhou *et al.* 1994) or specially designed chirped dielectric mirror sets (Stingl *et al.* 1995). For these pulse durations the bandwidths are relatively broad at 105nm but adequate gain bandwidth is available for the achievement of significantly shorter pulses provided the higher-order dispersions can be properly controlled.

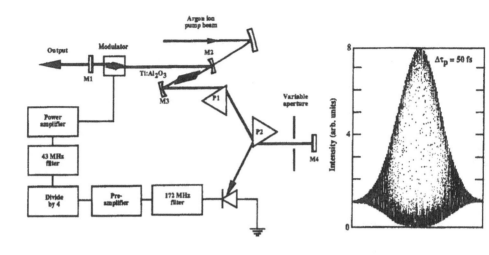

Figure 18. *Cavity configuration for the regeneratively initiated self-mode-locked Ti:Al₂O₃ laser. Also shown is a typical interferometric autocorrelation trace of the mode-locked output where the pulse duration was 40fs*

Although most of the initial research on the self-modelocking technique has involved titanium-doped sapphire as the gain medium, femtosecond pulse generation using other vibronic crystals has also been successfully achieved using this methodology. Important systems include Cr^{3+}:LiCaAlF₆ (LiKam Wa *et al.* 1992), Cr^{3+}:LiSrAlF₆ (Evans *et al.* 1992, Cr^{3+}:LiSrGaF₆ (Yanovsky *et al.* 1995, Cr^{4+}:forsterite (Seas *et al.* 1992) and Cr^{4+}:YAG (Sennarglu *et al.* 1993), where the latter two address the 1.2–1.3μm and 1.5–1.6μm spectral regions respectively. Pr^{3+}:LiYF₄ has also been self-modelocked to produce 400fs pulses in the red around 613nm (Ruan *et al.* 1995). The operational robustness of self-modelocked lasers in general is such that they afford excellent versatility in configurational aspects of their design. A good example of this is the cavity configurational enhancement that enables two independently tunable sequences of pulses to be produced simultaneously by a single self-modelocked laser. This was first demonstrated for the Ti:Al₂O₃ laser where three destructive approaches were evaluated (deBarros and Becker 1993, Dykaar and Darack 1993, Evans *et al.* 1993). The schematic of the system set up in our laboratory is shown in Figure 19. It is of similar basic design to that given earlier as Figure 15 but it now incorporates an additional prism and output coupler. Bandwidth limitations and tuning of each output beam is achieved by adjusting the variable slit apertures.

For a given set of cavity optics, the tunability of the two outputs was 700–820nm and 830–870nm. Interestingly, with this configuration several different modes of operation are possible. In the entirely regeneratively modelocked regime ~60ps pulses were produced whereas in the self modelocking regime the pulse durations were ~100fs. In

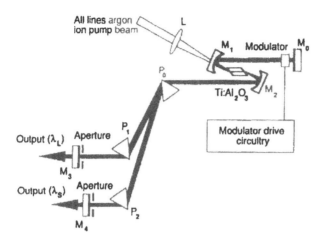

Figure 19. *Diagram showing the cavity configuration for the dual-wavelength self-mode-locked Ti:Al$_2$O$_3$ laser*

a third distinctive mode of operation, the pulse at one spectral band had femtosecond durations while those at the second band had picosecond durations. When the degree of synchronisation was measured using cross-correlation techniques, it was readily confirmed that the pulse sequences were produced with negligible relative timing jitter. The deduced value of 26fs for sech^2 profile pulses was considered to be a measurement limit (Evans *et al.* 1993). Indeed, in a subsequent work on a two-colour, self-modelocked Ti:Al$_2$O$_3$ laser (which shares design concepts with the above) it has been reported that the relative jitter is less than 2fs (Leitenstorfer *et al.* 1995). With average output powers of 100s of mW in each output, this type of laser is especially well suited to further improve the efficiency of difference-frequency mixing which provides femtosecond pulses that are widely tunable in the mid-to-far infrared regions.

Other cavity designs have been reported such that significantly higher pulse repetition frequencies can be achieved. An especially straightforward, though elegant, design (illustrated in Figure 20) was reported by Ramaswamy-Paye and Fujimoto (1994). Pulses ~50fs were produced at a repetition frequency of 385 MHz and related cavity designs enable this to be increased to beyond 1 GHz but for somewhat greater pulse durations.

Because the initiation of the self-modelocking process usually requires some mechanical perturbation (such as tapping a resonator-mirror mount) there has been considerable research effort directed at establishing the conditions that would enable the self-modelocking to self-start. Of the work reported on this topic to date, specific cavity design considerations (Cerullo *et al.* 1994) can afford greater practical tolerance than the use of either unidirectional ring resonators (Tamura *et al.* 1993) or an additional intracavity nonlinear element such as a ZnS component (Radzewicz *et al.* 1993). Within

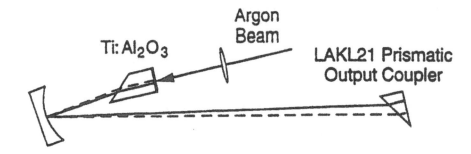

Figure 20. *Cavity configurations for high repetition-rate, femtosecond Ti:Al$_2$O$_3$ laser*

the modelling procedures developed by Cerullo and co-workers, a detailed theoretical framework is established and this permits the cavity design for self-modelocked lasers in particular to be systematically developed. In this approach a ray-transfer matrix analysis is used to determine the small-signal relative spot-size variation at an intracavity slit such that the higher peak-intensity regime associated with modelocked pulses can have a favoured cavity-stability status. This can be sufficient to satisfy a self-starting condition whereby the self-modulation operation is initiated spontaneously and this has been demonstrated experimentally by a number of research groups for a range of lasers since the original work of Cerullo *et al.* (1994).

By way of illustration, consider a four-mirror standing-wave resonator that is typical of that for a self-modelocked laser (see Figure 21(a)). The folding or focusing section of this cavity is of key importance where the separations x, z are of particular relevance. From the ABCD matrix analysis the small-signal relative spot-size variation, δ, at the slit, given by

$$\delta = \left(\frac{1}{w} \frac{dw}{dP} \right)_{P=0}$$

depends strongly upon the folding mirror separation (z) and the laser-crystal position (x). (w is the spot size and dw/dP is the rate of change of spot-size with beam intensity P). The analysis produces a plot of the type reproduced in Figure 21(b) for a Ti:Al$_2$O$_3$ laser where cavity stability data are complemented with a clear indication of the strongly-negative-δ regions where the intracavity beam is narrower for the high-peak-power modelocked pulses. An intracavity slit that is appropriately located within the resonator can therefore be used to discriminate in favour of the ultrashort-pulse regime. With suitable design, δ tends to negative infinity at the centre of a single, broad stability region and this requires that the cavity should be symmetrical about the laser crystal.

Cavity configurations that are optimised for the self-starting of self-modelocked, are of general interest and importance. Additionally, however, in the case of the all-solid-state versions, described in the next subsection cavity designs can be selected to suit

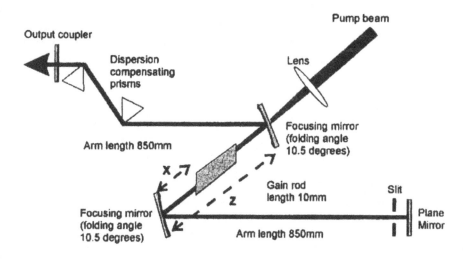

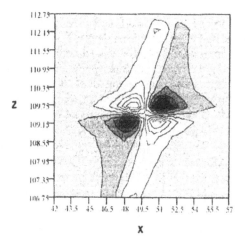

Figure 21. *Schematic of a typical 4-mirror cavity used in a self-modelocked Ti:sapphire laser and corresponding plot of spot-size variation δ as a function of separations x,z*

low threshold modelocking operations in particular.

3.5 All-solid-state, self-modelocked vibronic lasers

Considerable recent research effort has been devoted to the realization of all-solid-state femtosecond lasers that do not rely upon optical pumping by inefficient and expensive ion (Ar,Kr) or arclamp-pumped Nd:YAG CW lasers. The two approaches involve either diode-pumped minilaser pumping or direct diode laser pumping and representative examples of each type are included below.

From the viewpoint of using a frequency-doubled Nd:YAG (or YLF) diode-pumped minilaser as an alternative to a higher-power argon-ion laser for pumping a femtosecond Ti:Al$_2$O$_3$ laser a practical consideration becomes the achievement of a reduction in the threshold pumping power requirement for the self-modelocking regime. In contrast to the multi-watt pump power levels that are typical for Ar-ion pumped, femtosecond Ti:Al$_2$O$_3$ lasers, the selection of high quality laser crystals of optimised length and absorption parameters enabled a self-modelocking threshold to be demonstrated for a pump power as low as 400mW from a frequency-doubled Nd:YLF minilaser (Lamb *et al.* 1994). The experimental arrangement used in this work is shown in Figure 22 together with an autocorrelation trace (inset) for the 110fs pulses produced at a 700mW pump power level. Typical output powers at around 800nm are 20mW. Power-scaling of the green minilaser pump lasers to the >4W regime is currently underway in several laboratories and this is expected to lead to performance levels more typical of those of the ion-pumped counterparts.

Although a self-modelocked Cr:LiSrAlF$_6$ (Cr:LiSAF) laser can also produce sub-100fs pulses when pumped at >0.5W by a green Nd:YLF minilaser it is evident from its absorption spectrum (see Figure 23) that excitation is more efficient in the red spectral region. In fact, the availability of relatively high power (>0.5W) AlGaInP diode lasers makes this gain medium very well suited to direct diode pumping. Whereas, some of the initial self-modelocking studies of the diode-pumped Cr:LiSAF involved saturable absorbers for the initiation of the self-modelocking process (French *et al.* 1993, Kopf *et al.* 1994) more recent results relate to regeneratively-initiated arrangements (Dymott *et al.* 1994). The resonator designs are generally similar to those for the Ti-sapphire lasers where Brewster-angled laser crystals are used. With an emphasis on low-threshold self-modelocking operation for a Cr:LiSAF laser, a plane-plane laser-crystal geometry can be shown to be effective in minimising astigmatism in the gain medium.

In order to better match the pump laser beam to the confocal parameter of the beam in the Cr:LiSAF laser a self-injection-locked AlGaInP diode laser has been employed (see Figure 24). Indeed, with a pair of SF14 prisms for dispersion compensation, pulses of <100fs were produced for an incident pump power level of 120mW and modelocking could be established at a pump power as low as 88mW. A major contribution to this observed reduction of pump power for the self-modelocking operation was due to self-injection locking of the AlGaInP diode laser. By using the basic feedback scheme illustrated in Figure 24 the beam quality of the pump beam was substantially enhanced and thereby provided a much better spatial match to the beam conformation in the Cr:LiSAF crystal. The layout of the cavity for self-modelocking was pre-analysed using the ray-transfer matrix analysis due to Cerullo *et al.* (1994). It is of perhaps tutorial relevance to comment here on the particularly significant influence of intracavity dispersive effects on the self-modelocking process under the conditions of low power circulating

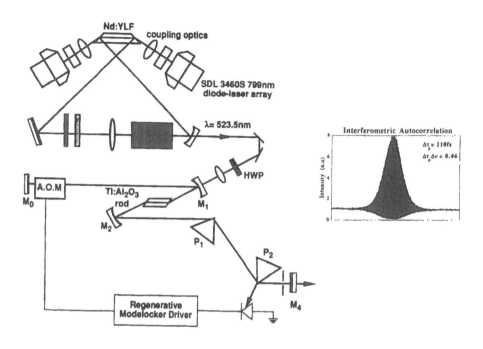

Figure 22. *Minilaser-pumped, femtosecond Ti:Al₂O₃ laser*

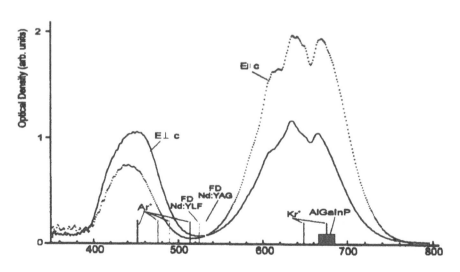

Figure 23. *Absorption spectra for Cr:LiSAF at room temperature.*

Wilson Sibbett

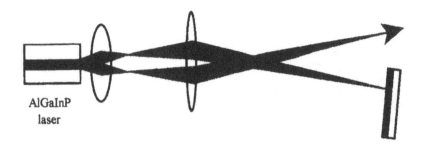

AlGaInP
laser

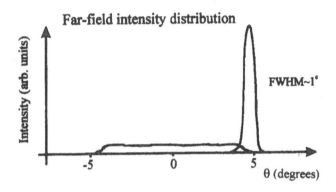

Figure 24. *Self-injection-locked AlGaInP diode laser used to pump self-modelocked Cr:LiSAF laser*

pulses. The duration of an ultrashort pulse in the laser cavity will be determined primarily by an interplay between intracavity dispersion and self-phase modulation. A dispersion length

$$L_D = \frac{T_0^2}{|\beta_2|}$$

(T_0 is the pulse duration, β_2 is the group delay dispersion) and a nonlinear length

$$L_N = \frac{1}{P_0\gamma}$$

(P_0 is peak pulse power and γ is the nonlinearity coefficient given by $\frac{n_2\omega_0}{cA_{\text{eff}}}$ (Agrawal 1989)). When L_D, L_N are significantly longer than the length of the laser cavity then the pulse behaviour can be considered using theory developed to describe optical soliton propagation in monomode optical fibres. The duration of such a soliton can be expressed as:

$$\tau_{\text{FWHM}} = \frac{4\ln(\sqrt{2}+1)}{L\gamma}\frac{|\beta_2|}{W} \tag{16}$$

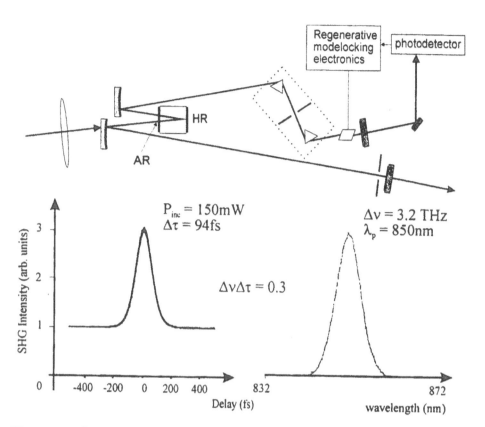

Figure 25. *Cavity configuration for the self-modelocked, diode-pumped Cr:LiSAF laser together with autocorrelation and spectral data*

where W is the pulse energy and L is the nonlinear interaction length (Agrawal 1989). From the relationship in Equation 16 for the solitonic pulse duration, it follows that longer pulses are to be expected at lower intracavity power levels. The generation of sub-100fs pulses requires that the group delay dispersion, β_2, be minimised. However, the phase shift, ϕ, in the laser cavity may be expressed as a Taylor series around the optical frequency ω_0 where:

$$\phi = \phi_0 + \beta_1(\omega - \omega_0) + \beta_2\frac{(\omega - \omega_0)^2}{2!} + \beta_3\frac{(\omega - \omega_0)^3}{3!} + ... \qquad (17)$$

This gives rise to the round-trip group delay, $\tau = \frac{d\phi}{d\omega}$ and we can define $\delta\tau = \tau - \beta_1$ such that the frequency-dependent relative group delay per cavity round trip is:

$$\Delta\tau = \beta_2(\omega - \omega_0) + \frac{\beta_3(\omega - \omega_0)^2}{2!} + \frac{\beta_4(\omega - \omega_0)^3}{3!} \qquad (18)$$

As attempts are made to reduce the second-order dispersive effects then the third-order dispersion (ie β_3) becomes increasingly significant. This third-order dispersion represents a linear variation of the group-delay dispersion across the bandwidth of a pulse and it can manifest itself as pulse broadening, the appearance of resonant sidebands and possibly the compromise of the modelocking process itself. A pulse cannot, in general, be shortened below a duration at which this third-order dispersion causes a significant change in the group delay dispersion across the pulse bandwidth—see Spielmann *et al.* 1994. This represents a limiting factor in attempts to demonstrate sub-100fs pulse generation at reduced levels of pump power and therefore intracavity laser power. It is thus vital to minimise the round-trip third-order dispersion. For instance, the negative third-order dispersion due to a prism pair can be used to balance that due to the gain medium and other intracavity elements in such a way that the round-trip third-order dispersion can be low or essentially zero—see Lemoff and Barty, 1993.

By compensating for higher-order dispersive effects it has been possible to demonstrate sub-10fs pulse durations in self-modelocked Ti:Al$_2$O$_3$ laser systems (Zhou *et al.* 1994, Stingl *et al.* 1995). For such broadband pulses the use of appropriately chirped multilayer dielectric mirrors is especially attractive because they produce an approximately constant negative group delay dispersion (Stingl *et al.* 1995) in contrast to prism pairs (Zhou *et al.* 1994). Also by applying the design criteria reported by Lemoff and Barty (1993), it has recently been demonstrated that pulses as short as 34fs can be produced by a diode-pumped, self-modelocked Cr:LiSAF laser (Dymott and Ferguson 1995). Additionally, low-threshold self-modelocking of a diode-pumped Cr:LiSAF laser has been demonstrated in our laboratory. (See Figure 25).

As already mentioned above, within the context of femtosecond pulse generation at low intracavity power levels, the third-order dispersive effects are of primary importance. By selecting LaKL21 prism material it has been possible to demonstrate in our labs that 60fs pulses can be produced for a diode pump power of 150mW (see Figure 26(a)) and 97fs pulse at pump powers of just 99mW (see figure 26 (b)). With further optimisations based on the existing knowledge base relating to cavity design criteria and intracavity dispersion control, it is likely that yet lower thresholds for self-modelocking will be demonstrated.

Ultracompact, frequency-tunable femtosecond laser sources are thus rapidly becoming a practical reality and taken together with ongoing developments of diode-based amplifier arrangements, such oscillator/oscillator-amplifier systems will be appropriate for a wide range of applications in the femtosecond regime. Given also that synchronously-pumped femtosecond optical parametric oscillators have been demonstrated at threshold pump powers as low as 50mW (Reid *et al.* 1995) the prospects of all-solid-state sources having near-to-mid infrared tunability appear especially promising. Moreover, the production of high optical quality colquirite-crystals such as Cr:LiSrGaF$_6$ afford the likelihood of power scalability because of their improved thermo-optical and thermo-mechanical properties. Research to date has already shown that sub-100fs pulses can be generated by a diode-pumped, self-modelocked Cr:LiSGAF laser—see Yanovsky *et al.* 1995.

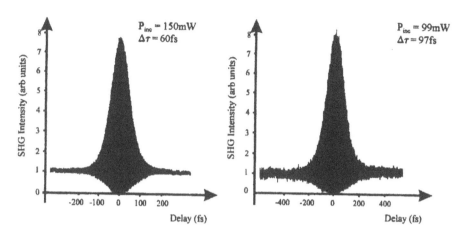

Figure 26. *Autocorrelation profiles for pulses produced by the diode-pumped Cr:LiSAF laser configuration of Figure 25.*

4   Conclusion

The underlying physical concepts by which ultrashort pulses can be produced by a variety of laser systems are now well understood. In modern systems where the optical Kerr nonlinearity is exploited either in a control cavity (ie coupled-cavity or additive-pulse modelocking) or in a single cavity (ie self-modelocking or Kerr-Lens modelocking) the access to the full tuning range offered by broadband gain media is especially relevant to time-resolved spectroscopic studies. Also with the generation of hypershort, high peak intensity pulses using complementary chirped-pulse amplification techniques (Strickland and Mourou 1985) it is now possible to obtain multi-terawatt powers from tabletop-sized femtosecond-pulse oscillator-amplifier combinations and these have already been impressively employed in novel scientific applications (Lemoff *et al.* 1994). It is therefore to be expected that self-modelocked lasers in particular will constitute a principal building block of many of the laser system implementations that will be used in future applications that will extend from X-ray pulse generation to time-domain spectroscopy in the mid-to-far infrared spectral regions.

References

Agrawal G P, 1989,*Nonlinear Fibre Optics, Acad Press Inc.*
Beaud P, 1990, *Opt Commun* **80** 31.
Blow K J and Wood D, 1988, *J Opt Soc Am B* **5** 629.
Brito-Cruz C H, Fork R L, Knox W H and Shank C V, 1986, *Chem Phys Lett* **132** 341.
Cerullo G, DeSilvestri S and Magni V, 1994, *Opt Lett* **19** 1040.
de Barros M R X and Becker P C, 1993, *Opt Lett* **18** 631.
de Barros M R X, Miranda R S, Jedju T M and Becker P C, 1995, *Opt Lett* **20** 480.

DiDomenico M, 1964, *J Appl Phys* **35** 2870.

Dykaar D R and Darack S, 1993, *Opt Lett* **18** 634.

Dymott M J and Ferguson A I, 1994, *Opt Lett* **19** 1988.

Dymott M J P and Ferguson A I, 1995, *Opt Lett* **20** 1157.

Evans J M, Spence D E, Burns D and Sibbett W, 1993, *Opt Lett* **18** 1074.

Evans J M, Spence D E, Sibbett W, Chai B H T and Miller A, 1992, *Opt Lett* **17** 1447.

Finch A, Chen G, Sleat W E and Sibbett W, 1988, *J Mod Opt* **35** 345.

Fork R L, Brito-Cruz C H, Becker P C and Shank C V, 1987, *Opt Lett* **12** 483.

Fork R L, Greene B I and Shank C V, 1981, *Appl Phys Lett* **48** 671.

Fork R L, Martine O E and Gordon J P, 1984, *Opt Lett* **9** 150.

French P M W, Kelly S M J and Taylor J R, 1990, *Opt Lett* **15** 378.

French P M W, Mellish R., Taylor J R. Delfyett P J and Florez L T, 1993, *Opt Lett* **18** 1934.

Grant R S and Sibbett W, 1981, *Opt Commun* **86** 177.

Grant R S, Su Z, Kennedy G T, Sibbett W and Aitchison J S, 1993, *Opt Let* **18** 1600.

Haung C P, Asaki M T, Backus S, Murnane M M, Kapteyn H C and H Nathel, 1992, *Opt Lett* **17** 61.

Johnson A M and Simpson W M, 1983, *Opt Lett* **8** 554.

Kane D J and Trebino R, 1993, *Opt Lett* **18** 823.

Kean P N, Zhu X, Crust D W, Grant R S, Langford N and Sibbett W, 1989, *Opt Lett* **14** 39.

Keller U, 'tHooft G W, Knox W H and Cunningham J E, 1991, *Opt Lett* **16** 1022.

Kennedy G T, Grant R S, Sleat W E and Sibbett W, 1993, *Opt Lett* **18** 208.

Knox W H, 1988, *IEEE J Quantum Electron* **24** 388.

Kopf D, Weingarten K J, Brovelli L R, Kaurp M and Keller U, 1994, *Opt Lett* **19** 2143.

Lamb K, Spence D E, Hong J, Yelland C and Sibbett W, 1994, *Opt Lett* **19** 1864.

Langford N, Grant R S, Johnston C I, Smith K and Sibbett W, 1989, *Opt Lett* **14** 45.

Leitenstorfer A, Furst C and Laubereau A, 1995, *Opt Lett* **20** 916.

Lemoff B and Barty C P J, 1993, *Opt Lett* **18** 57.

Lemoff B E, Barty C P J and Harris S E, 1994, *Opt Lett* **19** 569.

LiKamWa P, Chai B H T and Miller A, 1992, *Opt Lett* **17** 1447.

Malcolm G P A, Curley P F and Ferguson A I, 1990, *Opt Lett* **18** 1303.

Mark J, Liu Y L, Hall K L and Hans H A, 1989, *Opt Lett* **14** 48.

Mollenauer LF, 1985, *Handbook of Lasers* M L Stitch and M Bass, eds., North Holland Amsterdam.

Mollenauer L F and Stolen R H, 1984, *Opt Lett* **9** 13.

Mollmann K and Gellermann W, 1994, *Opt Lett* **19** 490.

Morin M and Piché M, 1989, *Opt Lett* **14** 1119.

New G H C, 1990, *Opt Lett* **15** 1306.

New G H C, 1983, *Rep Prog Phys* **46** 877.

Ouellette F and Piché M, 1986, *Opt Commun* **60** 99.

Reid D T, Ebrahimzadeh M and Sibbett W, 1995, *Appl Phys B* **60** 437.

Ruan S, Sutherland J M, French P M W, Taylor J R and Chai B H T, 1995, *Opt Lett* **20** 1041.

Radzewicz C, Pearsen G W and J S Krasinki, 1993, *Opt Commun* **102** 464.

Ramaswamy-Paye M and Fujimoto J G, 1994, *Opt Lett* **19** 1756.

Sarakura N and Ishida Y, 1992, *Opt Lett* **17** 61.

Seas A, Petricevic V and Alfano R R, 1992, *Opt Lett* **17** 937.

Sennarglu A, Pollock C R and Nathel H, 1993, *Opt Lett* **18** 826.

Spence D E, Evans J M, Sleat W E and Sibbett W, 1991, *Opt Lett* **16** 1762.

Spence D E, Kean P N and Sibbett W, 1991, *Opt Lett* **16** 42.

Spielmann Ch, Curley P F, Brabec T and Krausz F, 1994, *IEEE J Quantum Electron* **39** 1100.

Spielmann Ch, Krausz F, Brabec T, Wintner E and Schmidt A J, 1991, *Opt Lett* **16** 1180.

Spinelli I, Couillaud B, Goldblatt N and Negus D K, 1991, *Tech Digest Conf on Lasers and Electro-Optics* paperCPD97 583.

Stingl A, Lenzner M, Spielmann Ch, Krausz F and R Szipöcs R, 1995, *Opt Lett* **20** 603.

Strickland D and Mourou G, 1985, *Opt Commun* **56** 219.

Tamura K, Jacobson J, Ippen E P, Haus H A and Fujimoto J G, 1993, *Opt Lett* **18** 220.

Tucker R S, Koren U, Raybon G, Burrus C A, Miller B I, Koch T L and Eisenstein G, 1989, *Electron Lett* **25** 621.

Valdmanis J A, Fork R L and Gordon J P, 1985, *Opt Lett* **10** 131.

von der Linde D, 1972, *IEEE J Quantum Electron* **QE-8** 328.

Walker D R, Crust D W, Sleat W E and Sibbett W, 1992, *IEEE J Quantum Electron* **28** 289.

Yanovsky V P, Wise F W, Cassanho A and Jenssen H P, 1995, *Opt Lett* **20** 1304.

Zhu X, Kean P N and Sibbett W, 1989, *Opt Lett* **14** 1192.

Zhu X, Sleat W, Walker D and Sibbett W, 1991, *Opt Commun* **82** 406.

Zhou J, Taft G, Haung C-P, Murnane M M, Kapteyn H C and Christov I P, 1994, *Opt Lett* **19** 1149.

Applications of Ultrashort Optical Pulses

A Mysyrowicz

Laboratoire d'Optique Appliquée,
ENSTA - École Polytechnique, Palaiseau, France.

1 Introduction

The full impact that the recent progress in femtosecond laser technology will have on science and technology can hardly be overstated. Important advances have already been achieved in the understanding of ultrafast relaxation processes, particularly in physics, biophysics and physical chemistry, thanks to the availability of reproducible, ultrashort optical pulses, a few tens of femtoseconds in duration. Another venue, still in its infancy concerns the field of laser-matter interactions at very high intensities. Since the instantaneous peak power of an optical pulse corresponds to its energy divided by its duration, extremely high peak intensities can be achieved with modest energy requirements. For instance, a visible light pulse of duration 30fs (30×10^{-15} s), when focussed close to the diffraction limit, requires only 500mJ of energy to deliver a peak intensity of 10^{20}W/cm^2. Femtosecond lasers approaching such performances presently operate in several institutes at a repetition rate of 10Hz. Foreseeable progress in diode pumping schemes will allow an increase in their repetition rate by several orders of magnitude and make them true table-top systems. Perhaps the most important aspect of this new technology lies in the fact that, once optimised, such very high intensity femtosecond lasers will not require monstrous operating costs. Their dissemination in many academic institutions will allow progress in unforeseen directions.

A first likely impact of ultra-high intensity femtosecond lasers will be the extension of nonlinear optics to the XUV and X-ray domain. Intense, tunable coherent emission can be obtained in the XUV domain, up to 5nm, by high-order harmonic generation in gases or solids. Preliminary nonlinear experiments, involving one XUV photon produced in this way, and one visible photon, have already been accomplished. In the near future, true XUV nonlinear experiments will be performed, with all the photons involved in the XUV domain. One can compare this situation to the early sixties, when the first nonlinear experiments in the visible domain were reported.

Irradiation of solids with intense ultrashort optical pulses leads to the creation of dense plasmas, of near solid density, which are of great interest to plasma physicists. Intense sub-picosecond bursts of incoherent X-ray radiation are emitted from such dense microplasmas. The number of X-ray photons produced is already sufficient to envisage time-resolved X-ray diffraction studies, which allow the monitoring of structural transformation with very high time resolution. The main technological stumbling block at the present time concerns the fabrication of good quality X-ray optical elements, which would allow the collection and transport of the X-ray radiation to the target without compromising the pulse duration.

Equally exciting is the prospect of obtaining an X-ray laser. High intensity femtosecond lasers are particularly well suited as a pumping source for X-ray lasers, since the spontaneous radiative lifetime in the X-ray region is typically a few femtoseconds. Small scale X-ray lasers operating at several Hertz, based on femtosecond laser pumping, can be anticipated.

Intense femtosecond pulses may induce plasma oscillations of large amplitude in underdense plasmas. The associated electric field gradient can reach values of the order of GeV/m, leading to new concepts for compact particle accelerators.

At the highest achievable intensities ($10^{20}W/cm^2$), the electric field associated with the radiation largely exceeds the internal Coulomb field tying electrons to the atomic nucleus, offering an interesting challenge to theorists. (Instead of the usual situation in nonlinear optics, the atomic field is a perturbation against the strong applied radiation field). Electrons undergo a quiver motion with a stored oscillation energy comparable to or even exceeding the electron rest mass energy. A new regime sets in where the change of mass due to relativistic effects leads to a nonlinear response. Many interesting phenomena, including self-guided propagation of optical pulses through plasmas, high harmonic generation via multiphoton Compton scattering of visible photons with relativistic electrons (with the prospect of obtaining ultrashort γ-ray pulses) have been predicted.

The purpose of this chapter is to describe some of the problems encountered when entering the field of high intensity laser-matter interactions. It is by no means a fair or complete description of the present status of a fast evolving field, nor does it attempt to describe the most significant developments. Rather, it discusses some of the immediate issues a young scientist entering the field is likely to be confronted with. A first section describes the basic principles of femtosecond pulse amplification. In a second section, a new technique to measure the complex field amplitude of visible radiation is described. The determination of the phase and amplitude of the intense femtosecond pulse is an unavoidable issue which needs to be addressed from the start, if only from a very practical point of view. When the laser pulse propagates through different transparent media, such as air or optical windows, distortions of the beam profile occur due to the nonlinear Kerr effect. It is essential to have 'in situ' diagnostics of these effects, otherwise, unpleasant surprises may well occur concerning the performance of the pulse focussed on the target. Finally, in a third section, a brief discussion of coherent emission of XUV by high harmonic generation in gases and incoherent X-ray emission from solid targets is described, and a technique to measure the short wavelength emission with sub-picosecond accuracy is presented. In a way these measurements represent some of the precursors of nonlinear X-ray experiments.

2 Chirped amplification of femtosecond pulses in Ti:Sapphire

This section was produced with the assistance of A Antonetti, J P Chambaret, and P F Curley, Laboratoire D'Optique Appliquée, ENSTA École Polytechnique Centre de L'Yvette F-91120 Palaiseau France.

Considerable research effort is currently being directed towards the generation of intense pulses with a view to studying high intensity interactions of light with matter. The general approach to achieve high intensities is to amplify a short optical pulse to high energy using the technique of Chirped Pulse Amplification (CPA)(Maine *et al.* 1988), and to focus it on to a target. Light intensities of 10^{17}-10^{18}Wcm^{-2} using 100fs (1fs = 10^{-15}s) pulses can already be realised by current laser-amplifier technology (Barty *et al.* 1994, Zhou *et al.* 1994a, White *et al.* 1995 and Le Blanc *et al.* 1993), but with the advent of new ultrashort pulse laser oscillators, effort is now being directed towards designing CPA systems to achieve even higher intensities on to the target. In particular, the development of sub-50fs, multi-Joule, CPA systems is now underway, and in the near future it should be possible to offer experimentalists table-top laser systems producing intensities approaching 10^{20}-10^{21}Wcm^{-2}. These intensities will be applied to studies of light-matter interactions in a previously unexplored intensity regime. In particular, high harmonic generation in gas jets and plasma formation in solid and gas targets will be studied (Murnane *et al.* 1991, L' Huillier *et al.* 1992). These phenomena can however, be very sensitive to the shape and energy distribution of the excitation pulse. In the case of high density plasma generation, the presence of a pre-pulse or a low intensity pedestal associated with the main pulse can be seriously detrimental to plasma formation. Therefore we shall discuss the basic design considerations for short pulse, high energy CPA systems, and highlight elements which can strongly influence the pulse quality. The basic concept required to achieve high optical intensities on to a target is described by the expression:-

$$\text{Intensity delivered on to target} = \frac{\text{Pulse Energy}}{\text{Pulse Duration} \times \text{Focal Spot Area}}$$

Thus we need to consider the best method for generating an ultrashort optical seed pulse, find an efficient high gain amplifier design that maintains the pulse quality, and an appropriate scheme for focussing the pulse to a near diffraction limited spot.

2.1   Generation of ultrashort optical pulses

A detailed discussion on short pulse generation using mode-locked lasers is provided by Sibbett elsewhere in this book, so here we will consider only the points relevant to our discussion. In order to generate short laser pulses, it is necessary to have a broadband amplifier medium, and an element in the laser cavity which can introduce a time varying loss mechanism to produce a source of modulation. In practice, broadband gain media such as dyes or solid state materials such as Titanium doped sapphire are used for short pulse laser sources. To generate pulses below the picosecond level (less than 10^{-12}s), it is efficient to use a fast saturable absorber mechanism inside the laser cavity, where the transmission of the absorber is a function of the intensity of the pulse incident upon it. With the advent of Ti:sapphire, which exhibits a fluorescence bandwidth extending over 400nm, the possibility of generating ultrashort pulses directly from lasers has become feasible (Spielmann et al. 1994). During the competitive research period surrounding the development of Ti:sapphire a number of new techniques was discovered for the generation of short pulses, most notably Kerr Lens Mode-locking (KLM) (Spence et al. 1991). The discovery of KLM in Ti:sapphire, and subsequent research has allowed pulses as short as 10fs to be produced directly from a laser (Spielmann et al. 1994, Zhou et al. 1994b, Stingl et al. 1995). Lasers similar to these are now being used as the seed oscillators for the next generation of high power CPA systems.

2.2   Kerr lens mode-locking (KLM)

The generation of ultra-short mode-locked pulses using Ti:sapphire was revolutionised by the discovery of Kerr Lens Mode-locking (Spence et al. 1991). If we consider the laser material, the refractive index is given by the expression $n = n_0 + n_2 I$, where I is the optical intensity, and n_2 is the nonlinear refractive index (see Van Stryland, this volume). At sufficiently high intensities, for example at a beam focus inside a laser cavity or in a high power amplifier, it is possible to introduce a non-negligible intensity-dependent modification to the refractive index n. For a Gaussian beam profile, the intensity dependent modification of n produces an index profile in the material similar to that of a positive lens, causing the beam to self-focus (ie.Kerr Lensing). Inside a laser, this intensity dependent self-focussing produces an intensity dependent modification to the beam propagation as it passes around the cavity. By introducing an aperture or slit at another intracavity focus to select the self-focussed cavity mode, it is possible to force the laser into a short pulse state. The configuration of a typical KLM laser is shown in Figure 1. A simplified scheme demonstrating the role of the slit as a method for selecting the higher intensity pulse mode of operation is shown in Figure 2. Any low intensity light undergoes weaker self-focussing, and is suppressed by the slit. The intense self-focussed light then goes on to be preferentially amplified in the oscillator cavity.

2.3   Chirped pulse amplification

Having generated the short seed pulse, it is necessary to address the design considerations for the amplifier. Whilst in the design of short pulse laser oscillators, the nonlinear

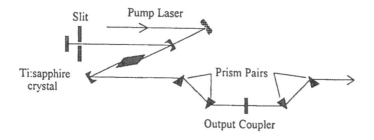

Figure 1. *Schematic description of a* KLM *laser*

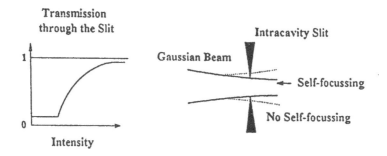

Figure 2. *(left) Transmission of light through a slit in the resonator as a function of intensity in the presence of self-focussing. (right) transverse beam profile with and without self-focussing.*

refractive index and the resulting self-focussing is an asset, it is certainly not the case for amplification. As the pulse intensity increases, the onset of effects resulting from the nonlinear refractive index can begin to distort the pulse both in the temporal and spatial domains. In amplification, there is a particular risk of self-focussing increasing to a level where catastrophic damage can occur to the amplifier medium. Hence the technique of chirped pulse amplification(CPA) was developed to reduce the peak pulse intensity during the amplification process. The principle of CPA is to stretch the pulse duration by several orders of magnitude, amplify the pulse to the required level, and then recompress the pulse to its original duration, albeit at a higher energy level (Maine

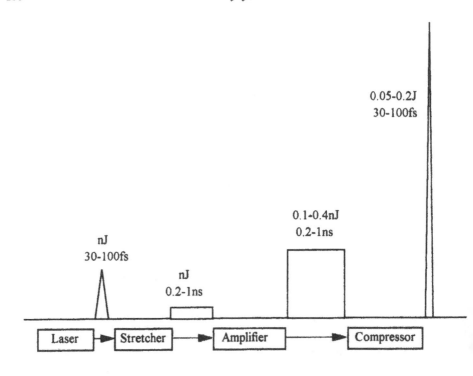

Figure 3. *Chirped pulse amplification* CPA

et al. 1988). A simplified schematic of the concept of CPA is shown in Figure 3. For efficient amplification of short pulses it is necessary to consider three main points. First, the amplifying medium used must exhibit sufficient bandwidth to support the seed pulse bandwidth. For high gain amplification, the attainable bandwidth can reduce significantly. For example, for a Gaussian lineshape, calculations reveal that for short pulse amplification in Ti:sapphire with an amplifier gain of order 10^6, narrowing of only a few percent is expected for an input pulse spectrum of 25nm (*i.e.* $\tau_{\text{pulse}} \approx 25$fs, at 800nm) with this increasing to 25% for a 40nm ($\tau \approx 17$fs) input spectrum, and up to almost 50% for a 100nm ($\tau \leq 10$fs) input spectrum. This narrowing can be circumvented to a certain extent by utilising a spectral mask in the pulse stretcher (Barty *et al.* 1994).

Secondly, the amplifier should be operated near the saturation fluence of the gain medium for efficient extraction of the energy stored in the amplifier. The saturation fluence is given by the expression, $F_{\text{sat}} = h\nu/\sigma_s$ where h is Planck's constant, ν is the laser transition frequency and σ_s is the emission cross-section. Solid-state materials such as Ti:sapphire, Alexandrite and Neodymium exhibit high saturation fluences of approximately 1–10Jcm^{-2}, and demonstrate efficient storage characteristics suitable for CPA.

Finally, as mentioned earlier, the peak intensity in the amplifiers should remain low, to avoid optically induced damage. A rule-of-thumb used in the design of amplifiers to quantify this problem is the B-integral. Representing the expected intensity dependent phase shift for a given amplifier configuration, the B-integral is defined by the expression,

$$B = \frac{2\pi}{\lambda} n_2 \int_0^L I(z)dz$$

where n_2 is the material nonlinear refractive index, and L is the material path length. In general it is best to minimise B, with values of $B < \pi$ representing the limit in CPA systems. If the B integral exceeds this value, the high spatial frequencies are amplified preferentially, and it becomes necessary to use spatial filters to remove them from the beam.

2.4 Discussion of pulse quality

Before discussing the various elements of CPA in more detail, it is necessary to consider the problem of pulse quality. For experimental applications, one needs to take into account not only the energy of the amplified pulse but also the quality of the recompressed pulse, particularly if any residual background, or long pedestal exists. In experiments, the presence of even a weak pedestal can dramatically affect the interaction process between the pulse and the target. For example, the ionisation of solid targets can occur at intensities of approximately 10^{12}Wcm^{-2}, therefore, if the amplified peak intensity is of order 10^{19}Wcm^{-2}, the pulse must exhibit a sharp rising edge over at least 7 orders of magnitude in intensity if pre-ionisation of the target is to be avoided. Such pre-ionisation of a target by a low intensity pedestal can produce a plasma plume which can act as a mirror to the main pulse as it approaches the target and thereby inhibit efficient transfer of the main pulse energy into the target. Unfortunately in CPA systems it is common to observe a background/pedestal in the recompressed pulse. This can arise for a wide variety of reasons.

In an effort to minimise the pedestal, we have examined the various components in CPA in order to identify the various contributions which can affect the level of the pedestal. Studies of the pulses generated by a KLM oscillator over a large dynamic range using an autocorrelator, have revealed that even an unoptimised laser can produce a seed pulse with a notable pedestal. It is therefore necessary to first characterise the seed oscillator, over a wide range of time durations to detect the presence of any residual pedestal. By optimising the KLM laser cavity it has been demonstrated that pulses as short as 30fs could be achieved with peak-to-pedestal dynamic ranges exceeding 6 orders of magnitude in intensity, limited by the detection (Curley *et al.* 1995).

To model the propagation of these short seed pulses through a stretcher/compressor system it was necessary to develop a ray tracing model which could reproduce the propagation of each spectral component of the pulse through the optical components. Our 2D ray tracing analysis of various stretcher designs revealed that a peak-to-pedestal (PTP) dynamic range greater than 10^7 is possible for ~50fs pulses. Therefore any pedestal in the pulse from the oscillator below this intensity level is unimportant.

We shall see later that whilst an unoptimised KLM oscillator seed pulse can limit the pulse quality, the main source of pedestals is due to spectral clipping and high order phase distortions introduced by the pulse stretcher.

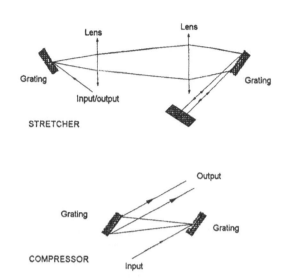

Figure 4. *Stretcher/Compressor design (top view). In the stretcher, reflecting mirrors elevate the beam so that input and output do not lie in the same horizontal plane.*

2.5   Pulse stretchers and compressors

To reduce the intensity during amplification we need to consider a system for stretching the pulse (Figure 4). For a 100fs pulse, stretching factors of order 10^3–10^4 are used to reduce the peak intensity in the amplifier. An optical system is used to introduce a large source of net positive Group Velocity Dispersion (GVD), and this is usually achieved using a telescope placed between a pair of diffraction gratings (Martinez 1987). In this scheme the higher spectral frequencies are diffracted by the first grating at an angle smaller than that of the lower frequencies. As a result the higher frequencies traverse a longer beam path, and therefore at the output of the stretcher the lower frequency components lead the higher components. This rising sweep of frequencies through the pulse is referred to as pulse chirping and produces in the time domain a stretching of the input pulse. The pulse can be recompressed by simply introducing a pair of diffraction gratings that supplies negative GVD to compensate the frequency chirp (Figure 4). In sub-100fs CPA systems, the design of the stretcher becomes important since even weak high order phase delays introduced at the stretching stage have to be recompensated at the recompression stage if the pulse quality is to be maintained. This is particularly important since uncompensated phase terms can result in a broadening of the recompressed pulse and introduce a residual pedestal. The main problems found with conventional pulse stretchers is that they incorporate refractive optics, ie. lenses. This inherently leads to additional chromatic aberration, which for pulses longer than 100fs is not a serious limitation. However, for pulses shorter than 100fs, spectral clipping, residual chromatic aberration, and high order phase distortion can lead to serious problems when the pulse is passed into the compressor for reconstruction of the original seed pulse. The problem of chromatic aberration can be avoided by using all reflective optics in the stretcher as we shall discuss in the next section. However spectral clipping represents a fundamental limitation, and for sub 50fs pulses it becomes necessary to make a trade-off between the desired pulse duration and the desired peak-to-pedestal (PTP) dynamic range. A given stretcher design has a finite bandpass, which is typically of order 100nm, limited by the finite size of the optics. For a 100fs pulse, (bandwidth ≈10nm, at 800nm), the effect of spectral clipping alone is not at a level sufficient to cause a serious problem in recompression. However, for a sub-50fs pulse, (ie. bandwidth >20nm, at 800nm) the clipping begins to introduce a notable pedestal ($\approx 10^{-7}$ level), which increases dramatically for shorter pulses. For an aberration free pulse stretcher the original seed pulse duration (at the full-width-half-maximum) can be reproduced down to 30fs, but the effect of finite spectral clipping can cause the pedestal level to change dramatically (Chambaret *et al.* 1995). Hence it is necessary for experimentalists to make a compromise between the desired duration and pedestal. For gas target interaction experiments, where a pedestal is not too problematic the CPA system can be optimised for the shortest pulses. In the case of solid target interactions, where a weak pedestal can seriously affect the studies, a high PTP range can be selected at the expense of a longer pulse duration.

2.6   Reflective stretcher design

For the amplification of sub-50fs pulses it is necessary to design a stretcher/compressor system with a minimum of aberration and residual high order phase distortions. These stipulations require the design of an all reflective stretcher to eliminate the chromatic aberration associated with lenses in conventional stretchers, and necessitate the reduction of the material dispersion through the amplifier to minimise problems of residual high order phase terms at the compression stage.

The advent of sub-100fs Ti:sapphire systems led to the development of the first all-reflective stretcher incorporating a cylindrical mirror pair, which was demonstrated to be able to support the stretching and recompression of sub-30fs pulses in a quintic-phase limited system.(Lemoff and Barty 1993). More recently, a new all reflective stretcher design has been proposed based on an Öffner Triplet (Chambaret *et al.* 1995, Öffner). The schematic of the system is shown in Figure 5 and incorporates a concentric mirror pair; one circular concave mirror and a rectangular convex mirror, radii of curvature ROC=1024 mm, and ROC=-512mm respectively. The main advantages of this configuration is that it is not only symmetric and exhibits no chromatic aberration, but, in addition, the precise relation between the mirror ROCs allows the spherical aberration and astigmatism of one mirror to be compensated by the other, thereby eliminating on-axis coma. In preliminary experiments this stretcher design has been demonstrated to support the stretching of a 30fs pulse up to 300ps and recompress back to 35fs. This stretcher design may prove invaluable in future CPA systems.

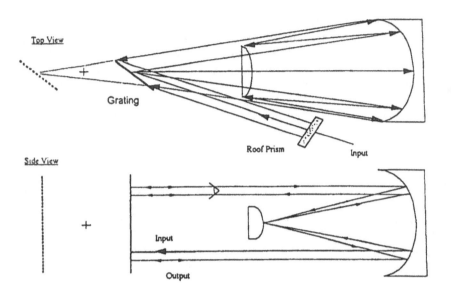

Figure 5. *All reflective optics of the stretcher design at LOA, ENSTA, with minimal aberration and astigmatism.*

2.7 Amplifier design

The typical pulse energy emitted from a KLM oscillator is of the order of a few nanojoules $(10^{-9}J)$, but this can be boosted towards the Joule level in a series of amplifier stages. In the design of each stage it is necessary to consider the minimisation of the B-integral, whilst still realising a high gain over a broad bandwidth. To reach energies of order 1mJ, (*i.e.* a gain of 10^6) two approaches exist:- regenerative amplification and multi-pass amplification, (see Figure 6). In the former case, the pulse is injected into a laser cavity, allowed to make multiple passes through the amplifier medium, and then the amplified pulse is extracted. Such regenerative systems are very efficient since it is possible to leave the pulse in the amplifier until the gain saturates, thus extracting all the available energy. To achieve the injection and ejection process, it is necessary to introduce into the laser cavity an optical element that can quickly turn the beam polarisation, together with a polarisation sensitive element. The pulse is injected by reflecting the beam off an intracavity polariser, and a Pockels cell is used to rotate the polarisation by 90° such that the pulse remains in the cavity. Once the pulse has been amplified to the required level, the beam polarisation is again rotated through 90° by the Pockels cell, and the pulse is reflected by the polariser and ejected from the cavity. Since the round trip losses are higher than in a multi-pass arrangement (see Figure 6), it is necessary

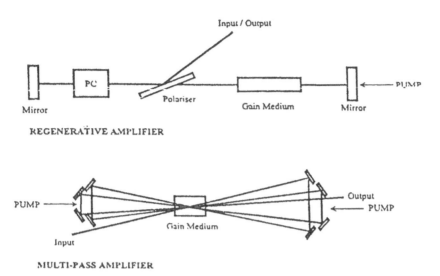

Figure 6. *Schematic description of regenerative and multipass amplifiers.*

to have a large number of passes in a regenerative cavity to realise the same gain. As a result, because the cavity incorporates the gain medium together with a Pockels cell and a polariser, the overall material path length experienced by the pulse is very long as compared to a simple multi-pass arrangement. As discussed before, the presence of large amounts of material dispersion and the increased problem of residual high order phase distortion can dramatically affect the final recompressed pulse contrast.

The use of regenerative amplification for short pulses has been demonstrated (Barty *et al.* 1994, Zhou *et al.* 1994, White *et al.* 1995); however a stretcher/compressor design capable of compensating high-order phase terms becomes necessary. Stretchers incorporating cylindrical optics have been used with a regenerative amplifier although the system design requires various combinations of glass materials to be inserted into the beam path in order to achieve a quintic phase limited system for a given pulse duration (Barty *et al.* 1994).

The advantage of using a regenerative scheme in a large amplifier chain is that the beam spot is 'cleaned' since it follows a resonant cavity mode. In addition, the regenerative amplifier provides a form of system isolation in that all components after the amplifier can be aligned by using the amplifier cavity mode, leaving the oscillator and stretcher free to be modified or optimised independently.

The alternative amplification scheme is to use a multi-pass amplifier arrangement (Figure 6), where the pulse is directed through an amplifier medium in a bow-tie arrangement. This scheme has the advantage that the transmission bandwidth is high, losses are low, fewer passes are required and so the material path length is reduced. This design is particularly suited for use with our Öffner reflective-optic stretcher, where the low material path length in the multipass amplifier allows recompression to be achieved to the quartic phase limit, by simply introducing a small adjustment in the separation and angle of the compressor gratings.

For amplification beyond a few millijoules, it is necessary to use additional multi-pass power amplifiers, which incorporate increasingly large diameter beam sizes to reduce the B integral. Various groups have demonstrated two stage amplification in Ti:sapphire to achieve energies of the order of 100–300mJ before compression, with pulse durations shorter than 50fs (Barty *et al.* 1994, Zhou *et al.* 1994). The design of a multipass system will be discussed in the following subsection.

2.8   Multipass amplifier chain

Since the goal is to achieve the shortest recompressed pulse with the maximum energy, our initial aim is to optimise our system to yield a sub-50fs recompressed pulse duration at the 200mJ level. A schematic of our CPA system is shown in Figure 7.

The amplification is divided into two multi-pass stages with an overall gain of approximately 10^8. The first amplifier stage is an 8-pass arrangement, which focusses the beam into a $5 \times 5 \times 10$mm long Ti:sapphire crystal using two confocal focussing mirrors (Georges *et al.* 1991). The focussing mirror on one side of the crystal has a lower radius of curvature as compared to the other, and as a result, the beam walks outward on the mirrors after each successive pass. A Pockels cell is positioned mid-way through the beam path in the amplifier and serves a dual purpose; to select a single pulse from the pulse train, and to eliminate the risk of the amplifier lasing independently. For a pump energy of 120mJ (532nm, 10Hz) using a Q-switched frequency-doubled YAG, the output from the first stage is typically 3mJ, which for an input pulse energy of approximately 0.5nJ represents a gain in excess of 10^6. The pulse is then sent through an isolation Pockels Cell and directed into a second 5-pass amplifier stage, which uses a $10 \times 10 \times 10$mm Ti:sapphire crystal. For a pump pulse energy of 1J (532nm, 10Hz) output energies in excess of 300mJ are achieved. Gain narrowing and saturation effects are observed to limit the spectral width to approximately 30nm (Figure 8), in agreement with modelling (LeBlanc and Salin 1995). From the Fourier Transform of the spectrum we deduce an optimised compressed duration of 25–30fs. With a grating compressor, the pulse is recompressed to a duration of 40–50fs, (Figure 8), with an energy of 180mJ.

From the autocorrelation a strong pedestal is evident in the recompressed pulse, however this is comparable to that obtained in preliminary experiments undertaken with just a stretcher-compressor system (Chambaret *et al.* 1995). The limiting factor in the generation of high quality pulses using an all-reflective stretcher system appears to be the surface quality of the mirrors. In these experiments the surface flatness was 5λ, but our ray tracing modelling of the stretcher-compressor design reveals the necessity for surface quality in excess of $\lambda/10$ in order to realise high dynamic ranges (Chambaret *et al.* 1995). The influence of surface quality is shown in Figure 9. The dynamic range increases, and the pedestal decreases with increasing mirror quality in the stretcher, with data shown for mirror surface quality of 5λ, 2λ, $\lambda/2$, and $\lambda/4$ respectively, for a 30fs recompressed pulse. In the very near future we hope to obtain a mirror with

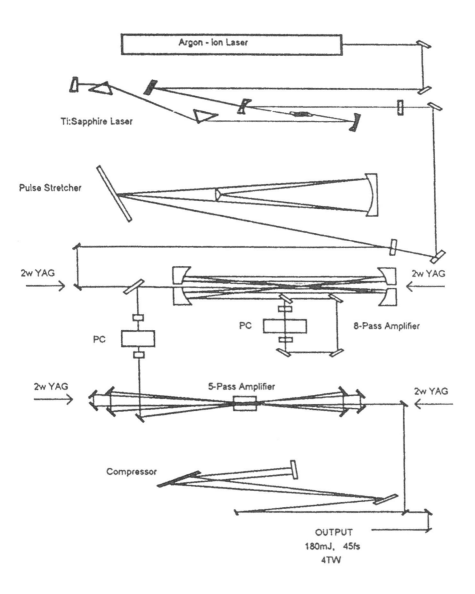

Figure 7. *Schematic description of the oscillator-*CPA *amplifier based on Ti:sapphire developed at LOA,ENSTA.*

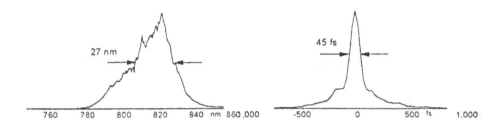

Figure 8. *(left) Spectrum of femtosecond pulse at the output of the chain shown in Figure 7. (right) Autocorrelation trace of the pulse at the output of the chain shown in Figure 7. The pulse duration (FWHM) is approximately 45fs.*

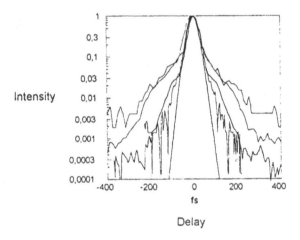

Figure 9. *High dynamic range cross-correlation trace w-2w of the fs pulse showing the influence of mirror quality in the stretcher in the pulse pedestal.*

a surface quality exceeding $\lambda/10$ which we anticipate should retain the pulse quality over at least six orders of magnitude for a 30fs pulse. We are currently expanding our system to incorporate a third amplifier stage, in order to realise amplification to 1J, in a sub-50fs, all solid-state Ti:sapphire system.

2.9   Beam propagation and focussing

After recompression of the pulse, it is necessary to both transport the beam to the target chamber, and then focus it close to its diffraction limit. With intense sub-100fs pulses the problem of nonlinear beam propagation can, just as in KLM oscillators, lead to

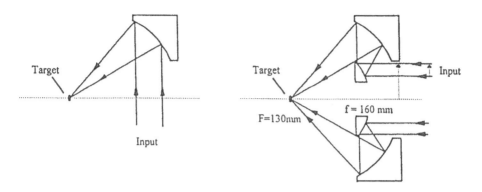

Figure 10. *(left) Schematic description of an off-axis parabolic mirror. (right) A Bowen telescope used for beam focussing.*

self-focussing and self-phase modulation. After amplification, pulses of approximately 100fs, and 50mJ energy with a beam diameter of a few centimetres can be sufficient to modify the refractive index of air. This leads to self-focussing, and can lead to a spectacular effect referred to as 'filamentation' or 'self-channelling' (Nibbering *et al.* 1995a, Braun *et al.* 1995). As the beam propagates, the radial change in the intensity dependent refractive index results in self-focussing, producing focussed intensities sufficient to ionise air and form a low density plasma. The low density plasma in turn leads to a defocussing effect which counteracts the ongoing Kerr-induced self-focussing. The resulting balance of self-focussing and defocussing results in self-channelling of the light into a pencil-like filament of order 100μm in diameter, which can propagate for several tens of metres. At present the confined energies are of order 1mJ, but in the future this phenomena could be exploited for high energy beam transport. It may prove possible to find a propagation medium where the spatial characteristics of the generated guiding mode could allow an increase in the confined energy, and provide an alternative method of delivering intense pulses on to target without the need for additional optical elements.

To avoid the problems of nonlinear propagation altogether, it is necessary to place both the compressor, and the beam propagation tubes under vacuum. To achieve a near diffraction limited spot, an appropriate focussing scheme has to be employed. Focussing lenses cannot be used because of the problem of chromatic aberration. Also, at high intensities the beam would catastrophically self-focus in a lens, hence a reflective optic focussing system must be devised. The most widely used scheme is an off-axis parabola (Figure 10). With careful alignment, spot sizes of order 5–10μm can be realised, although parabolic mirrors have the disadvantage that since they are produced by diamond polishing, the beam can suffer more than 40% loss by diffraction effects. In addition off-axis parabolic mirrors exhibit a very limited angular field acceptance, making alignment very critical. An alternative scheme is a Bowen Telescope, which incorporates spherical mirrors, is off-axis aplanetic and demonstrates a large acceptance angle (Figure 10).

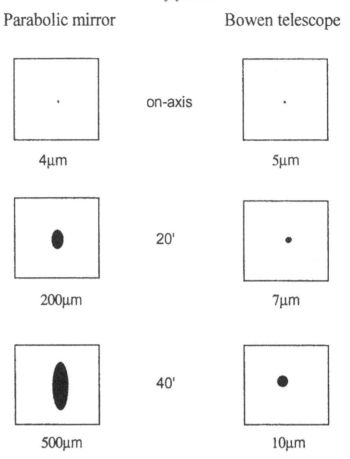

Figure 11. *Comparison of beam spot obtained with off-axis parabolic mirror and Bowen telescope at focus, for different angles of incidence with respect to mirror size.*

A comparison of the beam focussing possible with a Bowen Telescope and an off-axis parabola is shown for increasing incidence angle in Figure 11. With this scheme, intensities of 10^{20}–10^{21}Wcm^{-2} could soon prove possible on a compact, table top, all solid-state system. One final point is worth mentioning with regard to pulse quality. As the pulse energies increase it may prove necessary to use active methods to clean the amplified pulse and improve the PTP dynamic range after compression. This can be achieved presently by frequency doubling the pulses or by incorporating 'active' cleaning using a plasma mirror. In the latter scheme, a pre-pulse is used to deliberately form a plasma, which in turn reflects the main pulse, now with a much faster rising edge. The efficiency, and reproducibility of this technique have yet to be fully characterised.

2.10 Summary

In summary the next few years will see CPA amplifier systems producing intensity levels approaching 10^{22}Wcm^{-2} . Such systems should take the field of high energy plasma interactions into a new regime, giving experimentalists the opportunity of making compact X-ray lasers, and to undertake experiments at previously unattainable energy levels and temporal resolution. The coming years could thus lead to many new discoveries in the field of high energy/density physics.

3 Phase retrieval of femtosecond pulses and the measurement of nonlinear refractive index

This section was produced with the assistance of M A Franco, E T J Nibbering and B S Prade, Laboratoire d'Optique Appliquée, ENSTA–École Polytechnique.

When dealing with very high intensity lasers, a knowledge of the phase of the optical pulse is imperative. Precise characterisation of the pulse temporal shape may lead to improvements in laser oscillators. It may also be desirable to analyse how much a pulse is influenced during chirped-pulse amplification (CPA) by effects such as gain narrowing, gain saturation as well as self-phase modulation (SPM) and group velocity dispersion (GVD) (Ditmire *et al.* 1995). These effects become more pronounced with pulses of duration below 100fs (Zhou *et al.* 1994a). The design of pulse shapers can also take advantage of an analysis of the amplitude and phase at the output of these devices (Weiner *et al.* 1992). Equally important is a reliable 'in situ' measurement of the wave front after pulse transport from the laser system to the target. Phase shifts due to transmission through transparent media such as air, optical elements or windows may alter the expected characteristics in terms of peak intensities on target. To illustrate this point we recollect that a plane parallel piece of glass in the path of an intense optical beam of Gaussian transverse intensity profile acts as a focusing lens since it induces a phase retardation on the central part of the beam, due to the increase of the refractive index (nonlinear optical Kerr effect). A good knowledge of the nonlinear Kerr coefficient of different transparent media is therefore essential. Finally the outcome of various types of experiments may also benefit from a knowledge of the pulse characteristics on the target. One can think of experiments where intense laser beams are self-guided through nonlinear media (Borisov *et al.* 1994). Such channeling effects depend on the input phase profile of the optical pulse. Another category of phase-sensitive experiments concerns the coherent control of matter wave packets in chemical physics (Warren *et al.* 1993, Kohler *et al.* 1995).

Several approaches to the problem of phase retrieval have been developed during the last decade. For instance, one can measure the frequency shift in a pulse by cross-correlating the pulse with a spectral slice of the pulse (Chu *et al.* 1995). Frequency-Resolved Optical Gating (FROG) (DeLong *et al.* 1994, DeLong and Trebino 1994) is based on spectrally and time-resolved pulse autocorrelation, after which the pulse temporal profile is calculated from the FROG-trace using iterative phase-retrieval routines. Recently we proposed and demonstrated a different approach in phase retrieval based on spectral analysis after nonlinear propagation (Prade *et al.* 1994, Nibbering *et al.* 1995b). In this method one measures spectra of the optical pulse both before and after propagation through a transparent medium of known Kerr-like nonlinear response. An iterative phase retrieval algorithm is then used to adjust an initially arbitrarily chosen phase and pulse energy until the calculated spectra pertaining to these parameters match the experimental results. One obtains the peak intensity of the incident pulse as well as its time variation by associating the retrieved phase with the input spectrum and by reconstructing the pulse in the time domain. Spectral analysis after nonlinear propagation can be used as an experimental method to determine the sign and value of the nonlinear refractive index n_2. Several methods have been developed addressing different situations and with corresponding merits (Chase and van Stryland 1994). The method relying on nonlinear spectral analysis is at the same time simple, fast, accurate and applicable to a large variety of solids and liquids (Nibbering *et al.* 1995c). Here we summarise some of its properties and extend it to the determination of the nonlinearity of air.

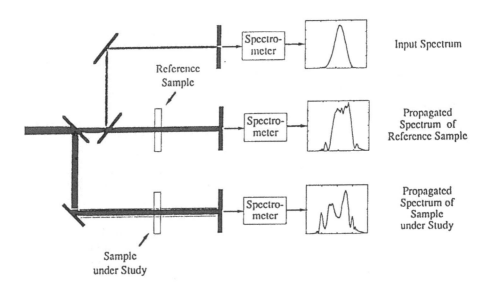

Figure 12. *Experimental set-up for measurements of the pulse phase and intensity and the nonlinear refractive index of materials. For details see text.*

3.1   Phase retrieval

Figure 12 shows the experimental setup used to demonstrate the potential of the method. Part of the optical pulse to be analysed crosses a thin plate of fused silica (Herasil). An optical ray of constant intensity is selected with a pinhole placed after the nonlinear medium. The spectrum of this ray is recorded with a spectrograph and associated detector. A similar ray but without the nonlinear sample provides the reference spectrum.

The algorithm used to recover the information on the phase and amplitude of the op-

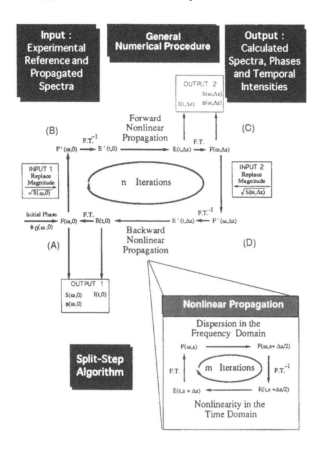

Figure 13. *General numerical procedure for pulse retrieval. The upper scheme shows the basic algorithm, while in the lower right corner the split-step algorithm is depicted that takes into account both group-velocity dispersion and self-phase modulation.*

tical pulse from the spectra is shown schematically in Figure 13. $S(\omega, 0)$ and $S(\omega, z)$ represent the experimental power spectra measured before and after propagation through the nonlinear medium respectively. Similarly, $F(\omega, 0)$ and $F(\omega, z)$ are the calculated amplitude spectra before and after propagation through the nonlinear medium.

The numerical procedure depicted in Figure 13 operates as follows. One uses the two functions $S(\omega, 0) = |F'(\omega, 0)|^2$ and $S(\omega, \Delta z) = |F'(\omega, \Delta z)|^2$ as input and one seeks the initial complex amplitude spectrum $F(\omega, 0)$ or the initial pulse $E(t, 0)$. Input 1 injects the square root of the measured reference spectrum, denoted as $\sqrt{S(\omega, 0)}$. The algorithm is initialised by assuming an associated flat phase $\phi_0(\omega, 0)$. However, we have verified that the introduction of a different (arbitrary) initial phase leads to the same final result. Substituting the initial experimental spectrum, we have $F_0'(\omega, 0) = \sqrt{S(\omega, 0)} \exp[i\phi_0(\omega, 0)]$ in (B). After reconstruction of the pulse in the time domain by an inverse Fourier transformation, one calculates the field amplitude corresponding to transmission through a transparent nonlinear medium of thickness Δz. The Fourier transform of the propagated pulse after the sample is then computed, leading to a calculated propagated spectrum $F(\omega, \Delta z)$ at (C). After $|F(\omega, 0)|$ is replaced by the measured propagated spectrum (Input 2) while the acquired phase $\phi(\omega, \Delta z)$ is retained in (D), the backward propagation is performed using the relation

$$E(t, 0) = E(t, \Delta z) \exp\left\{-i\frac{2\pi n_2}{\lambda}I(t, 0)\Delta z\right\} \tag{1}$$

The procedure calculates the pulse at its original location (A), but now with a phase content closer to the one able to reconstitute the propagated spectrum. The experimental magnitude of the reference spectrum is then substituted again for the calculated one, but the phase is retained. The iterative procedure is repeated until the phase profile is stable. The algorithm provides, in the form of a numerical array, the phases and the spectra as well as the temporal profiles of the reference and propagated beams at Output 1 and Output 2 of Figure 13 respectively.

Convergence of the simulated reference and propagated spectra to the measured ones is achieved by repeating the iterations for different magnitudes of the experimental power spectra. This is equivalent to changing the energy W of the pulse. The magnitude of the nonlinear effect is thus adjusted for every run of the algorithm. Minimisation of a distance error function will result in the best reconstituted spectra.

After fine adjustment of the laser intensity, the procedure leads to an excellent reconstruction of the reference and propagated spectra. The calculated phase is then expected to give the best approximation to the true phase. Of course this fine adjustment of the pulse energy can be omitted if the pulse energy density at the position of the diaphragm is known precisely.

Figure 14 shows the results of a characterisation of the output of the Ti:sapphire oscillator-amplifier chain at LOA. Single-shot spectra were recorded before and after propagation through 3mm of Hersasil (fused silica with $n_2 = 3.21 \times 10^{16} \mathrm{cm}^2/\mathrm{W}$). A typical run of the numerical procedure on a desk computer requires about 50 iterations. Characteristically, the routine converges rapidly within 10 iteration rounds after which a fine adjustment occurs. The reconstituted spectra (represented by dots) clearly match the experimental ones (as depicted by the solid lines). The B-integral has a maximum value equal to about 2. The dashed line is the calculated initial phase. We sometimes observed phase jumps at frequencies corresponding to the wings of the spectra. We have confirmed that these are artifacts of our basic algorithm. The phase jumps are a consequence of the fact that the phase is undefined for those frequencies where the spectral density is zero. The erroneous phase changes are introduced if, during one iteration, a region of the calculated spectrum happens to be of zero intensity. For more details on the properties of the method, the reader should consult Nibbering et al (1995b).

Preliminary studies have been performed with spectra recorded by Salin at LOA using an amplified laser operating at KHz repetition rates with a pulse duration of 50fs. Results of the analysis are shown in Figure 15. This example shows that the method also works for shorter pulses.

Note that group velocity dispersion (GVD) may become important with shorter pulses since the associated spectrum is larger. Inclusion of GVD in the routine may

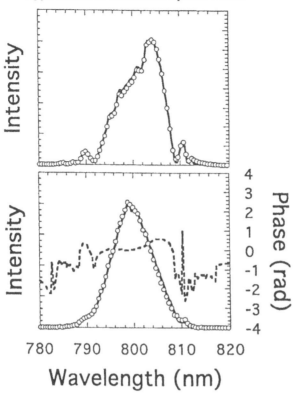

Figure 14. *Result obtained with the output of a 45mJ, 120fs Ti:sapphire laser amplifier chain. The lower part shows the input spectrum at the exit of the compressor, while the upper part demonstrates the spectrum after nonlinear propagation through 3mm of Herasil. The solid lines denote the experimental spectra while the circles represent the calculated ones. The pulse spectral phase at the exit of the compressor is depicted as the dashed line.*

become necessary for uv pulses or ultrashort visible pulses (less than 20fs in duration), even if very thin fused silica plates are used. A GVD compatible algorithm is also necessary if one wants to determine the phase of much weaker pulses, *e.g.* at the output of amplified femtosecond dye laser systems or (cavity-dumped) solid-state femtosecond oscillators or if the pulse has to propagate through a significant thickness of nonlinear material,such as an optical fibre. In the general case both GVD and SPM have to be taken into account at the same time. A split-step algorithm is one of the possible approaches to resolve the problem of propagation through a medium that is both nonlinear and dispersive (see Figure 13). Numerical simulation of the nonlinear propagation process then requires a greater effort. One can start the procedure at Input 1 by using an intensity spectrum $S(\omega, 0)$ and a spectral phase $\phi(\omega, 0)$ at position $z = 0$ for the complex spectral amplitude $F(\omega, 0)$. [Alternatively one can start the procedure in the time domain at Input 2 using an intensity temporal profile $I(t, 0)$ and a temporal phase $\phi(t, 0)$.] First one takes into account only the linear dispersion of the medium for an interval $\Delta z/2$ using Equation 1. For a following interval Δz, one introduces the

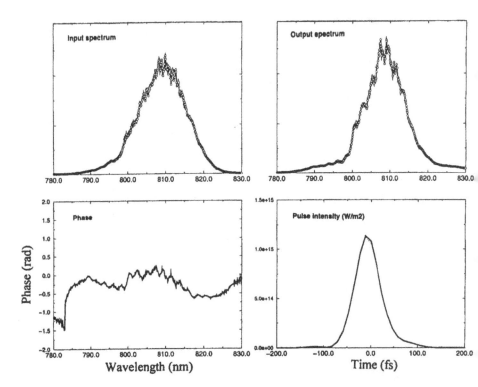

Figure 15. *Results obtained at LOA with a 50fs, 1KHz, 1W average output power Ti:sapphire laser amplifier chain. The upper part shows the spectrum before and after propagation through Herasil. The lower left part shows the phase at the exit of the compressor, while the lower right part depicts the calculated pulse intensity profile.*

nonlinear effect of SPM while discarding GVD. This process is then repeated for n steps. This method, as schematically depicted in Figure 13, necessitates the implementation of the Fast Fourier Transform (FFT) algorithm in the numerical procedure. Figure 16 demonstrates that the general numerical procedure including GVD works well for the analysis of spectra transmitted through a thick glass plate.

Finally we come back to the question of absolute intensity calibration of the optical pulse. Measurement of the spatial profile can be obtained by exploring the beam transverse profile with a pinhole. Experiments at LOA show that the intensity profile of the beam extracted from the algorithm applied to different parts of the beam agree with the intensity beam profile measured independently with a beam profile analyser. The intensity integrated over the pulse time profile yields a pulse energy in agreement with the measured value for both the cases shown in Figures 14 and 15.

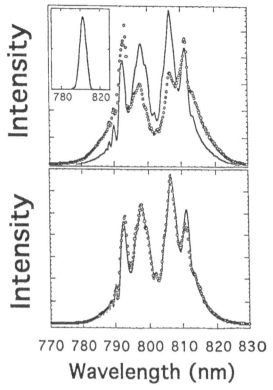

Figure 16. *Propagated spectrum of 150fs pulses through 1cm of BK7 glass for an intensity of 1.2×10^{11} W/cm^2 (solid line). The B-integral is equal to about 3. The upper part shows the calculated result (dots) when only a Kerr-like nonlinearity is taken into account. The lower part shows the result (dots) when group-velocity dispersion is included in the calculations. Only the latter result is a stable solution for our numerical procedure. The inset shows the reference spectrum.*

3.2 Measurement of nonlinear refractive index

The phase retrieval method described above yields the quantity $I(t)n_2z$, where z is the sample thickness. To obtain information on the pulse evolution, one assumes that the nonlinear response of the medium is instantaneous and of the form

$$n(I) = n_0 + n_2 I(t)$$

with n_2 a known quantity. On the other hand, if the pulse time evolution is known it is a simple matter to determine n_2 for different materials from an analysis of the pulse spectrum after transmission through the transparent nonlinear medium to be studied. In a single laser shot one records simultaneously the laser spectra both after transmission through the material to be probed and through a reference material of known n_2 as well as the laser spectrum before transmission (see Figure 12). In a first step, the phase retrieval algorithm is applied to the reference material in order to determine the pulse time profile I(t). In a second step the method is applied to the determination of n_2 from

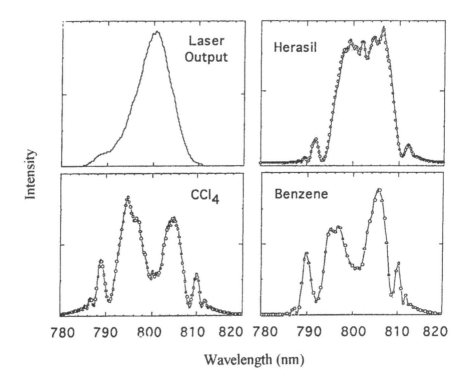

Figure 17. *Propagated spectra of 150fs pulses through three transparent materials (solid lines) together with the calculated spectra (dotted traces). The left upper curve shows the output of the laser set-up. For herasil the maximum intensity $I_{max} = 240\text{GW}/\text{cm}^2$ and the material thickness $z = 3mm$. The liquids carbon tetrachloride and benzene were probed using a fused silica cell with $z = 1mm$, with $I_{max} = 120\text{GW}/\text{cm}^2$. In the case of benzene a finite response time of $t = 75fs$ had to be used in the calculation.*

the test material making use of the known laser intensity $I(t)$. An attractive feature of the present method is that the iterative routine performs well even with 'imperfect' pulses, in which the phase profile as a function of frequency is not flat. Calculations indicate that nonlinear materials with negative and positive values of n_2 lead to different spectra. Therefore the method can reveal the sign of n_2. In addition, the method is well suited to the determination of n_2 in solids and liquids as well as gases.

We have applied this method to determine n_2 in a number of solids and liquids. Figure 17 shows the spectra of optical pulses with 24mJ pulse energies and a 150fs duration after propagation through different representative materials. The high-intensity pulses were delivered by a 100mJ Ti:Sapphire oscillator-amplifier system using the chirped-pulse amplification technique in a multipass configuration. The far-field spatial profile of the parallel, nearly Gaussian, beam (FWHM 15mm) was monitored with a CCD-camera in order to check that self-focussing did not occur during propagation. In order to eliminate uncertainties in laser intensities related to the transverse spatial profile of the beam, the single-shot far-field spectra of the pulses with and without non-linear materials were recorded by a spectrometer after transmission through a 1mm diameter diaphragm placed at the centre of the beam. Typical values obtained by the method are shown in the following table.

Compound	Manufacturer	$n_2(10^{-20}\text{m}^2/\text{W})$ $\lambda = 804nm$	reponse time τ (fs)
$MgF_2(o)$	Melles Griot	1.15 ± 0.20	
Ruby (Ec)		4.15 ± 0.30	
Suprasil	TecOptics	2.82 ± 0.30	
Herasil	TecOptics	3.21 ± 0.24	
BK7 glass	TecOptics	3.75 ± 0.30	
Water	distilled	5.7 ± 0.5	
CCl_4	Merck,Uvasol	19 ± 3	
Methanol	Merck, p.a.	6.7 ± 0.5	27 ± 1
Benzene	Merck, p.a.	24 ± 1	75 ± 10
CS_2	Merck, p.a.	150 ± 8	145 ± 10
Air		$3.0 \times 10^{-3} \pm 0.3$	70 ± 10

Table 1. *Estimated n_2 values of transparent materials obtained with the ENSTA-method*

In the analysis of the measurements on liquids, the influence of the cell containing the liquid must be taken into account. In situations where the only dephasing mechanism is due to the nonlinear response of the medium, the contributions of the cell windows and the liquids are additive and commutable with $\Phi_{\text{tot}} = \Phi_1 + \Phi_2 + \Phi_3$, where the total dephasing factor Φ_{tot} is the sum of the dephasing through the first cell window (Φ_1), the liquid (Φ_2) and the second cell window (Φ_3). It is therefore sufficient to use the modulated spectra after propagation through the empty cell as the reference (or input spectrum) for the calculations on the liquids. An interesting feature of the method is that one can also obtain some information on the response time of the optical-Kerr effect. Indication of a non-instantaneous n_2 can be given by visual inspection of the spectra after nonlinear propagation. For a finite response time of n_2, the leading edge of the response is steeper than the trailing part. For a positive n_2 it results in an enhanced red-part of the spectrum. Such a spectral red-shift is readily observed in Figures 17 and 19. Spectra after propagation through methanol and benzene exhibit more pronounced red components than through water and carbon tetrachloride. Clearly, no adequate spectral reconstruction can be expected from the phase retrieval algorithm if a delayed response is not included in the Kerr-nonlinearity. In fact we could not obtain convergence of the algorithm for spectra obtained after propagation through the liquids methanol, benzene and CS_2 by using an instantaneous Kerr-law. The algorithm was consequently modified by assuming a response time of the refractive index of the form:

$$\Phi(t) = \frac{2\pi z}{\lambda_0} \int_{-\infty}^{t} \frac{n_2 I(t)}{\tau} \exp\left\{-\frac{t-t'}{\tau}\right\} dt' \tag{2}$$

A single-sided exponential time response for the refractive index has been introduced

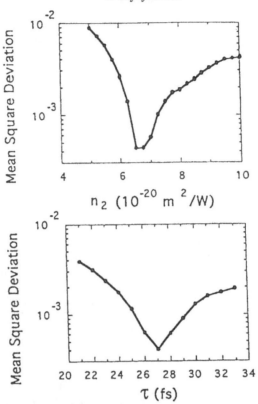

Figure 18. *The accuracy of the method is demonstrated for the case of methanol. In the upper curve the mean square deviation is plotted for different values of n_2, while the dependence to the finite response time t is shown in the lower curve.*

previously by several authors to describe the response of various liquids. Excellent results are obtained by introducing Equation 2 in the algorithm (with τ an adjustable parameter) for some liquids. A demonstration is given in Figure 17 for the cases of methanol and benzene. Figure 18 shows the mean square deviation (distance) between the calculated and measured output spectra of methanol plotted both as a function of n_2 and of the parameter τ. It gives the precision of the value of τ.

A similar analysis has been performed for carbon disulfide yielding a value $\tau = 145$fs. However the results were less satisfactory in the sense that the mean square deviation was much larger. This is probably due to the fact that the single-sided exponential law of Equation 2 is not adequate to describe the dynamics of this molecule (Etchepare *et al.* 1985, Etchepare *et al.* 1988, McMorrow *et al.* 1988).

The method can be applied to cases of large propagation distances as long as the spatial profile does not change significantly due to normal diffraction or due to self-focusing. One can thus measure nonlinearity of materials with very low n_2 values such as gases. There is an advantage of choosing a non-focussed situation, since it avoids the problem of possible competing processes, such as multiphoton ionisation, in the focal area. Ionisation may seriously complicate the situation since it brings additional

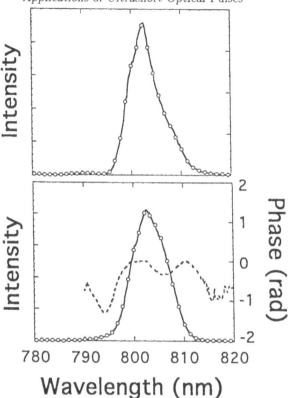

Figure 19. *The propagated spectrum of 150fs pulses through 7.5m air (solid line) together with the calculated spectrum (dotted trace) is shown in the upper part. The lower part shows the corresponding output of the laser set-up together with the pulse spectral phase (dashed curve). The value for n_2 is 3.0×10^{-23} W/cm^2, but a finite response time of $t = 70fs$ had to be used in the calculation.*

changes in the SPM-spectrum (see Shaw *et al.* 1993). Figure 19 shows spectra after propagation through the atmosphere. Note that a red-shift occurs, as in benzene. In air one has to take into account also finite-response time, due to a nuclear contribution to the nonlinearity from rotational and vibrational transitions of the molecules of oxygen and nitrogen (Hellwarth *et al.* 1990, Nibbering *et al.* 1995d) The extracted value for n_2 and of the response time is indicated in Table 1.

3.3 Conclusions and prospects

We have shown that an important modification of the spectrum of intense femtosecond pulses occurs during propagation through ordinary transparent materials. Using an inverse propagation algorithm it is possible to determine the phase and amplitude of the ultrashort optical pulses in a single laser shot by analysis of the pulse spectra before and after passage through the transparent medium, provided the nonlinear response of the material is known. The potential of the method for the characterisation of ultrashort optical pulses is demonstrated using experimental results obtained with a Ti: sapphire oscillator-amplifier chain. The same procedure can also be used to characterise quickly and accurately the nonlinear refractive index of materials in the solid, liquid or gaseous state. Moreover, the algorithm can be adapted to include the effects of a finite response of the nonlinear medium. A value of the noninstantaneous response time has been obtained for a number of liquids as well as air.

4   Sub-picosecond XUV pulses; generation and characterisation by two-colour ATI.

This section was produced with the assistance of: P Agostini, Service des Photons, Atomes et Molécules CEA Saclay, France; H G Muller, FOM Institute for Atomic and Molecular Physics, the Netherlands; D von der Linde, Institut für Laser and Plasmaphysik, Universität Essen, Germany.

The recent advances in compact, high-intensity, femtosecond laser sources have opened the road to a wealth of new investigations and applications. Among the most exciting, are the schemes for generating intense, vacuum UV-(soft) X-Ray pulses via laser plasma (LP) or *high-order harmonic generation* (HOHG). Such pulses are potentially extremely short and can be used for a variety of *time-resolved* experiments in physics, chemistry, biophysics etc .... They are, obviously, difficult to characterise. Direct measurement of subpicosecond pulses (in the XUV as well as in the visible) is beyond the possibilities of electronic devices and one must rely on indirect methods for the cross-correlation with a well-characterised pulse.

Two-colour Above-Threshold Ionisation (ATI), mixing one IR and one XUV pulse provides such a scheme. The basic idea is the following: an atom is ionised by the XUV pulse *in the field of an intense IR pulse*. The IR field modifies the energy of the photoelectron and creates sidebands in the energy spectrum which can be monitored as a function of the delay between the two pulses. Under certain conditions, this yields the desired cross-correlation function.

In this section, we will first give some background on ATI especially in the strong-field limit in which an approximation known as 'Simpleman Theory' applies (van den Heuvell and Muller 1988). This theory very simply explains some important characteristics of the HOHG in gases. Next we briefly review the different methods used to produce XUV pulses with a special emphasis on HOHG. The next sub-section reviews the experimental results on time-measurements. Finally, the last sub-section will review new proposals to extend the production and measurement of XUV pulses to the *sub-femtosecond* range as well as perspectives for novel applications.

4.1   ATI in the strong-field limit

Multiphoton ionisation is not usually limited to the strict minimum of photons $N_0 = E_0/\hbar\omega$ required to overcome the ionisation energy E_0. Instead the atom can absorb excess-photons above the threshold, a process known as ATI. At moderately high intensities (up to 10^{13} Wcm^{-2}), the corresponding photoelectron energy spectrum comprises a (small) number of discrete lines separated by the photon energy $\hbar\omega$. In the limit of very low frequencies or very high intensities, however, it is more convenient to think of this process as two step. The first step takes the electron into the continuum through the interaction with the electromagnetic field and the Coulomb potential. In the second step, the interaction with the Coulomb field is neglected and the interaction with the electromagnetic field is treated classically (Di Mauro and Agostini 1995).

In linear polarisation, the classical motion of an electron in the field is separable into an oscillatory motion, with an associated average kinetic energy called the ponderomotive energy (denoted as U_p) and a drift motion which depends upon the *phase ϕ* of the field at the instant, t_0, the electron is placed in the continuum. The ponderomotive energy U_p is defined in atomic units as $I/4\omega^2$ where I and ω are the laser intensity and frequency, respectively. For 780nm photons, $U_p = 5.7$eV at 10^{14} W/cm^2. The electron energy spectrum is then simply given by the classical kinetic energy, averaged over one optical cycle and its envelope is determined by the probability distribution of ϕ. This is the essence of the 'Simpleman's theory' (van den Heuvell and Muller 1988). It is easy to see for instance that the maximum *average* kinetic energy is $2U_p$, while the maximum *instantaneous* kinetic energy at the origin (after $1/2$ cycle) is $3.2U_p$ (Kulander *et al.* 1993, Corkum 1993). The latter quantity is intimately related to the harmonics cut-off, as discussed in Section 4.2.

In circular polarisation, it can be shown that the drift motion has a constant velocity $\mathbf{v} = e\mathbf{E}/m\omega$ perpendicular to $\mathbf{E}$ at t_0 where $\mathbf{E}$ is the electric field of the wave. Thus the electron exits the field in a direction that is directly related to the phase. The quantum equivalent of this motion is the angular momentum transferred to the electron by the photons. An important consequence is that the electron does not return to its parent core. This is the key idea behind recent proposals (Ivanov *et al.* 1995) for producing and measuring sub-femtosecond pulses (see Section 4.5).

Two-colour ATI

Of special interest for this chapter is the case a two-'colour', XUV-IR, ATI process. The electron is dropped into the continuum by the XUV photon and subsequently interacts with the IR intense field. Since the ionisation rate is small compared to the IR frequency, and the XUV pulse is normally much longer than the IR period, the phase ϕ is uniformly distributed over 2π. Furthermore, the initial velocity $T_0 = E_0 - \hbar\omega_{xuv}$ is large. The Simpleman's Theory then predicts that the maximum kinetic energy will be $\sqrt{8T_0U_p}$ (Schins *et al.* 1994). This simple expression works surprisingly well even for relatively low values of U_p, provided T_0 is large enough, as shown in Section 4.4. Of course, the quantum (photon) aspects are lost in this classical description. Actual electron energy spectra display discrete sidebands (separated by the IR photon energy) around the central energy T_0 which are comprised in the classical envelope.

ATI as a cross-correlation method

One off-shoot of the two-colour ATI is that it provides a method for measuring the duration of ultrashort XUV pulses by cross-correlation by monitoring the amplitude of the first sideband as a function of the timing between the XUV and IR pulses. Theoretically, the problem has been treated for the cases of both phase-incoherent (Cionga *et al.* 1993) and phase-coherent (Veñiard *et al.* 1995) fields. In general, the expression for the intensity $I_{\pm 1}(t)$ expected in the first sidebands (corresponding to the absorption or emission of one IR photon following the ionisation by the XUV photon) as a function of time delay t between the pulses can be written as

$$I_{\pm 1}(t) = \int dt' I_{xuv}(t'-t) I_{IR}(t'), \tag{3}$$

where $I_{xuv}(t)$ and $I_{IR}(t)$ represent the intensities of the XUV and optical fields, respectively. Assuming that the IR pulse shape is known, then $I_{xuv}(t)$ is obtained by a de-convolution procedure. Two examples of such measurements are described in Section 4.4.

4.2   Generation of ultrashort XUV pulses

IR radiation can be coherently or incoherently converted into the XUV/X range with reasonable efficiencies through a number of mechanisms. Since the electromagnetic field ($\mathbf{E}$, $\mathbf{B}$), strongly couples to electrons only, the primary mechanism is electron acceleration by the Lorentz force $(e/m)(\mathbf{E} + (\mathbf{v} \times \mathbf{B})/c)$. Even in a free-electron gas, the non-linear term $(\mathbf{v} \times \mathbf{B})/c$ is able to generate harmonics of the incident frequency, provided the electron density has a non-zero gradient (Shen 1984). In the case of dense targets, accelerated electrons may interact with phonons or with inner-shell electrons. XUV radiation is produced from blackbody radiation or atomic inner-shell transitions. In the case of low density targets, XUV results from fluorescence from multiply-charged ion excited states and HOHG. In the following subsections these various mechanisms are briefly reviewed.

Laser plasma

When an intense, ultrashort laser pulse is focussed on a dense target (solid or liquid), the medium is rapidly heated to very high temperatures and retains a density close to the solid density (Strickland and Mourou 1985, More *et al.* 1988) since the hydrodynamic expansion of the created plasma is negligible during the laser pulse. Emission of electromagnetic energy occurs through blackbody radiation, Bremmsstrahlung, inner-shell recombination. Depending on the amount of energy deposited by the laser pulse, the emitted spectrum extends from a few hundred eV to a few hundred keV with overall efficiency ranging from 0.1% to 1%. The duration of the XUV pulse is strongly dependent on which mechanism is responsible for it, which in turn, depends on the characteristics of the primary pulse. However, as a general thumb rule, the shorter the X-ray wavelength, the shorter the pulse duration.

HOHG on solid metallic targets

A laser pulse sent at nearly-grazing incidence on to the surface of a solid target generates harmonics in the direction of the specular reflection (Figure 20).

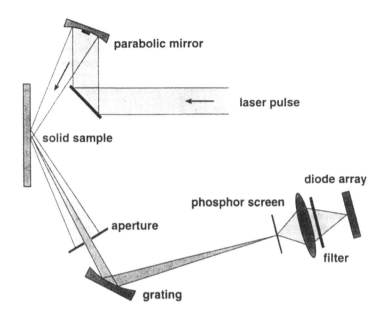

Figure 20. *Schematic of the experimental arrangement for measuring harmonic generation from solid surfaces*

We will distinguish two cases:

(1) The IR pulse intensity on the target is below the damage threshold, *i.e.* the temperature is below the melting point. Harmonic generation then occurs essentially through the steep density gradient at the surface. For instance, for the second-order current density $\mathbf{J}(2\omega)$ the dominant term is proportional to $\nabla \rho \cdot \mathbf{E}$ (where ρ is the solid electron density) which implies that only the component of $\mathbf{E}$ normal to the surface generates the harmonic (Shen 1984). In practice, generation by the component parallel to the surface is very weak. Harmonics of IR to UV radiation have been observed under such conditions (Frakas *et al.* 1992).

(2) At very high intensities (typically in the 10^{16}-10^{17} Wcm^{-2} range) a thin layer of metal is ionised very rapidly. However, if femtosecond pulses are used, then there is virtually no expansion of the plasma. Under such conditions, harmonic generation occurs at the plasma-vacuum interface, just as in the previous case. Electrons are driven back and forth by the field across the density gradient undergoing a strongly anharmonic motion which gives rise to even and odd harmonics. Recent experiments (Figure 21) (Kohlweyer *et al.* 1995, von der Linde *et al.* 1995) have clearly demonstrated HOHG, and wavelengths as short as 45nm have been generated (von der Linde *et al.* 1995) (18th harmonic of 800nm). These observations extend to the near visible in pioneering experiments (Carman *et al.* 1981) using long pulses from an IR CO_2 laser in which the 46th harmonic could be detected. In all cases, a cutoff has been observed which roughly

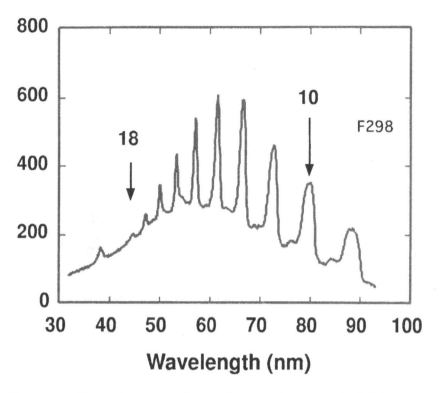

Figure 21. *Harmonic spectrum from a glass sample showing the 18th harmonic.*

corresponds to the value of the plasma frequency, in agreement with earlier theoretical models (Bezzerides *et al.* 1981). However, in the case of recent observations by von der Linde *et al.* (1995), this agreement could be spurious. More experiments are required to explore the cut-off behaviour at ultra-high intensities. Note that in both cases, even and odd harmonics are generated, at variance with HOHG in free atoms, as discussed in the next Section.

Harmonic Generation in Gases

HOHG in gases offers an alternative method of producing ultrashort pulses of XUV radiation and has attracted a great deal of interest in the last few years (L' Huillier *et al.* 1992). A strong IR pulse irradiating free atoms, generates a nonlinear polarisation P_{NL}. Due to the inversion symmetry of the Hamiltonian, all even Fourier components of P_{NL} vanish. In the perturbative limit, the harmonic amplitudes rapidly decrease with the order. On the contrary, in strong fields, the spectrum becomes essentially flat beyond, say the 9th harmonic, (the so-called 'plateau' region) and stops abruptly at a 'cut-off' harmonic. An example of this behaviour is shown in Figure 22 (Wahlström *et al.* 1993). The conversion efficiency in the plateau and its extension depend on the gas and the intensity. The energy conversion efficiencies range from 10^{-6} for xenon to 10^{-8} for neon. The conversion saturates when the degree of ionisation of the medium

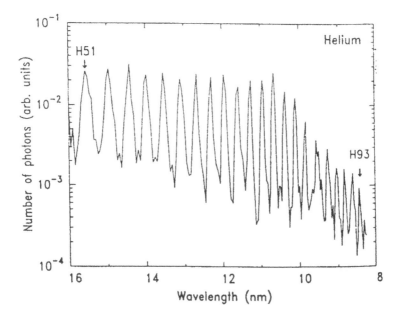

Figure 22. *Harmonic spectrum in helium at 10^{15} $W.cm^{-2}$ after Wahlström et al 1993).*

becomes high due to the relatively lower nonlinear susceptibility of the ions and the influence of a high density of free electrons on the focussing of the fundamental beam. The HOH emission is *coherent* and is confined in a small solid angle. This makes the HOH one of the brightest available sources in the VUV domain. The highest harmonic reported so far is the 143rd of a Nd laser corresponding to a photon energy of 168eV or a wavelength of 7.4nm (Perry and Mourou 1994). The shortest wavelength generated by HOHG is 6.7nm, the 37th harmonic of a KrF (248nm) laser in rare gas ions (Preston *et al.* 1995).

Simpleman's model of HOHG in gases

Experiments show that the cut-off (the highest order generated) depends linearly on intensity and more specifically, scales approximately as $3U_p + (IP)$ where (IP) is the ionisation potential of the atom. The ATI Simpleman's theory, extended to high-order harmonic generation (Kulander *et al.* 1993, Corkum 1993) immediately predicts this law: classically, the electron in the continuum, oscillates at the fundamental frequency and each time its trajectory brings it in the vicinity of the nucleus, it is accelerated by the Coulomb field. Its trajectory is then slightly deformed by small kinks whose Fourier spectrum contains the harmonics. The characteristics of the electron harmonic motion depend on the initial conditions, *i.e.* the initial phase and velocity. It can be shown that the maximum kinetic energy at the origin, for an electron initially at rest at the origin is 3.17 U_p and is obtained for an initial phase of about 17 degrees. Therefore, the cut-off is understood as the energy generated by those electrons which recombine with the core at the maximum kinetic energy:

$$\text{cutoff} = \text{max kinetic energy} + \text{ionisation potential} \qquad (4)$$

This is actually observed in experiment (allowing for some modifications due to propagation effects) and is a major success of this model. Quantum calculations using the quasi-classical approximation have recently confirmed and made the classical result precise (Lewenstein 1994).

With circular polarisations, as discussed above, the drift motion prevents any return of the electron to the core. This is the classical extension to non-perturbative regimes of angular momentum conservation which forbids, in the perturbative regime, harmonic generation by circularly polarised photons. It turns out that harmonic generation is extremely sensitive to the ellipticity of the fundamental light (Dietrich *et al.* 1994). This property can be used to generate sub-femtosecond pulses as discussed in the last sub-section.

Duration of HOH

In the perturbative limit, the nonlinear polarisation at the harmonic frequency $(2q+1)\omega$ is proportional to the fundamental field to the power $2q + 1$. This implies that the harmonic pulse duration is much shorter than the fundamental pulse. In the nonperturbative regime in which HOH are generated, there is no simple expression for the nonlinear polarisation. However, numerical calculations show that the q-th harmonic intensity scales as the fundamental intensity to a power $q_{\text{eff}} \approx 15$, independent of the harmonic order. Again, this implies a duration shorter than that of the fundamental. Assuming a Gaussian fundamental pulse of 150fs FWHM, a duration of about 40fs is expected.

Streak cameras have been used to separate the harmonic spectra from recombination radiation emitted by the target gas after the pulse (Kohlweyer *et al.* 1995), but their resolution is limited to 1ps. On the other hand, measurements on the basis of cross-correlation is limited only by the duration of the probe pulse. Such measurements using optical and XUV photons have been reported by van Woerkom *et al.* (1989) and Schins *et al.* (1994) (Figure 23), and have been used to determine the temporal profile of laser-generated X-ray pulses near 90eV and 250eV respectively.

4.3   Experiments

In this Section we summarise two recent experiments which realise the principle introduced in Section 4.2 (Schins *et al.* 1994, Schins *et al.* 1995a). They both rely on photoelectron spectrometry, combined to a pump-probe arrangement suitable for the cross-correlation measurements. They differ by the method used to produce XUV radiation. In the first one, broadband XUV pulses are produced by a gallium plasma generated by focusing a 4mJ, 150fs, 780nm pulse on the surface of a liquid gallium sample. In the second, HOH are generated in argon and focussed in the target gas by a spherical multilayer mirror. In both, the laser is a CPA Ti:S femtosecond laser delivering 50mJ of energy in, linearly polarised pulses at 790nm (Le Blanc 1993).

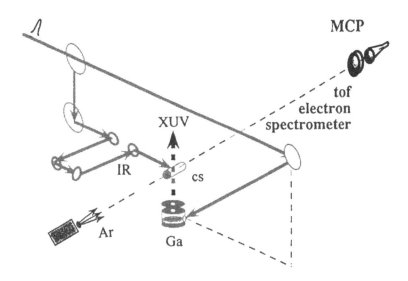

Figure 23. *Experimental set-up for the cross-correlation measurement of a sub-picosecond* XUV *pulse from a gallium LP (see Schins et al 1994).*

Measurement of sub-picosecond, broadband (250–400eV) pulses

Two-colour ATI, as described above demands XUV photons of well-defined energy T_0. This is difficult to achieve without changing the pulse duration. There is however a way around this requirement, namely to detect an Auger decay (in the strong IR field). Auger decay is a well-characterised effect resulting from inner-shell vacancies: if an inner-shell electron is removed by photo-ionisation, the core-hole ion may decay through a process in which one valence electron fills the vacancy while a second is simultaneously ejected into the continuum with a well-defined energy, *independent* of the incoming photon energy (thus eliminating the need of selecting an XUV band). Argon atoms irradiated by photons above the L-edge (250eV) decay to the doubly charged ion through a number of well-identified Auger transitions. Studies of the $L_{2,3}$; $M_{2,3}$; $M_{2,3}$ line and its modifications by the strong IR pulse (Schins *et al.* 1994, Schins *et al.* 1995b) (Figure 24), have lead to the following results; (i) as predicted, sidebands at multiple energies of the IR photons appear on both sides of the bare Auger line. (ii) the number of sidebands is correctly predicted by the $\sqrt{8T_0U_p}$ of Simpleman's formula, and (iii) the amplitudes of the n-th sidebands are given by the expression:

$$J_n(a,b) = \sum_{l=-\infty}^{\infty} J_{n+2l}(a)J_l(b), \tag{5}$$

with $a = \boldsymbol{\alpha}\cdot\mathbf{k}$ and $b = U_p/2\omega$. Here $\boldsymbol{\alpha}$ is the amplitude of the periodic motion of the free electron in the linearly polarised IR dressing field. In the direction of polarisation, $\alpha = E/\omega^2$ (in atomic units), ω the frequency of the IR dressing field, and k is the wavevector of the electron in the dressing field. The experiment is described in Schins *et al.* (1994). Here we just point out that the broadband XUV beam and the IR laser

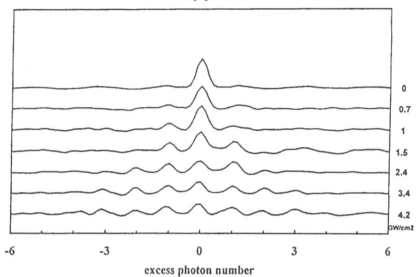

Figure 24. *Electron energy spectra for different laser intensities displaying up to four sidebands around the central lines. Spectra have been deconvoluted from the Auger line structure (after Schins et al 1994).*

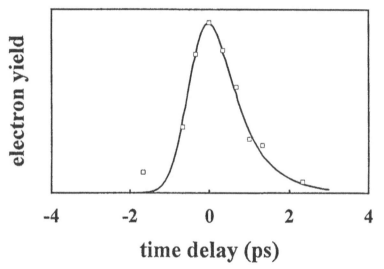

Figure 25. *Cross-correlate signal as a function of the time delay between the XUV and IR pulses (after Schins et al 1994).*

beams cross at right angles in the electron spectrometer.

Applying the method outlined above, the cross-correlation of the XUV and IR pulses is obtained and yields a duration of 700fs (FWHM) for the XUV (Figure 25). Due to the right-angle crossing, the pulses are also convoluted with the geometrical aperture of the spectrometer, which, in the present case limits the resolution of the measurement. This drawback, as well as some problems connected with the structure of the Auger line, are

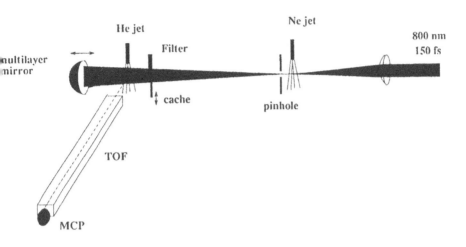

Figure 26. *Experimental set-up for the cross-correlation measurement of 21-st harmonic generated in argon and measured by two-colour ATI in helium. More details can be found in Schins et al. 1995a.*

removed in the experiment described in the following sub-section.

Measurement of femtosecond, 33ev harmonic pulses

Photoionisation of helium with a well-defined energy photon provides a structureless line. Using co-linear XUV and IR beams suppresses the convolution with the spectrometer aperture. It is thus possible to solve the two problems of the previous set-up. This is achieved with a multilayer mirror which selects a narrow band around harmonic 21 (33eV), produced in argon, and focusses both the harmonic and the infrared beams to the same spot (Figure 26). The time overlap is also automatically insured with this set-up. Helium atoms were ionised, resulting in ATI spectra which consisted of combined peaks scaling as the cross-correlation function of the two pulses. This could be mapped out by varying the time delay between the two pulses. A time delay was introduced on the IR pulse by interposing thin calibrated plates of fused silica (20, 30, 50, 110 and 500μm thick) while the intensity was controlled by a mask movable under vacuum. Using this technique, we have carried out the first measurement of the duration of high-order harmonic pulses with sub-picosecond resolution (Schins *et al.* 1995a). A value of 40fs was obtained by deconvoluting the cross-correlation measurement, assuming a squared hyperbolic secant shape for the IR pulse. This result is much shorter than the IR pulse, and consistent with the theory (L' Huillier *et al.* 1992, Antoine *et al.* 1995). Note however that with this set-up, the IR pulse could only be delayed with respect to the harmonic allowing us to measure only the trailing edge of the XUV pulse

4.4   Extension to the sub-femtosecond range and perspectives

The high sensitivity of harmonics to the ellipticity of the light led Ivanov *et al.* (1995) to suggest a method of generating sub-femtosecond pulses. The basic idea is to generate the harmonics of a femtosecond pulse whose polarisation changes with time, circular at the beginning, linear for a short time at the top of the pulse and circular again. Harmonics would then be created only for a very short time interval near the maximum of the pulse. A pulse with a time-dependent polarisation can be constructed in principle from two linearly polarised pulses with perpendicular E vectors and slightly different frequencies.

There has also been a proposal to characterise the sub-femtosecond pulses by using a sort of streak camera based on the phase dependence of the two-colour ATI (Ivanov *et al.* 1995). We have seen above that the characteristics of photoelectron drift motion (velocity, direction) depend on the phase ϕ. Suppose photoelectrons are created by a sub-femtosecond pulse in the field of a low frequency laser (say CO_2). Then the time interval during which they are dropped in the low frequency field maps a corresponding interval of phase ϕ. The resulting velocity distribution can be recorded for instance using a velocity-dependent deflection and a position-sensitive detector thus providing a measurement with sub-femtosecond resolution.

Harmonic generation in gases is limited by the cutoff and the saturation intensity. HOHG in ions (Preston *et al.* 1995) will certainly extend the useful range but the efficiency is expected to be low. According to theoretical models, HOHG from solid-density plasmas with a steep density-gradient is expected to have a cut-off at the plasma frequency of the high-density region. However, recent simulations (Gibbon) show that HOHG could be generated well beyond this limit if the intensity reaches the range of relativistic oscillation velocities 10^{18} Wcm^{-2}. Recently, the authors have become aware of a first demonstration carried out with the VULCAN laser at Appleton Rutherford Laboratory where the 70-th harmonic of 1.05μm at a wavelength of 15nm was observed (Moustaizis *et al.* 1995).

Finally, combining the currently available high-reflectivity, normal incidence mirrors and the techniques of HOHG, the intensity of soft X-ray sources could be boosted to 10^{11}–10^{12} Wcm^{-2} in the near future. This should be sufficient to observe non-linear X-ray processes such as two-photon inner-shell ionisation.

References

Antoine P, Piraux B and Maquet A, 1995, *Phys Rev A* **51** R1750.
Barty C P J, Gordon C L and Lemoff B E, 1994, *Opt Lett* **19** 1442.
Bezzerides B, Jones R D and Forslund D W, 1981, *Phys Rev Lett* **49** 202.

Borisov A B, Shi X, Karpov V B, Korobkin V V, Solem J C, Shiryaev O B, McPherson A, Boyer K and Rhodes C K, 1994, *J Opt Soc Am B* **11** 1941.

Braun A, Korn G, Liu X, Du D, Squier J and Mourou G, 1995, *Opt Lett* **20** 73.

Carman R L, Rhodes C K and Benjamin R F, 1981, *Phys Rev A* **24** 2649.

Chambaret J P, Rousseau P, Curley P, Cheriaux G, Grillon G and Salin F, 1995, *Conference on Lasers and Electro-Optics,Baltimore* Paper CFD5.

Chase L L and Van Stryland E W, 1994,*Handbook of Laser Science and Technology, Supplement 2: Optical Materials, (CRC, Boca Raton).*

Cheriaux G, Rousseau P, Salin F, Chambaret J P, Walker B and Dimauro L, 1995, *Opt Lett* (submitted).

Chu K C, Heritage J P, Grant R S, Liu K X, Dienes A, White W E and Sullivan A, 1995, *Opt Lett* **20** 904.

Cionga A, Florescu V, Maquet A and Taoeb R, 1993, *Phys Rev A* **47** 1830.

Corkum P, 1993, *Phys Rev Lett* **71** 1994.

Curley P F, Darpentigny G, Cheriaux G and Chambaret J P, 1995, *Conference on Lasers and Electro-Optics,Baltimore* Paper CThI50.

DeLong K W and Trebino R, 1994, *J Opt Soc Am A* **11** 2429.

DeLong K W, Trebino R and Kane D J, 1994, *J Opt Soc Am B* **11** 1595.

Dietrich P, Burnett N H, Ivanov M, and Corkum PB, 1994, *Phys Rev A* **50** R3585.

DiMauro L F and Agostini P, 1995, *Advances in Atom Mol Phys*, (Academic Press,New York)

Ditmire T, Nguyen H and Perry M D, 1995, *Opt Lett* **20** 1142.

Etchepare J, Grillon G, Antonetti A and Orszag A, 1988, *Physica Scripta* **T23** 191.

Etchepare G, Grillon G, Thomazeau A, Migus A and Antonetti A, 1985, *J Opt Soc Am B* **2** 649.

Frakas G, Toth C, Moustaizis D, Papadogiannis N, Fotakis C, 1992, *Phys Rev A* **46** R3605.

Georges P, Estable F, Salin F, Poizat J P, Grangier P and Brun A, 1991, *Opt Lett* **16** 144.

Gibbon P, Harmonic generation from solid-vacuum interface irradiated at high laser intensities(to be published).

Hellwarth R W, Pennington D M and Henesian M A, 1990, *Phys Rev A* **41** 2766.

Ivanov M, Corkum P B, Zhuo T and Bandrauk A, 1995, *Phys Rev Lett.* **74** 2933.

Kohler B, Krause J L, Raksi F, Wilson K R, Yakovlev V V, Whitnell R M and Yan Y J , 1995, *Acc Chem Res* **28** 133.

Kohlweyer S, Tsakiris G D, Walhstrvm C-G, Tillman C and Mercer I, 1995, *Opt Commun* **117** 431.

Kulander K C, Schafer K J and Krause J L, 1988,in *Super-Intense Laser-Atom Physics*, eds B Piraux, A L'Huillier and K Rzazewski, (Plenum Press,New York).

Larsson C G, Persson A, Starczewski T, Svanberg S, Salihres P, Balcou P and L'Huillier A, 1993, *Phys Rev A* **48** 4709.

Le Blanc C, Grillon G, Chambaret J P, Migus A and Antonetti A, 1993, *Opt Lett* **18** 140.

Le Blanc C and Salin F, 1995, *CLEO* Paper CWF46.

Lemoff B E and Barty C P J, 1993, *Opt Lett* **18** 1651.

Lewenstein M, Balcou P, Ivanov M Y, L'Huillier A and Corkum P B, 1994, *Phys Rev A* **49** 2117.

L'Huillier A, Lompri L A, Manfray G and Manus C, 1992,in *Atoms in Intense Laser Fields*, ed M Gavrila, (Academic Press,New York).

McMorrow D, Lotshaw W T, Kenney-Wallace G A, 1988, *J Quantum Electron* **QE-24** 443.

Maine P, Strickland D, Bado P, Pessot M and,Mourou G, 1988, *IEEE J Quantum Electron* **QE-24** 398.

Martinez O E, 1987, *IEEE J Quantum Electron* **QE-23** 59.

More R M, Zinamon K H, Warren K H, Falcone R and Murnane M, 1988, *J Phys* **C7** 43.

Moustaizis S D, Bakarezos M, Danson C N, Dangor A E,Dyson A, Fews P, Gibbon P, Lee P,
    Loukakos P, Neely D, Norreys P, Walsh B, Wark J S, Zepf M, Zhang J, 1995,
    Euroconference on Generation and Applications of Ultrashort X-ray pulses, Pisa, Italy,
Murnane M M, Kapteyn H C, Rosen M D and Falcone R W, 1991, *Science* **251** 531.
Nibbering E T J, Curley P F, Grillon G, Franco M A, Prade B S and Mysyrowicz A, 1995a,
    Opt Lett (submitted).
Nibbering E T J, Franco M A, Prade B S, Grillon G, Chambaret J P and Mysyrowicz A,
    1995b, *J Opt Soc Am B* (in press).
Nibbering E T J, Franco M A, Prade B S, Grillon G, Le Blanc C and Mysyrowicz A, 1995c,
    Opt Commun (in press).
Nibbering E T J, Grillon G, Franco M A, Prade B S, Mysyrowicz A, 1995d, (in preparation).
Offner A, *US Patent 3,748.015.*
Perry MD and Mourou G, 1994, *Science* **264** 917.
Prade B S, Schins J M, Nibbering E T J, Franco M A and Mysyrowicz A, 1994,
    Opt Commun **113** 79.
Preston S G, Sanpera A, Zepf M, Blyth W J, Wark J S, Key M H, Burnett K, Nakai M, Neely
    D and Offenberger A A, 1995, *Phys Rev Lett* (submitted).
Schins J M, Breger P, Agostini P, Constantinescu R C, Muller H G, Bouhal, Grillon G,
    Antonetti A and Mysyrowicz A, 1995a, *J Opt Soc Am* (in press).
Schins J M, Breger P, Agostini P, Constantinescu R C, H G, Muller, H G, Grillon G, Antonetti
    A and Mysyrowicz A, 1995b, *Phys Rev A* (in press).
Schins J M, Breger P, Agostini P, Constantinescu R C, H G, Muller, H G, Grillon G, Antonetti
    A and Mysyrowicz A, 1994, *Phys Rev Lett* **73** 2180.
Shaw M J, Hooker C J and Wilson D C, 1993, *Opt Commun* **103** 153.
Shen Y R, 1984, *The Principles of Nonlinear Optics*, (Wiley, NewYork).
Spence D E, Kean P N, Sibbett W, 1991, *Opt Lett* **42** 16.
Spielmann Ch, Curley P F, Brabec T and Krausz F, 1994, *IEEE J Quantum Electron* **QE-30**
    1100 and references therein.
Stingl A, Lenzer M, Spielmann Ch, Krausz F and Szipocs R, 1995, *Opt Lett 20* **12** 1399.
Strickland D and Mourou G, 1995, *Opt Commun* **56** 219.
van Linden van den Heuvell H B and Muller H G, 1988, *Multiphonon Processes*, eds S D
    Smith and P L Knight, (Cambridge University Press)
van Woerkom L D, Freeman R R, Davey S, Cooke W E and McIlrath T J, 1989,*J Mod Optics*
    36 1817.
Véniard V, Taoeb R and Maquet A, 1995, *Phys Rev Lett* **74** 4161.
von der Linde D, Engers T, Genke G, Agostini P, Grillon G, Nibbering E, Mysyrowicz A and
    Antonetti A, 1995, *Phys Rev A* **52** 25.
Warren W S, Rabitz H and Dahleh M, 1993, *Science* **259** 1581.
Weiner A M, Leaird D E, Patel J S and Wullert J R, 1992,
    J Quantum Electron **QE-28** 908.
White W E, Sullivan A, Proce D F, Bonlie J and Stewart R, 1995, Conference on Lasers and
    Electro-optics CLEO, *PDP: CPD45-1.*
Zhou J, Huang C-P, Shi C, Murnane M M and Kapteyn H C, 1994a, *Opt Lett* **19** 126.
Zhou J, Taft G, Huang C, Murnane M, Kapteyn H and Christov I P, 1994b,
    Opt Lett 19 **15** 1151.

Ultrafast Imaging Techniques

J Reintjes

US Naval Research Laboratory
Washington DC, USA

1 Introduction

Time-gated imaging provides a view of the world well beyond the one available to the unaided eye. Details of the motion of creatures or objects and scientific studies of explosions, shock fronts and other interactions are among the things that become visible with imaging systems with sufficiently short time resolutions. Various technologies have been employed in producing time-gated images ranging from simple mechanical shutters to the sophisticated laser-based systems of today. As advances in technology have provided capability of gating at progressively shorter times, new regimes of measurement and a new class of objects and interactions have become available for observation and study.

Time-gated imaging involves the simple concept of restricting the light used in forming an image of an object according to its time history. This can be done either by using pulsed illumination to limit the duration of the illumination source, or a shutter at the recording medium, as illustrated in Figure 1. Technological advances have been made for both techniques over time. The mechanical shutters of early imaging systems, although still very much in use today, have advanced to electronically-controlled shutters, and most recently to light-controlled shutters. Pulsed sources have progressed from the flash powder of the nineteenth century through strobe lights to the laser sources of today. A summary of approximate capabilities of various shutters and pulsed light sources is given in Tables 1 and 2.

Type	Minimum time (s)	Characteristics
Mechanical	$10^{-3} - 10^{-4}$	Convenient to use Ambient light source Inherent lower time limit
Electronic	10^{-10}	Electron tube – Framing
	10^{-12}	Electron tube – Streak
Electro optic	2×10^{-10}	Limited angular acceptance Low on/off contrast
Optical	10^{-14}	Lower time limit unconstrained Based on nonlinearities

Table 1. *Shutter characteristics*

Type	Minimum time (s)	Characteristics
Flashlamp	$10^{-8} - 10^{-6}$	Xenon strobe
Laser	10^{-8}	Q-Switch
	10^{-12}	Modelocked
	6×10^{-15}	Femtosecond

Table 2. *Pulsed illumination source characteristics*

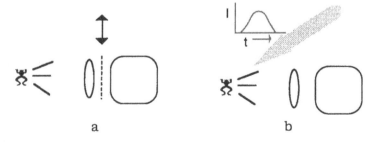

a   b

Figure 1. *Time gated imaging techniques: a) shutter at the recording medium; b) pulsed illumination.*

Figure 2. *Sequence of pictures obtained with a shutter speed of 0.002s showing a galloping horse with four feet off the ground (far left). From Muybridge, 1887. Reprinted with permission.*

1.1 Mechanical shutters

Although we have a tendency to think of the 'ultrafast' in the title of this chapter as unique to the new technology of our era, the concept of 'ultrafast' has always existed and has always been ascribed to the most advanced technology of the time. Early gating systems used mechanical shutters to open and close the exposure to film. An example of an early 'ultrafast' image obtained in this way is the famous picture (Figure 2) of a galloping horse made by Edward Muybridge (Muybridge 1887). Muybridge developed a camera with a shutter speed of 0.002s to settle a bet as to whether a horse ever has all four feet off the ground when galloping. While their shutter times are slow by current standards these pictures illustrate an important aspect of technological development - its application to obtaining information that was previously unavailable. Mechanical shutters of course remain in use today. However they have a practical limit on the speed that they can attain caused by the need to accelerate the material of the shutter to velocities near the speed of sound. For example, a 1ms exposure with 35mm film requires the shutter element to move a distance of the order of 24mm in 0.001s, which is about 1/10 of the speed of sound. To increase the speed of mechanical shutters to the next order of magnitude would require moving the mechanical components at close to supersonic speeds.

1.2 Strobe lamps

A major advance in high speed imaging was made in the 1930's when Edgerton introduced *electronic strobe photography* (Edgerton 1931). This advance reduced gating times to the microsecond range and opened up an entirely new domain to the realm of stop-action photography. The main technological advance by Edgerton was not in developing microsecond flash sources. Indeed, microsecond and nanosecond pulsed sources had been known for some time in the form of spark gaps. Rather his advance was in devising a practical method for producing enough light to expose film in the microsecond times possible with his flash lamps. His pictures are famous for their ability to show new aspects of the world around us, ranging from the familiar to things that were

 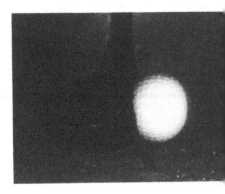

Figure 3. *Electronic strobe pictures of (a) multiple exposures of a golf swing and (b) a golf ball at the moment of impact. From Kayafas, 1987; copyright The Harold E Edgerton 1992 Trust, courtesy Palm Press.*

indistinguishable prior to his inventions. Early examples were images of a splashing milk drop and a hummingbird's wings frozen in flight; in Figure 3 we see a multiple exposure of a golf swing and golf ball at the moment of impact (Kayafas 1987). There are now innumerable scientific applications of these techniques.

Edgerton's technology changed the way high speed imaging was done. Prior to his development, ambient light was used for illumination and the film exposure was controlled by the shutter. In his approach the film was exposed continuously in a dark room and the duration of exposure was controlled by the duration of the light flash. His technology concentrated on the development of pulsed electronic flash units based on xenon flash lamps. These are now routinely available as the electronic flash unit used by amateur and professional photographers. Further advances of this technique has followed the progression of pulsed light sources, first to the nanosecond range with *Q-switched lasers*, then to the picosecond regime with *mode-locked lasers*, and to the present day with *femtosecond pulses*.

1.3 Electronic shutters

Progress has also been made in the technology of fast shutters. Although the minimum exposure time of mechanical shutters is limited as we discussed before, electronic techniques have allowed shutters to advance in two forms: *framing cameras* and *streak cameras* as illustrated in Figure 4. Both techniques involve the use of electron imaging tubes to convert the optical signal to an electron beam and to reform the image on a phosphor. In each case the time-gating is achieved by pulsing a voltage that controls the electrons that are emitted from the cathode. In the case of a framing camera, a two dimensional image is obtained at the phosphor over an area of the display. Time

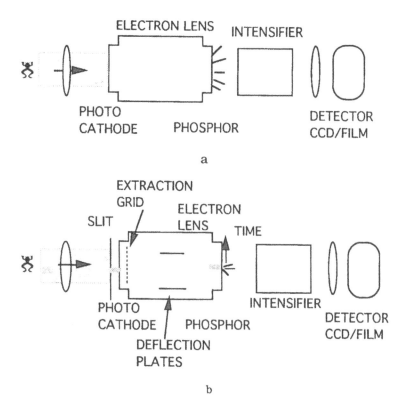

Figure 4. *(a) Schematic illustration of a framing camera; (b) schematic illustration of a streak camera.*

gating is obtained by pulsing the accelerating voltage between the photocathode and the anode, providing a stop-action picture of the type familiar from flash photography. Framing cameras with exposure times of several nanoseconds have been available for many years. Recent advances allow single exposures to be made in the 100 picosecond range (Hares 1987). This has been accomplished by using a gated microchannel plate (MCP) that is capacitively coupled to the photocathode, as shown in Figure 5, along with an avalanche transistor switch. Gate times of the order of 100ps have been obtained with switching voltages up to 3kV, although only about 200V actually appears across the MCP. The camera is capable of two dimensional pictures and the use of multiple cameras allows a time sequence of multiple images to be obtained. The system has been used to image expanding plasmas in laser fusion experiments (Ripin *et al.* 1993), and for range-gated underwater imaging (McLean *et al.* 1995).

Streak cameras use a small slit at the entrance to the camera and sweep the image of the slit across the phosphor with a deflecting voltage. One dimension of the resulting image records time while the second dimension carries one-dimensional image

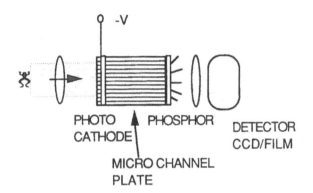

Figure 5. *Schematic illustration of a high speed framing camera with a capacitively coupled photocathode. Adapted from Hares, 1987*

information about the object. Two dimensional information can be built up by scanning the object in the direction perpendicular to the slit. The chief limitation on time resolution is the blurring of the image due to variation in arrival time of the electrons at the phosphor, caused by, among other things, the thermal spread of the electrons leaving the cathode. The resolution time is reduced by constructing the streak tube with an extraction grid close to the cathode to raise the electrons to their final velocity as quickly as possible, minimising the effects of thermal velocity spread. Streak cameras are available today with resolution times of the order of 1ps, and sub-picosecond cameras have been developed by using two dimensional scan techniques.

1.4   Optical gating techniques

The progression of high-speed imaging to ever shorter times has gone from shuttered systems to pulsed illumination to shutters again as each technology has reached a practical limit. In order to advance beyond the limits of electronic shutters another change in technology is required. This has been provided in recent years by the development of optically gated systems in which the selection of light from an illuminating signal for imaging is controlled by a second light beam. All techniques used for optical gating involve *nonlinearities*. Many of them involve nonlinear optical effects. Some, however, involve interference between signal and gating optical beams at a photodetector, with the nonlinearity being provided by the square-law response of the detector. The time resolution that can be achieved with these techniques is much shorter than can currently be achieved with electronic switching.

Pulsed optical gating

Optically gated imaging systems can be divided into two categories: *pulsed gating*, in which the shutter time is determined by the duration of an optical pulse; and *correlation*

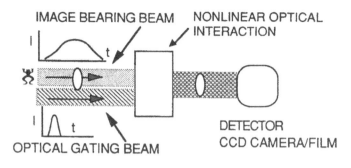

a

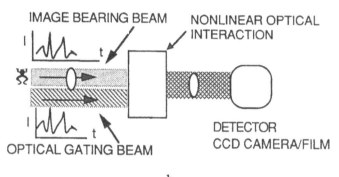

b

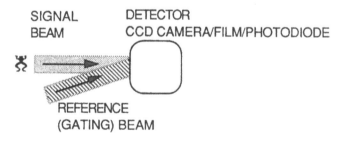

c

Figure 6. *(a) Schematic illustration of a system used for pulsed optical gating; (b) schematic illustration of system used for nonlinear optical correlation gating; (c) schematic illustration of system used for interference correlation gating.*

or *coherence gating*, in which the shutter time is determined by the duration of mutual coherence between the illumination signal and the gating signal. A typical configuration used for pulsed optical gating is shown in Figure 6a. Here the nonlinear interaction occurs between a signal beam carrying the image and a gating beam, resulting in a third beam which carries the gated image. The third beam may or may not be at the same wavelength as the signal beam and may or may not propagate in the same direction. Some of the nonlinear interactions that have been used for time-gated imaging are listed in Table 3.

Optical Kerr effect
Stimulated Raman amplification
Second harmonic generation
Sum frequency generation
Third harmonic generation
Four wave mixing
Coherent anti-Stokes Raman scattering
Parametric amplification

Table 3. *Some nonlinear optical interactions used for time-gated imaging*

The minimum time resolution for the pulsed nonlinear optical gates is determined by a combination of the optical pulse duration and the time constant of the interaction under use. As a class they are potentially capable of operation with time gates as short as 6 femtoseconds, the limit of current pulsed-laser technology, although not all interactions can be used at this short a time. Depending on the particular interaction and medium involved, the limiting time can be in the femtosecond to picosecond time regime or longer. To date pulsed gated systems have been demonstrated at time resolutions down to about 100 femtoseconds.

We can make a few general comments about the operational requirements of a pulsed optical gate. The equation describing the *output* of the nonlinear optical gate is:

$$\nabla_{\perp}^2 A_{\text{out}}(L,r,t) - 2ik\left(\frac{\partial A_{\text{out}}(L,r,t)}{\partial z} + \frac{1}{v}\frac{\partial A_{\text{out}}(L,r,t)}{\partial t}\right) = f\left(A_{\text{in}}(0,r,t), A_{\text{ref}}(z,r,t)\right) \quad (1)$$

where the optical fields of the signal and gate beam are defined as

$$E_{\text{sig}}(z,r,t) = \frac{1}{2}\left(A_{\text{sig}}e^{i(\omega_{\text{sig}}t - k_{\text{sig}}z)} + \text{c.c.}\right),$$

$$E_{\text{gate}}(z,r,t) = \frac{1}{2}\left(A_{\text{gate}}e^{i(\omega_{\text{gate}}t - k_{\text{gate}}z)} + \text{c.c.}\right),$$

r describes the *transverse spatial variation* that carries image information on the signal, the medium extends from $z = 0$ to $z = L$ and f is a nonlinear function that describes the appropriate interaction. The signal intensity emerging from a proper pulsed imaging gate satisfies the relation

$$I_{\text{out}}(L,t) = K I_{\text{in}}(0,t) I_{\text{gate}}^n(t) \quad (2)$$

where I is the intensity of the appropriate signal. This relation indicates simply that the time dependence of the gated signal is determined by the product of the image signal and the gate signal raised to the power n, with n determined by the specific interaction.

In order to minimize distortion of the gated image, the spatial variation of the gated signal should be determined entirely by the spatial variation on the input beam. Generally speaking this means that the gating beam should be uniform over the interaction region, but there are some interactions that do not require even this.

The nonlinear optical interactions occur over an extended interaction volume. It is not required that the spatial variation of the signal at the exit of the gate be the same as that at the entrance. All that is required is that any diffraction that takes place on the image bearing beam within the nonlinear medium be the same in the presence of the gating interaction as it is without it. Then the image can be faithfully reconstructed with lenses or other imaging optics.

Correlation (coherence) gating

Correlation or coherence gating is similar in principle to pulsed gating, except for the specific interaction, pulsed formats and detectors as illustrated in Figure 6b. Coherence gating is typically done with long pulse or CW broad band optical signals, for example, radiation from pulsed dye lasers or superluminescent diodes. In coherence gating, the gated signal is determined by a *zero time correlation integral* between the amplitudes of the signal and a gate:

$$S(r) = K \left| \int A_{\text{sig}}(z, r, t) \, A^*_{\text{gate}}(z, r, t) \, dt \right|^2 \tag{3}$$

Several types of nonlinear optical interaction have this property, although interestingly enough, the more common ones of non resonant sum and difference frequency mixing and harmonic generation do not exhibit this property. The time resolution is determined by the coherence time of the gating beam, and not its pulse duration. As a result it is possible to obtain picosecond or subpicosecond gate times with reference signals that last for nanoseconds or are even CW. This can be very important for many applications such as medical imaging or materials inspection in which the level of illumination of the sample is limited by optical damage. In such cases the number of photons that can be used to illuminate the sample with short pulses is limited. If attenuation of the signal is large it can be very hard to obtain enough signal photons to be detected within the duration of an ultrashort pulse.

Coherence gating can also be carried out by allowing the signal and gating beams to interfere at the detector as shown in Figure 6c. The resulting signal is given by

$$S = K \int I_{\text{total}} \, dt = K \int |A_{\text{sig}}(z, r, t) + A_{\text{ref}}(z, r, t)|^2 \, dt \tag{4}$$

$$= K \left[\int |A_{\text{sig}}(z, r, t)|^2 \, dt + \int |A_{\text{ref}}(z, r, t)|^2 \, dt + 2 \text{Re} \int A_{\text{sig}}(z, r, t) \, A^*_{\text{ref}}(z, r, t) dt \right] \tag{5}$$

The first two terms in Equation 5 represent constant background terms while the last term contains the gated signal information. Although the first two terms do not contribute to the desired gated signal, they can contribute shot noise that affects the minimum detectable signal.

Interference gating can be realized either in the time or spatial domains. In the spatial domain, the technique is the same as pulsed holography and provides two dimensional images directly. It can be used not only with the correlation gating described here using broad band long duration signals, but also with short pulsed sources. Interference gating in the time domain is typically accomplished by modulating the gating signal in time and then detecting the resulting component at the difference frequency. Single point images are obtained directly with this method, and two dimensional images can be built up by scanning either the illumination spot or the sample.

1.5   Characteristics of optically gated systems

The characteristics of optically gated imaging systems that are of interest are

- minimum gate time
- overall image exposure time
- leakage/gate contrast
- minimum detectable signal level
- number of resolution elements
- minimum resolvable spatial dimension
- imaging fidelity/distortion

Minimum gating time refers to the minimum time in which a signal can be selected for detection from a longer signal which one desires to suppress. *Overall image exposure time* refers to the time required to build up enough signal to detect or the time to scan an image if scanning is employed. The minimum gating time is determined by the specific interaction and light source used, while exposure time is determined also by the detector sensitivity, the illumination power available or usable, and the efficiency or insertion loss of the gate. For the systems considered here the gating time is in the ultrafast domain, while overall exposure time can be fractions of a second or more depending on the system.

Leakage refers to how much light is transmitted through the gate when it is 'off'. In general we can assume that the light input to the gate is composed of light we wish to detect in a short time interval and light we wish to suppress. The better the suppression and lower the leakage the better. Leakage determines how good the gate is at discriminating in favour of the desired signal and can determine which of various gating mechanisms are preferable in various situations. *Minimum signal level* is generally determined by the noise properties of the gate and the optical detector. In some cases it is determined entirely by the optical gate, and in others entirely by the optical detector. Generally speaking, the lower the minimum signal level the better, but it is not always necessary to operate at the minimum signal level. In many applications the signal intensity is severely attenuated, therefore the ability to detect signals as low as possible is important. The minimum signal level can also have a bearing on the overall exposure time.

The *minimum detectable signal level* is determined by the noise characteristics of the system. For all systems the absolute minimum noise level is due to photon noise on the incoming signal, although this may not be reached if there are other noise sources.

The signal photon noise may be described by Poisson statistics:

$$\Delta n_{\text{photon}} = \sqrt{n_{\text{photon}}} \tag{6}$$

where n_{photon} is the number of photons detected. In modern systems it is common to use low noise CCD cameras for image recording. Current CCD cameras have quantum efficiencies near 80% and noise levels of the order of 5 to 8 photoelectrons, or 10 photons, rms, and typically 5–8 times higher peak to peak. The rms value is applicable if images from several CCD exposures are combined, while the peak to peak values are appropriate if a single CCD exposure is used. For the numbers quoted above, the photon noise on the gated signal is comparable to the detector noise when the number of detected photons is of the order of 2500–6400 per pixel for single shot detection, and 100 photons per pixel for multiple shot detection. If the gating mechanism has an insertion loss, still higher signal levels are required from the illuminated object. For signal levels larger than these, the noise will be dominated by photon noise on the incoming signal, and the S/N is given by the square root of the number of detected photons. For signal levels smaller than these, the noise is determined primarily by the detector properties, although for systems that employ heterodyne detection the signal to noise ratio can be determined by the signal photon noise even at low signal levels.

The spatial properties of most optical systems are characterisable by the number of resolution elements, and not necessarily the minimum spatial resolution. This is because the minimum spatial resolution is determined by a diffraction spot diameter. The size of the diffraction spot can be adjusted with magnification or demagnification, but this is usually accompanied by a change in the overall size of the image. As a result it is the number of resolution spots that is a good number to compare for different optical imaging systems.

The *number of resolution elements* can be determined by the recording medium, the nonlinear optical interaction or both. Limitation on the number of resolution elements caused by the nonlinear interaction arise because the interaction takes place over an extended interaction length. To determine the effect of the interaction geometry on the nonlinear interaction we consider the system shown in Figure 7. The number of transverse modes in the interaction volume can be shown with simple box normalisation to be given by (Duncan *et al.* 1992)

$$N_{\text{modes}} = \left(\frac{A}{\lambda L}\right)^2 \tag{7}$$

where A is the effective cross sectional area of the gate beam, and λ is the signal wavelength. For commonly encountered situations, the pump beam has a finite area but does not have a perfectly constant profile. In these cases, an effective area must be used which is generally smaller than the one calculated using the pump beam radius. If we define the *Fresnel number* of the interaction as

$$F = \frac{A}{\lambda L} \tag{8}$$

the number of modes is given by $N_{\text{modes}} = F^2$. In some situations the number of resolvable spots can be made equal to the number of modes. Thus, roughly speaking,

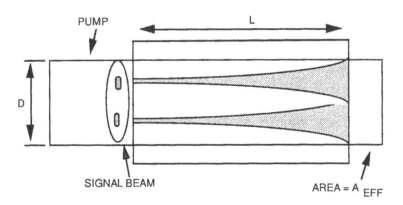

Figure 7. *Illustration of concepts used to determine the number of resolution spots in a nonlinear optical imaging system.*

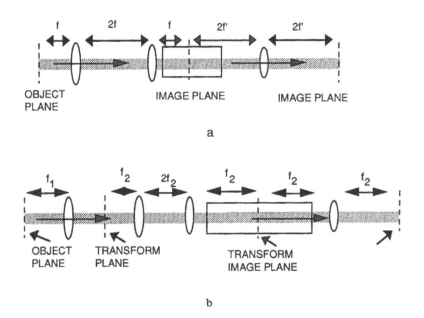

Figure 8. *(a) Schematic illustration of configuration used for direct image gating; (b) schematic illustration of configuration used for Fourier gating.*

a nonlinear interaction of area A and length L will provide an image with resolution of $n \times n$ spots, where $n = F = A/\lambda L$.

The image can be brought into the nonlinear medium either directly, by imaging the object into the gate medium and reimaging onto the detector (Figure 8a), or by forming the Fourier transform of the object in the medium and reconstructing the image at the detector (Figure 8b). Direct image amplification in principle can provide higher resolution, but can have limited field of view. For Fourier amplification, the higher spatial frequencies are cut-off by the finite gate beam diameter, limiting the resolution, but providing a full field of view.

2 Nonlinear optical interactions

The concept of all optical gating implies the ability of one light signal to affect another and, as such, lies naturally within the realm of nonlinear optics. Many different types of nonlinear optical interaction can be used for image gating: see Table 3.

All of the techniques listed in Table 3 can be used for pulsed gating and some can also be used for correlation gating. Some of the interactions involve amplification, and some involve wavelength conversion. The interactions currently being used most commonly for gated imaging are the optical Kerr effect, stimulated Raman amplification and coherent anti-Stokes Raman scattering.

2.1 Optical Kerr effect

The optical Kerr gate is based on birefringence induced in a nonlinear material by a strong optical field (Duguay and Mattick 1971, Wang *et al.* 1991, Ho *et al.* 1993). It can be used for pulse gating, but not for correlation gating. A typical arrangement for image gating with the optical Kerr effect is illustrated in Figure 9. It consists of a nonlinear material that exhibits optically induced birefringence arranged between crossed polarisers. The image is carried by a signal beam which passes through the nonlinear material and both polarisers. The gating pulse is brought in on a second beam that can be either collinear with the signal beam or at a small angle. The gating pulse is generally polarized at 45° to the polarization direction of the image beam and produces an intensity-dependent anisotropic change in the refractive index of the nonlinear material in the gate. The resulting birefringence causes a portion of the signal beam to have its polarisation rotated. That part of the image beam with rotated polarisation is transmitted through the analysing polariser and is detected with a suitable two dimensional detector, *e.g.* film or CCD camera.

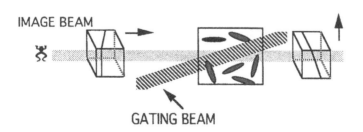

IMAGE BEAM

GATING BEAM

Figure 9. *Schematic diagram of an Optical Kerr imaging gate.*

There is no requirement for the wavelength of the signal and gate beams to be related. In fact if they are not the same, it is easier to separate them using a colour filter after the gate because the gating signal is usually much stronger than the image signal. In principle. any material with a nonlinear coefficient of refractive index can be used for the Kerr gate. The power drive requirements, insertion loss and gating times are dependent on the particular medium. We will illustrate this for a CS_2 gate, the most commonly used gate material.

CS_2 is a linear molecule with an anisotropic polarisability. In the presence of a linearly polarised optical field, the molecules tend to align along the polarisation direction. For pulse durations longer than the reorientational relaxation time of 2.3ps, we can consider the response to be instantaneous, with the refractive index given by:

$$n(t) = n_0 + n_2\, I(t) \tag{9}$$

where n_0 is the linear refractive index and n_2 is the nonlinear refractive index.

For CS_2 the change in index for light polarised parallel to the gate pulse is twice as great as the change in index for light polarised perpendicular to the gate:

$$\Delta n_\| = -2\Delta n_\perp \tag{10}$$

This difference results in *optically induced birefringence*. It occurs because molecules that reorient in the presence of the gate field contribute to an increase in refractive index parallel to the gate polarisation, while molecules that are initially oriented perpendicular to the direction of propagation contribute to a decrease of index for polarisation perpendicular to the gate polarisation. The factor of 2 arises because molecules that are initially oriented along the direction of propagation do not contribute to the decrease in index of the perpendicular polarisation but do contribute to the increase in index parallel to the gate polarisation.

If we consider the configuration shown in Figure 9 where the signal light is initially polarised at 45° to the direction of gate polarisation, the signal intensity transmitted through the analysing polariser is given by

$$I_T = I_{in}\, \sin^2\left[k\left(\Delta n_\| - \Delta n_\perp\right) L/2\right] \tag{11}$$

where L is the length of the gate and $k = 2\pi/\lambda$ is the wave vector of the signal light. For gate transmissions of the order of 10% or less we can use the small parameter expansion of the sine function to obtain

$$I_T = \left(\frac{3\pi n_0 L}{2\lambda} n_2\right)^2 I_{in}\, I_{gate}^2(t) \tag{12}$$

which is the same form as in Equation 2 with $n = 2$.

Thus, for gate pulses longer than the relaxation time of the nonlinear material the Kerr gate provides an instantaneous gate for the incoming signal. For shorter pulses, the gate has an integrating response with an exponential decay. As pointed out earlier the response time for CS_2 is 2.3ps, the shortest time available in common liquids. Other choices such as nitrobenzene, toluene or alcohol have longer times— up to a few tens of picoseconds.

It is generally desirable to operate the gate with minimum insertion loss, which would require operation at the level of a half wave plate with relatively high drive power. On the other hand, if some insertion loss can be accepted, then lower drive power can be used. To get a feel for the required values, we consider our CS_2 gate with $n_2 = 2.8 \times 10^{-14}$ cm^2/W. For 10% gate transmission we calculate a drive power of 1.6×10^8 W/cm^2 for a gate 1cm in length at a wavelength of 1μm. Operation of Kerr gates with tranmissions up to 60% have been described (Wang *et al.* 1993).

The minimum detectable signal level for Kerr gates is usually determined by the CCD noise properties which were discussed earlier. The gate contrast is determined primarily by leakage through the crossed polarisers. High quality crystal polarisers can have extinctions up to 10^6 although somewhat smaller values are usually achieved in practice because of imperfections in optics placed between the polarisers. Usually however, the contrast is always above 10^3. Contrast can also be increased by using multiple Kerr gates in series (Wang *et al.* 1993).

Spatial resolution is determined either by the resolution limits of the camera or resolution limits of the Kerr gate. The resolution limits of the Kerr gate can be adjusted by varying its length and area. For example, for an image containing about 500×500 resolution elements (comparable to a CCD camera) the Kerr gate described above would have to have a Fresnel number of

$$F = 500 = \frac{A}{\lambda L}. \tag{13}$$

At a wavelength of 1μm and a length of 1cm, the area required is 0.05cm^2, corresponding to a diameter of 2.5mm. For geometries with larger Fresnel numbers the number of resolution elements is determined by the detector.

2.2 Stimulated Raman gating

Pulsed gating can also be accomplished with stimulated Raman amplification. A typical arrangement is shown in Figure 10 (Duncan *et al.* 1991a). Here the image is carried by the Stokes beam and the Raman pump pulse serves as the gating pulse. Raman amplification is described by the equations

$$\frac{\partial A_S(z,\tau)}{\partial z} = -i\kappa_2 A_L(z,\tau) Q^*(z,t) \tag{14}$$

$$\frac{\partial Q^*(z,\tau)}{\partial t} + \frac{1}{T_2} Q^*(z,t) = -i\kappa_1 A_L^*(z,t) A_S(z,t) \tag{15}$$

where Q is the normal coordinate of the material excitation, T_2 is the dephasing time of the Raman transition and κ_1 and κ_2 are coupling constants. For most cases important to time-gated imaging, the Raman process will be in the transient regime with the pulse duration $t_p \ll T_2$ and the amplified Stokes intensity is given by

$$I_S(z,t) \propto \kappa_1 \kappa_2 I_L(t) \left[\int_{-\infty}^{t} \frac{E_L^*(t')E_S(0,t')I_1\left(\sqrt{4\kappa_1\kappa_2 z[p(t)-p(t')]}\right)}{\sqrt{p(t)-p(t')}} dt' \right]^2 \tag{16}$$

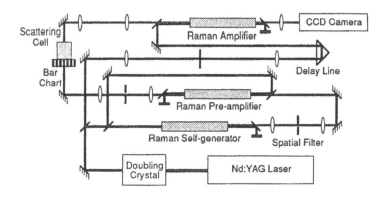

Figure 10. *Schematic illustration of a pulsed Raman imaging gate. (Duncan 1991)*

where

$$p(t) = \int_{-\infty}^{t} I_{\mathrm{L}}(t')dt'$$

is the integrated energy of the pump pulse and I_1 is a modified Bessel function.

The incident Stokes field is amplified only when the pump pulse is present, resulting in a pulsed amplification gate. When the signal level is low enough to require quantum effects to be taken into account the field amplitudes in Equations 14 and 15 are replaced by creation and annihilation operators (Raymer and Mostowski 1981, Duncan *et al.* 1991b). Thus the Raman interaction amplifies the complex field amplitude, or the complex probability amplitude for the quantum regime. As a result, the Raman amplifier will maintain the features of an image, even if the image structure diffracts within the amplifier, provided structure from the pump beam does not transfer to the Stokes beam.

The pump beam can have structure in the amplitude or phase or both. Because the amplification is dependent on the square of the pump amplitude, structure in the pump phase will not imprint directly on the Stokes beam. However, amplitude structure will imprint directly. In addition, if phase structure in the pump can diffract to amplitude structure, it too will imprint on the Stokes beam, distorting the image. Thus it is desirable to have a smooth pump beam. This can be accomplished by spatially filtering or apodizing a beam from a high quality laser. Alternatively, intensity averaging can also be used to smooth out pump intensity structure over the Raman interaction length (Reintjes *et al.* 1986).

Since the Raman gain is usually high enough to raise the detected signal level well above the noise of a CCD camera, the noise level in a Raman amplifier is determined by the quantum fluctuations that are responsible for spontaneous Raman scattering. Analysis shows that these can be taken as arising from either the fluctuations of the material polarisation, the fluctuations in the Stokes field or both (Raymer and Mostowski 1981, Duncan *et al.* 1991b). The equivalent noise level can be shown to be $h\nu/2$ per mode of the amplifier, corresponding to one photon per mode, with the number of modes given by (Duncan *et al.* 1992)

$$N = \left(\frac{A}{\lambda L}\right)^2 \Delta\nu\Delta t \qquad (17)$$

where $\Delta\nu$ is the effective bandwidth of the amplified Stokes field. The first factor in Equation 17 describes the number of transverse spatial modes and the second factor describes the temporal modes. The number of temporal modes, $N_t = \Delta\nu\Delta t$, is determined by the relation of the pulse duration to the Raman dephasing time, T_2. N_t is near unity for interactions in the extreme transient regime, $t_p \ll T_2$. For long pulse interactions that satisfy the condition $t_p \gg t_{SS} = G_{SS}T_2$, the number of temporal modes is given approximately by $N_{temp} \approx t_p/t_{SS}$ and is much greater than unity.

The number of spatial modes is given by the square of the Fresnel number as discussed before

$$N_{spatial} = F^2 = \left(\frac{A_{gate}}{\lambda_S L}\right)^2 \qquad (18)$$

where A is the effective area of the gate beam and λ_S is the wavelength of the Stokes beam. Comparison of Equations 6 and 18 shows that the minimum noise level in a Raman amplifier is achieved in the extreme transient regime for which $t_p \ll T_2$ and the noise level is one photon per spatial mode. For example, in hydrogen gas at 20 atmospheres of pressure, $T_2 \approx 350$ps. Gating systems with pulse durations shorter than this will exhibit minimum noise. The Fresnel number of Raman amplifiers is usually somewhat limited and so there is also an element of Fourier filtering in the amplifier. The gate contrast is determined by the amplification level since unamplified Stokes light is passed through at its incident level to the detector. Stable Raman gains can be obtained up to values of the order of 10^9 providing potential for a contrast of nine orders of magnitude.

As a numerical example typical of laboratory experiments, we consider a hydrogen amplifier pumped with a single 30ps pulse from a doubled Nd:YAG laser at 532nm. The hydrogen pressure is 20atm, the beam size is 3mm and $T_2 = 350$ps. The pump Fresnel number is 15 and the number of resolvable spots is about $15\times15 = 225$. An amplification of 10^6 is obtained for a pump pulse energy of 10mJ. An example of an image of bars on a resolution chart obtained with direct image amplification from such an amplifier is shown in Figure 11 (Duncan *et al.* 1992). At low illumination levels, there is grain in the images due to photon noise, which produces modulation in the bright areas of the image, but no photons in the dark regions. This type of behaviour can be interpreted as interference between the quantum fluctuations and the field of the signal, or as photon noise due to statistical variations of *Glauber coherent states*. The structure in the image gradually becomes less noticeable as the signal level increases. As with most other systems, averaging over multiple shots can be effective in reducing the noise.

2.3 Correlation (coherence) gating

Correlation gating is accomplished by selecting out a portion of the signal based on its degree of correlation with the gating pulse. The most effective techniques for this

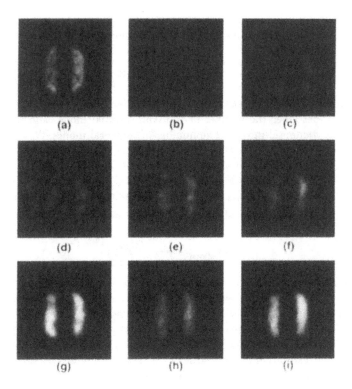

Figure 11. *Example of images of a 0.316 mm wide bars from a bar chart obtained with Raman Fourier image amplification for various seed-Stokes input levels: (a) Seed Stokes, no amplification; (b) no seed Stokes; (c) seed Stokes = 70 photons; (d) seed Stokes = 106 photons; (e) seed Stokes = 160 photons; (f) seed Stokes = 280 photons; (g) seed Stokes = 600 photons; (h) seed Stokes = 2400 photons; (i) seed Stokes = 1.5 × 10⁶ photons. From Duncan et al. 1992.*

purpose are based on field correlations, in which the detected signal is determined by the field correlation integral:

$$S = K \left| \int |A_{\text{sig}}(r,t)| \, |A^*_{\text{gate}}(r,t)| e^{i(\phi_{\text{sig}} - \phi_{\text{gate}})} \, dt \right|^2 \tag{19}$$

Many nonlinear optical interactions, as well as square law optical detectors, can be used for correlation gating. Suitable nonlinear optical interactions involve an integrating response that can be obtained with a resonant interaction or other slow response. Examples of nonlinear optical interactions that can be used to give correlation gating are stimulated Raman scattering, coherent anti-Stokes Raman scattering, certain types of resonant four wave mixing and photorefractive interactions. Certain classes of common nonlinear interactions, however, do not produce this type of response. An example is second harmonic generation with nonresonant interactions—as in a crystal such as KDP.

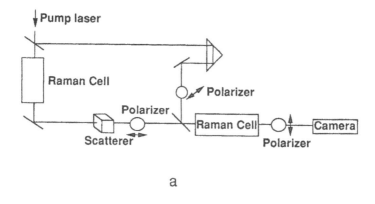

a

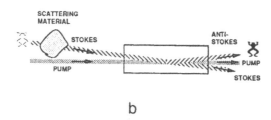

b

Figure 12. *Schematic illustration of correlation gates using (a) stimulated Raman scattering and (b) coherent anti-Stokes Raman scattering. From Reintjes et al 1993.*

Here the second harmonic signal is proportional to the intensity correlation integral:

$$S = K \left| \int I_{\text{sig}}(r,t) \, I_{\text{ref}}^*(r,t) \, dt \right|^2 \tag{20}$$

The intensity correlation integral gives a much poorer level of discrimination against unwanted light than does the field correlation integral of Equation 19 because the cancellations resulting from non-correlated phase reversals that are present in the field correlation integral are not present in the intensity correlation integral. As a result second harmonic generation in non-resonant materials, as well as sum or difference frequency conversion, can be used for pulsed gating, but not for correlation gating.

An arrangement that is suitable for correlation gating with stimulated Raman scattering is shown in Figure 12a (Reintjes *et al.* 1993, Bashkansky *et al.* 1994), and one using coherent anti-Stokes Raman scattering is shown in Figure 12b. The stimulated Raman coherence gate is similar to the pulsed Raman amplifier gate shown in Figure 10, except that crossed polarisers have been added and long pulse broad band radiation is used. For correlation gating, the pump and Stokes beams consist of relatively long signals with bandwidths that are wider than the Raman linewidth:

$$\Delta\nu_L \gg \Delta\nu_R = \frac{1}{\pi T_2} \tag{21}$$

Because of the integrating response of the Raman effect with broad band excitation, the exponential gain is determined by the steady state gain coefficient and the average pump intensity, while the average amplified Stokes intensity is proportional to the square of the field correlation integral between the pump and incident Stokes field amplitudes (Bashkansky *et al.* 1994),

$$\bar{I}_S(L) = \bar{I}_S(0)|R|^2 \exp(g_{SS}\bar{I}_L L) + \bar{I}_S(0)\left(1 - |R|^2\right) \tag{22}$$

where $\bar{I}$ is the average intensity over the laser pulse, g_{SS} is the steady-state Raman gain and R is the normalized Stokes-pump cross field correlation function defined as:

$$|R|^2 = \frac{\left|\int A_S(t) A_L^*(t)\,dt\right|^2}{\left|\int A_S(t) A_S^*(t)\,dt\right|^2 \left|\int A_L(t) A_L^*(t)\,dt\right|^2} \tag{23}$$

If the gain in the Raman amplifier is high enough, it can be used for correlation gating in the same way as for pulsed gating. Generally however, maintaining such high gains requires large pump energies in the longer pulses associated with correlation gating. Lower gains can be used with the cross polarisation technique shown in Figure 12a. In materials such as hydrogen, Raman gain occurs primarily only for Stokes light that is polarised parallel to the pump. In this arrangement, the Stokes signal transmitted through the second polariser is given by

$$I_T = I_S(0)\left(\frac{A-1}{2}\right)^2 \tag{24}$$

where A is the field amplification factor

$$A = \sqrt{|R|^2 \exp(g_{SS}\bar{I}_L L) + 1 - |R|^2} \tag{25}$$

This configuration has been given the name CARP (coherent amplifying Raman polarisation) gate (Bashkansky *et al.* 1994).

In addition to quantum noise, the correlation gate also has a leakage contribution from Stokes light that is not correlated with the pump but receives amplification. The primary contribution to this light comes from finite bandwidth effects. By expanding the mode expressions to include the lowest order bandwidth term, we can estimate the gate contrast against unwanted components of the Stokes signal to be (Bashkansky *et al.* 1994)

$$C = R(\tau = 0)\frac{6}{\pi^2}\frac{\left[\exp(g\bar{I}_L z/2) - 1\right]^2}{gIL\left(\frac{\Gamma}{\Delta}\right)\left(\frac{\Gamma}{\Delta\omega}\right)} \tag{26}$$

where $\bar{I}_L$ is the average gate intensity, $\Delta\omega$ is the mode spacing of the gate radiation and where $\Gamma = 1/T_2$. In practice, we have achieved contrast levels of 10^3–10^4 and contrasts of 10^6 or more are possible in principle with suitable designs. Similar behaviour can be derived for the CARS gate (Bashkansky *et al.* 1994, Bashkansky and Reintjes 1993, Bashkansky and Reintjes 1994).

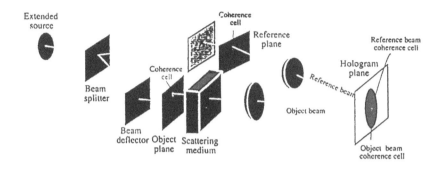

Figure 13. *Schematic diagram of holographic image gating. From Leith et al 1993.*

2.4 Interference gating

Correlation gating can also be done by interfering the image signal and gate signal on a square law detector. Spatial interference corresponds to pulsed holography, while temporal interference corresponds to heterodyne or homodyne detection.

Holography

Holography for time-gated imaging has been demonstrated by several many (Spears *et al.* 1989, Leith *et al.* 1991, Valdmanis and Abramson 1991, Chen *et al.* 1991, Leith *et al.* 1993). A typical setup is shown in Figure 13 and is similar to standard holographic configurations. For broadband interactions, interference fringes are present only when the signal is coherent with the reference. If only part of the signal is coherent with the reference, then holographic fringes are produced only for that part of the signal that is coherent with the reference, selecting that part of the signal for detection. Subsequent reconstruction of the hologram produces an image of the object illuminated with the selected portion of light. Light that is not coherent with the reference forms a background that does not contribute to the reconstructed image. It can however, contribute to noise that reduces the fringe contrast, and hence to the ultimate sensitivity of the technique. This technique can be used for direct two dimensional imaging and, with scanning of the relative timing of the reference light, three dimensional imaging.

When used with pulsed illumination, fringes are formed only when the reference pulse is present. In this situation, the holographic image is formed only by light that arrives simultaneously with the gating pulse, while any other light contributes to back-

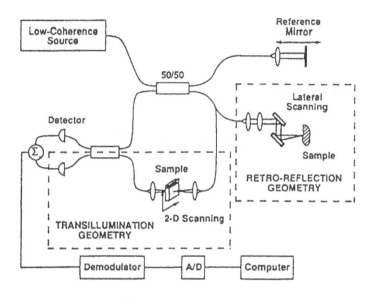

Figure 14. *Schematic diagram of apparatus for low coherence tomography. From Izatt et al 1993.*

ground noise and reduction of fringe contrast. Variation of the relative timing of the signal and reference can be used to change the part of an object that is imaged. Resolution is determined by the duration of the pulse for pulsed illumination or the coherence length (roughly the inverse of the bandwidth) for broad band illumination. Such a system has been demonstrated for light in flight photography for examination of three dimensional objects (Abramson 1978, 1991). By shearing the reference pulse across a film plane, holograms of objects taken at different positions on a single pulse were obtained. The various spatial positions appear at different places on the film. By viewing the objects in sequence it is possible to obtain a rendering of a light pulse as it propagates over an object. A low light level holographic detection method has also been demonstrated by recording the interference pattern on a CCD array, transferring to a computer and performing the reconstruction electronically, greatly reducing the signal level needed for detection (Leith *et al.* 1991).

Low coherence tomography

Another approach to coherence gating is termed 'low coherence tomography' (Izatt *et al.* 1993, Huang 1991, Gelikonov *et al.* 1995). In this approach, a single spatial mode broad band source is used with a fibre interferometer, as shown in Figure 14.

Again the time resolution is given by the bandwidth of the source. In this arrangement, interference occurs between the gate signal passed through the reference arm of the interferometer, and the image signal that is transmitted through, or reflected from, the object under study. Single point information is obtained directly. Two dimensional image information can be obtained by scanning the illumination spot or the sample laterally, and depth information can be obtained by changing the reference timing of the gate pulse. The resulting signal is detected in essentially a heterodyne mode, with only that part of the signal that is temporally coherent with the reference contributing a signal. Additional discrimination against unwanted portions of the signal light is obtained by modulating the gate pulse in time, either by dithering the mirror in the reference arm or by using the Doppler shift produced by the depth scanning motion. Detection is then done in a narrow bandwidth region offset from the laser frequency. Discrimination of 94-db has been reported with a superluminescent diode (Izatt *et al.* 1993) and 135-db with a Ti:sapphire laser source (Hee *et al.* 1993).

3 Applications

Much of the work in high speed imaging has traditionally been directed at 'stop-action' photography. As gating times extend to the picosecond scales and shorter, the added benefit for such applications becomes limited. For example, an object in motion to be photographed in green light has a minimum resolvable displacement of the order of 0.5μm. If we wish to make optimal use of a 1 picosecond gating system, the object would have to move at a speed of $5 \times 10^7 \mathrm{cm\ s^{-1}}$, two orders of magnitude faster than the speed of sound. Speeds of this order of magnitude can be found in expanding plasmas from laser targets, but longer pulses are usually adequate to freeze the motion of most physical objects. As a result, although stop-action photography can be done with the ultrafast time-gates described here, there are few things that make optimal use of their capabilities. Two new applications that have received attention in recent years, and that do take advantage of the ultrafast gating capability, are imaging through turbid materials and range-gated imaging with microscopic depth resolution. These applications have uses in areas such as medical imaging, underwater imaging and materials inspection.

3.1 Imaging in turbid materials

When light propagates through turbid materials, images of objects that are part of, embedded in, or viewed through the medium are degraded because of the multiple path scatter. Once the multiple scattering becomes sufficiently well developed, the light looses all memory of its original direction or location and the ability to image any object within or behind such media is severely limited. One method of obtaining improved image resolution is to restrict the light received at the detector with time-gated imaging. The concept behind the use of time-gated imaging is that light that is multiply scattered propagates over longer paths, and arrives at the detector later than light that is scattered less. By selecting out light that arrives early, one can reduce the scattered component, improving the resolution of the image. Imaging can be done

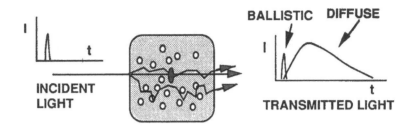

Figure 15. *Schematic diagram showing unscattered and multiply scattered light passing through a scattering medium. The qualitative time distribution for the various components is shown at the right.*

in both transmission (in which it is the earliest arriving light that is detected) or in backscatter (in which light that is scattered or reflected from a specific depth within the material is selected). Other methods that can be used include spatial filtering to restrict the highly scattered light, and analysis of the time evolution of the scattered light to infer the presence of hidden objects. Photon density waves are also being used for imaging into scattering materials (Chance *et al.* 1993, O'Leary *et al.* 1992, Knuttel *et al.* 1992). Descriptions of these various techniques are given in the issue of *Optics and Photonics News* cited in the references. Here the discussion will be limited to the time-gated techniques.

Light that is transmitted through the medium has an unscattered component, often called *ballistic light* (Wang *et al.* 1991) and diffuse light as illustrated in Figure 15. The ballistic light is transmitted through the material with minimum delay, while the diffuse light is spread out in time. Ballistic light allows imaging with as much spatial resolution as the large scale optical properties of the medium allow. Its intensity is attenuated exponentially according to the law

$$I_{\text{ballistic}} = I_0 \, e^{-\mu_s L} \tag{27}$$

where μ_s is the scattering coefficient of the material. Ballistic scattering will give the highest amount of spatial resolution possible in such imaging applications, provided there is enough intensity for detection. Depending on the sensitivity of the method being used, attenuations from e^{-25} up to e^{-40} can be worked with. The highest attenuation reported to date for which images are obtained is about e^{-33} (Duncan *et al.* 1991a).

Diffuse light is described by the diffusion equation

$$\nabla^2 \psi = \frac{1}{D} \frac{\partial \psi}{\partial t} \tag{28}$$

where ψ is the power flux of the signal beam and D is a diffusion coefficient given by $D = cL_t/3n$, where n is the refractive index of the medium and L_t is the transport length defined in Equation 33. The approximations that lead to this equation assume that the scattering has caused the loss of all memory of original position, direction and phase of the light so that the optical signal can be described by the intensity rather than the field amplitude. Analytic solutions for diffuse light have been obtained only for isotropic scatterers, *i.e.* scatterers that are small compared to the optical wavelength. For this situation the point spread function (PSF) for light emerging from a sample of length L has the form (Moon *et al.* 1993)

$$\text{PSF} = \frac{1}{\pi R^2} \exp(-r^2/R^2) \tag{29}$$

where

$$R = \sqrt{4Dt} \tag{30}$$

When the scatterers are comparable to, or larger than, the wavelength, the scattering is dominantly forward. In this case it is common to treat the propagation in the diffusion regime as if the scattering were isotropic, but with a modified scattering coefficient

$$\mu'_s = \mu_s(1 - g) \tag{31}$$

where g is given by

$$g = \int S(\theta) \cos\theta \, d\theta \tag{32}$$

The anisotropic scattering material is commonly characterized by the transport length, L_t, defined as

$$L_t = \frac{1}{\mu'_s} = \frac{L_s}{1 - g} \tag{33}$$

where L_s is the scattering length given by $L_s = 1/\mu_s$. The magnitude of g lies between 0 for isotropic scatterers and 1 for extremely anisotropic scatterers. Materials such as biological tissue and sea water typically have values of g near 0.9.

Time-gated imaging - Examples

Many of the techniques described earlier have been used to obtain time-gated imaging in various materials. Here we will give examples of the different types of images that have been obtained. A more complete description can be found in the references.

Low coherence tomography has been used to photograph the human eye (Izatt *et al.* 1993, Hee *et al.* 1995) and to study human membranes (Gelikonov *et al.* 1995). Typical arrangements are as shown previously in Figure 14. A depth resolved photograph of the human eye is shown in Figure 16a, with various elements of the eye indicated (Izatt *et al.* 1993). Since the eye does not scatter much, the detection is in the ballistic regime, with what scattering is present providing the signal that is detected. This technique has also been used to probe fine detail of the retina, showing structure in the foveal region (Figure 16b)(Izatt *et al.* 1993). Studies of human membrane tissue have been reported by Gelikonov *et al.* (1995) in which certain types of cancer are able to be detected earlier than with conventional methods.

The optical Kerr gate has been used to study imaging in a variety of materials (Ho *et al.* 1993). An example of ballistic imaging with an optical Kerr gate is shown in Figure 17. Here a bar chart with 0.25mm bars was photographed through 55mm of 0.2% intralipid solution. A Kerr-Fourier gate was used with a Nd:glass laser at 1.054μm

(a)   (b)

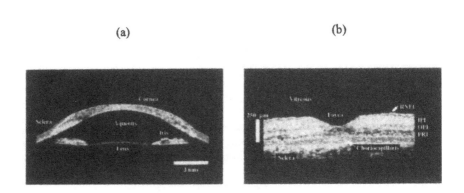

Figure 16. *(a) Time gated image of the human eye obtained with low coherence tomography; (b) Low coherence tomography image of the retina of the human eye. From Izatt et al. 1993.*

with a pulse duration of 8ps. The material had a scattering attenuation of 10^{-10}. Other studies using Kerr gates are summarized in the article by Ho *et al.* (1993).

Holographic imaging of wires placed behind 6mm of flesh from the human hand is shown in Figure 18 (Leith *et al.* 1993). These images were obtained using a 100mW dye laser with a coherence time of 300fs.

Examples of imaging with the stimulated Raman pulsed gate are shown in Figures 19 and 20 (Duncan *et al.* 1991a, Mahon *et al.* 1993). In Figure 19, a sequence of images obtained at different gate timings is shown (Mahon *et al.* 1993). The material was a 5cm long solution of non-dairy creamer. The image is obscured before and after the optimal timing, but the bars of the resolution chart are evident at the optimal timing. Attenuations up to e^{-33} have been obtained with this system. The image in Figure 20 was obtained with Fourier imaging with the Raman system (Mahon *et al.* 1993). A series of 10 light and dark bars each 0.125μm wide can be resolved with this system.

Imaging with multiply scattered light

When scattering attenuation becomes greater than about e^{-20} to e^{-40} depending on the sensitivity of the system, the unscattered ballistic light becomes too weak to detect, and imaging must be done with multiply scattered light. Time gated imaging using diffuse light consists of selecting light near the front of the diffuse peak, before the full diffuse spreading has a chance to develop (see Figure 17)(Wang *et al.* 1991, Ho *et al.* 1993).

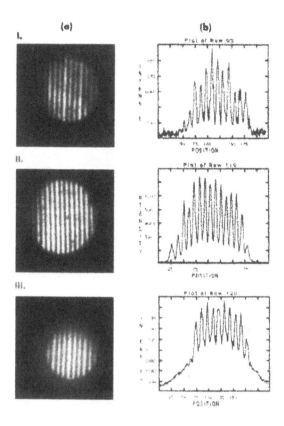

Figure 17. *Images of 0.25mm wide bars obtained through about 55mm of 0.2% intralipid solution with an optical Kerr-Fourier gate; I no scatterer; II scatterer present; III photograph with spatial filter only. From Ho et al. 1993.*

Imaging in this regime has been done with streak cameras (Hebden *et al.* 1991, Hebden and Wong 1993, Hebden and Delpy 1994), an example of which is shown in Figure 21 (Hebden and Delpy 1994) A series of transparent and dark spheres 8mm in diameter were placed in a suspension of 1.27μm polystyrene spheres with a *g* value of about 0.9. The system was illuminated with pulses from a Ti:sapphire laser a few picoseconds in duration. The images were recorded with a streak camera with 15ps resolution.

The pulsed Raman gate has been used to study the resolution of images that can be obtained in the diffuse regime (Moon *et al.* 1993). The image resolution in the diffuse regime is related to the radius of the point spread function given in Equation 30. This radius increases with time during the pulse. The difficulty is that the highest resolution is obtained at the very earliest part of the image for which very few photons are available. As time progresses, the intensity increases, but the spatial resolution

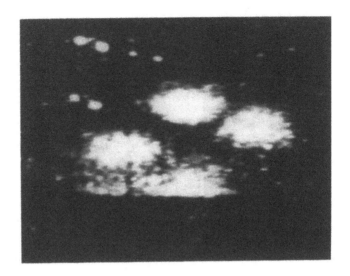

Figure 18. *Holographic image of wires placed behind the human hand. From Leith et al. 1993.*

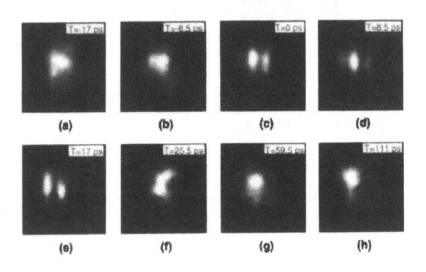

Figure 19. *Sequence of images of 0.316mm wide bars obtained through a solution of non dairy creamer with attenuation of e^{-33} with a pulsed Raman gate for various reference timings. From Mahon et al. 1993.*

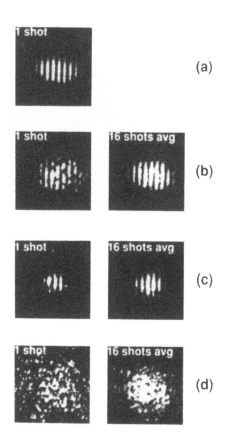

Figure 20. *Image obtained of 0.1255m wide bars obtained with a Raman gate with Fourier imaging through a 5cm long cell of non-dairy creamer with an attenuation of e^{-24} for various time delays; (a) no scatterer; (b) relative time = -16ps; (c) relative time = 0ps; (d) relative time = 24psec. From Mahon et al. 1993.*

decreases. A compromise is needed to make use of the fewest number of photons possible and to accept some broadening in spatial resolution. Calculations have shown that, for attenuations of 10^{15}, the spatial resolution, defined as the point spread function at the exit plane of an object located at the entrance plane, is about 1/5 of the total sample thickness regardless of the scattering properties (Moon *et al.* 1993). For a sample embedded in the middle of a sample, resolution will be about 1/10 of the thickness.

Measurements of the radius of the point spread function for materials of various length and scattering strength are shown in Figure 22 (Moon *et al.* 1993). Here the line representing a radius that is 1/5 of the sample length is also shown. The agreement between the line and the data confirms the prediction. Further calculations show that this relation holds approximately for attenuations between 10^{-10} and 10^{-20}.

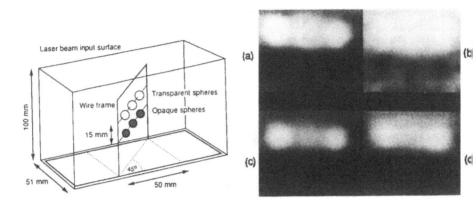

Figure 21. *Images of spheres obtained with a streak camera in the multiple scatter regime. From Hebden and Delpy 1994.*

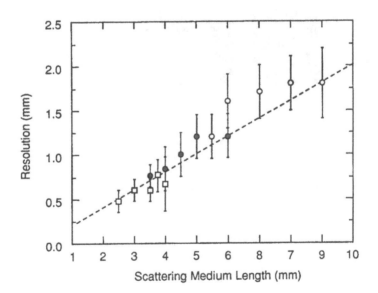

Figure 22. *Measurement of the width of an opaque edge viewed through intralipid solutions of various transport lengths and sample lengths. Squares: $L_t = 0.33mm$ crosses: $L_t = 0.5mm$; diamonds: $L_t = 1mm$. The line corresponds to a resolution that is 1/5 of the sample length. From Moon et al. 1993.*

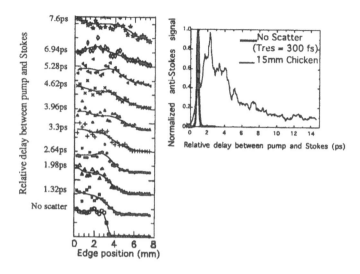

Figure 23. *Images of an opaque edge sandwiched inside 15mm of raw chicken obtained with a CARP Raman correlation gate with 250fs resolution. Inset shows correlation trace with and without scatterer present showing the ballistic peak and the delayed peak from multiple scattering.*

Measurements obtained with the CARP correlation gate (Figure 12b) of an opaque edge sandwiched in the middle of 15mm of raw chicken are shown in Figure 23 for various delays between the signal and reference. The time resolution was about 250fs. The definition of the edge is highest for short times for which the imaging is ballistic. For delays of a picosecond or longer, the ballistic light is no longer present and the imaging is done with multiply scattered light corresponding to the delayed, broad correlation peak shown in Figure 23b. The edge definition gradually degrades as the time delay increases as shown in Figure 23a.

If the scattering is sufficiently anisotropic it might be expected that better image information can be obtained at earlier times than predicted by the diffusion regime (Wang *et al.* 1991, Wang *et al.* 1993, Ho *et al.* 1993). Properties of light propagation in this regime have been discussed in Bucher (1993), Arnush (1972), Mooradian *et al.* (1979), and Zege *et al.* (1991).The spatial-resolution limits in this intermediate regime have been examined theoretically using a Markov chain calculation to model the probabilities of scatter at each point (Moon and Reintjes 1994). The results predict regime in which the resolution can be increased by factors of 2 to 3 over the diffuse limit. Measurements of the point spread function for scatterers with large *g* values are shown in Figure 24 (Moon *et al.* 1995). Increased resolution compared to the diffuse limit at ballistic scattering attenuations up to e^{-100} was seen at *g* values greater than 0.95. The transition to the diffusion regime usually occurred after 2–3 transport lengths.

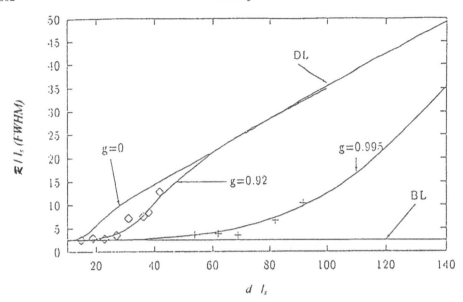

Figure 24. *Measurement of the width of an opaque edge obtained in the intermediate scattering regime showing the transition from ballistic to diffuse for various g values. From Moon et al. 1995.*

3.2 Materials inspection

The CARP gate has been used for materials inspection to find subsurface flaws in highly scattering material (Bashkansky and Reintjes 1995). An example is shown in Figure 25. The material used was silicon nitride ceramic, which is black to the eye and consists of tubules of the order of $3\mu m$ in diameter. A saw cut across one corner was made in the sample as shown in Figure 25a to provide a defect at different depths. The illumination beam was scanned across the image and at each point the gate timing was varied to probe the depth of the material. An example of the resulting trace is shown in Figure 25b, where a strong return from the surface is seen, along with a weaker return from the subsurface defect.

3.3 Underwater imaging

Propagation through sea water is dominated by a combination of scattering and absorption. Imaging can be expected to be carried out primarily with the unscattered ballistic component. Again typical g values are of the order of 0.9. Imaging is usually done in a reflection mode and the timing between the reference and the signal is used to determine the distance of the object being photographed. Time-gated imaging can improve the image quality by rejecting light that is scattered back into the camera. Time-gated imaging has been carried out previously with several systems in the nanosecond range. Recently ultrafast gating techniques have been used for this purpose. An example of an object photographed with a framing camera with 250ps resolution is shown in Figure 26 through about 6.5 scattering lengths one way over a

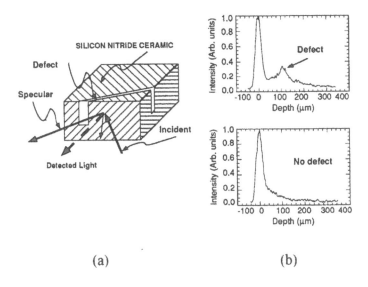

(a)　　　　　　　　　　　　　(b)

Figure 25. *Detection of a subsurface flaw in a ceramic; (a) diagram of sample; (b) Trace of time gated reflection as a function of depth at two different positions. From Bashkansky et al. 1995.*

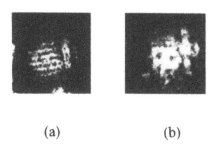

(a)　　　　　　　　(b)

Figure 26. *Images obtained through 6.5 scattering lengths of water/maalox solution with a framing camera with 250ps resolution (a) and 5ns resolution (b).*

total distance of 10m (McLean *et al.* 1995). The improvement provided by time-gating can be seen by comparing the image obtained with a time resolution of 5ns.

A three dimensional reconstruction of an object obtained with a pulsed Raman gate with 30ps resolution is shown in Figure 27 (Moon *et al.* 1994). The object was

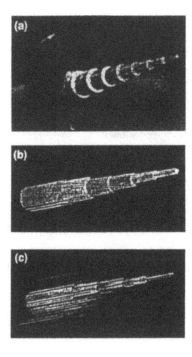

Figure 27. *Three dimensional reconstruction of an object through 8.5 scattering lengths of a water/Ropaque solution obtained with a Raman pulsed imaging gate; (a) original object; (b) three dimensional reconstruction of object with no scattering present; (c) three dimensional reconstruction through scattering material. From Moon et al. 1994.*

imaged through about 8.5 attenuation lengths over a path length of 55cm in a solution of Ropaque in water. The three dimensional reconstruction was made by varying the relative timing of the gate and signal on successive laser shots and combining the images.

4   Acknowledgements

The author would like to acknowledge numerous colleagues who have made major contributions to much of the work described in this chapter, including Dr M D Duncan, Dr R Mahon, Prof L L Tankersley, Dr M Bashansky, Dr J A Moon, Dr P Battle, Dr C L Adler, and Dr R H Lehmberg. In addition he would like to acknowledge the numerous colleagues who have allowed reproduction of their work: Dr J G Fujimoto, Dr R R Alfano, Dr E Leith, Dr J A Valddmanis, Mr E A McLean and Dr J Hebden.

References

Abramson N, 1978, Light in Flight Recording by Holography, *Opt Lett* **3** 121.

Abramson N,1991, Time Reconstructions in Light in Flight Recording by Holography,*Appl Opt* **30** 1242.

Arnush D, 1972, Underwater Light-Beam Propagation in the Small-Angle-Scattering Approximation, *J Opt Soc Am* **62** 1109.

Bashkansky M, Adler C L, Battle P, and Reintjes J, 1994, Nonlinear optical field cross correlation techniques for imaging through a strong scattering medium, *SPIE* **2241** 228.

Bashkansky M, Battle P R, Mahon R, and Reintjes J,1995, Subsurface Defect Detection in Ceramic Materials Using Ultrafast Optical Techniques,*22nd Ann.Rev. of Prog. in Quantitative Nondestructive Evaluation,Seattle.*

Bashkansky M, and Reintjes J, 1993, Imaging Through a Strong Scattering Medium with Nonlinear Optical Field Cross-correlation Techniques,*Opt Lett* **18** 2132.

Bashkansky M, and Reintjes J,1994, Image Upconversion Using Coherent anti-Stokes Raman Scattering, *IEEE J Quant Elect* **QE-30** 318.

Bucher E A, 1973, Computer Simulation of Light Pulse Propagation for Communication through Thick Clouds, *Appl Opt* **12** 2391.

Chance B, Kang K, and Sevick E, 1993, Photon diffusion in breast and brain: spectroscopy and imaging, *Opt Photonics News* **4** 9.

Chen H, Chen Y, Dilworth D, Lopez J, Masri R, Rudd J, and Valdmanis J, 1991, Two-dimensional Imaging Through Diffusing Media Using 150fs Gated Electronic Holography Techniques, *Opt Lett* **16** 487.

Duncan M D, Mahon R, Tankersley L L, and Reintjes J,1991a, Time-gated Imaging Through Scattering Media Using Stimulated Raman Amplification, *Opt Lett* **16** 1868.

Duncan M D, Mahon R, Tankersley L L, and Reintjes J,1991b,Spectral and Temporal Characteristics of Spontaneous Raman Scattering in the Transient Regime, *J Opt Soc Am* **B8** 300.

Duncan M D, Mahon R, Tankersley L L, and Reintjes J,1992, Low-light Level, Quantum-noise-limited Amplification in a Stimulated Raman Amplifier, *J Opt Soc Am* **B9** 2107.

Duguay M A and Mattick A T,1971, Ultrahigh-speed Photography of Picosecond Light Pulses and Echoes, *Appl Opt* **10** 2162.

Edgerton H E, 1931, Stroboscopic Moving Pictures, *Elect.Eng.(Am.I.E.E.)* **50** 327.

Gelikonov V M, Gelikonov G V, Kuranov R V, Pravdenko K I, Sergeev A M, Feldshtein F I, Khanin Ya I, and Shabanov D V, 1995, Coherent Optical Tomography of Microscopic Inhomogeneities in Biological Tissues, *JETP Lett.* **61** 159.

Hares J D, 1987, Advances in sub-nanosecond shutter tube technology and applications in plasma physics, *SPIE* **831** X-rays from Laser Plasmas, **831** 165.

Hebden J C, and Delpy D T, 1994, Enhanced Time-resolved Imaging with a Diffusion Model of Photon Transport, *Opt Lett* **19** 311.

Hebden J C, Kruger R A, and Wong K S,1991, Time-resolved Imaging Through a highly Scattering Medium, *Appl Opt* **30** 788.

Hebden J C, and Wong K S,1993, Time-resolved Optical Tomography, *Appl Opt* **32** 372.

Hee M R, Izatt J A, Jacobson J M, and Fujimoto J G, 1993, Femtosecond Transillumination Optical Coherence Tomography, *Opt Lett* **18** 950.

Hee M R, Izatt J A, Swanson E A, Huang D, Schuman J S, Lin C P, Puliafito C A, and Fujimoto J G, 1995, Optical coherence tomography of the human retina, *Archives of Opthalmology* **113** 325.

Ho P P, Wang L, Liang X, Galland P, Kalpaxis L L, and Alfano R R, 1993, Snake ligh
tomography: ultrafast time-gated snake light 2D and 3D image of objects in biomedica
media, *Opt Photonics News* **4** 23.

Huang D, Swanson E A, Lin C P, Schuman J S, Stinson W G, Chang W, Hee M R, Flotte T
Gregory K, Puliafito C A, and Fujimoto J G, 1991, Optical Coherence Tomography,*Scienc*
254 1178.

Izatt J A, Hee M R, Huang D, Swanson E A, Lin C P, Schuman J S, Puliafito C A, and Fujimoto J G, 1993, Micron-resolution biomedical imaging with optical coherence tomography
Opt Photonics News **4** 14.

Kayafas G, ed,.1987,Stopping Time: The Photographs of Harold Edgerton, (Abrams, New
York)

Knuttel A, 1992, Spatial Localisation of Absorbing Bodies by Interfering Diffusive Photor
Density Waves, *Appl Opt* **32** 381

Leith E, Arons E, Chen H, Chen Y, Dilworth D, Lopez J, Shih M, Sun P C, and Vossler G
1993, Electronic holography for imaging through tissue, *Opt Photonics News* **4** 19.

Leith E, Chen H, Chen Y, Dilworth D, Lopez J, Masri R, Rudd J, and Valdmanis J,1991
Appl Opt **30** 4204.

McLean E A, Burris H R, and Strand M P, 1995, Short pulse range gated optical imaging ir
turbid water, *Appl Opt* **34** 4343

Mahon R, Duncan M D, Tankersley L L, and Reintjes J, 1993, Time-gated imaging througt
dense scatterers with a Raman amplifier, *Appl Opt* **32** 7425.

Moon J A, Mahon R, Duncan M D, and Reintjes J, 1993, Resolution limits for imaging througt
turbid media with diffuse light, *Opt Lett* **18** 1591.

Moon J A, Mahon R, Duncan M D, and Reintjes J,1994, Three Dimensional Reflective Image
Reconstruction Through a Scattering Medium Using Time Gated Raman Amplification
Opt Lett **19** 1234.

Moon J A, and Reintjes J, 1994, Image resolution by use of multiply scattered light,
Opt Lett **19** 521.

Moon J A, Battle P, Bashkansky M, Mahon R, Duncan M D, and Reintjes J, 1996, Achievable
spatial resolution of time resolved transillumination imaging systems which utilize multiply
scattered light, *Phys Rev E* January.

Mooradian G C, Geller M, Stotts L B, Stephens D A, and Krautwald R A,1979, Blue-green
pulsed propagation through fog, *Appl Opt* **18** 429.

Muybridge E, 1987, Animal Movements, (Chapman and Hall, London).

O'Leary M A, Boas D A, Chance B, and Yodh A G, 1992, Refraction of Diffuse Photon density
Waves, *Phys Rev Lett* **69** 2658.

Raymer M G, and Mostowski J, 1981, Stimulated Raman Scattering: Unified Treatment of
Spontaneous Initiation and Spatial Propagation, *Phys Rev A* **24** 1980.

Reintjes J, Duncan M D, Mahon R, Tankersley L L, Bashkansky M, Moon J A, Adler C L
and Prewitt J M S, 1993, Time-gated imaging with nonlinear optical Raman interactions.
Opt Photonics News **4** 28.

Reintjes J, Lehmberg R H, Chang R S F, Duignan M T, and Calame G, 1986, Beam Cleanup
with stimulated Raman Scattering in the Intensity Averaging Regime, *J Opt Soc Am* **B3**
1408.

Ripin B H, Huba J D, McLean E A, Manka C K, Peyser T, Burris H R, and Grun J, 1993
Sub-Alfvenic Plasma Expansion, *Phys.Fluids* **B5** 3491.

Spears K G, Serafin J, Abramson N H, Zhu X, and Bjelkhagen H, 1989, Chrono-coherent
Imaging for Medicine, *IEEE Trans. Biomed.Eng.* **36** 1210.

Valdmanis J A, and Abramson N H, 1991, Holographic imaging captures light in flight, *Laser
Focus World*, February edition p 111.

Wang L M, Ho P P, and Alfano R R, 1993a, Double-stage Picosecond Kerr Gate for Ballistic Time-gated Optical Imaging in Turbid Media, *Appl Opt* **32** 535.

Wang L M, Ho P P, and Alfano R R, 1993b, *Appl Opt* **32** 5043.

Wang L M, Ho P P, Liu C, Zhang G, and Alfano R R, 1991, Ballistic 2-D Imaging Through Scattering Walls Using an Ultrafast Optical Kerr Gate, *Science* **253** 769.

Zege E I, Ivanov A P, and Katsev I L, 1991, Image Transfer Through a Scattering Medium, (Springer Verlag, Berlin)

General reference for imaging

Opt Photonics News, 1993, **4** No.3 (October edition).

Fibre Nonlinearities

Pavel V Mamyshev

AT & T Bell Laboratories
New Jersey, USA

1 Fibres: structure, modes, dispersion, losses

The progress made in the technology of fabricating silica-based glass optical fibres has ensured that the transmission losses are now less than 1dB/km over a fairly wide range of the near-infrared wavelength region (Figure 1). The optical losses are determined only by the fundamental properties of the fibre materials. Electronic transitions from valence to conduction bands cause absorption at short wavelengths (in the far-ultraviolet region), which decays exponentially with increasing wavelength. Absorption from lattice vibrations occur in the infrared, also decaying exponentially with decreasing wavelength. Rayleigh scattering is the third intrinsic loss phenomena. The Rayleigh loss decays with wavelength as λ^{-4}. The sum of these three intrinsic effects results in a 'transmission window', with a minimum loss (less than 0.2dB/km) for silica-based fibres in the spectral region of $\lambda \sim 1.5$–1.6μm.

The simplest optical fibre consists of a light-guiding core (radius a) with a refractive index n_{co} and a reflecting cladding with a refractive index $n_{cl} < n_{co}$. The transverse distribution of the optical field in the fibre is determined by the competition between the diffraction and refraction processes. The fibre mode structure is determined by the dimensionless frequency parameter V:

$$V = \frac{2\pi a}{\lambda} (n_{co}^2 - n_{cl}^2)^{1/2} \tag{1}$$

where λ is the light wavelength in vacuum. Different fibre LP_{lm} modes have different

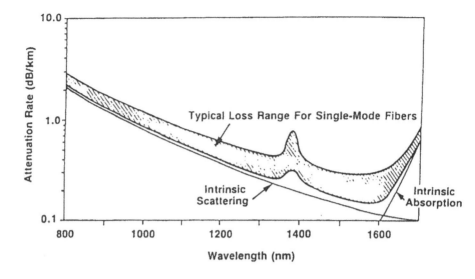

Figure 1. *Spectral dependence of loss in a silica-based fibre.*

propagation constants k_{lm}:

$$k_{lm} = \frac{2\pi}{\lambda}\left(n_{cl} + b_{lm}\Delta n\right) \qquad (\Delta n \equiv n_{co} - n_{cl}). \qquad (2)$$

Figure 2 shows the near-field patterns of different LP_{lm} modes and the dependencies of the normalised propagation constant b_{lm} of the LP_{lm} modes on the dimensionless frequency parameter V (Gloge 1971). One can see that the difference between the refractive indices of the core and the cladding Δn, the core radius a, and the optical wavelength determine the fibre propagation regime: if $V < 2.4$, only one mode can propagate in the fibre, whereas for $V > 2.4$ the fibre is a multimode one. Typical values of the parameters of silica-based single-mode fibres are: $2a \sim 3$–$10\mu m$ and $\Delta n \sim 10^{-2}$–10^{-3}.

It should be mentioned that even 'single-mode' fibres are usually not truly single-mode. An ideal circular 'single-mode' fibre can support two independent degenerate modes of orthogonal polarisation. However, real 'single-mode' fibres have internal imperfections, such as slightly elliptical cores, and can experience external perturbations which cause the two modes to propagate with different phase (and group) velocities. This effect is referred to as *modal birefringence*. The effect is further complicated by external perturbations that can cause coupling between the two modes. These effects can lead to *polarisation mode dispersion* (PMD) and instabilities in output polarisation. Fibres used for telecommunications should have values of PMD as low as possible. A real fibre can be viewed as a set of randomly oriented wave plates. The PMD of a low-birefringent fibre is a statistical value, measured in $psec/(km^{1/2})$. State of the art fibre fabrication technology has ensured that the polarisation mode dispersion parameter of single-mode, low-loss telecommunication fibres is less than $0.1psec/(km^{1/2})$.

The fact that different modes have different propagation constants is known as *intermode dispersion*. In turn, each mode is subject to a *chromatic dispersion*. A very

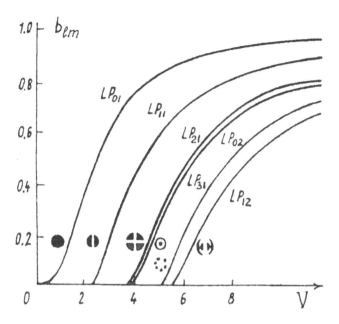

Figure 2. *The near-field patterns and dependencies of the normalised propagation constant b_{lm} on the dimensionless frequency parameter V for different LP_{lm} fibre modes.*

important role in (linear and nonlinear) light propagation in fibres is played by the *group velocity dispersion* $D_{lm}(\lambda)$:

$$
\begin{aligned}
D_{lm}(\lambda) &\equiv \frac{\mathrm{d}}{\mathrm{d}\lambda}\left(\frac{1}{v_{\mathrm{gr}}}\right) = -\frac{2\pi c}{\lambda^2}\frac{\mathrm{d}^2 k_{lm}}{\mathrm{d}\omega^2} \\
&= -\frac{\lambda}{c}\frac{\mathrm{d}^2 n}{\mathrm{d}\lambda^2} + \frac{\Delta n}{\lambda c}\frac{\mathrm{d}^2(Vb_{lm})}{\mathrm{d}V^2}
\end{aligned}
\tag{3}
$$

Different kinds of nonlinear phenomena can be observed depending on the sign and magnitude of the group velocity dispersion. As one can see from Equation 3, the fibre group velocity dispersion depends not only on the material dispersion (first term), but also on the waveguide dispersion (second term). Silica glass has a 'zero-dispersion' wavelength λ_0 near 1.3μm ($D(\lambda_0)=0$). The dispersion is negative for $\lambda < \lambda_0$ and positive for $\lambda > \lambda_0$. It is very important for many applications given the present fibre fabrication technology that by changing the refractive index profile of the fibre (and, therefore, changing the waveguide contribution to the dispersion), one can shift the 'zero-dispersion' wavelength of single-mode fibres up to $\lambda \sim 2\mu$m (Figure 3). This means that one can have low-loss, single-mode silica-based fibres with different signs and magnitudes of group velocity dispersion in the spectral region $\lambda > 1.3\mu$m (*i.e.* where fibres have minimum optical loss).

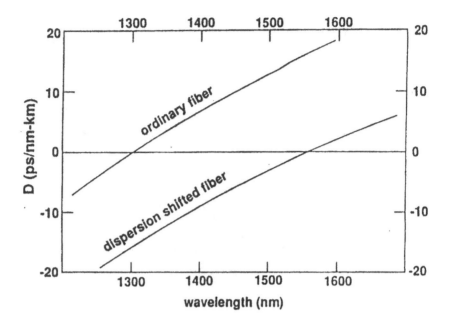

Figure 3. *Spectral dependence of the dispersion D for a standard fibre and for a dispersion-shifted fibre.*

2 Nonlinear properties of fibres: Comparison with bulk media

A wide variety of nonlinear phenomena in silica-based optical fibres are determined mainly by the third-order nonlinear susceptibility $\chi^{(3)}$. The second-order susceptibility $\chi^{(2)}$ should vanish since silica glass is an isotropic medium (note that this is not exactly true, because, as we have mentioned above, fibres usually have some birefringence).

It should be noted that silica glass is a fairly weak nonlinear optical material. The third-order nonlinear susceptibility of silica glass is two to three orders of magnitude less than that of traditional nonlinear materials, such as CS_2. One can therefore ask the question: why are the nonlinear processes so important in fibres? It would seem that pump power thresholds for the observation of nonlinear processes in optical fibres should be several orders of magnitude larger than in traditional materials. However, the efficiency of the nonlinear effects in fibres is determined not by the high value of nonlinear coefficients, but by the waveguide nature of light propagation in fibres. The nonlinear process efficiency is usually determined not only by the nonlinear coefficient, but also by the product of the pump power density and the interaction length (let us call this product A). We shall now calculate the values of A for the cases of a bulk nonwaveguide medium (A_{nw}) and of fibre (A_f) at pump powers P less than the critical self-focusing power P_{sf}. In the case of a bulk medium, the experimental geometry is as follows: pump light is focused by a lens into a nonlinear medium to form a spot of radius r_0 (Figure 4a). Because of diffraction the size of the spot increases with the

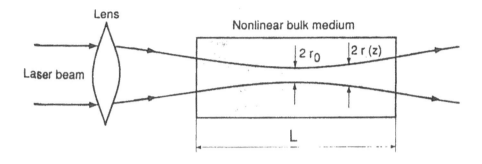

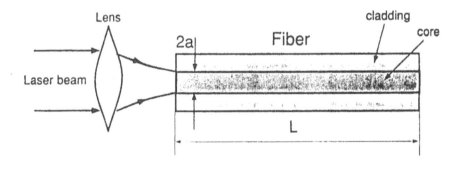

Figure 4. *Experimental geometry used for laser pumping nonlinear (a) bulk media and (b) optical fibres.*

propagation distance z according to the formula:

$$r(z) = r_0\sqrt{1 + (z\lambda/2\pi n r_0)^2} \tag{4}$$

The product A_{nw} is

$$A_{nw} = \int_{-L/2}^{L/2} \frac{P}{\pi r^2(z)}\, dz < \int_{-\infty}^{+\infty} \frac{P}{\pi r^2(z)}\, dz = P\frac{2\pi n}{\lambda} \tag{5}$$

where L is the length of the medium. It therefore follows that in the case of an infinitely long bulk nonlinear medium, the product of the pump power density and the interaction length is independent of the degree of focusing, *i.e.* it is independent of the beam waist r_0 (because a reduction of r_0 increases the diffraction divergence of the light). The parameter A_{nw} is determined only by the pump power P.

In the case of a fibre, the effects of diffraction are compensated completely by the refraction in the reflecting cladding, so that the transverse dimensions of radiation inside the fibre are constant along its length L and the product A_f is defined as follows (Figure 4b):

$$A_f = PL/(\pi a^2) \equiv (P/\pi a)\,(L/a) \tag{6}$$

Comparing this expression with that for A_{nw} and bearing in mind that the radius of a single-mode fibre a is of the order of λ, we find that for a given pump power the advantage in the case of a fibre is of the order of L/a:

$$A_f \sim A_{nw}\,(L/a) \tag{7}$$

The maximum fibre length L is limited, in principle, only by the losses in the fibre and can reach tens of kilometers, and then $L/a \sim 10^9$–10^{10}. Such a great advantage, with respect to the value of A_{nw}, compensates for the small nonlinear coefficients of silica glass and significantly reduces the threshold for the observation of nonlinear processes in silica fibres (sometimes to milliwatt pump powers). This result is due more to the waveguide nature of the propagation of radiation than to the great lengths of fibres. It should be stressed that we are comparing the cases of a fibre of finite length and of an unbounded bulk medium.

2.1   Effective fibre length

Equations 6 and 7 assume that the fibre has no loss. Nevertheless, in real fibres the pump intensity $I(z) \equiv P(z)/(\pi a^2)$ decreases with distance z due to the fibre loss:

$$I(z) = I_0\, e^{-\alpha z}\ ; \tag{8}$$

where α is the fibre loss coefficient. Efficiency of nonlinear processes is determined by

$$A_f = \int_0^L I_0\, e^{-\alpha z}\, dz = I(0)\,\frac{1 - e^{-\alpha L}}{\alpha} \equiv I_0\, L_{\text{eff}}\,. \tag{9}$$

where the effective fibre length is

$$L_{\text{eff}} = \frac{1 - e^{-\alpha L}}{\alpha} \tag{10}$$

One can see that for the case of 'short' fibres (when $\alpha L \ll 1$), the effective length equals the fibre length $L_{eff} = L$. For the case of long fibres ($L \to \infty$, $\alpha L \gg 1$),

$$L_{eff} \to \frac{1}{\alpha} \tag{11}$$

It follows from Equations 9–11 that for 'short' fibres ($L < 1/\alpha$) the efficiency of nonlinear processes is determined by the whole length of the fibre, while for the case of $L > 1/\alpha$, the nonlinear processes are effective only in the first $1/\alpha$ length of the fibre and the light propagation through the rest of the fibre is essentially linear. Typical example:

$$\text{for} \quad \alpha = 0.048 \text{ km}^{-1} \ (0.21 \text{ db/km}) \qquad L_{eff}(\infty) = 1/\alpha = 20.83 \text{ km}$$

Reduction in the threshold power of nonlinear processes in fibre greatly extends the capabilities of nonlinear fibre optics, because it is then possible to use low power lasers that operate at high repetition rates and which are tunable over a wide spectral range.

The transition from bulk nonlinear media to fibre not only quantitatively reduces the threshold powers of nonlinear processes, but also has certain qualitative advantages. It is well-known that the self-focusing of radiation is frequently the main negative factor at high laser radiation powers—in particular, it prevents the utilization of a number of nonlinear effects. This is due to the fact that the self-focusing of a radiation beam usually makes the beam highly inhomogeneous in space. Self-focusing is also frequently accompanied by optical damage to the medium.

Therefore, one of the most important advantages of fibre is that a number of nonlinear processes can occur effectively in the absence of self-focusing of radiation—which is fundamentally impossible in the case of bulk nonwaveguide media. This will be discussed in greater detail when the effects of self phase-modulation are considered.

Another important advantage of fibre is the ability to realise (by varying the fibre parameters, such as the fibre diameter $2a$ and the profile of the fibre refractive index) both single-mode and multimode propagation conditions. This is manifested most strikingly in the case of stimulated four-photon processes.

3 Stimulated four-photon processes

In a stimulated four-photon process (also referred to as *four-wave mixing*), two pump photons of frequencies ω_{p1} and ω_{p2} dissociate into Stokes and anti-Stokes photons: $\omega_{p1} + \omega_{p2} \to \omega_s + \omega_a$. For the case of $\omega_{p1} = \omega_{p2}$, it follows from the law of energy conservation that $\Delta\omega = \omega_a - \omega_p = \omega_p - \omega_s$, *i.e.* the Stokes and anti-Stokes frequency shifts are equal. However, the conservation law applies not only to the energy, but also to the momentum of photons. This means that in order for the four-photon process to be efficient, the phase-matching conditions should be satisfied, *i.e.* the phase mismatch

$$\Delta k \equiv k_s + k_a - k_{p1} - k_{p2} \tag{12}$$

should be zero. (Here k_s, k_a and k_p are wave vectors.)

In an isotopic bulk medium with non-zero second-order material dispersion

$$k_2 = \frac{\lambda^3}{2\pi c^2} \frac{\mathrm{d}^2 n}{\mathrm{d}\lambda^2} \neq 0 \qquad \left(k_2 \equiv \frac{\partial^2 k(\omega)}{\partial \omega^2} \right)$$

it is not possible to achieve phase matching in a collinear interaction: $\Delta k = k_2 \Delta\omega^2 \neq 0$. Phase matching in media of this kind is attainable only when the waves interact at certain angles (this is known as *vector phase matching*), but then the interaction length is clearly short and the process as a whole is ineffective.

In fibre optics, there are methods for compensating the material dispersion in order to achieve phase matching in four-photon processes—both in multimode and single-mode fibres. The efficiency of the pump energy transformation into the energy of the Stokes anti-Stokes components can reach tens of percent in fibres. This means that fibre parametric oscillators can be used effectively as tunable wavelength converters of laser light.

To achieve the phase-matching condition $\Delta k = 0$ in fibres, the material dispersion

$$\Delta k_{\mathrm{mat}} = \frac{\lambda^3}{2\pi c^2} \frac{\mathrm{d}^2 n}{\mathrm{d}\lambda^2} \Delta\omega^2$$

can be compensated by:

- **Intermode dispersion** (Stokes, anti-Stokes and pump waves propagate in different fibre modes).

- **Fibre birefringence**

- **Intramode dispersion**

- **Nonlinear refractive index** (modulational instability)

We consider these methods of achieving phase-matching conditions in four wave mixing processes below.

3.1   Intermode dispersion

As was pointed out above (see Equation 2), for the case of multimode fibres, different fibre modes have different propagation constants k (intermode dispersion). Therefore, if the pump, Stokes, and anti-Stokes waves participating in the four-photon process represent different fibre modes, the material dispersion can be compensated by the intermode dispersion and one can ensure phase matching for a certain frequency shift $\Delta\omega$ (Stolen 1975, Lin and Bosch 1981, Dianov *et al.* 1982).

Stimulated four-photon processes in fibres can be observed in the following experimental setup (Figure 5). Radiation representing the first or second harmonic of a Q-switched Nd:YAG laser is coupled by a lens into a low-mode fibre of length from 10cm to 10m. The radiation coupled out of the fibre by an exit lens is focused, after reflection by a diffraction grating, onto a screen or photographic film. By varying both

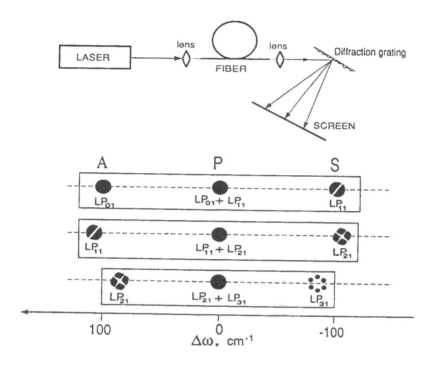

Figure 5. *(a) Experimental set-up for the observation of parametric processes in optical fibres and (b) the spectral images of near-field patterns at the fibre output for different four-photon processes.*

the angle at which the pump radiation is coupled into the fibre and the degree of focusing (which makes it possible to excite preferentially specific groups of modes in a fibre), various mode combinations of the Stokes anti-Stokes generation can be obtained. Stimulated four-photon processes become effective when the power reaches a value of the order of 1kW. Figure 5 shows the images of various four-photon processes expanded as a spectrum on a screen. The frequency shifts for various combinations of pump modes can reach 5500cm^{-1}. Tuning of the frequency shift for a specific mode combination may occur as a result of a change in the fibre parameters, such as the core radius and Δn. Note that as one can see from Figure 5, stimulated four-photon processes can be used as means for selective excitation of fibre modes.

Knowing the experimental values of the frequency shifts, one can determine the dispersive properties of fibres - such as the differences between the propagation constants of various modes and between their group delays. By way of example, we can cite our experimental observations,(with the aid of four-photon processes), of the lifting of degeneracy between the LP_{11} modes with orthogonally distributed lobes, associated with a slight ellipticity of the core (Dianov *et al.* 1982). Figure 6 illustrates the process

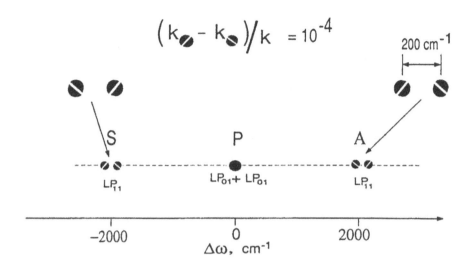

Figure 6. *Spectral image of the near-field fibre output for the case of the four-photon process when both pump photons propagate in the LP_{01} mode. The experiment shows that the propogation constant of the LP_{11} mode with different lobe orientations is different.*

when both pump photons propagate in the LP_{01} mode, with the Stokes and anti-Stokes photons in the LP_{11} mode. The frequency shift is about 2000cm^{-1}, but, as a result of a difference in their propogation constants (amounting to $\delta k/k \sim 10^{-4}$), the frequency shift for the LP_{11} modes with different directions of lobes differs by 200cm^{-1}. As one can see, four-photon processes make it possible to readily detect fairly fine effects.

The experimentally determined coherence length of the four-photon processes phased-matched due to the inter-mode dispersion in multi-mode fibres can reach tens of metres. This leads to a high efficiency of the pump energy conversion (tens of percent). The maximum coherence length is probably determined by imperfections in the fibre (*i.e.* the fibre parameters vary slightly along the length).

3.2 Fibre birefringence

A method for attaining phase matching in single-mode fibres involves the use of bire-fringence which occurs in certain types of fibres, *i.e.* the use of the difference between the refractive indices for two specific directions of polarisation of radiation. Thus, if the pump, Stokes, and anti-Stokes radiation propagate in different polarisation modes of a fibre, the phase matching can be achieved at a certain frequency shift which depends on the fibre birefringence (Stolen *et al.* 1981). The experimental frequency shifts in the Stokes anti-Stokes generation can be used to determine the birefringence of a fibre. Note that as we have already mentioned above, single-mode birefringent fibres are actually two-mode fibres. That is why the phase matching in birefringent single-mode fibres can be considered as a particular case of phase matching in multi-mode fibres.

3.3 Intramode dispersion

In addition to the above method of ensuring phase matching in four-photon processes in multi-mode fibres by compensation of the material dispersion with the intermode dispersion, there are also methods for achieving phase matching in single-mode fibres. The phase mismatch of a four-photon process with a frequency shift $\Delta\omega$ is:

$$\Delta k = k_2 \Delta\omega^2 \tag{13}$$

Here k_2 is the fibres second-order dispersion, which is proportional to the fibre group velocity dispersion D (see Equation 3). The coherence length of the four-photon process is:

$$L_{\text{coh}} = \frac{\pi}{\Delta k} = \frac{\pi}{k_2 \Delta\omega^2} \tag{14}$$

In the wavelength region of $k_2=0$ the coherence length L_{coh} is large, which means that the four-photon processes efficiency is very high. The dispersion k_2 of a single-mode fibre consists of the material and waveguide (or *intramode*) components (see Equation 3). By changing the fibre parameters (for example, the fibre core radius and/or Δn), one can change the waveguide dispersion. As a result, the zero fibre dispersion point (wavelength where k_2 of the fibre equals zero) can be shifted from $\lambda \sim 1.3\,\mu$m (the material zero dispersion point) up to $\lambda \sim 2\mu$m. This means that for any desired wavelength in the region 1.3μm $< \lambda < 2\,\mu$m, one can get a fibre with $k_2(\lambda)=0$.

Note that in the region of $k_2=0$ the phase matching conditions are satisfied for a broad continuum spectrum of frequency shifts $\Delta\omega$, starting with $\Delta\omega=0$. Note also that for optical transmission systems operating near the zero dispersion point of the fibre, four-photon processes are one of the main factors deteriorating the signal transmission.

In some cases, the waveguide dispersion can have not only the second-order term, but higher-order terms as well. These higher-order dispersion terms (of even orders) add additional terms to the expression for the phase mismatch (Equation 13). In some cases, the combination of second-order and higher-order dispersion can lead to the phase matching condition for certain frequency shifts. Stokes anti-Stokes generation with frequency shifts up to thousands of cm^{-1} have been achieved in this way (Lin *et al.* 1981).

3.4 Nonlinear refractive index—modulational instability

Phase mismatch caused by the fibre dispersion can also be compensated by the effect of cross-phase modulation. Let us consider propagation of a strong pump wave with intensity I_p and weak Stokes and anti-Stokes waves. Due to the Kerr effect, the refractive index at the pump wavelength is:

$$n_p(I_p) = n_p + n_2 I_p \tag{15}$$

Here n_p is the 'linear' refractive index at pump wavelength, and n_2 is the *Kerr coefficient* (also called the *nonlinear refractive index*). It is very important to note that the nonlinear change of the refractive index for waves with frequencies different from ω_P (due to the cross-phase modulation effect) is twice as big as the self-modulation:

$$n_{s,a}(I_p) = n_{s,a} + 2n_2 I_p \tag{16}$$

The magnitudes of the pump, Stokes and anti-Stokes wave vectors are:

$$k_p = \frac{2\pi}{\lambda}[n_p + n_2 I_p] \tag{17}$$

$$k_{s,a} = \frac{2\pi}{\lambda}[n_{S,A} + 2n_2 I_p] \tag{18}$$

Therefore, the phase mismatch is:

$$\Delta k \equiv k_s + k_a - k_p - k_p = k_2 \Delta\omega^2 + \frac{2\pi}{\lambda} 2n_2 I_p; \tag{19}$$

One can see from Equation 19 that the phase matching condition is satisfied for a frequency shift $\Delta\omega_{MI}$:

$$\Delta\omega_{MI}^2 = -\frac{2\pi}{\lambda}\frac{2n_2 I_p}{k_2} \tag{20}$$

Since the nonlinear refractive index coefficient is positive in silica-based fibres (with $n_2 \sim 3 \times 10^{-16} \mathrm{cm}^2/\mathrm{W}$ (Kim *et al.* 1994)), this kind of phase matching can only be achieved in the spectral region of *negative* k_2. Such phase matching leads to a so-called *modulational instability effect* (Hasegawa 1984). Due to this effect, the Stokes anti-Stokes spectral components with frequency shifts $0 < \Delta\omega < \Delta\omega_{MI}\sqrt{2}$ experience exponential gain. The intensity gain coefficient $G_{MI}(\Delta\omega)$ is:

$$G_{MI}(\Delta\omega) = 2\Delta\omega\,|k_2|\left[\frac{2\pi n_2 I}{\lambda\,|k_2|} - \frac{\Delta\omega^2}{4}\right]^{1/2} \tag{21}$$

Maximum gain takes place for the frequency shift $\Delta\omega = \Delta\omega_{MI}$:

$$G_{MI}^{\max} = \frac{4\pi n_2 I}{\lambda} \tag{22}$$

Note that the maximum modulational instability gain does not depend on the magnitude of the fibre dispersion, it depends only on the pump intensity. In the time domain, the modulational instability effect manifests itself as an exponential growth of the amplitude modulation of a CW wave propagating through the fibre.

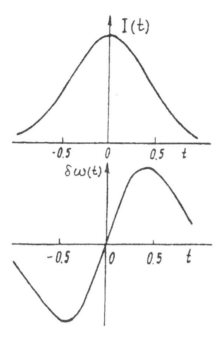

Figure 7. *Envelope of the pulse intensity I(t) and the time-dependent carrier frequency deviation (frequency chirp) δω(t) caused by the effect of self-phase modulation.*

4 Self-phase modulation

During the propagation of a light pulse with an intensity envelope $I(t)$ (t is time) in a nonlinear medium characterised by an intensity-dependent refractive index $n(I) = n_0 + n_2 I(t)$, different parts of the pulse experience different additional nonlinear phase shifts (Stolen and Lin 1978):

$$\Phi_{nl}(t) = \frac{2\pi}{\lambda} \, n_2 I(t) L \tag{23}$$

Here L is the length of the medium. A change of phase in time means a carrier frequency deviation:

$$\delta\omega(t) = -\frac{\partial \Phi}{\partial t} = -\frac{2\pi}{\lambda} \, n_2 \frac{\partial I(t)}{\partial t} L \tag{24}$$

One can see from Equation 24 that the pulse acquires a frequency modulation (*chirp*), the front pulse edge becomes red-shifted in frequency, and the trailing edge becomes blue-shifted in frequency (Figure 7). The pulse also becomes spectrally broadened. If the input pulse is a transform-limited Gaussian pulse (or sech^2 pulse) with spectral bandwidth (FWHM) $\Delta\nu_{in}$, the degree of spectral broadening for $\Phi_{nl}(0) > 2$ is:

$$\frac{\Delta\nu_{out}}{\Delta\nu_{in}} \sim \Phi_{nl}(0) \tag{25}$$

Here $\Phi_{nl}(0)$ is the maximum nonlinear phase shift in the pulse peak. Substituting the value of LI for a bulk nonwaveguide medium (Equation 5) into the above expression, we obtain:

$$\left[\frac{\Delta\nu_{\text{out}}}{\Delta\nu_{\text{in}}}\right]_{\text{nw}} < \frac{2\pi}{\lambda} P \frac{2\pi n_0}{\lambda} \tag{26}$$

When the pulse power P in the above expression is less than the critical self-focusing power for the medium $P_{\text{sf}} = \lambda^2/(8\pi n_2 n_0)$, we find:

$$\left[\frac{\Delta\nu_{\text{out}}}{\Delta\nu_{\text{in}}}\right]_{\text{nw}} < \frac{\pi}{2} \tag{27}$$

This means that in a bulk nonwaveguide nonlinear medium it is not possible to achieve significant spectral broadening up to the appearance of self-focusing. By contrast, one can easily obtain spectral broadening of laser pulses (as large as several hundred times) due to the self-phase modulation effect in optical fibres. The effect of self-phase modulation in fibres is widely used for laser pulse compression.

5 Stimulated Raman scattering

Stimulated Raman scattering (SRS) represents the scattering of light caused by molecular vibrations. When a strong (pump) wave with frequency ω_p and a weak Stokes-shifted (red-shifted) wave with frequency $\omega_s = \omega_p - \Delta\omega$ propagate in a medium at the same time, the Stokes wave can experience exponential gain if the frequency difference $\Delta\omega$ corresponds to the frequency of the medium molecular vibrations:

$$I_s(z) = I_s(0) \exp\left[g_R(\Delta\omega)I_p z\right] \tag{28}$$

Here I_p, I_s are the wave intensities, z is the distance and $g_R(\Delta\omega)$ the Raman gain coefficient. The Raman gain curve $g_R(\Delta\omega)$ of silica fibres is shown in Figure 8. One can see that, in practice, the Raman gain spectrum stretches from 0 to 1000cm^{-1}. It consists of a number of inhomogeneously broadened lines of vibrational resonances corresponding to different modes of the lattice vibrations. The maximum gain is obtained for a frequency shift of 440cm^{-1}: $g_R(\Delta\nu = \Delta\omega/(2\pi c) = 440\text{cm}^{-1}) = 0.92\times10^{-11}\text{cm/W}$ ($\lambda \sim 1\,\mu\text{m}$) for linear polarisation of the pump. For the case of a non-polarisation preserving fibre, the gain coefficient averaged over all polarisations is half of that value. The Raman gain coefficient is inversely proportional to the wavelength λ.

If SRS originates from spontaneous noise, SRS reaches saturation (*i.e.* the Stokes wave intensity becomes equal to the pump intensity $I_s = I_p$) when (Smith 1972):

$$g_R I_p z \sim 16 \tag{29}$$

Equation 29 determines a 'threshold' pump intensity for SRS in a fibre of length L:

$$I_p^{\text{threshold}} = \frac{16}{g_R L} \tag{30}$$

When the Stokes wave reaches its saturation, it depletes the pump wave. With further propagation, the Stokes wave becomes strong enough to act as a pump for the next

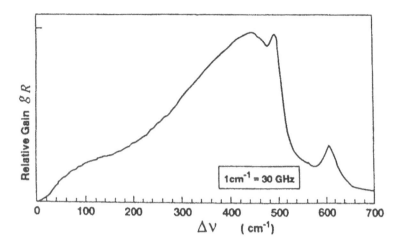

Figure 8. *Dependence of the Raman gain coefficient g_R on the frequency difference between the pump and the Stokes waves.*

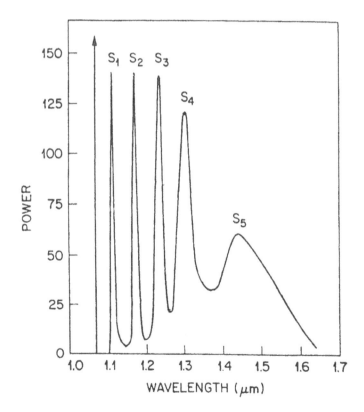

Figure 9. *Spectrum of cascaded stimulated Raman scattering when pumping at $\lambda_p = 1.064 \mu m$.*

Stokes component. One can easily obtain many orders of cascaded SRS generation (Figure 9). The spectral shift between SRS components 440–460cm^{-1} corresponds to the maximum of the Raman gain spectrum.

It has been demonstrated that the Raman gain in optical fibres can be used for amplification of optical signals (Stolen 1980, Dianov *et al.* 1987b, 1995, Grubb 1995). One can build very efficient Raman lasers by placing a fibre inside a resonator (Stolen 1980, Golovchenko *et al.* 1990a, Grubb *et al.* 1995). In a recent experiment (Grubb *et al.* 1995) with a fifth-order cascaded Raman laser operating at 46% slope efficiency, a diode-pumped output of 1.5W was obtained at 1484nm with a 2nm spectral width.

When considering the Raman interaction of the pump and Stokes waves, we did not take into account the anti-Stokes wave at frequency $\omega_a = \omega_p + \Delta\omega$. Nevertheless it has been shown both theoretically and experimentally that parametric interaction of the pump, Stokes and anti-Stokes waves can suppress the exponential Raman gain (Golovchenko *et al.* 1990b, Vertikov *et al.* 1991, Mamyshev *et al.* 1992, Vertikov and Mamyshev 1992). When the parameter $\Delta k/(g_R I_p) \to 0$, the exponential Raman gain coefficient also tends to zero. (Here g_R is the Raman gain for the case when the parametric interaction is negligible and $\Delta k = k_2 \omega^2$ is the phase mismatch for the process $\omega_p + \omega_p \to \omega_s + \omega_a$).

6 Stimulated Brillouin scattering

Stimulated Brillouin Scattering (SBS) represents the scattering of light by acoustic phonons. Similar to the case of stimulated Raman scattering, SBS manifests itself as exponential amplification of the Stokes wave in the field of the pump wave. Only the backward SBS is observed in fibres. The frequency shift of SBS ($\Delta\nu_{SBS} = \omega_p - \omega_s)/(2\pi)$) is inversely proportional to the pump wavelength and for $\lambda = 0.53\mu m$ it amounts to 32GHz in silica fibres (Stolen 1979). The SBS gain coefficient is more than two orders of magnitude higher than the Raman gain coefficient, $g_{SBS} = 4.5 \times 10^{-9}$cm/W. The 'threshold' pump power for SBS can be as small as several milliwatts. However, the SBS gain spectrum is very narrow ($\sim$100MHz). For the case of spectrally broad pumping, the SBS efficiency decreases as the ratio of the SBS linewidth to the pump spectral bandwidth. Moreover, for the case of pulsed pumping, the interaction length between the Stokes and pump waves is short because of the backward nature of the Brillouin scattering. As a result, stimulated Raman scattering frequently predominates over Brillouin scattering. Nevertheless, SBS can cause serious problems in fibre communication lines, even in soliton communication systems, where picosecond pulses are used (Mamyshev *et al.* 1991).

The effect of phase-conjugation at SBS has been observed in multimode fibres (Basiev *et al.* 1982). This means that such fibres can be used as phase-conjugation mirrors in laser light amplifiers.

7 Self-phase modulation and group-velocity dispersion. The non-linear Schrödinger equation (NSE)

We have already considered the self-phase modulation (SPM) effect without taking into account the effect of the fibre dispersion. However, the dispersion plays an extremely important role in the nonlinear dynamics of light propagation in a fibre. The nonlinear dynamics of optical pulse propagation in fibres depends qualitatively on the sign of the fibre group velocity dispersion—equal to $\partial^2 k/\partial\omega^2$ and abbreviated to GVD. As we have already discussed in Section 5, the self-phase modulation effect broadens the propagating pulse spectrally and produces frequency modulation of the pulse (chirp), so that the front edge of the pulse becomes down-shifted (red-shifted) in frequency and the trailing edge becomes up-shifted (blue-shifted) in frequency. This chirped pulse will tend to self-compress in the time-domain if the group velocity of blue spectral components is higher than that of red spectral components (the case of negative GVD). But for the case of positive GVD this pulse will tend to self-broaden (faster than in a linear case). Let us estimate a typical fibre length (a self-action length $z_{\rm sa}$) at which the pulse width starts to change significantly due to the combined action of self-phase modulation and group velocity dispersion (self-compression for the case of negative GVD and self-broadening for the case of positive GVD). Assume that the initial pulse shape is

$$E(z=0, \tau) = E_0 \, \text{sech}(\tau/t_0) \tag{31}$$

The full width at half maximum FWHM is $\tau_0 = 1.763 t_0$. The spectral bandwidth of this pulse can be estimated as $\Delta\omega_0 \sim 1/t_0$. Initially neglecting the dispersion effects, we estimate the pulse spectral $\Delta\omega$ broadening in the distance $z_{\rm sa}$ (see Equation 25) to:

$$\Delta\omega = \frac{2\pi}{\lambda} \, n_2 \, |E_0|^2 \, z_{\rm sa} \tag{32}$$

Now, neglecting the fibre nonlinearity, we assume that the temporal spread of spectral components $\Delta\omega$, occurring in the distance $z_{\rm sa}$, is equal to the initial pulse width t_0:

$$\Delta\omega \left| \frac{\partial^2 k}{\partial\omega^2} \right| z_{\rm sa} = t_0 \tag{33}$$

Now we can find the self-action length $z_{\rm sa}$ from Equations 32 and 33:

$$z_{\rm sa} = \left[t_0^2 \left| \frac{\partial^2 k}{\partial\omega^2} \right|^{-1} \left(\frac{2\pi}{\lambda} \, n_2 \, |E_0|^2 \right)^{-1} \right]^{1/2} = [z_{\rm d} \, z_{\rm nl}]^{1/2} \tag{34}$$

Here the dispersion length $z_{\rm d}$ represents the fibre length at which the pulse starts to broaden significantly in the linear case:

$$z_{\rm d} = t_0^2 \left| \frac{\partial^2 k}{\partial\omega^2} \right|^{-1} \tag{35}$$

The nonlinear length $z_{\rm nl}$ represents the fibre length at which the pulse spectrum starts to broaden significantly due to the self-phase modulation effect, without taking into account the dispersion effect:

$$z_{\rm nl} = \left[\frac{2\pi}{\lambda} \, n_2 \, |E_0|^2 \right]^{-1} \tag{36}$$

For a detailed analysis of the nonlinear light propagation through a fibre, one has to solve a nonlinear wave equation. A theoretical analysis of the combined action of refractive index nonlinearity and fibre dispersion is usually done in the frame of the *Nonlinear Schrödinger* equation. The electric field in the fibre is considered to be transverse (this is a good approximation for fibres with a small difference between the refractive indices of the core and the cladding Δn). The electric complex field Ψ is represented as:

$$\Psi(x,y,z) = \eta(x,y)\,\tilde{E}(z,t)\,\exp\left[i(\omega t - kz)\right] \tag{37}$$

where $\eta(x,y)$ represents the normalised transverse spatial distribution of the mode field in a linear approximation (the light power is considered to be below the critical self-focusing power), $\eta(0,0) = 1$; $\tilde{E}(z,t)$ is the slowly varying amplitude of the electric field, z is the coordinate along the fibre axis, t is time; ω is the light carrier frequency, and k is the propagation 'constant'. The analysis is usually carried out in the second-order approximation of the dispersion theory, *i.e.* the terms up to the second-order are kept when the propagation constant $k(\omega)$ is expanded into a Taylor series. In this case, one can get the following equation (nonlinear Schrödinger equation (NSE): Zakharov and Shabat 1971, Hasegawa and Tappert 1973) for the effective electric field $E(z,t)$ averaged over the transverse coordinates:

$$\frac{\partial E}{\partial z} = -\frac{i}{2}\frac{\partial^2 k}{\partial \omega^2}\frac{\partial^2 E}{\partial \tau^2} + \frac{in_2}{n_0}k_0\,|E|^2\,E - \alpha E \tag{38}$$

Here $\tau = t - z/v_{\text{gr}}$ is the time the light is travelling, $v_{\text{gr}} = \partial\omega/\partial k$ is the light group velocity, and α is the linear fibre loss constant. The effective electric field is given by

$$E(z,t) = \tilde{E}(z,t)\left[\frac{\langle\eta^4\rangle}{\langle\eta^2\rangle}\right]^{1/2},$$

where the angular brackets denote integration over transverse coordinates. The effective electric field is related to the light power $P(z,t)$ as follows: $|E(z,t)|^2 = P(z,t)/S_{\text{eff}}$, where the effective mode area of the transverse cross section of the mode area is $S_{\text{eff}} = \langle\eta^2\rangle^2/\langle\eta^4\rangle$.

Equation 38 describes the combined action of self-phase modulation (second term on the right side of the equation) and dispersion of group velocities (first term on the right side). In many cases, it describes the nonlinear evolution of picosecond pulses well and even that of subpicosecond pulses. But in some cases, especially in the spectral region of negative group velocity dispersion (group velocity v_{gr} decreases as wavelength λ increases), one has to add additional terms to the equation to take into account higher-order nonlinear and dispersive effects (see section 12). It should be noted that there are no principal differences between the effect of self-phase modulation, four-wave mixing and modulational instability. All of them are associated with the nonlinear refractive index nonlinearity. This means, in particular, that the nonlinear Schrödinger equation 38 describes all these effects.

8   Fibre-grating pulse compression

We have already discussed above that the self-phase modulation effect broadens a spectrum and produces frequency modulation (chirp) of the pulse propagating through a

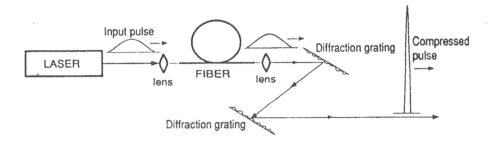

Figure 10. *Experimental set-up for the fibre-grating compression of laser pulses.*

fibre (see Section 4). If such a chirped pulse is transmitted through a dispersive delay line with negative group velocity dispersion, the blue spectral components catch up with the red ones and the pulse can therefore be compressed into a duration determined by the inverse width of its (broadened) spectrum. The most popular dispersive delay line consists of two diffraction gratings (Figure 10). This technique of *fibre-grating pulse compression* was mainly designed for fibres with a positive group velocity dispersion. Depending on the length of the fibre, one can distinguish two basic regimes of pulse spectral broadening due to the self-phase modulation in the fibre. When the fibre length L is much less than the self-action length z_{sa}, the fibre dispersion is of no importance. This case is called the 'dispersion-free' regime of SPM. In the case of the 'dispersive' regime of SPM, the fibre length is of the order of, or longer than, z_{sa}. Let us consider the pulse compression using these two regimes of SPM in more detail.

8.1 Pulse compression using 'dispersion-free' SPM

In this case, the frequency chirp of the pulse is described by Equation 24 (Figure 7). The pulse spectrum broadens according to Equation 25, and the achievable pulse compression ratio (with an optimum dispersion of the dispersive delay line) is:

$$\frac{\tau_0}{\tau_{comp}} \sim 1 + 0.57 \, \Delta\Phi_{nl}(0) \tag{39}$$

According to Equation 39, the pulse compression ratio increases linearly with the fibre length and with the pulse peak power. Nevertheless, stimulated Raman scattering arising in the fibre results in the conversion of the pulse energy into the energy of SRS generation which distorts the linearity of the frequency chirp, and in turn hinders the pulse compression (Dianov *et al.* 1984a,b, 1985b). This means that SRS limits the fibre length and the allowable pulse power in the experiment, and hence limits the pulse compression ratio. If we put the expression for the 'threshold' SRS intensity into Equation 23 for $\Delta\Phi_{nl}(0)$, we get a 'critical' value of $\Delta\Phi_{nl}(0)$ attainable due to SPM at the SRS 'threshold':

$$\Delta\Phi_{cr}(0) = 32\pi n_2/(\lambda g_R) \tag{40}$$

Note that this expression does not depend on either the fibre parameters or the wavelength (since g_R is inversely proportional to λ), but only on the fibre material. For silica fibres and linear light, the polarisation Equation 40 gives $\Delta\Phi_{cr} \sim 28$, and for the non-polarisation preserving fibre $\Delta\Phi_{cr} \sim 50$. This means that, according to Equation 39, the ultimate compression factors (limited by SRS) are 17 and 30, correspondingly. This agrees well with available experimental results.

Note that we assumed that the Raman gain coefficient did not depend on the pump power or the fibre dispersion. Nevertheless, parametric interaction can suppress the Raman gain (see Section 5) and thus increase the allowable pulse compression ratio. A two-fold increase in the maximum degree of compression was achieved experimentally by using parametric suppression of stimulated Raman scattering. The compression ratio of pulses from an actively Q-switched and mode-locked Nd:YAG laser ($\lambda = 1.064\mu m$), $\tau_0/\tau_{comp} = 30$, was obtained in a polarisation-preserving fibre in a regime of parametric suppression of the Raman gain, while without parametric interaction the maximum compression ratio was 16 (Vertikov *et al.* 1991).

Thus, in the case of 'dispersion-free' self-phase modulation, stimulated Raman scattering limits the compression factor, but does not limit the power or energy of the compressed pulses (to increase the critical SRS power, one should decrease the fibre length, Equation 30). Limitations on the power of pulses under compression are determined only by the threshold for optical damage of the fibre and the self-focusing of light.

As one can see from Figure 7, the frequency chirp of the pulse after 'dispersion-free' SPM is linear only in the pulse centre, while the pulse wings have opposite signs to the chirp. For this reason, the compressed pulse has a low-intensity pedestal. This pedestal can be suppressed using the effects of nonlinear fibre birefringence (Stolen *et al.* 1982, Nikolaus *et al.* 1983, Mollenauer *et al.* 1983, Halas and Grischkovsky 1986, Dianov *et al.* 1989a). Due to this effect, the state of polarisation of light in the fibre is intensity-dependent. One can adjust the input polarisation of the pulses under compression so that at the fibre output, the states of polarisation of the high-intensity pulse centre and of the low-intensity wings are orthogonal. By putting a polariser at the fibre output, one can suppress the wings (which contribute to the pedestal of the compressed pulse) and hence increase the quality of the compressed pulse (Dianov *et al.* 1989a).

8.2 Pulse compression using 'dispersive' SPM

When the fibre length is of the order of the self-action length z_{sa}, the fibre dispersion plays an important role. When the parameter

$$N = \left[\frac{z_d}{z_{nl}}\right]^{1/2} \tag{41}$$

is small, the nonlinear effects in the fibre are of no importance and the pulse temporally broadens during the propagation due to the dispersion. The pulse spectral bandwidth does not change significantly. If $N > 1$ the nonlinearity becomes important and the pulse spectrum broadens due to SPM. Due to the combined action of SPM and positive GVD, the pulse temporal width broadens faster than for the case of linear dispersive broadening. During the pulse broadening, the pulse intensity decreases, therefore, the efficiency of the self-phase modulation decreases with distance. At a fibre length

$$z_{opt} = 2.5\, z_{sa} \tag{42}$$

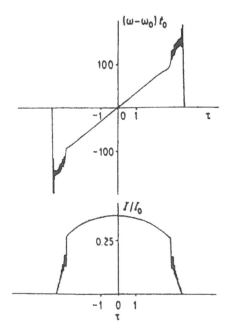

Figure 11. *The pulse intensity envelope and pulse frequency chirp caused by the 'dispersive' regime of self-phase modulation in a fibre. The fibre length is optimal $z = z_{opt}$ for subsequent pulse compression in the dispersive delay line.*

the pulse temporal profile becomes nearly rectangular. It is very important for the subsequent pulse compression that the frequency chirp is linear within the entire pulse (Figure 11). This then makes it possible to obtain high-quality compressed pulses with extremely low pedestals (Grischkovsky and Balant 1982, Tomlinson *et al.* 1984, Golovchenko *et al.* 1988). This length of fibre is optimal for high-quality pulse compression. The degree of possible pulse compression is:

$$\frac{\tau_0}{\tau_{comp}} \sim 0.63 \left[\frac{z_d}{z_{nl}}\right]^{1/2} \tag{43}$$

Optimal dispersion of the delay line is:

$$\left(\frac{dt}{d\nu}\right)_{opt} \sim -6.47 \, c\tau^2 \left[\frac{z_d}{z_{nl}}\right]^{-1} \tag{44}$$

The use of fibres with length $L > z_{opt}$ for subsequent pulse compression is undesirable, because with further propagation through the fibre, the spectral bandwidth of the pulse does not significantly increase any more, but the linearity of the chirp may break down.

One can see from Equations 42 and 43 that the optimum fibre length depends on the pulse intensity, pulse width and value of the fibre dispersion. The pulse compression ratio increases (as the square root) with the pulse intensity. Similar to the case of the 'dispersion-free' SPM, the maximum pulse power coupled into the fibre is limited by the appearance of SRS. Nevertheless, the regime of SRS is different because of the group velocity dispersion (Stolen and Johnson 1986, Dianov *et al.* 1986a,1989a). The group velocities of the pulse under compression (pump) $v_{gr,p}$ and of the Stokes SRS wave $v_{gr,s}$ are different. The interaction length between the pump and SRS pulses is the *walk-off length*:

$$L_{\text{walk-off}} = \frac{\tau_0}{\left| \dfrac{1}{v_{gr,p}} - \dfrac{1}{v_{gr,s}} \right|} \sim \frac{\tau_0}{\left(\dfrac{\partial^2 k}{\partial \omega^2} \right) \Delta\omega_{\text{SRS}}} \tag{45}$$

where $\Delta\omega_{\text{SRS}}$ is the frequency shift between the pump and the Stokes waves. (For silica fibres, $\Delta\omega_{\text{SRS}}/(2\pi c) = 440\text{cm}^{-1}$.) The SRS walk-off length is usually less than the optimum fibre length for the pulse compression. By putting expression 45 into Equation 30, we find the critical (SRS 'threshold') pulse power:

$$P_{\text{cr}}\tau_0 = 16 S_{\text{eff}} \frac{\Delta\omega_{\text{SRS}}}{g_R} \frac{\partial^2 k}{\partial \omega^2} \tag{46}$$

Note that the product $P_{\text{cr}}\tau_0$ is the energy of the input pulses. This means that the above expression estimates the critical ('threshold') pulse energy for SRS in long fibres. As one can see the SRS threshold energy does not depend on the fibre length when the fibre length is longer than the walk-off length. Analysis of Equation 46 shows that the critical pulse energy has the same order of magnitude (a few nanojoules) over the entire transparent spectral region of single-mode silica fibres. The exception is the case where the pump and Stokes wavelengths are symmetrical with respect to the zero dispersion wavelength λ_0 (where the GVD vanishes). In this case, $v_{gr,p} = v_{gr,s}$, $L_{\text{walk-off}} \to \infty$ and $P_{\text{cr}}\tau_0 \to 0$ (Dianov *et al.* 1986a).

Combining Equation 43 and Equation 46, we can estimate the minimum critical width of the compressed pulses:

$$\tau_{\text{comp,cr}} = 0.25 \left[\frac{g_R \lambda}{\Delta\omega_{\text{SRS}} n_2} \tau_0 \right]^{1/2} \tag{47}$$

Note that this expression depends neither on the fibre parameters nor on the wavelength (since g_R is inversely proportional to λ), but only on the fibre material. Substituting numerical values for silica fibres, we get:

$$\tau_{\text{comp,cr}} = 0.05 \left[\tau_0 \right]^{1/2} \tag{48}$$

where $\tau_{\text{comp,cr}}$ and τ_0 are expressed in picoseconds.

Note that in some cases of the 'dispersive' SPM regime, the pulse compression can be obtained with high stability in the presence of SRS (Weiner *et al.* 1988, Kuckartz *et al.* 1988, Fursa *et al.* 1992). The limitation (Equation 46) on the energy of the *compressed* pulses is still valid for this case.

Summarizing this section, one can distinguish the following features of pulse compression using the 'dispersion-free' and 'dispersive' regimes of SPM. For the case of the 'dispersion-free' regime of SPM, the effect of stimulated Raman scattering limits the *degree* of pulse compression ($\tau_0/\tau_{\text{comp}}$ <17–30) and does not limit the power and energy of the compressed pulses. Compression of pulses with energy up to $\sim 10\mu J$ was realised experimentally using single-mode fibres (Dianov *et al.* 1989a). The 'dispersive' regime of SPM gives a higher quality of compressed pulses. SRS limits the maximum *energy* of the compressed pulses by several nJ, and does not limit the compression ratio. The minimum achievable duration of the compressed pulses is proportional to the square root of the initial pulsewidth. Pulses as short as 6fs were obtained by compression of 40fs pulses (Fork *et al.* 1987). Compression ratios as high as 110–160 of 60–90ps pulses were obtained in single-stage configurations (Dianov *et al.* 1987a, Kuckartz *et al.* 1988, Fursa *et al.* 1992). It should be noted that when the 'dispersive' regime of SPM is used, high-quality pulse compression is possible even when the initial pulse shape is not symmetrical (Golovchenko *et al.* 1988).

9 Solitons

Let us consider propagation of pulses in the region of negative group velocity dispersion. When the parameter N (Equation 41) is small, *i.e.* $N \ll 1$, the nonlinearity (the second term on the right side of Equation 38) is of no importance. The pulsebroadens temporally due to the dispersion as it propagates through the fibre, while the pulse spectral bandwidth does not change. In the opposite case, if we neglect the dispersion term in NSE (Equation 38), the pulse width does not change with the propagation, but the pulse spectrum broadens due to the SPM effect (see Equations 23 to 25). Nevertheless, there is a regime when these two effects (SPM and GVD) compensate for each other (the second term on the right side of Equation 38). It happens when the parameter N equals 1:

$$N = \left[\frac{z_d}{z_{nl}}\right]^{1/2} = 1\,.$$

In this case, due to the combined action of SPM and GVD, the pulse propagates through the fibre without change to its spectral and temporal shapes. This regime is called the *soliton regime of propagation* (Hasegawa and Tappert 1973, Mollenauer *et al.* 1980). Qualitatively, the soliton propagation regime can be understood as follows (Figure 12). Assume that we have a chirp-free pulse. The self-phase modulation broadens the pulse spectrum and produces a frequency chirp: the front edge of the pulse becomes red-shifted and the trailing edge - blue-shifted. When negative GVD is applied to this chirped pulse, the red spectral components are delayed in time with respect to the blue ones. This means that the blue spectral components shift in time to the front pulse edge, while the red spectral components move to the trailing edge. When the nonlinearity is applied again, it shifts the frequency of the front edge to the red spectral region and up-shifts the frequency of the trailing edge. This means that the 'blue' front edge becomes 'green' again, the 'red' trailing edge also becomes 'green' and the pulse spectrum bandwidth narrows to its original width. Of course, in reality the effects of SPM and GVD act at the same time so that the pulse spectral and temporal widths don't change during the propagation. The only net effect is a (constant within the entire pulse) phase shift of 0.5 radians per z_d. The soliton condition $N=1$ ($z_d = z_{nl}$) can be rewritten as:

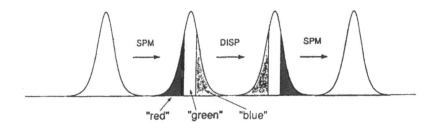

Figure 12. *Qualitative explanation of the soliton regime of pulse propagation (see text).*

$$|E_\theta|^2 = \lambda \left| \frac{\partial^2 k}{\partial \omega^2} \right| \frac{1}{2\pi n_2 t_0^2} . \tag{49}$$

Equation 49 gives the relationship between the soliton parameters (pulsewidth τ(FWHM), equal to $1.763t_0$, and pulse peak intensity $|E_0|^2$) and the fibre dispersion.

Soliton pulses are extremely robust. The combined action of the fibre nonlinearity and dispersion picks up the soliton out of whatever reasonable pulse is launched into the fibre (*i.e.* when the initial pulse parameters are not too far from Equation 49), and discards the residue as dispersive wave radiation. The soliton pulse is robust with respect to perturbations of the fibre parameters (mode area and dispersion) and is stable in fibre transmission lines where fibre loss is compensated by lumped amplifiers. The robustness of solitons makes them extremely promising for use in long distance high-bit-rate transmission, optical switching and optical processing.

These soliton pulses are also referred to as *fundamental solitons* or *first-order solitons*. When N (Equation 41) equals an integer number greater than 1, this corresponds to higher-order soliton (*N-soliton or multi-soliton*) pulses. The power of multi-soliton pulses is N^2 times higher than that of fundamental solitons. This means that at the initial stage of propagation, the nonlinearity predominates over the dispersion, and the pulses broaden spectrally and self-compress in the time domain. The propagation evolution of multi-soliton pulses is fairly complex. The pulse spectrum and temporalshapes restore at a fibre length equal to the soliton period z_0 (Mollenauer *et al.* 1983):

$$z_0 = \frac{\pi}{2} z_d \tag{50}$$

The fibre length for the maximum self-compression of multisoliton pulses is (Dianov *et al.* 1986b):

$$z_{opt} = \left[0.5 + \frac{1.7}{N} \right] z_{sa} \tag{51}$$

One can see that for large N, the optimal self-compression length $z_{opt} \to 0.5 z_{sa}$. The degree of self-compression of multi-soliton pulses increases with N:

$$\frac{\tau_0}{\tau_{comp}} \sim 4.1 \left[\frac{z_d}{z_{nl}} \right]^{1/2} = 4.1\,N \tag{52}$$

By using the multi-soliton pulse compression technique, one can get very high compression ratios. Compression ratios of 27 (Mollenauer *et al.* 1983), 22 and 110 (Dianov *et al.* 1984c) have been achieved experimentally. Nevertheless, the quality of the compressed pulses produced using this technique is poor. The compressed pulses have a wide low-intensity pedestal with the energy in the pedestal increasing with the compression ratio.

High-quality pulse compression can be achieved when adiabatic compression of fundamental solitons is used (Dianov *et al.* 1989b, Chernikov and Mamyshev 1991). Equation 49 can be rewritten as:

$$t_0 = \left|\frac{\partial^2 k}{\partial \omega^2}\right| \frac{1}{|E_0|^2 t_0} \frac{\lambda}{2\pi n_2} \tag{53}$$

One can see from Equation 53 that the soliton pulsewidth t_0 is proportional to the fibre dispersion and inversely proportional to the soliton energy $(S_{\text{eff}}|E_0|^2 t_0)$. This means that a fundamental soliton can be adiabatically compressed

- by adiabatic amplification of the soliton in the fibre or

- if the soliton propagates through a special fibre with the dispersion $\left|\frac{\partial^2 k}{\partial \omega^2}\right|$ decreasing with fibre length and/or

- in fibres with a mode area S_{eff} decreasing with fibre length.

The soliton compression is adiabatic if the change of fibre parameters or change of soliton energy is small on the soliton dispersion length. The first experiments on adiabatic soliton compression in fibres with slowly decreasing dispersion are described elsewhere (Dianov *et al.* 1989b, Chernikov and Mamyshev 1991), along with experiments on soliton compression by adiabatic amplification (Smith and Mollenauer 1989). Note that the fibre mode area usually (but not necessarily) decreases with distance in fibres with slowly decreasing dispersion (Bogatyrev *et al.* 1991). Another method of adiabatic soliton compression in axially uniform fibres is based on the combined action of higher-order dispersion and Raman self-scattering effects (Mamyshev *et al.* 1993).

As we have already mentioned above, a soliton can be generated in a fibre if a pulse with parameters close to those of Equation 49 is launched into the fibre. During the transformation of the initial pulse, a nonsoliton component in the form of dispersive wave radiation is usually generated along with the soliton. Hasegawa (1984) suggested a method for high-bit-rate pulse train generation based on the *induced modulational instability effect*. This method was realised by Tai *et al.* (1986). Nevertheless, such a method does not permit the generation of stable trains and the quality of the pulses is poor. The pulses generated by this technique are not fundamental solitons. Mamyshev *et al.* (1991) described a method for generating extremely high-quality high-repetition-rate stable trains of fundamental soliton pulses from CW dual-frequency sinusoidal beat signals. When these signals propagate through a fibre with effective adiabatic amplification (it could be a fibre with actual amplification or a fibre with decreasing dispersion along the fibre length), the new spectral components are generated. It is important to note that due to the joint action of the effective amplification, the fibre dispersion and nonlinearity, the phases of the spectral components are self-adjusted so that in the time domain the signal reshapes into a high-quality soliton pulse train without nonsoliton components (Figure 13). Soliton trains with repetition rates of 60–200GHz have been generated by this method (Mamyshev *et al.* 1991, Chernikov *et al.* 1992). Another method for generating high-quality high-repetition-rate pulse trains of fundamental solitons is based on the combined action of the induced modulational instability and the Raman self-scattering effect (Mamyshev *et al.* 1990).

9.1   The Raman self-scattering (self-frequency shift) effect

Since its discovery the Raman self-scattering effect (RSS) (Dianov *et al.* 1985a) has attracted much attention because of its important role in nonlinear light propagation in optical fibres (Mitschke and Mollenauer 1986, Gordon 1986, Golovchenko *et al.* 1989, Stolen *et al.* 1989, Blow and Wood 1989, Mamyshev and Chernikov 1990). The RSS effect consists of an energy exchange between spectral components *within a single pulse spectrum* due to the Stimulated Raman scattering effect. The Raman gain spectrum in silica fibres stretches practically from 0 to 1000cm^{-1}. If the spectrum of a propagating pulse is broad enough, the red spectral components of the pulse spectrum can be amplified in the field of the blue spectral components of the pulse—due to the Raman gain. The RSS effect results in a shift of the propagating pulse spectrum to the Stokes spectral region. In particular, RSS causes the *soliton self-frequency shift effect*: the spectrum of a fundamental soliton shifts as a whole to the Stokes region during propagation through a fibre. The rate of the mean soliton frequency shift is inversely proportional to the fourth order of the soliton pulsewidth (Gordon 1986). The molecular vibrations, on which the light is self-scattered, are excited by various pairs of spectral components of the light. The excitation is efficient if all the pairs excite the vibrations in phase. The latter takes place if all the spectral components are in phase. That, in turn, is satisfied in a nonlinear dispersive media for fundamental soliton pulses only. This exceptional feature of fundamental solitons makes the RSS effect very 'selective' with respect to solitons.

This 'selectivity' of the RSS effect with respect to solitons is used in a very efficient method for high-quality femtosecond soliton formation (Dianov *et al.* 1985a, Mamyshev 1991, Mamyshev 1992). When light radiation propagates through a fibre in the negative group velocity dispersion spectral region, the RSS effect 'selects' only the soliton component from the radiation. As a result, the soliton component shifts to the Stokes spectral region, the nonsoliton component being unshifted. At the fibre output, the soliton component is spectrally separated from the rest of the radiation spectrum, and so solitons can now be easily obtained by spectral selection of the Stokes components from the spectrum. In particular, one can generate femtosecond fundamental solitons from multisoliton picosecond pulses. The method permits one to obtain single soliton pulses if multisoliton pulses of the order of $N < 13$ are used as the input radiation (Mamyshev 1991). If the order of input multisoliton pulses is too high, or a noise radiation is used (for example, a Stokes component of cascaded SRS), a random sequence of fundamental solitons is obtained by this method (Mamyshev 1991, Golovchenko *et al.* 1991).

The nonlinear Schrödinger equation does not describe the RSS effect. In the next section, we will discuss a model which takes into account the Raman self-scattering effect and the higher-order nonlinear and dispersive effects (Mamyshev and Chernikov 1990).

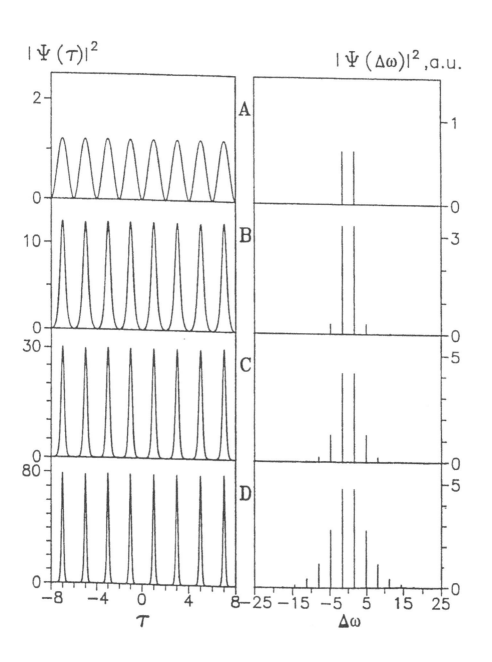

Figure 13. *Intensity (left) and spectrum (right) evolution of a dual-frequency beat signal in a fibre with effective amplification. During propagation, the signal reshapes into a stable, high-quality train of fundamental solitons (Mamyshev et al. 1990).*

10   Raman and higher-order nonlinear and dispersive effects. The modified NSE

Light propagation in a single-mode fibre with third-order nonlinearity is described by the scalar wave equation for the electric field $E(t, r)$:

$$\Delta E(t, r) - \frac{1}{c^2}\frac{\partial^2 D(t, r)}{\partial t^2} = \frac{4\pi}{c^2}\frac{\partial^2 P^{(3)}(t, r)}{\partial t^2}, \tag{54}$$

where $\Delta = \partial^2/\partial z^2 + \Delta_\perp$ is the Laplacian operator. The linear electric induction D and the nonlinear third-order polarisation $P^{(3)}$ are:

$$D(t, r) = \int \varepsilon(t', r)\, E(t - t', r)\, dt', \tag{55}$$

$$P^{(3)}(t, r) = E(t, r) \int R(t')\, |E(t - t', r)|^2\, dt', \tag{56}$$

Here $\varepsilon(t)$ and $R(t)$ are linear and nonlinear response functions of the medium, respectively. $R(t)$ consists of practically instantaneous electronic and delayed Raman responses. The propagating electric field $E(t, r)$ can be presented as a superposition of monochromatic waves:

$$E(t, r) = \int E(\omega, z)\, U\left(\omega, r_\perp \exp[ik(\omega)z - i\omega t]\right) d\omega, \tag{57}$$

where $U(\omega, r_\perp)$ is the transverse fibre mode profile, $U(\omega, 0, 0) = 1$; $k(\omega) = n_{\mathrm{eff}}(\omega)\omega/c$ is the mode propagation constant, and $n_{\mathrm{eff}}(\omega)$ is the effective index. Substituting Equations 55–57 into Equation 54, we obtain:

$$E(\omega, z)\left[\Delta_\perp + \varepsilon(\omega, r_\perp)\frac{\omega^2}{c^2} - k^2(\omega)\right] U(\omega, r_\perp) + U(\omega, r_\perp)\left[\frac{\partial^2}{\partial z^2} + 2ik(\omega)\frac{\partial}{\partial z}\right] E(\omega, z)$$

$$= -\frac{4\pi\omega^2}{c^2} \int d\omega' \int d\omega''\, E(\omega', z)\, U(\omega', r_\perp)\, E(\omega'', z)\, U(\omega'', r_\perp)$$

$$\times E^*(\omega' + \omega'' - \omega, z) U(\omega' + \omega'' - \omega, r_\perp)\chi^{(3)}(\omega - \omega')\exp\left(i\Delta k\, z\right); \tag{58}$$

Here

$$\varepsilon(\omega, r_\perp) = \int \varepsilon(t, r_\perp)\exp\left(i\omega t\right) dt\,;$$

$$\Delta k = k(\omega') + k(\omega'') - k(\omega' + \omega'' - \omega) - k(\omega),$$

and

$$\chi^{(3)}(\omega) = \int R(t)\exp\left(i\omega t\right) dt$$

is the third-order susceptibility. The third-order susceptibility $\chi^{(3)}(\Delta\omega)$ is a complex value, the real part of which is responsible for parametric and self-phase modulation effects, and the imaginary part of which is responsible for the Raman effect.

The first term in Equation 58 is zero, because $U(\omega, r_\perp)$ is the fibre mode. One can show that the term

$$\frac{\partial^2 E(\omega, z)}{\partial z^2}$$

is much smaller than the others in Equation 58, and we neglect it. On averaging over transverse coordinates $r_\perp$, Equation 58 is reduced to:

$$\frac{\partial E(\omega, z)}{\partial z} = \frac{i2\pi\omega}{cn_{\text{eff}}(\omega)} \int d\omega' \int d\omega'' G(\omega, \omega', \omega'')$$
$$\times E(\omega', z) E(\omega'', z) E^*(\omega' + \omega'' - \omega, z) \chi^{(3)}(\omega - \omega') \exp{(i\Delta k\, z)}, \quad (59)$$

where

$$G(\omega, \omega', \omega'') = \frac{\int U(\omega', r_\perp) U(\omega'', r_\perp) U(\omega' + \omega'' - \omega, r_\perp) U(\omega, r_\perp)\, d^2 r_\perp}{\int U^2(\omega, r_\perp)\, d^2 r_\perp}$$

By assuming a Gaussian mode profile $U(\omega', r_\perp)$, G can be estimated as

$$G(\omega, \omega', \omega'') = [\, S(\omega)\, S(\omega')\, S(\omega'')\, S(\omega' + \omega'' - \omega)\,]^{1/4} / S(\omega),$$

where

$$S(\omega) = A(\omega)/A(\omega_0),$$

and the effective mode area $A(\omega)$ is:

$$A(\omega) = \frac{[\int U^2(\omega)\, d^2 r_\perp]^2}{\int U^4(\omega)\, d^2 r_\perp}.$$

Expanding $k(\omega)$ about ω_0 and correcting to the mth order, we obtain:

$$k(\omega) = k_0 + k_1 \Delta\omega + \frac{k_2 \Delta\omega^2}{2} + \frac{k_3 \Delta\omega^3}{6} + \ldots + \frac{k_m \Delta\omega^m}{m!},$$

$$\Delta\omega \equiv \omega - \omega_0, \qquad k_l \equiv \left. \frac{\partial^l k(\omega)}{\partial \omega^l} \right|_{\omega = \omega_0} \quad (60)$$

.By substituting

$$\tilde{E}(\Delta\omega, z) = \frac{1}{2} S^{1/4}(\omega)\, E(\omega, z)\, \exp{(ik(\omega)z - ik_0 z)}$$

(where $k_0 \equiv k(\omega_0)$) and Equation 60 into Equation 59 we obtain:

$$\frac{\partial \tilde{E}(\Delta\omega, z)}{\partial z} - i\left(k_1 \Delta\omega + \frac{k_2 \Delta\omega^2}{2} + \ldots + \frac{k_m \Delta\omega^m}{m!} \right) \tilde{E}(\Delta\omega, z)$$

$$= \frac{i2\pi\omega_0}{cn_{\text{eff}}(\omega) \sqrt{S(\omega)}} \left(1 + \frac{\Delta\omega}{\omega_0} \right) \int d\omega \int d\omega'' \tilde{E}(\Delta\omega', z)\, \tilde{E}(\Delta\omega'', z)$$

$$\times \tilde{E}^*(\Delta\omega' + \Delta\omega'' - \Delta\omega, z)\, \chi^{(3)}(\Delta\omega - \Delta\omega') \quad (61)$$

Equation 61 is the nonlinear propagation equation in the spectral domain. By Fourier transforming Equation 61, we can obtain the propagation equation in the time domain:

$$\left[\frac{\partial}{\partial \xi} - i^{1/2} \frac{\partial^2}{\partial \tau^2} - \gamma \frac{\partial^3}{\partial \tau^3} \right] \Psi(\tau, \xi) \quad (62)$$

$$= i\Psi(\tau,\xi)\int |\Psi(\tau - \tau',\xi)|^2 F(\tau')d\tau' - \sigma\frac{\partial}{\partial\tau}\left[\Psi(\tau,\xi)\int |\Psi(\tau - \tau',\xi)|^2 F(\tau')\,d\tau'\right].$$

Equation 62 is written in dimensionless form, for $m=3$. Here

$$\xi = z/z_d;\quad z_d = t_0^2/|k_2|;\quad \tau = (t - k_1 z)/t_0;$$

$$\gamma = k_3/(6k_2 t_0);\quad \Psi(\tau;\xi) = \tilde{E}(\tau,\xi)/E_0;$$

$$\sigma = 1/(\omega_0 t_0) - \frac{\partial}{\partial\omega}\left[\ln(n_{\text{eff}}(\omega)\sqrt{S(\omega)})\right]_{\omega=\omega_0};$$

$$E_0^2 = 2c|k_2|/(\omega_0)n_2 t_0^2)\quad\text{and}\quad F(\tau) = R(\tau)/\left[\int R(\tau')d\tau'\right].$$

Equation 61 contains a term that is proportional to $\Delta\omega/\omega_0$. This term (and the last term in Equation 62) describes the Stokes losses associated with the material excitation during the Raman process (*i.e.* the equation takes into account that the photon energy is proportional to its frequency). This term also describes the frequency dependence of the nonlinearity (*i.e.* it describes the fact that the nonlinear term is proportional to the light frequency). This term describes the *self-steepening effect* (DeMartini *et al.* 1967), and is also known as the *shock term*. It should be noted that in a vast majority of papers, the shock term is overestimated by a factor of two (see, for example, Tzoar and Jain 1981, Anderson and Lisak 1983, Bourkoff *et al.* 1987, Dianov *et al.* 1989c), and only the real part of the nonlinear susceptibility is taken into account. The imaginary Raman part of the shock term describes the Stokes losses. We will show later (Section 12) the importance of the imaginary Raman part of the shock term for femtosecond soliton propagation.

It should be emphasized that Equations 61 and 62 conserve the photon number of the signal:

$$\frac{\partial}{\partial z}\left[\int \frac{n_{\text{eff}}(\omega)\,S(\omega)|E(\omega,z)|^2}{\omega}\,d\omega\right] = 0 \tag{63}$$

11   Response functions

Equations 61 and 62 contain a nonlinear response function. In order to solve such equations, one has to determine this function from the available experimental data on the nonlinear refractive index coefficient n_2 and the Raman gain curve $g_R(\Delta\omega)$ (Stolen *et al.* 1989, Golovchenko *et al.* 1989).

The third-order nonlinear susceptibility of fused silica in the near infra-red and visible spectral regions can be represented as a sum of a nonresonant electronic non-linear susceptibility and a resonant nonlinear susceptibility associated with molecular vibrations (Raman susceptibility):

$$\chi^{(3)}(\Delta\omega) = \chi_{NR}^{(3)} + \chi_R^{(3)}(\Delta\omega) \tag{64}$$

Here $\chi_{NR}^{(3)}$ is suggested to be real and independent of $\Delta\omega$ (this is a good approximation for visible and infra-red spectral regions, because the electron absorption lies in the ultra-violet region). The imaginary part of the resonant susceptibility is antisymmetric in $\Delta\omega$. It determines the Raman gain coefficient:

$$g_R(\Delta\omega) = \mathrm{Im}\left[\chi_R^{(3)}(\Delta\omega)\right]\frac{4\pi\omega_0}{cn_0} \tag{65}$$

The spectral dependence of the Raman gain coefficient for fused silica $g_R(\Delta\omega)$ is known from literature (see Figure 8), so using the Kramers-Krönig relations, the real part of the Raman contribution to the nonlinear susceptibility can also be obtained (Figure 14). From the expression for the nonlinear refractive index

$$n_2 = \frac{2\pi}{n_0}\left[\chi_{NR}^{(3)} + \mathrm{Re}(\chi_R^{(3)}(0))\right] \tag{66}$$

and from the known value of n_2, one can find $\chi_{NR}^{(3)} = 4.5\,\mathrm{Re}(\chi_R^{(3)}(0))$.

As a result we have calculated the spectral dependence of the third-order nonlinear susceptibility from the known experimental data on n_2 and $g_R(\Delta\omega)$. Now we can solve the nonlinear propagation equation (Equation 61) in the spectral domain. If we want to work in the time domain, we must calculate the nonlinear response function, which is the inverse Fourier transform of the nonlinear susceptibility. After the inverse Fourier transformation of Equation 64, we obtain the response function $F(t)$:

$$F(\tau) = (1 - \beta)\,\delta(\tau) + \beta f(\tau) \tag{67}$$

The first term, which is proportional to the delta-function $\delta(\tau)$, is determined by the electron contribution, while the second term is determined by the Raman contribution. The functions $F(t)$ and $f(t)$ are normalised so that their integrals are a unit. From $\beta/(1-\beta) = \mathrm{Re}(\chi_R^{(3)}(0))/\chi_{NR}^{(3)}$, we obtain the parameter $\beta=0.18$. Figure 15 shows the Raman response function of fused silica $f(t)$.

For an adequate description of nonlinear light propagation through optical fibres, one should use the actual Raman gain curve (and the response function corresponding to it). Nevertheless most theoretical papers, in fact, use different approximated forms of the response function. As a rule, the degree of accuracy of these approximations is not discussed. In connection with this, it is worthwhile considering some of the approximations used in the literature and comparing their accuracy for different experimental situations. For simplicity and clearness, in this section we shall not take into account the higher-order dispersive and nonlinear effects (*i.e.* the last term in Equation 62 will be neglected). In this case, taking into account Equation 67, the propagation equation is simplified to:

$$\left[\frac{\partial}{\partial\xi} - i^{1/2}\frac{\partial^2}{\partial\tau^2} - \gamma\frac{\partial^3}{\partial\tau^3}\right]\Psi(\tau,\xi)$$

$$= i(1 - \beta)|\Psi(\tau,\xi)|^2\Psi(\tau,\xi) + i\beta\Psi(\tau,\xi)\int|\Psi(\tau - \tau',\xi)|^2 f(\tau')\,d\tau' \tag{68}$$

Expanding $|\Psi(\tau-\tau',\xi)|^2$ using a Taylor series around time τ and neglecting the second- and higher-order terms, we obtain the 'linear approximation' equation (Gordon 1986):

$$\left[\frac{\partial}{\partial\xi} - i^{1/2}\frac{\partial^2}{\partial\tau^2} - \gamma\frac{\partial^3}{\partial\tau^3}\right] = i|\Psi(\tau,\xi)|^2\Psi(\tau,\xi) + it_R\Psi(\tau,\xi)\frac{\partial|\Psi(\tau,\xi)|^2}{\partial\tau} \tag{69}$$

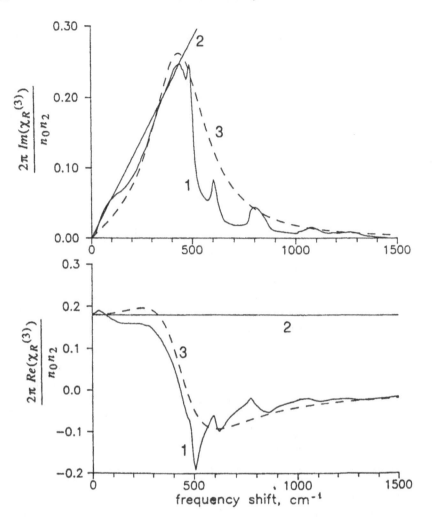

Figure 14. *Dependencies of the (a) imaginary and (b) real parts of the Raman non-linearity as a function of $\Delta\omega$. The imaginary part is antisymmetric in $\Delta\omega$, while the real part is symmetric. Curve 1 is the actual Raman curve, curve 2 is the linear approximation and curve 3 is the Lorentzian approximation.*

where the parameter t_R is determined as:

$$T_R = t_R t_0 = t_0 \beta \int f(\tau)\tau d\tau = \frac{2\pi}{n_0 n_2}\left[\frac{d(\mathrm{Im}(\chi_R^{(3)}(\Delta\omega)))}{d(\Delta\omega)}\right]_{\Delta\omega=0} \tag{70}$$

Equation 69 in fact suggests a linear dependence of the Raman gain coefficient on the spectral shift and an independence of the real part of $\chi_R^{(3)}(\Delta\omega)$ from the spectral shift $\Delta\omega$. One can see that the parameter T_R is proportional to the derivative of the actual Raman gain curve at the point of $\Delta\omega=0$. When using the linear approximation in practice, it is usually suggested that the Raman gain is a straight line which goes

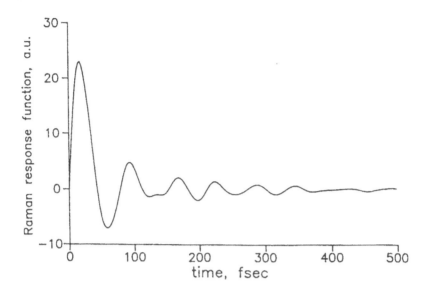

Figure 15. *The Raman response function $f(t)$ of fused silica.*

through the zero point and the point of maximum value of the actual Raman curve (see Figure 14a, curve 2) (Gordon 1986). In this case, the parameter $T_R = t_R t_0 \sim 2.8 - -3\text{fs}$. (Note that in some papers (Haus and Nakazawa 1987, Islam *et al.* 1989, Agrawal 1990), the parameter T_R was considered to be $T_R = 6\text{fs}$. We believe that this is a two-fold overestimated value, because when calculating T_R (Haus and Nakazawa 1987), the *intensity* Raman gain coefficient was substituted for the *field* Raman gain coefficient.) The first drawback of the linear approximation is obvious: it suggests a strong Raman interaction among spectral components separated by more than 500cm^{-1}, while in fact this interaction is almost absent (see Figure 14). It is also obvious that the linear approximation cannot describe a cascaded SRS. The linear approximation does not take into account the spectral dependence of the real part of the Raman nonlinear susceptibility $\text{Re}(\chi^{(3)}(\Delta\omega))$ (it is considered to be frequency-independent). As a result, this model does not describe, in particular, the effect of the increase of 'pulse area' for short solitons (Stolen and Tomlinson 1992, Mamyshev and Chernikov 1992).

Let us consider the effect of the soliton pulse area increase in more detail. According to the NSE model, the soliton parameters and the fibre parameters are related to each other through Equation 49. This means that the pulse area of NSE solitons (defined as follows) equals a unit:

$$\frac{|E_0|^2 \, t_0^2 n_2 2\pi}{\lambda \left| \dfrac{\partial^2 k}{\partial \omega^2} \right|} = 1 .$$

In fact, the NSE model suggests the soliton spectrum to be narrow enough so that parametric interaction (self-phase modulation effect) is described by the nonlinear refractive index n_2 (remember that n_2 represents the nonlinear susceptibility for *zero frequency shift* $\Delta\omega$, see Equation 66). For short pulses, however, the assumption that the pulse

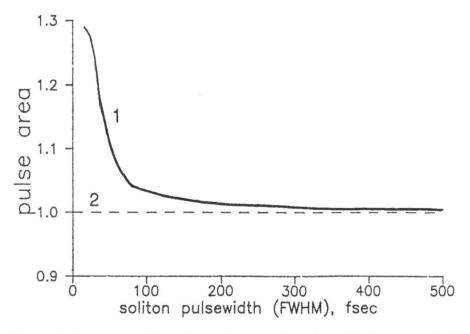

Figure 16. *Dependence of the soliton pulse area on the pulse width. For curve 1 (solid), the actual Raman curve is used, while for curve 2 (dashed), the linear approximation of the Raman curve is used.*

spectrum is narrow is no longer valid. When a more accurate relation (Equation 68) is used (taking into account the actual dependence of $\mathrm{Re}(\chi^{(3)}(\Delta\omega))$), the soliton pulse area (formally determined above) becomes greater than 1 for short pulses (Figure 16). Qualitatively, this can be explained as follows: the 'effective n_2' decreases when the soliton spectrum becomes broader, and to compensate for this decrease, the soliton pulse intensity increases.

Note that the spectral dependence of the real part of the nonlinear susceptibility also affects the process of modulational instability (Golovchenko *et al.* 1990b).

Another approximation of the Raman gain curve used in the literature is approximation by a Lorentzian line (Blow and Wood 1989). The normalised Raman response function in the Lorentzian approximation is:

$$f(t) = \frac{\tau_1 + \tau_2}{\tau_1 \tau_2^2} \exp(-t/\tau_2) \, \sin(t/\tau_1) \tag{71}$$

Parameters τ_1=0.0122ps and τ_2=0.032ps are chosen to fit the actual Raman line in the most optimal way (see Figure 14a, curve 3). The Lorentzian approximation describes the Raman gain for high frequency shifts and the increase of the 'pulse area' of short soliton pulses much better than the linear approximation.

Consider now how different approximations describe the rate of the soliton self-frequency shift (SSFS) $d\omega_{\mathrm{mean}}/dz$ caused by the RSS effect. For the 'linear' approxima-

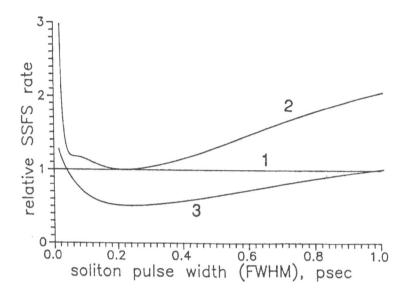

Figure 17. *Dependence of the relative soliton self-frequency shift rate (determined as $(\mathrm{d}\omega_{\mathrm{mean}}/\mathrm{d}z)/(\mathrm{d}\omega_{\mathrm{mean}}/\mathrm{d}z)_{\mathrm{actual}})$ on the soliton pulse width. Curve 1—actual Raman curve is used in simulations; curve 2—linear approximation; curve 3—Lorentzian approximation.*

tion, the SSFS rate is:

$$\frac{\mathrm{d}\omega_{\mathrm{mean}}}{\mathrm{d}z} = -\frac{8}{15} T_{\mathrm{R}} \left| \frac{\partial^2 k}{\partial \omega^2} \right| t_0^{-4} \tag{72}$$

For the other models, the SSFS rate can be calculated by solving Equation 68 numerically, or by using the following formula (Gordon 1986):

$$\frac{\mathrm{d}\omega_{\mathrm{mean}}}{\mathrm{d}z} = -\frac{2.7\pi}{t_0^4} \left| \frac{\partial^2 k}{\partial \omega^2} \right| \int \frac{x^4 \, \mathrm{Im}(\varepsilon(x/t_0)) \, dx}{\sinh^2 (x\pi/3.526)(x/t_0)} \tag{73}$$

where

$$\varepsilon(\omega) = \beta \int f(t) \exp (i\omega t) \, dt \,.$$

(Note that this expression does not take into account the effect of the pulse area increase.) Note also that Equation 72 represents a particular case of Equation 73. Figure 17 shows dependencies of normalised SSFS rates calculated in different approximations as a function of the soliton pulsewidth $\tau_0 = 1.763t_0$. The SSFS rates are normalised on the SSFS rate calculated using the Raman curve: relative SSFS rate $\equiv (\mathrm{d}\omega_{\mathrm{mean}}/\mathrm{d}z)/(\mathrm{d}\omega_{\mathrm{mean}}/\mathrm{d}z)_{\mathrm{actual}}$. The degree of deviation of the curves from unity shows the degree of inaccuracy of the models. One can see that for short soliton durations, the linear approximation gives considerably overestimated values for the SSFS rate. This fact is explained as follows: short soliton durations correspond to broad spectra. As mentioned above, the linear approximation considerably overestimates the Raman interaction among spectral components separated from each other by 500cm^{-1} or more. The Lorentzian approximation also gives overestimated values for the Raman gain for large frequency shifts (though to less of a degree than the linear approximation), so it also overestimates the SSFS rate for short (<40fs) solitons.

The SSFS rate of long solitons (with narrow spectral bandwidth) is determined by the slope of the curve $\text{Im}(\chi_R^{(3)}(\Delta\omega))$ near zero frequency shifts $\Delta\omega$. One can see from Figure 14a that the slope of curve 2 (linear approximation) is greater than that of the actual Raman curve (curve 1). As a result, the linear approximation gives an over-estimated SSFS rate for solitons with pulse durations of more than 300fs (Figure 17). The Lorentzian approximation gives underestimated values of the Raman interaction $\text{Im}(\chi_R^{(3)}(\Delta\omega))$ for frequency shifts $\Delta\omega/(2pc)$ in the approximate interval 20–300cm^{-1} (Figure 14). This explains why the Lorentzian model considerably underestimates the SSFS rate of solitons with pulse durations in the 40fs–1ps range (Figure 17, curve 3). Note that this underestimation is about two-fold in the region of 100–500fs. One can see that in some cases utilization of the linear approximation is more preferable to the Lorentzian one—particularly for the case of a single isolated fundamental soliton. How-ever, in some other situations, the linear approximation can give qualitatively incorrect results (for example, as already mentioned, it cannot describe a cascaded SRS).

Some other papers which provide a theoretical description of the RSS effect in fibres should be mentioned. A relaxation-type equation for the nonlinear addition to the re-fractive index was used in Dianov *et al.* (1989c) and in Haus and Nakazawa (1987) to describe the RSS effect. Similar to the linear approximation, this model gives consider-ably overestimated values of the Raman amplification for large spectral shifts, and does not describe the actual spectral dependence of the real part of the nonlinear susceptibil-ity. Such an approach, which uses a relaxation-type equation for the nonlinear addition to the refractive index, seems to be inappropriate for the case of silica glass—because the nonlinear index consists of electron and Raman parts. Moreover, the 'relaxation' model causes misunderstanding in the physical interpretation of the phenomenological parameters of the model. In fact, in the paper by Dianov *et al.* 1989c, the parameter T_R was regarded as the relaxation time of the electronic nonlinearity. But as one can see from Equation 70, T_R is not the relaxation time of the electronic nonlinearity: T_R is determined mainly by the Raman nonlinearity. Moreover, it is not a relaxation time at all, but rather just a parameter which has the dimension of time.

In the paper by Afanasyev *et al.* (1990), a system of two coupled equations—a NSE-type equation for the electric field envelope, and a classical oscillator-type equation for the molecular vibrations—was used to describe the RSS effect. Such an approach is, in fact, identical to the Lorentzian model. Nevertheless it should be noted that in this paper, Afanasyev *et al.* did not take into account the fact that the nonlinear index n_2 contains not only the electronic part, but also the Raman part. As a result the electronic nonlinearity was overestimated by about 20%.

12   Higher-order effects

Let us consider the importance of the influence of higher-order nonlinear and dispersive effects. To demonstrate the importance of the higher-order effects described by the last term in Equation 62, first consider the propagation of a 40fs (FWHM) soliton-like pulse in a lossless dispersion-flattened fibre (the third-order and higher-order dispersion

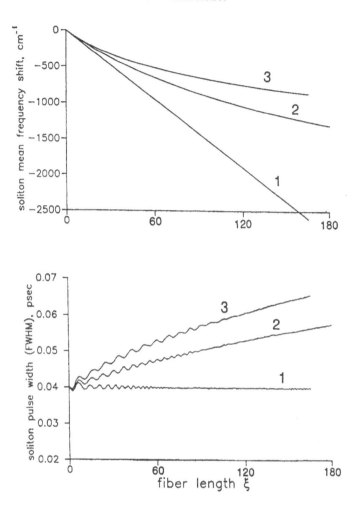

Figure 18. *Evolution of the soliton mean frequency shift and the soliton pulse width as a function of the propagation distance ξ. Model 1 does not take into account the higher-order effects (the last term in Equation 62 is neglected); model 2 takes into account the higher-order effects but assumes that the fibre mode area is frequency-independent ($S(\omega)$=const in Equation 62); model 3 takes into account all higher-order effects including the spectral dependence of the fibre mode area.*

is zero) using different models (Figure 18, curves 1–3). Model 1 does not take into account the higher-order effects (that is, the last term in Equation 62 is neglected). Model 2 takes into account the higher-order effects but assumes the fibre mode area to be independent of the wavelength ($S(\omega)$ = const in Equation 62). Model 3 takes into account all the higher-order nonlinear effects (including the fibre mode area dependence on wavelength). The actual Raman gain curve is used in all three cases. One can see the great importance of higher-order effects. Without taking into account these effects (model 1), the soliton pulsewidth remains unchanged during the pulse propagation, the mean frequency being shifted to the Stokes region linearly with length—due to the RSS effect. Nevertheless, the self-frequency shift should lead (via higher-order effects, models 2 and 3) to a decrease of the nonlinearity (in comparison with the dispersion) and to the Stokes losses associated with material excitation. As a result, the pulse broadens, and the soliton self-frequency shift rate decreases considerably. For example, the models give the following results at a fibre length of $\xi=120$ (for a fibre dispersion $D = 10\text{ps/nm/km}$, this distance corresponds to a fibre length of 4.8m):

- **Model 1**: the pulsewidth remains unchanged, $\tau(\text{FWHM})=40\text{fs}$ and the mean frequency shift $\Delta\omega_{\text{mean}}=1900\text{cm}^{-1}$;

- **Model 2**: $\tau(\text{FWHM})=53\text{fs}$, $\Delta\omega_{\text{mean}}=1038\text{cm}^{-1}$;

- **Model 3**: $\tau(\text{FWHM})=61\text{fs}$, $\Delta\omega_{\text{mean}}=747\text{cm}^{-1}$.

One can conclude that to describe nonlinear broadband (about 100cm^{-1} and more) light propagation, and to describe the evolution of solitons that spectrally shift $\sim100\text{cm}^{-1}$ or more during their propagation, it is absolutely necessary to take into account the higher-order nonlinear effects.

The effect of third-order dispersion on short soliton propagation is also important. When the mean frequency of the soliton shifts to the Stokes spectral region due to the RSS effect, the second-order dispersion at the carrier frequency changes due to the third-order dispersion. In the adiabatic approximation, without taking into account the higher-order nonlinear effects (in the frame of the 'linear' model, Equation 69), this leads to the following dependence of the soliton pulse width as a function of the propagation distance z:

$$t_0(z) = t_0(0)\left[1 + \frac{32\,T_{\text{R}}\,z}{15\,t_0^4(0)}\frac{\partial^3 k}{\partial\omega^3}\right]^{1/4} \tag{74}$$

To simulate real experimental situations where broad light spectra are involved, one should take into account the higher-order nonlinear and dispersive effects at the same time (Equation 62). Note that in some cases the dispersion effects of the fourth and higher orders need to be taken into account (Mamyshev *et al.* 1993).

Appendix: typical pump power levels needed to observe nonlinear processes in fibres

In this section, we will estimate the pump power levels that are required to observe different nonlinear processes in silica optical fibres. As a typical example, we take a single-mode non-polarisation-preserving fibre with an effective mode area $S_{\text{eff}}=50\mu\text{m}^2$ at a wavelength $\lambda=1.55\mu\text{m}$. The fibre loss coefficient is considered to be $\alpha=0.048\text{km}^{-1}$ (0.21db/km), which corresponds to an effective length $L_{\text{eff}}=20.8\text{km}$ (see Equations 10 and 11).

Stimulated Raman scattering

The P_R is determined here as the CW pump power required to observe saturated SRS at a given fibre length:

$$g_R \frac{P_R}{S_{eff}} L = 16$$

(see Equation 29). Using the value of $g_R = 3.5 \times 10^{-12}$ cm/W for the non-polarisation-preserving case, we get

$$P_R = 1.1 \, W$$

For the case of polarisation-preserving fibres, the Raman gain coefficient is two times higher, and the P_R is twice as small.

Stimulated Brillouin scattering

Similar to the case of SRS, the SBS 'threshold' power P_{SBS} is determined from:

$$g_{SBS} \frac{P_{SBS}}{S_{eff}} L = 20$$

Taking the value of $g_{SBS} = 2.25 \times 10^{-9}$ cm/W, we get

$$P_{SBS} = 2.1 mW$$

For the polarisation-preserving fibre, this value is twice as small.

Self-phase modulation

The 'self-phase modulation power' P_{SPM} is determined here as the power which corresponds to a nonlinear phase shift of one rad at the fibre output:

$$\frac{2\pi}{\lambda} n_2 \frac{P_{SPM}}{S_{eff}} L = 1$$

Using the averaged value of n_2 for the non-polarisation-preserving case,

$$[n_2]_{average} = \frac{8}{9} [n_2]_{linearpol} = 2.6 \times 10^{-16} cm^2/W \, ,$$

we get

$$P_{SPM} = 22 mW$$

Modulational instability

The modulational instability 'threshold' power P_{MI} is determined from Equation 22:

$$\frac{4\pi}{\lambda} n_2 \frac{P_{MI}}{S_{eff}} L = 16$$

Using $n_2 = 2.6 \times 10^{-16} cm^2/W$, we get

$$P_{MI} = 180 \, mW$$

Solitons

The parameters of fundamental solitons are determined by Equation 49. For the case of a soliton with a pulse width of $\tau_0(\text{FWHM})=1.763t_0=20\text{ps}$, and a fibre dispersion of $D=0.5\text{ps/nm/km}$, the soliton peak power is 2.4mW, while the soliton dispersion length (see Equation 35) is $z_d=2001\text{km}$. Note that the soliton peak power scales as

$$\frac{D}{\tau_0^2}$$

(*i.e.* is inversely proportional to the dispersion length z_d). For example, for a soliton with $\tau_0=200\text{fs}$ in a fibre with $D=15\text{ps/nm/km}$, the soliton peak power is 720W and the dispersion length is 0.67m.

References

Afanasyev V V, Vysloukh V A, and Serkin V N, 1990, *Opt Lett* **15** 489.

Agrawal G P, 1990, *Opt Lett* **15** 224.

Anderson D and Lisak M, 1983, *Phys Rev A* **27** 1393.

Basiev T T, Dianov E M, Karasik A Ya, Luchnikov A V, Mirov S B, and Prokhorov A M, 1982, *Pis'ma Zh Eksp Teor Fiz* **36** 85 [*JETP Lett* **36** 104].

Blow K J and Wood D, 1989, *IEEE J Quantum Electron* **25** 2665.

Bogatyrev V A, Bubnov M M, Dianov E M, Kurkov A S, Mamyshev P V, Prokhorov A M, Rumyantsev S D, Semeonov V A, Semeonov S L, Sysoliatin A A, Chernikov S V, Gurianov A N, Devyatykh G G, and Miroshnichenko S I, 1991, *J Lightwave Technol* **LT-9** 561.

Bourkoff E, Zhao W, Joseph R I, and Christodoulides D N, 1987, *Opt Commun* **62** 284.

Chernikov S V and Mamyshev P V, 1991, *J Opt Soc Am B* **8** 1633.

Chernikov S V, Mamyshev P V, Dianov E M, Richardson D J, L\,ming R I, and Payne D N, 1992, *Sov Lightwave Commun* **2** 161.

DeMartini F, Townes C H, Gustafson T K, and Kelley P L, 1967 *Phys Rev* **164** 312.

Dianov E M, Zakhidov E A, Karasik A Ya, Mamyshev P V, anu Prokhorov A M, 1982, *Zh Eksp Teor Fiz* **83** 39 [*Sov Phys JETP* **56** 21].

Dianov E M, Karasik A Ya, Mamyshev P V, Onischukov G I, Prokhorov A M, Stel'makh M F, and Fomichiv A A, 1984a, *Kvantovaya Elektronika (Moscow)* **11** 1078 [*Sov J Quantum Electron* **14** 726].

Dianov E M, Karasik A Ya, Mamyshev P V, Onischukov G I, Prokhorov A M, Stel'makh M F, and Fomichiv A A, 1984b, *Pis'ma Zh Eksp Teor Fiz* **39** 564 [*JETP Lett* **39** 691].

Dianov E M, Karasik A Ya, Mamyshev P V, Onischukov G I, Prokhorov A M, Stel'makh M F, and Fomichev A A, 1984c, *Pis'ma Zh Eksp Teor Fiz* **40** 148 [*JETP Lett* **40** 903].

Dianov E M, Karasik A Ya, Mamyshev P V, Prokhorov A M, Serkin V N, Stel'makh M F, and Fomichev A A, 1985a,*Pis'ma Zh Eksp Teor Fiz* **41** 242 [*JETP Lett* **41** 294].

Dianov E M, Karasik A Ya, Mamyshev P V, Prokhorov A M, and Serkin V N, 1985b, *Zh Eksp Teor Fiz* **89** 781 [*Sov Phys JETP* **62** 448].

Dianov E M, Ivanov L M, Karasik A Ya, Mamyshev P V, and Prokhorov A M, 1986a, *Zh Eksp Teor Fiz* **91** 2031 [*Sov Phys JETP* **64** 1205].

Dianov E M, Nikonova Z S, Prokhorov A M, and V N Serkin, 1986b, *Pis'ma Zh Tekhn Fiz* **12** 756 [*Sov Techn Phys Lett* **12** 311].

Dianov E M, Karasik A Ya, Mamyshev P V, Prokhorov A M, and Fursa D G, 1987a, *Kvantovaya Elektronika (Moscow)* **14** 662 [*Sov J Quantum Electron* **14** 415].

Dianov E M, Mamyshev P V, Prokhorov A M, and Fursa D G, 1987b, *Pis'ma Zh Eksp Teor Fiz* **46** 383 [*JETP Lett* **46** 482].

Dianov E M, Ivanov L M, Mamyshev P V, and Prokhorov A M, 1989a, *IEEE J Quantum Electron* **25** 828.

Dianov E M, Ivanov L M, Mamyshev P V, and Prokhorov A M, 1989b, in *Topical Meeting on Nonlinear Guided-Wave Phenomena: Physics and Applications, Houston, Texas, Technical Digest Series* **2** 157 (OSA, Washington DC).

Dianov E M, Grudinin A B, Khaidarov D V, Korobkin D V, Prokhorov A M, and Serkin V N, 1989c, *Fibre and Integrated Optics* **8** 61.

Dianov E M, Abramov A A, Bubnov M M, Shipulin A V, Semionov S L, Schebuniaev A G, Gurianov A N, and Khopin V F, 1995, in Optical Amplifiers and their Applications topical meeting, *OSA Technical Digest Series* **18** 189.

Fork L R, Brito Cruz C H, Becker P C, and Shank C V, 1987, *Opt Lett* **12** 483.

Fursa D G, Mamyshev P V, and Prokhorov A M, 1992, *Sov Lightwave Commun* **2** 59.

Gloge D, 1971, *Appl Opt* **10** 2252.

Golovchenko E A, Dianov E M, Mamyshev P V, and Prokhorov A M, 1988, *Opt Quantum Electron* **20** 343.

Golovchenko E A, Dianov E M, Karasik A Ya, Mamyshev P V, Pilipetskii A N, and Prokhorov A M, 1989, *Kvantovaya Elektronika (Moscow)* **16** 592 [*Sov J Quantum Electron* **19** 391].

Golovchenko E A, Mamyshev P V, Prokhorov A M, and Fursa D G, 1990a, *J Opt Soc Am B* **7** 172.

Golovchenko E A, Mamyshev P V, Pilipetskii A N, and Dianov E M, 1990b, *IEEE J Quantum Electron* **26** 1815.

Golovchenko E A, Mamyshev P V, Pilipetskii A N, Dianov E M, and Prokhorov A M, 1991, *J Opt Soc Am B* **8** 1626.

Gordon G P, 1986, *Opt Lett* **11** 662.

Grischkovsky D and Balant A C, 1982, *Appl Phys Lett* **41** 1.

Grubb S G, 1995, in *Optical Amplifiers and their Applications topical meeting*, *OSA Technical Digest Series* **18** 186.

Grubb S G, Strasser T, Cheung W Y, Reed W A, Mizrahi V, Erdogan T, Lemaire P J, Vengsarkar A M, DiGiovanni D J, Peckham D W, and Rockney B H, 1995, in *Optical Amplifiers and their Applications topical meeting*, *OSA Technical Digest Series* **18** 197.

Halas N J and Grischkovsky D, 1986, *Appl Phys Lett* **46** 823.

Hasegawa A and Tappert F, 1973, *Appl Phys Lett* **23** 142.

Hasegawa A, 1984, *Opt Lett* **9** 288.

Haus H A and Nakazawa M, 1987, *J Opt Soc Am B* **4** 652.

Islam M N, Sucha G, Bar-Joseph I, Wegener M, Gordon J P, and Chemla D S, 1989, *J Opt Soc Am B* **6** 1149.

Kim K S, Stolen R H, Reed W A, and Quoi K W, 1994, *Opt Lett* **19** 257.

Kuckartz M, Schultz R, and Hard H, 1988, *J Opt Soc Am B* **5** 1353.

Lin C and Bosch M A, 1981, *Appl Phys Lett* **38** 479.

Lin C, Reed W A, Pearson A D, and Shang H T, 1981, *Opt Lett* **6** 493.

Mamyshev P V and Chernikov S V, 1990, *Opt Lett* **15** 1076.

Mamyshev P V, Chernikov S V, Dianov E M, and Prokhorov A M, 1990, *Opt Lett* **15** 1365.

Mamyshev P V, 1991, *Izv Acad Nauk, SerPhys* **55** 374 [*Bull Acad Sci USSR, Phys Ser* **55** No. 2].

Mamyshev P V, Chernikov S V, and Dianov E M, 1991, *IEEE J Quantum Electron* **27** 2347.

Mamyshev P V and Chernikov S V, 1992, *Sov Lightwave Commun* **2** 97.

Mamyshev P V, Vertikov A P, and Prokhorov A M, 1992, *Sov Lightwave Commun* **2** 73.

Mamyshev P V, 1992, in *Optical Solitons—Theory and Experiment*, chapter 8, ed Taylor J R (Cambridge University Press, UK).

Mamyshev P V, Wigley P G J, Wilson J, Stegeman G I, Semenov V A, Dianov E M, and Miroshnichenko S I, 1993, *Phys Rev Lett* **71** 73.

Mitschke F M and Mollenauer L F, 1986, *Opt Lett* **11** 659.

Mollenauer L F, Stolen G P, and Gordon G P, 1980, *Phys Rev Lett* **45** 1095.

Mollenauer L F, Stolen R H, Gordon J P, and Tomlinson W J, 1983, *Opt Lett* **8** 289.

Nikolaus B, Grischkovsky D, and Balant A C, 1983, *Opt Lett* **8** 189.

Smith R G, 1972, *Appl Opt* **11** 2489.

Smith K and Mollenauer L F, 1989, *Opt Lett* **14** 751.

Stolen R H, 1975, *IEEE J Quantum Electron* **11** 100.

Stolen R H and Lin C, 1978, *Phys Rev A* **17** 1448.

Stolen R H, 1979, Nonlinear properties of optical fibres, in *Optical Fibre Telecommunications* p125 (Academic, New York).

Stolen R H, 1980, *Fibre and Integrated Optics* **1** 21.

Stolen R H, Bosch M A, and Lin C, 1981, *Opt Lett* **6** 213.

Stolen R H, Botineau J, and Ashkin A, 1982, *Opt Lett* **7** 512.

Stolen R H and Jonhson A M, 1986, *IEEE J Quantum Electron* **22** 2154.

Stolen R H, Gordon J P, Tomlinson W J, and Haus H A, 1989, *J Opt Soc Am B* **6** 1159.

Stolen R H and Tomlinson W J, 1992, *J Opt Soc Am B* **9** 565.

Tai K, Tomita A, Jewell J L, and Hasegawa A, 1986, *Appl Phys Lett* **49** 236.

Tomlinson W J, Stolen P H, and Shank C V, 1984, *J Opt Soc Am B* **1** 139.

Tzoar N and Jain M, 1981, *Phys Rev A* **23** 1266.

Vertikov A P, Mamyshev P V, and Prokhorov A M, 1991, *Sov Lightwave Commun* **1** 363.

Vertikov A P and Mamyshev P V, 1992, *Sov Lightwave Commun* **2** 119.

Weiner A M, Heritage J P, and Stolen R H, 1988, *J Opt Soc Am B* **1** 364.

Zakharov V A and Shabat A B, 1971, *Zh Eksp Teor Fiz* **61** 118 [*Sov Phys JETP* **34** 62].

Electromagnetically Induced Transparency

Malcolm H Dunn

University of St Andrews, Scotland

1 Introduction

Following much theoretical speculation, the first experiments on electromagnetically induced transparency (EIT) were carried out in the Steve Harris group at Stanford University in 1991 (Boller *et al.* 1991, Field *et al.* 1991). Just what this remarkable effect is may be best appreciated by going back a little way and examining the ideas behind conventional absorption, because this is the context in which we must place the effect of induced transparency. The idea of resonant absorption is one that is widely familiar. A transition in which the lower level (often a ground state) is more heavily populated than the upper level exhibits absorption of radiation whose frequency corresponds closely to the energy gap between the states (see Figure 1). The energy lost by the radiation field through absorption is taken up by the atoms when they are excited from the lower to the upper state. If the atoms then lose this energy of excitation in some random way, for example by undergoing spontaneous emission where the radiation is emitted in all directions and not just in the direction of propagation of the radiation that was initially absorbed, then the absorption process is sustained through many cycles of the atom absorbing a photon and emitting it subsequently in a random direction compared to the direction of the original radiation. If the radiation is not too strong, the population of the atoms in the lower state is little changed, and the absorption coefficient is then independent of the radiation intensity. Associated with the absorption profile of the medium, there is also a dispersion profile (*i.e.* variation of refractive index of the

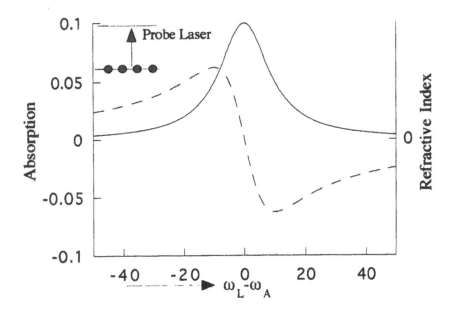

Figure 1. *Frequency dependences of the absorption (solid line) and dispersion (dashed line) associated with a transition from a populated lower state. ω_L is the frequency of the radiation, and ω_A that associated with the energy gap of the transition.*

medium with frequency), and this effect, known as anomalous dispersion, is also shown in Figure 1. The dispersion relations (Kramers-Kronig relations) tell us how to derive the refractive index frequency profile from the absorption profile. The two are related through deep principles to do with causality. These relations are of course universal, and do not just apply to the particular absorption profile that we are considering explicitly here. An important point to keep in mind is that if we change one, e.g. absorption, then we also change the other, *i.e.* dispersion, in a defined and predictable fashion.

As the radiation field intensity is progressively increased, a point is reached where the excitation (absorption) process begins to dominate the random decay processes that establish the status quo. Atoms do not return quickly enough to the lower state, and a significant population begins to build in the upper state (see Figure 2), with the consequence that the absorption coefficient is thereby reduced. What is being described here is the saturation of the absorption. If the intensity is sufficiently high, a point can be reached where the population of the upper state begins to approach that of the lower state, when the absorption approaches zero, *i.e.* the absorption is said to be bleached. It should be apparent from this argument that it is not possible to do other than equalise the populations of the upper and lower states through the pumping effects of radiation acting directly between the states. However, if some additional excitation process to the upper state is brought into play, then it becomes possible to have more atoms in the upper state than the lower state, when the transition exhibits optical gain, and we move into the regime of an active gain medium appropriate to the case of the laser.

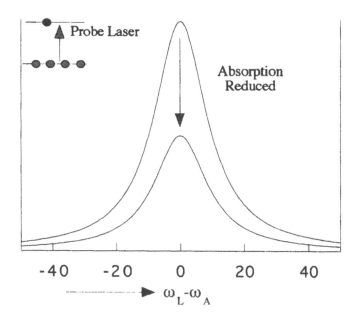

Figure 2. *Effect of increased radiation intensity on lower state population and hence absorption profile.*

The point of the discussion thus far is to emphasise firstly that absorption is reduced through population transfer, from the lower to the upper state of the transition, and secondly that this reduction in absorption becomes significant when the intensity exceeds some threshold value (the saturation intensity). Of course, to overcome absorption requires the expenditure of power in order to keep the population transfer in place, and to this extent there is always some absorption going on. However, for intensities in excess of the saturation intensity, it represents a small fractional loss of the high intensity of radiation then present.

This preamble enables us to appreciate more fully the remarkable nature of EIT. The process is illustrated in Figure 3. What it entails is the following. If a second radiation field is applied to the atom as shown, so as to couple the upper level of the first transition to a third level (three level atom), then, under the correct conditions, the absorption experienced on the original transition can be switched off, in a situation where, and these are the key points, firstly, there is no population transfer (*i.e.* the population of the lower level remains essentially unchanged), and secondly, the effect occurs for all intensities of the radiation field coupling the lower two levels. (We will refer to this field as the probe field, *i.e.* the field that probes the absorption, whereas the field coupling the upper pair of levels will be referred to as the coupling field.) Since there is little population transfer (there must of course be some taking place, for if the coupling field is to have any effect it must have population or more precisely amplitude probability in the states with which it can interact), then the process is a 'non-dissipative' one, in that neither the probe laser nor the coupling laser experience significant absorption.

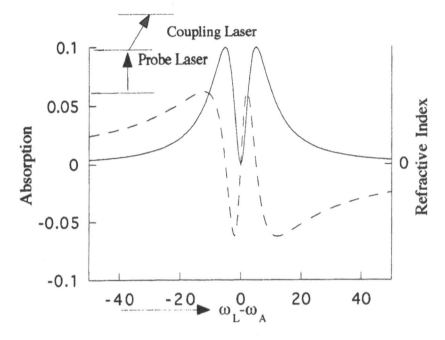

Figure 3. *Frequency dependences of the absorption (solid line) and refraction (dashed line) as seen by the probe laser when the upper state of the transition is coupled by the coupling laser field to a third state. The deep dip in the absorption profile is due to electromagnetically induced transparency.*

The effect is therefore totally different to the process of saturated absorption.

So far, we have considered what is known as the cascade scheme for EIT, in that the third level lies above the first two levels considered, so that one transition (the coupling transition) cascades into the other transition (the probe transition). However, there are other possibilities in which EIT can arise, and a further two (there are more) are shown in Figure 4, which introduces the idea of the lambda scheme and the vee scheme.

In so far as the absorption profile experienced by the probe laser is modified by the coupling laser, it follows that the dispersion (refractive index) profile is also modified, since the two are intimately related through the Kramers-Kronig dispersion relations. This is shown in Figures 1 and 3, where the change produced by the EIT process in the resonant absorption profile is reflected in a corresponding change in the anomalous dispersion profile. We will come back to discuss these ideas later on when we consider the process of electromagnetically induced focusing (EIF) (see Section 6).

From EIT it is not a great step to postulate the idea of inversionless lasing. If the absorptive effects of the lower state population can be diminished through the EIT process, then it is reasonable to speculate that any population deposited in the upper state (through some incoherent excitation process, for example) could lead to net gain occurring on the transition, even when the upper state population is less than the lower

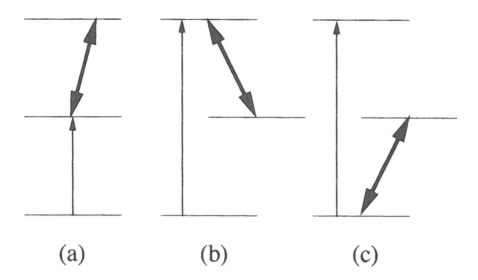

Figure 4. *Various level schemes in which it is possible to observe electromagnetically induced transparency. (a) cascade scheme, (b) the lambda scheme, and (c) the vee scheme. The heavy double-ended arrow designates the coupling laser field, and the light single-ended arrow the probe laser field.*

state population, *i.e.* in the absence of a population inversion. In order to confirm that such an effect could arise, it will be necessary to explore the contribution that an atom in the upper state can make to the amplification of the probe radiation field when the coupling field is also present. This point we will also return to later on (see Section 7). The concept of the inversionless laser is illustrated in Figure 5.

So far, we have described the effect of EIT in a phenomenological way, but have not yet given any hint of an explanation as to how it arises in terms of basic physical principles. The conventional, and indeed more powerful way from an analytical viewpoint, in which to proceed is to set up a density matrix model for the three level atom and to proceed from there. Such an approach makes it easy to average over the whole ensemble of atoms involved in the process, so allowing for atoms being excited randomly in time to the different states involved, and for the random decay of populations and coherences. However, although we will set up such a model later on (see Section 4), such an approach tends to obscure physical insight into what exactly is going on. For this reason, we are going to first of all adopt an approach that involves following the time evolution of a single three-level atom, created in a particular state at some defined time. By averaging over the atom's effective lifetime, we can calculate its net contribution to the radiation field (either the probe or the coupling field). By following how the three state amplitudes of the single atom evolve in time, and how they thereby contribute to the polarisations that drive the radiation fields, we can see how the process of EIT works. All this insight is lost in the density matrix modelling of the process, since although EIT emerges from the density matrix equations, it is extremely difficult to see why indeed it should.

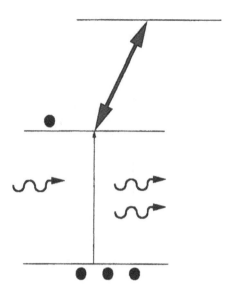

Figure 5. *An inversionless laser. Despite there being fewer atoms in the upper laser state (one dot) than the lower laser state (three dots), in the presence of the coupling laser field, designated by the heavy double-ended arrow, gain is experienced on the transition.*

However, before we go on to develop our single atom model in Section 3, we will first of all review a number of processes related to, but preceding in historical development, the discovery of EIT, as well as describing some of the classic EIT experiments that have now been reported.

2 Brief Historical Review

An early example of a situation in which absorption can be made to disappear in the presence of a lower state population is that of Fano interference, identified by Fano (1961). The situation involved in this case is that of photoionisation, as shown in Figure 6a. Photons photoionise from some lower (ground) state to the ionisation continuum. However, if the absorption cross-section is explored in the vicinity of an auto-ionising state embedded within the continuum, then a zero can occur in the absorption (and hence ionisation) process at particular detunings from exact resonance with the auto-ionising state, as shown in Figure 6b. The underlying explanation of this process is that the two routes to ionisation that are now present, namely directly to the continuum or via the auto-ionisation state, can interfere destructively under the appropriate conditions, so that no ionisation thereby occurs, and hence no absorption, despite the population remaining in the ground state.

In 1976, what has now become known as the 'Pisa' experiment was carried out (Arimondo *et al.* 1976, Alezetta *et al.* 1976, 1979). This is another experiment that involves interfering pathways leading to the absence of absorption. The generic situation

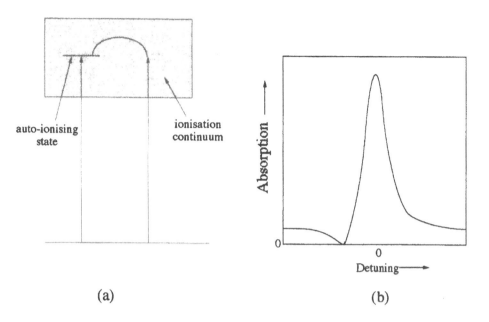

(a)   (b)

Figure 6. *Fano interference. (a) Energy states involved, (b) absorption as a function of detuning showing minimum (zero) in the absorption profile.*

is illustrated in Figure 7. Here there are two lower states, close enough together so as for both to be populated. Two radiation fields are present, one coupling one of the lower states to an upper state, the other field coupling the other lower state to the same upper state. Each radiation field acting separately would be expected to transfer lower state population to the upper state, the presence of which could be monitored through the occurrence of spontaneous fluorescence (perhaps to some other level) from the upper level. The finding of the 'Pisa' experiment was that when both radiation fields were present, a situation could arise under appropriate conditions when no fluorescence was observed to occur at all, even though each field induced fluorescence when applied separately. Again the explanation of the effect is that the two possible excitation pathways, now from different lower states but to the same upper state, interfere destructively in a quantum-mechanical sense, so as to lead to no net absorption occurring, hence no population being transferred to the upper state, and hence no fluorescence being observed. A particular variant of the experiment carried out by Gray, Whittley and Stroud (1978) is illustrated in Figure 8, and is based on the D1 line in sodium, with hyperfine splitting leading to the two lower states involved (F=1, 2). Theoretical and experimental results are illustrated in Figure 9. Here one of the lasers generating one of the radiation fields is held at a constant frequency, in this case on exact resonance with the transition, while the second laser is tuned to scan the frequency of the second radiation field across the other transition. The absence of fluorescence at the appropriate detuning, in this case zero, from the absorption line centre is clearly seen.

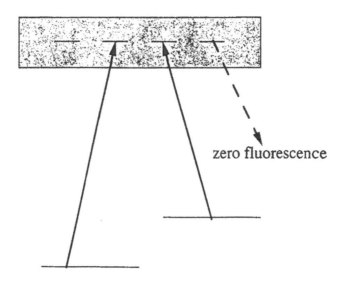

Figure 7. *Energy levels involved in population trapping - the 'Pisa' experiment*

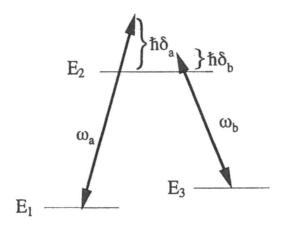

Figure 8. *Energy level scheme involved in the experiment of Gray, Whittley and Stroud.*

There is an important conclusion that we can reach in this particular type of experiment. This is that a coherently trapped population is set up in the ground states so that no population transfer takes place to the excited state. What exactly we mean by this statement will become clearer when we consider the single atom model of such

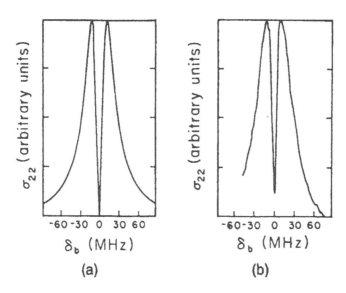

Figure 9. *Fluorescence as monitored from the state E2 for the case where one of the lasers is held on exact resonance, and the second laser is detuned. (a) theoretically predicted, (b) experimentally observed. From Gray, Whittley and Stroud (1978).*

effects later on (see Section 3). However, what is important to realise is that the coherence referred to here is within each atom, and not necessarily between atoms. In other words each and every atom is driven by the presence of the two radiation fields into a state which is a suitable coherent superposition of the amplitudes associated with the two ground states (the relative phases of the two amplitudes being important), and this state is then unaffected by the presence of the two radiation fields together, so that no probability transfer to the upper state takes place. We can look on this situation as one where the amplitude transfer from one of the lower states to the upper state, mediated by the appropriate radiation field, is of the same magnitude, but in exact antiphase to the amplitude transfer from the other lower state, as mediated by the other radiation field. The net effect is for no net amplitude transfer to the upper state to occur, and hence for the atom to remain trapped in the coherent superposition of the lower states. When looking at the collection of atoms as a whole, we can talk of a trapped population being present in the two lower states, but the important thing to recognise is that this is really a situation that occurs at the intra-atom level.

The first reported experiments in EIT were those of Harris and co-workers at Stanford (Boller *et al.* 1991). The initial experiments were carried out in strontium vapour with pulsed lasers providing the radiation fields, and involved auto-ionising transitions in a lambda arrangement. These were followed shortly thereafter by EIT experiments in lead vapour (Field *et al.* 1991) involving only bound states, and these are the ones described here. The energy level scheme is illustrated in Figure 10, and can be seen to be a cascade scheme. The laser used to generate the coupling radiation between the $^3D_1$ state and the $^3P_1$ state was a Nd:YAG laser operating at 1064nm, while the laser used to generate the probe radiation between the ground state $^3P_0$ and the $^3P_1$ state was a

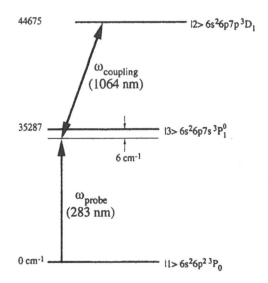

Figure 10. *Energy level scheme in neutral lead appropriate to the experiments of Field et al. (1991).*

frequency-doubled dye laser tuneable around 283nm. Typical experimental results are shown in Figure 11. The transmission of the probe radiation through the lead vapour is initially of the order of e^{-14}, but increases to e^{-4} in the presence of the coupling radiation, despite the high ground state population of lead atoms still being present in this latter case. As we shall see shortly, one way of looking at this cascade-type EIT experiment is in terms of population trapping discussed above. The combination of the probe and coupling radiation results in a trapped population occurring between the top-most state ($^3D_1$) and the ground state ($^3P_0$), or putting it more correctly, in each and every atom a coherent superposition state, involving these two states, is established by the two radiation fields. The wavefunction amplitudes generated in the intermediate state through the presence of the radiation fields coupling to the other two states making up the superposition state are of equal magnitude but of opposite sign, so that destructive interference occurs, and hence no net transfer of amplitude takes place. In other words there is a trapped population and transparency results on both the probe and coupling transition. The close relation between EIT and population trapping effects will be appreciated from this discussion. Of course, the maintenance of the coherence established between top and bottom states is vital to the EIT process, and anything that tends to diminish this coherence, for example by up-setting the phase relation established between the two states, will diminish the effectiveness of the EIT process. The relevance of population trapping to other EIT schemes will be discussed more fully later, but direct comparison of the 'vee' population trapping scheme described above (the 'Pisa' experiment) and the 'vee' EIT scheme will be readily appreciated.

The intimate relation between absorptive and refractive effects has already been stressed, and it is apparent that in so far as EIT alters the absorptive properties of a medium, it will also alter the refractive properties of that medium. This leads to

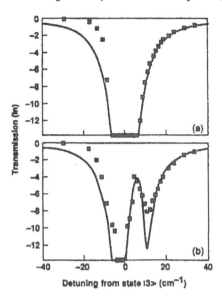

Figure 11. *Transmission through lead vapour as a function of probe laser detuning; (a) in the absence of the coupling laser field, and (b) with the coupling laser field present. From Field et al. (1991).*

some very interesting possibilities, as has been pointed out by Scully (1991). From Figure 1 it will be realised that in the usual case of anomalous dispersion, the largest change in refractive index due to the presence of the transition occurs at a frequency where there is still high absorption (in fact the absorption coefficient has only fallen to half its peak value by the point of maximum dispersion is reached for the case of a Lorentzian absorption profile). However, the dispersion profile is modified by the EIT process corresponding to the changes in the absorption profile. As has been pointed out by Scully, it is now possible, in a suitable EIT scheme, to experience high dispersion (point B on the dispersive curve in Figure 12) in the complete absence of absorption (point C on the absorption curve in Figure 12). It thus becomes possible, at least in principle, to contemplate a medium, of high particle density, that shows no significant absorption but huge refractive index. Such a medium has been called "phaseonium" by Scully.

3 State Amplitude Evolution in a Three Level Atom

We now develop a simple three level atom model to describe the effects discussed above. In this model we describe the time evolution of the state amplitudes for the case of a single atom subject to two, near-resonance radiation fields. From these amplitudes we can of course readily deduce the associated probability densities, which hence allows us to infer the state populations for a collection of such atoms, and also the polarisations associated with the atom, and hence its effect on the radiation fields themselves (*i.e.* absorption/gain/dispersion). This approach is particularly useful in gaining an insight

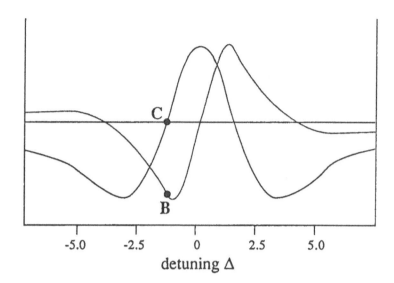

Figure 12. *Absorption (solid line) and dispersion (dotted line) in 'phaseonium', from Scully (1991). At the zero absorption point (C), there is a substantial contribution to the refractive index (B).*

into the origins of EIT at a fundamental level, and this is the main reason for us pursuing it here. However, such an approach becomes somewhat clumsy when averaging over large numbers of atoms excited at different times, as occurs in real situations, or when doppler broadening, for example, is to be taken into account. Then it is more appropriate to work in terms of the density matrix description. The density matrix is a more powerful computational tool when ensemble averages are involved, but a good deal of the insight is lost inside the black-box thereby created.

The three level atom is illustrated in Figure 13. The three levels have energy eigenvalues E_1 (=0), E_2 and E_3, associated eigenfunctions ψ_1, ψ_2, and ψ_3 respectively, and associated spontaneous decay rates (to other levels outside the system not explicitly considered) γ_1, γ_2, and γ_3 respectively. The coupling strength of the radiation fields between the two pairs of states are Ω_c for the coupling laser (between states 2 and 3), and Ω_p for the probe laser (between states 1 and 2), where the Ω's are the associated angular Rabi frequencies.

We proceed in the usual way by writing the wavefunction of the atom subject to the radiation fields as an expansion in terms of the unperturbed eigenfunctions, but where the associated coefficients are allowed to be time-dependent, namely

$$\Psi = a_1(t)\psi_1 + a_2(t)\psi_2 \exp(-i\omega_{21}t) + a_3(t)\psi_3 \exp(-i\omega_{31}t) \tag{1}$$

where ω_{21} and ω_{31} are the angular frequencies of the resonant transitions. We look for the above as a solution of the time-dependent Schrodinger equation

$$H\Psi = i\hbar \frac{\partial \Psi}{\partial t} \tag{2}$$

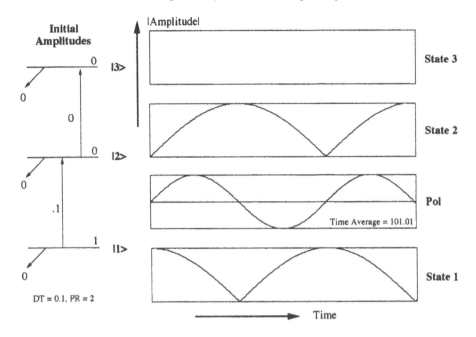

Figure 13. *Rabi flopping between states 1 and 2.*

where the Hamiltonian, H, is made up of two parts, namely that associated with the isolated atom, but including the phenomenological damping introduced through the decay rates γ_1, γ_2, and γ_3, and that describing the additional energy of the atom as a result of its interaction with the two time-varying radiation fields. This latter is treated in terms of the electric dipole approximation. The total Hamiltonian is hence

$$H = H_0 + exE_p \cos(\omega_p t) + exE_c \cos(\omega_c t), \tag{3}$$

where ω_p and ω_c are the angular frequencies and E_p and E_c the electric field amplitudes of the probe radiation field and the coupling radiation field respectively, e is the electronic charge, and x the (generalised) position of an electron in the atom. We proceed in the usual way by substituting for Ψ and H in equation (2) using equations (1) and (3) respectively. We then seek to recover three coupled time-dependent equations for a_1, a_2, and a_3 by multiplying through the resultant equation by ψ_1^*, ψ_2^*, and ψ_3^* respectively and integrating over the whole of space. We make the rotating wave approximation (neglect of high frequency terms) in the usual way, and further assume that the two radiation fields are on exact resonance with their associated transitions, i.e. that $\omega_p = \omega_{12}$, and $\omega_c = \omega_{32}$. (It is straightforward to generalise the expressions we obtain to the non-resonant case as well, but the simpler form of the equations in the case of exact resonances are adequate for our subsequent purpose here of elucidating the fundamental nature of EIT).

We hence obtain the three coupled equations describing the state amplitudes a_1, a_2, and a_3 respectively

$$\dot{a}_1 = -i\Omega_p a_2 - \gamma_1 a_1 \tag{4}$$

$$\dot{a}_2 = -i\Omega_p a_1 - i\Omega_c a_3 - \gamma_2 a_2 \tag{5}$$

$$\dot{a}_3 = -i\Omega_c a_2 - \gamma_3 a_3 \tag{6}$$

where

$$\Omega_c = \frac{E_c D_{23}}{2\hbar}$$

is the Rabi frequency (angular) for the coupling laser, and

$$\Omega_p = \frac{E_p D_{21}}{2\hbar}$$

is the Rabi frequency (angular) for the probe laser. D_{12} and D_{23} are the associated dipole matrix elements for the two transitions.

These three coupled equations are the basis of our further discussion. Although a physically meaningful analytical solution for these three equations is not possible, it is a straightforward process to solve them numerically for the situations of interest here. This we have done using well-known numerical analysis techniques implemented through a PC-based programme (A compiled version of this programme is available on disc for use in PC systems, for further details see the end of this chapter.) In addition to the requirement to follow the time evolution of the state amplitudes of the three level atom, as given by Equations 4-6 above, we also need to describe the influence of the atom on the two radiation fields. This we do by way of the associated polarisation terms

$$P_c = D_{23} \int_{-\infty}^{t} dt_0 \, a_2^*(t, t_0) \, a_3(t, t_0) \tag{7}$$

$$P_p = D_{21} \int_{-\infty}^{t} dt_0 \, a_1^*(t, t_0) \, a_2(t, t_0) \tag{8}$$

where the associated absorption or gain is given by $Im(P)/\epsilon_0 E$

We are now in a position to explore the basic physics of EIT, but first of all we review our ideas about the way in which the above equations describe the conventional absorption process (*i.e.* what happens when we switch-off the coupling field).

Initially, though, we give a brief introduction to the diagrams that we will use in this next section. Typical of these diagrams is Figure 13. Here the moduli of the state amplitudes a_1, a_2, a_3, are plotted as a function of time as the interaction of the three level atom with the two radiation fields proceeds. The squares of these amplitudes are of course the probabilities of finding the atom in the particular states. Also plotted as a function of time is the imaginary part of the polarisation for the transition between state 1 and state 2, namely the probe transition. The definition of this in terms of the appropriate state amplitudes, is given by Equation 9. The time averaged value of this quantity over the effective lifetime of the atom is a measure of the total net effect of the atom on the radiation field. A negative value corresponds to net gain and a positive value to net absorption. The value of the time average is written on the diagram. The energy level diagram on the left-hand side is used to designate the random decay rates of the three states (values placed at the ends of the short arrows leaving the states), through processes either radiative or collisional that are not explicitly considered here, together with the coupling strengths associated with the two radiation fields (values placed within the appropriate arrows connecting the states), and the initial conditions (initial state amplitudes, with respect to both their magnitudes and phases) of the three level system (values placed above the states).

We first of all consider the case where the atom is initially in the lowest state (state 1), and is subject to just a single radiation field coupling state 1 to state 2. The atom is assumed not to decay to any other states. The influence of the field is to cause the atom to transfer progressively from state 1 to state 2, *i.e.* the atom is excited, during which time the polarisation indicates that absorption of radiation takes place. However, once the atom has transferred entirely to state 2, *i.e.* when the amplitude for state 2 becomes unity, then as time evolves beyond this point, the atom begins to transfer back to state 1 again, during which time the polarisation indicates that emission of radiation is taking place, *i.e.* the atom exhibits optical gain as expected. The process keeps on repeating indefinitely in so far as the atom has no where else to which it can go apart from transferring between the two energy states. What is being described here is Rabi flopping between the two states connected by the radiation field. The net effect on the radiation field, when averaged over a considerable period of time, is zero, since the absorption at one half of each flopping period is cancelled by the gain at the other half.

In order for there to be net absorption, the atom must be able to leave the system from the upper state (state 2), having first absorbed a photon from the radiation field in order to reach that state from the lower state (state 1). This situation is illustrated in Figure 14, where the decay rate from state 2 is now finite. As the amplitude of state 2 builds up in time, as a result of the radiation field transfering the atom from state 1, this decay damps away the amplitude of the state, more or less before there is an opportunity for the field to transfer the atom back to state 1 again. The time average of the polarisation over the effective lifetime is now positive, indicating a situation of net absorption. When the atom decays away from the three levels being considered explicitly here, it ceases to influence the radiation field. However, within an ensemble of atoms, others are constantly being created in the lower state, so ensuring that the absorption process continues throughout time. What we are hence describing here is the well-known process of absorption. However, what the present analysis does clearly indicate is the importance of relative decay rates in determining the magnitude of absorption.

If we repeat the analysis but with the atom initially in state 2, and with the major decay rate being from state 1 (the lower state), then as expected, the overall effect of the atom on the radiation field is one of gain. The atom enters the process in the upper state and leaves by the lower state, the energy thereby released going towards amplifying the radiation field.

Having reviewed the situation with regard to a two level atom interacting with a single radiation field, we now proceed to consider the situation appertaining to EIT, where two radiation fields interact with a three level atom.

In Figure 15 the three level atom is subject to a coupling field between states 2 and 3, which has a Rabi frequency that is three times that of the probe radiation field coupling state 1 to state 2. The atom is initially in the lowest state 1, and decay from the system is through state 2, as before. The first thing to note is that the amplitude of state 1 changes very little. Despite the presence of a probe radiation field of the same magnitude as in our previous example, the presence of the coupling field has

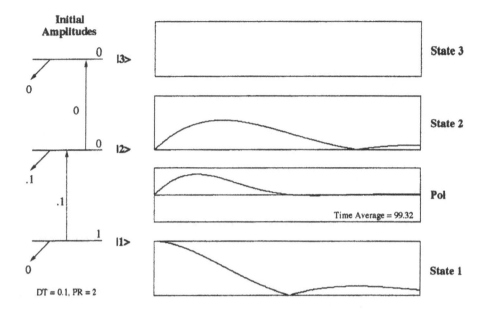

Figure 14. *Conventional absorption. Atom enters by state 1 and ultimately leaves by state 2, with the net effect being one of absorption. The positive time average of the polarisation is indicative of absorption.*

diminished the rate of amplitude transfer from state 1 to state 2. The time average of the polarisation associated with the transition from state 1 to state 2 is now much smaller, namely around 5 units rather than the 99 units in the previous case. This of course means that the absorption associated with the transition is much smaller than previously, and what we are seeing here is the occurrence of EIT on the transition from state 1 to state 2 in a cascade energy level scheme.

A notable feature of Figure 15 is that as time progresses the amplitude (and hence associated population) seems to become trapped in state 1, and also, but to a much lesser extent, in state 3. In other words a coherent superposition state involving the states 1 and 3 is formed, and this superposition state is decoupled from the radiation field. Thus we see the relation between EIT and the phenomenon of population trapping which is the underlying explanation of the 'Pisa' experiment discussed above (see section 2). The two are synonomous, at least in the cascade scheme for EIT. Because the atom evolves rapidly into a trapped superposition state, with only a much reduced probability of decay from state 2 compared to previously, its probability of removing a photon from the probe radiation field is much reduced, *i.e.* absorption is much reduced.

Examination of Figure 15 suggests that the final relative amplitude of state 3 compared to state 1 (around 1 to 3) is in inverse ratio to the relative Rabi frequencies associated with the coupling radiation field (1 unit) compared to the probe field (0.3 units). That this is in fact the case is confirmed by the behaviour of the atom shown in Figure 16. Here we have set up at the start a coherent superposition state in which the

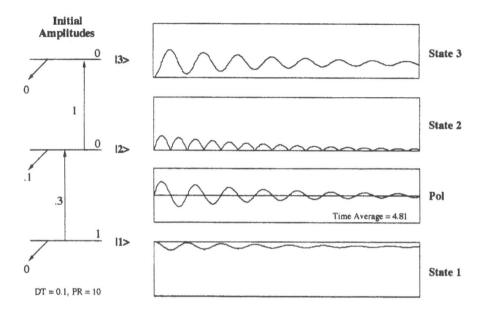

Figure 15. *Three level atom in the presence of two radiation fields. Coupling strength between states 2 and 3 is 1.0, and between states 1 and 2 is 0.3. Decay rate from state 2 is 0.1.*

initial relative amplitudes of states 3 and 1 are 0.6 and 0.8 respectively, these being in inverse ratio to the Rabi frequencies of the coupling and probe lasers (0.8 units to 0.6 units). Under these initial conditions nothing happens! There is no absorption of the probe radiation (or the coupling radiation for that matter), and the state amplitudes do not change with time. Since no amplitude builds up in state 2, the only state by which the atom can decay to other non-specified states, the atom is effectively trapped in that it is fully decoupled from the radiation fields (but only provided that both fields are simultaneously present). Examination of the basic equations (see Equations 4–6) involved here highlights the reason for this state of affairs. Amplitude is being fed into state 2 from state 3 by the coupling field (as described by the second term on the right hand side of equation 5) which over a given time interval is exactly equal in magnitude but in anti-phase with the amplitude that is being fed into state 2 from state 1 by the probe field (as described by the first term on the right hand side of Equation 5). As a result, the net effect is a zero increase in the amplitude of state 2. This perspective highlights the viewpoint that the origin of EIT effects is through the destructive interference of alternative excitation pathways, which shows its commonality with the process of Fano interference.

It will be noted from Figure 16 that the amplitude of state 3 is in anti-phase to that of state 1. This selection is most important if the integrity of the superposition state is to be preserved in the presence of the radiation fields. This is clearly shown in Figure 17, where the phase of state 3 has been perturbed by 90°. As a result, amplitude builds up in state 2, and hence partly leaks away from the 3 state system, before the

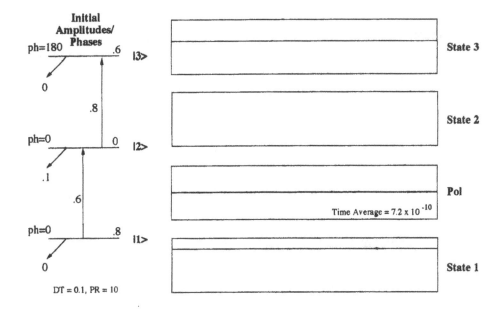

Figure 16. *Trapped populations due to coherent superposition of states 1 and 3.*

condition of a trapped population is re-established. However, as a result of the loss, through state 2, of amplitude from state 1, probe radiation is absorbed, as is shown by the non-zero time average of the polarisation. From this we conclude that the coherence established between state 3 and state 1 is crucial to the occurrence of EIT, and that processes that destroy this coherence diminish the effectiveness of the EIT process (in the nomenclature of the density matrix to be discussed in Section 4, this coherence is described by ρ_{31}).

So far, we have considered the occurrence of EIT in a cascade scheme where state 3 is above state 2 which is above state 1. Other schemes are possible and these also can be described by our single atom model.

Figure 18 can be used to describe the lambda scheme if we regard state 3 as lying below state 2, and close to state 1.The two lower states (states 1 and 3) have zero decay rates away from the 3-state scheme, whereas state 2, which is now the upper state of the lambda scheme has a finite decay rate. The probe laser couples from state 1 to state 2 as before, and the coupling laser couples from state 3 to state 2, likewise as before. It is apparent that under the conditions appertaining to Figure 18, a trapped population is rapidly established involving states 1 and 3. This results in little decay occurring from state 2 (none after the superposition state is established), and hence very little net absorption (around 2 units) being observed on the probe transition, *i.e.* EIT is observed on this transition. As before the crucial coherence in this case is that established between states 1 and 3.

The vee scheme for EIT may be explored through Figure 19. In this case state 1 must be regarded as being of greater energy than state 2, and placed close to state

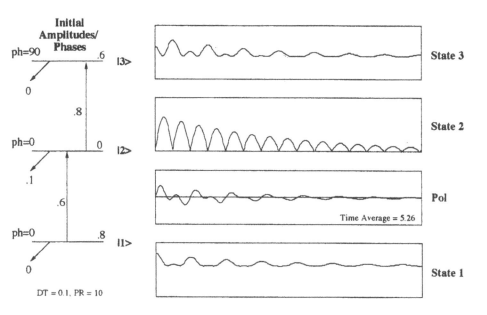

Figure 17. *Effect of perturbing the initial phase of state 3 relative to state 1 by $\pi/2$.
The trapped populations are damped away, but eventually re-establish a steady-state.*

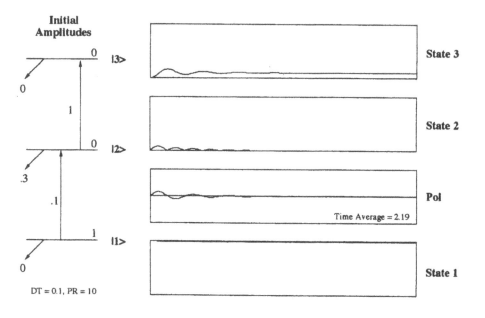

Figure 18. *Lambda EIT scheme. State 3 is to be regarded as lying below state 2 and
close to state 1.*

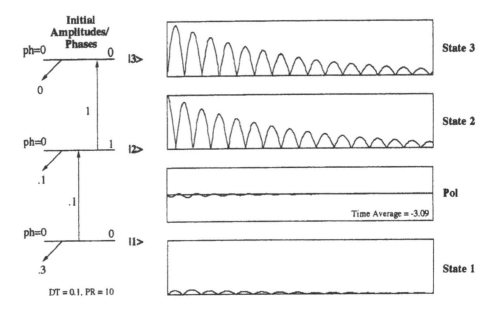

Figure 19. *Vee EIT scheme. State 1 is to be regarded as lying above state 2 and close to state 3.*

3. State 2 then becomes the lowest state (ground state). The probe radiation couples between state 2 and state 1, while the coupling radiation couples state 2 to state 3. It is apparent from Figure 19 that EIT is also observed in this case in that there is little amplitude transfer to state 1 from state 2, and hence the associated time average of the polarisation on the probe transition is small (3 units, but note that although this has a negative value in this case, it still corresponds to net absorption since now state 1 is above state 2). However it is apparent that in the case of the vee scheme, a trapped population state is not formed. Amplitude rapidly oscillates between states 2 and 3 due to the coupling field. This leads to a cancellation of the amplitude transfer between states 2 and 1 because the rapidly changing amplitude in state 2 rapidly cycles the sign of the polarisation on the transition from 2 to 1 from negative to positive, and hence this polarisation averages over time to be close to zero. Whereas in the absence of the coupling field, absorption of the probe field would allow the atom to leave by the more rapid decay route from state 1, with net absorption, this route is now effectively blocked and the atom leaves the 3 state system via the lowest state, state 2, even though this has a lower decay rate than state 1. In so doing, the atom leaves the 3 state system without having extracted a photon from the probe radiation field, and absorption is thereby decreased. No trapped populations are involved in this case, and EIT results from destructive interference between the direct pathway from state 2 to state 1, and that going from state 2 to state 1 via state 3.

The simple 3 state model, when solved numerically, provides considerable insight into the origins of EIT in both cascade, lambda and vee schemes. Whereas in the former two, the origin of EIT is as the result of a trapped population (a coherent superposition state) that is decoupled from the radiation fields, this is not so in the last scheme where amplitude beating and quantum interference effects occur instead.

We now introduce a comprehensive and analytically powerful approach based on the density matrix. We will demonstrate how this approach may be applied to describe the cascade scheme for EIT, but extension to the other schemes is straightforward.

4   The density matrix description.

The density matrix is useful for calculating the expectation values of operators when the precise wavefunction of the whole system is not known. Although the microscopic wavefunction of an individual atom may be precisely specified, in considering the effects of the whole medium, a summation needs to be made over all the atoms making up the medium, and hence over their wavefunctions. These atoms start to evolve at different times (*i.e.* are excited to specific states at different times), and may involve coherences between the states with a distribution of phases, etc. The density matrix allows average macroscopic properties of the medium to be calculated without the requirement of precisely specifying the exact wavefunction of each and every atom, but instead by working in terms of appropriate averages. This description is particularly appropriate when random decay processes, such as spontaneous emission, and random excitation processes need to be taken into account, and also when the damping of coherences is involved. As previously, in the case of our single atom system, the atomic wavefunction is written as a superposition of the unperturbed states of the isolated atom, but the coefficients are allowed to be functions of time to allow for effects of the perturbations (in this case the radiation fields). Hence we have:

$$\Psi(r,t) = \sum_n C_n(t)\psi_n(r,t)$$

The density matrix elements are then defined as ensemble averages (*e.g.* over the creation time of the states), thus we have

$$\rho_{nm} = \overline{C_m^* C_n}$$

When $m = n$, the element, which is then referred to as an on-diagonal element, describes a level population; when $m \neq n$, the element, which is then refered to as an off-diagonal element, describes a coherence between the two states involved, which is maintained after averaging over all the atoms. Certain of the coherences drive the optical field amplitudes. In addition such effects are fundamental in the occurrence of EIT as we have already seen in our consideration of the single atom.

The time evolution of the density matrix is described by the Liouville equation, namely

$$\frac{\partial \rho_{ij}}{\partial t} = -\frac{i}{h}[H,\rho]_{ij} = -\frac{i}{h}\sum_{k=1}^n (H_{ik}\rho_{kj} - \rho_{ik}H_{kj})$$

where H is the Hamiltonian. This equation follows from the time-dependent Schrodinger equation.

We now include damping (decay) terms, so the equation becomes

$$\frac{\partial \rho_{ij}}{\partial t} = -\frac{i}{\hbar}[H,\rho]_{ij} - \Gamma_{ij}\left(\rho_{ij} - \rho_{ij}^0\right)$$

Note that there are off-diagonal as well as diagonal damping terms. The latter describe the decay of the state populations (amplitudes) as previously, whereas the former describe the decay of the coherences. The importance of coherences to the EIT process has already become apparent from our previous discussion of the single atom case.

We write the Hamiltonian in terms of that for the isolated atom, H_0, and the additional energy, H_i, arising from the interaction of the electric fields of the applied optical radiation fields with the electron(s) of the atom, thus we have:

$$H = H_0 + H_i = H_0 - \mu_{ij}E(t)$$

The induced optical polarisation is given by:

$$P(t) = N\mu_{ij}\rho_{ij}$$

The susceptibility can be deduced from the polarisation, and written in terms of real and imaginary parts :

$$P(t) = \epsilon_0(\chi' - i\chi'')E$$

The imaginary part of the susceptibility, which describes the optical gain or absorption, is hence given by:

$$\chi''(\omega) = -\frac{2N\mu_{ij}^2}{\epsilon_0\hbar}\frac{\mathrm{Im}(\rho_{ij})}{\Omega_R}$$

where $\Omega_R = \mu_{ij}E/2\hbar$, is the Rabi frequency induced by the optical field, and the pair of states i and j are those appropriate to the particular resonance under consideration.

We now consider the case of the cascade configuration for EIT with parameters as described in the Figure 20.

The atomic selection rules lead to the following relations for the dipole matrix elements:

$$\begin{aligned}
\mu_{12} &= \mu_{21} = \langle 1|\mu|2\rangle \neq 0 \\
\mu_{23} &= \mu_{32} = \langle 2|\mu|3\rangle \neq 0 \\
\mu_{13} &= \mu_{31} = \langle 1|\mu|3\rangle = 0
\end{aligned}$$

The electric fields of the two coherent optical fields applied to the atom are described by:

$$E_j(z,t) = \frac{E_j^0}{2}\{\exp\left[i\left(\omega_j t - k_j z\right)\right] + \mathrm{c.c.}\}, \qquad j = 1,2$$

and have associated Rabi frequencies:

$$\Omega_1 = \frac{\mu_{12}E_1^0}{2\hbar}, \quad \Omega_2 = \frac{\mu_{23}E_2^0}{2\hbar},$$

A particular off-diagonal decay rate includes a dephasing term that describes phase fluctuations between the associated states involved in the off-diagonal element. This is

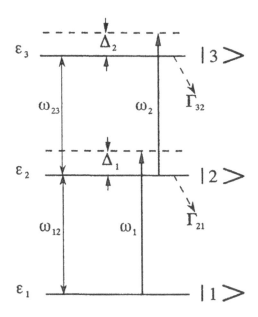

Figure 20. *A cascade three level atom*

in addition to the contribution to the off-diagonal decay associated with the decay of the populations themselves, so that we have

$$\gamma_{ij} = \frac{1}{2}(\gamma_{ii} + \gamma_{jj}) + \gamma_{\text{deph}}$$

We now define the detunings between the atomic resonances and the optical fields by:

$$\Delta_1 = \omega_1 - \omega_{12} - k_1 v_z$$
$$\Delta_2 = \omega_2 - \omega_{23} - k_2 v_z$$

These include the effect of the Doppler shift caused by the atom moving with the z-component of velocity, v_z .

The oscillations at the corresponding optical frequencies of the off-diagonal elements are removed by the substitutions:

$$\rho_{12} = \rho_{12}\exp[i(\omega_1 t - k_1 z)]$$
$$\rho_{23} = \rho_{23}\exp[i(\omega_2 t - k_2 z)]$$
$$\rho_{13} = \rho_{13}\exp[i\{(\omega_1 + \omega_2)t - (k_1 + k_2)z\}]$$

The application of the Liouville equation then yields the following six simultaneous equations for the different elements of the density matrix

$$\dot{\rho}_{11} = i\Omega_1(\hat{\rho}_{21} - \hat{\rho}_{12}) + \Gamma_{21}\rho_{22}$$

$$\dot{\rho}_{22} = i\Omega_1(\hat{\rho}_{12} - \hat{\rho}_{21}) + i\Omega_2(\hat{\rho}_{32} - \hat{\rho}_{23}) - \Gamma_{21}\rho_{22} + \Gamma_{32}\rho_{33}$$

$$\dot{\rho}_{33} = i\Omega_2(\hat{\rho}_{23} - \hat{\rho}_{32}) - \Gamma_{32}\rho_{33}$$

$$\dot{\hat{\rho}}_{12} = -i(\Delta 1 - i\gamma_{12})\hat{\rho}_{12} + i\Omega_1(\rho_{22} - \rho_{11}) - i\Omega_2\hat{\rho}_{13}$$

$$\dot{\hat{\rho}}_{23} = -i(\Delta 2 - i\gamma_{23})\hat{\rho}_{23} + i\Omega_2(\rho_{33} - \rho_{22}) + i\Omega_1\hat{\rho}_{13}$$

$$\dot{\hat{\rho}}_{13} = -i(\Delta_1 + \Delta_2 - i\gamma_{13})\hat{\rho}_{13} + i\Omega_1\hat{\rho}_{23} - i\Omega_2\hat{\rho}_{12}$$

These six complex equations can be converted into nine real equations by making the substitutions

$$\hat{\rho}_{ij} = \rho_{ij}^r + i\rho_{ij}^i$$

$$\hat{\rho}_{ij}^* = \rho_{ij}^r - i\rho_{ij}^i$$

The population conservation between the three states implies that

$$\rho_{11} + \rho_{22} + \rho_{33} = 1$$

which removes one degree of freedom from the system.

In the steady state all the time derivatives can be set equal to zero when the system is described by the following set of equations expressed in matrix form

$$
\begin{bmatrix}
-\Gamma_{21} & \Gamma_{32} & 0 & 0 & 0 & -2\Omega_1 & 0 & 2\Omega_2 \\
0 & \Gamma_{32} & 0 & 0 & 0 & 0 & 0 & -2\gamma_2 \\
0 & 0 & \gamma_{12} & 0 & 0 & -\Delta_1 & -\Omega_2 & 0 \\
0 & 0 & 0 & \gamma_3 & 0 & -\Omega_2 & -(\Delta_1+\Delta_2) & \Omega_1 \\
0 & 0 & 0 & 0 & \gamma_{23} & 0 & \Omega_1 & -\Delta_2 \\
-2\Omega_1 & -\Omega_1 & \Delta_1 & \Omega_2 & 0 & \gamma_{12} & 0 & 0 \\
0 & 0 & \Omega_2 & (\Delta_1+\Delta_2) & -\Omega_1 & 0 & \gamma_{13} & 0 \\
\Omega_2 & -\Omega_2 & 0 & -\Omega_1 & \Delta_2 & 0 & 0 & \gamma_{23}
\end{bmatrix}
\begin{bmatrix}
\rho_{22} \\
\rho_{33} \\
\rho_{12}^r \\
\rho_{13}^r \\
\rho_{23}^r \\
\rho_{12}^i \\
\rho_{13}^i \\
\rho_{23}^i
\end{bmatrix}
=
\begin{bmatrix}
0 \\
0 \\
0 \\
0 \\
0 \\
-\Omega_1 \\
0 \\
0
\end{bmatrix}
$$

The detunings experienced by the different atoms are changed by the Doppler effect, so an average must be made over the velocity distribution $f(v_z)$, namely

$$\overline{\rho_{ij}} = \int_0^\infty f(v_z)\rho_{ij}(v_z)dv_z.$$

Solution of the matrix equation and the averaging over the Doppler profile are carried out through appropriate computer routines. An example of the calculation of the off-diagonal element, ρ_{12}, which describes the interaction of the atom with the probe radiation field for a cascade system is illustrated in Figure 21.

In Figure 22 we show a set of experimental results for the absorption on the designated transition in rubidium vapour under similar conditions to those corresponding to the theoretical calculations. The occurrence of the EIT transmission window is clearly observable.

The lambda configuration shown in Figure 23 can be analysed in a similar way. The density matrix equation for this situation can readily be derived following procedures

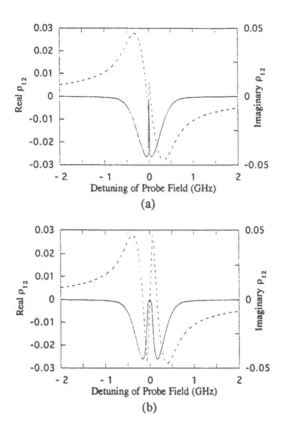

Figure 21. *Real (solid line) and imaginary (dashed line) components of ρ_{12} for the cascade configuration with a coupling field strength of (a) 0.04GHz, and (b) 0.16 GHz.*

similar to those discussed above, when we obtain

$$
\begin{bmatrix}
0 & \Gamma_{32} & 0 & 0 & 0 & 0 & 0 & 2\Omega_2 \\
0 & -\Gamma_{32}-\Gamma_{31} & 0 & 0 & 0 & 0 & -2\Omega_1 & -2\Omega_2 \\
0 & 0 & \gamma_{12} & 0 & 0 & \Delta_2-\Delta_1 & -\Omega_2 & -\Omega_1 \\
0 & 0 & 0 & \gamma_{13} & 0 & -\Omega_2 & -\Delta_1 & 0 \\
0 & 0 & 0 & 0 & \gamma_{23} & \Omega1 & 0 & -\Delta_2 \\
0 & 0 & \Delta_1-\Delta_2 & \Omega_2 & -\Omega_1 & \gamma_{12} & 0 & 0 \\
-\Omega_1 & -2\Omega_1 & \Omega_2 & \Delta_1 & 0 & 0 & \gamma_{13} & 0 \\
\Omega_2 & -\Omega_2 & \Omega_1 & 0 & \Delta_2 & 0 & 0 & \gamma_{23}
\end{bmatrix}
\begin{bmatrix}
\rho_{22} \\
\rho_{33} \\
\rho_{12}^r \\
\rho_{13}^r \\
\rho_{23}^r \\
\rho_{12}^i \\
\rho_{13}^i \\
\rho_{23}^i
\end{bmatrix}
=
\begin{bmatrix}
0 \\
0 \\
0 \\
0 \\
0 \\
0 \\
-\Omega_1 \\
0
\end{bmatrix}
$$

where

$$
\Delta_1 = \omega_1 - \omega_{13} - k_1 v_z, \qquad \Omega_1 = \frac{\mu_{13}E_1^0}{2\hbar}, \qquad \Omega_2 = \frac{\mu_{13}E_2^0}{2\hbar}
$$

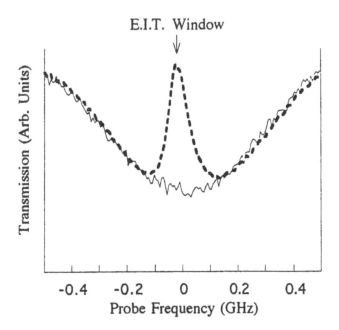

Figure 22. *Experimentally observed EIT in rubidium vapour. Solid curve is in the absence of the coupling field, while the dashed curve is with the coupling field present.*

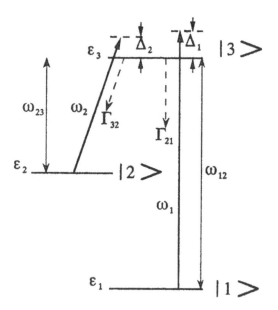

Figure 23. *Lambda EIT scheme.*

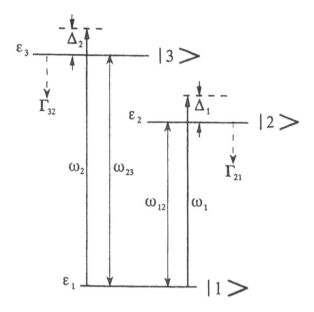

Figure 24. *Vee EIT scheme.*

Likewise the vee scheme, which is illustrated in Figure 24, can be modelled within the density matrix representation to yield the following matrix equation

$$
\begin{bmatrix}
\Gamma_{21} & 0 & 0 & 0 & 0 & 2\Omega_1 & 0 & 0 \\
0 & \Gamma_{31} & 0 & 0 & 0 & 0 & 0 & 2\Omega_2 \\
0 & 0 & \gamma_{12} & 0 & 0 & -\Delta_1 & -\Omega_2 & 0 \\
0 & 0 & 0 & \gamma_{23} & 0 & \Omega_2 & \Delta_1 - \Delta_2 & \Omega_1 \\
0 & 0 & 0 & 0 & \gamma_{13} & 0 & \Omega_1 & -\Delta_2 \\
2\Omega_1 & \Omega_1 & -\Delta 1 & \Omega_2 & 0 & -\gamma_{12} & 0 & 0 \\
0 & 0 & \Omega_2 & \Delta_2 - \Delta_1 & -\Omega_1 & 0 & \gamma_{23} & 0 \\
\Omega_2 & 2\Omega_2 & 0 & \Omega_1 & -\Delta_2 & 0 & 0 & -\gamma_{13}
\end{bmatrix}
\begin{bmatrix}
\rho_{22} \\
\rho_{33} \\
\rho_{12}^r \\
\rho_{23}^r \\
\rho_{13}^r \\
\rho_{12}^i \\
\rho_{23}^i \\
\rho_{13}^i
\end{bmatrix}
=
\begin{bmatrix}
0 \\
0 \\
0 \\
0 \\
0 \\
\Omega_1 \\
0 \\
\Omega_2
\end{bmatrix}
$$

where

$$
\Delta_2 = \omega_2 - \omega_{13} - k_2 v_z, \qquad \Omega_2 = \frac{\mu_{13} E_2^0}{2\hbar}, \qquad \Omega_1 = \frac{\mu_{13} E_1^0}{2\hbar}
$$

5 EIT in Rubidium

As an example of an experimental system in which EIT effects can be demonstrated, we consider the case of rubidium vapour, and in particular explore the cascade scheme. The energy level scheme is shown in Figure 25. The ground state is a $5S_{1/2}$ state which is split into two hyperfine levels for each of the two isotopes of rubidium that are present, namely ^{85}Rb and ^{87}Rb. The intermediate level is the $5P_{3/2}$ level which

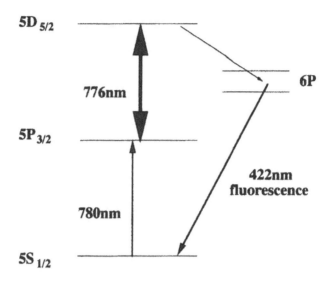

Figure 25. *Cascade EIT scheme in rubidium vapour.*

is hyperfine split into four states for each of the two isotopes. The probe field is scanned over the $5S_{1/2}$ to $5P_{3/2}$ transition. The coupling field is applied between the $5P_{3/2}$ level and the upper level shown, which is the $5D_{5/2}$ level. This latter level is split into six hyperfine levels for each of the two isotopes. In the experiments the coupling field frequency is held fixed while the probe field frequency is scanned. To minimise the effects of Doppler broadening, the probe and coupling fields are counter-propagated and have close to the same wavelength (780nm for the former and 776nm for the latter). Electromagnetically induced transparency occurs at every two photon resonance point, and is superimposed on the Doppler broadened profile of the probe field absorption, as shown in Figure 26. Because of the counter-propagating beam geometry which minimises Doppler broadening, and provided the coupling laser is unfocused and operated at low power ($\sim$ 100mW) to prevent excessive power broadening, the induced transparency mirrors the hyperfine level structure, as is apparent in Figure 26, which is for the case of ^{87}Rb. Figure 27 shows the effects of progressively increasing the power of the coupling laser from 5mW to 300mW, and the scan of the probe laser frequency in this case encompasses both the ^{87}Rb and ^{85}Rb isotopes. As the power is increased, power broadening washes out the hyperfine structure, but the EIT effect becomes significantly more pronounced, with the induced transparency approaching 100%. It should be borne in mind that the probe field used here is of very low intensity, so that there is negligible population transfer out of the ground state.

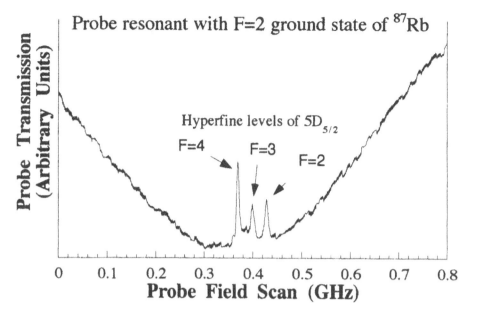

Figure 26. *EIT in rubidium vapour at low coupling field power for the case of the isotope* ^{87}Rb.

6 Electromagnetically Induced Focussing (EIF)

As a result of the change in the absorption profile that results from electromagnetically induced transparency, there is an associated change in the dispersion (refractive index) profile experienced by the probe laser. This effect is shown in Figure 21. It has an interesting consequence in so far as the effect is dependent on the intensity of the coupling field. The radiation field from a laser oscillating on single transverse mode changes smoothly across the beam profile in Gaussian fashion. Hence, if such a laser is used as the coupling laser, the refractive index experienced by the probe laser exhibits smooth radial change. Under appropriate circumstances, therefore, the coupling laser can be used to write a lenslike profile into the medium which is then experienced by the probe laser. It is of course a distributed lens extending throughout the volume of the medium lying within the coupling laser field. Depending on the detunings, this profile can result in either focusing or defocusing effects. We have called the effect electromagnetically-induced-focusing (EIF) (Moseley *et al.* 1995). It is important to realise that it occurs close to the position where EIT has rendered the vapour transparent, so there is little or no disipation on the probe laser transition. (Since so little population is transferred to the intermediate state, this applies to the coupling laser as well). The principles of the EIF effect are summarised in Figure 28. In Figure 29, the effect of detuning on the nature of the induced lens is explored for the case of rubidium vapour (probe laser wavelength 780nm, coupling laser wavelength 776nm). Throughout the figure, the coupling laser is maintained on exact resonance with the $5P_{3/2}$ to $5D_{5/2}$ transition, but the probe laser is progressively detuned from resonance with the $5S_{1/2}$ to $5P_{3/2}$ tran-

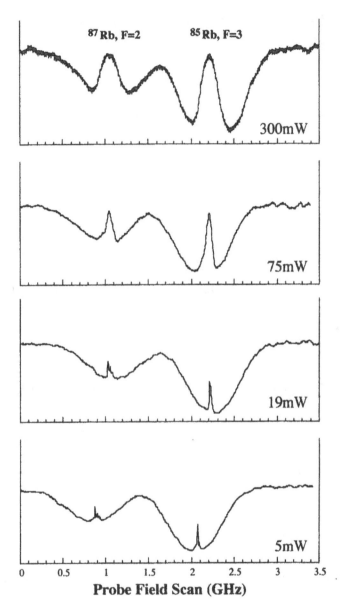

Figure 27. *EIT in rubidium vapour showing the effect of increasing the coupling field power from 5mW to 300 mW.*

sition. At zero detuning, the probe laser experiences a wide, but hard-edged aperture due to the EIT effect, and zero lensing (*i.e.* the refractive index is unity across the whole profile). When the probe laser is detuned by 50MHz from the resonance condition, the medium behaves as a diverging lens and the EIT aperture is reduced in size. Finally,

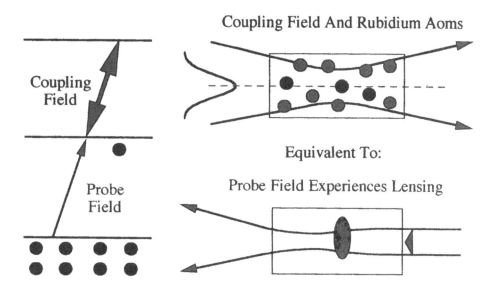

Figure 28. *Principles of electromagnetically induced focusing. The coupling field prepares the rubidium atoms so that the probe field experiences a lensing effect.*

when the detuning reaches 150MHz, the medium behaves as a converging lens, and the EIT aperture now appears small, but with a soft edge. A comprehensive theoretical treatment of the EIF effect must consider both the distributed nature of the lens as well as the diffraction effects associated with the aperture (Moseley *et al.* 1996). The results of such an analysis are shown in Figure 30 where the size of the emerging probe beam is plotted as a function of detuning. The dashed horizontal line corresponds to the case of zero lensing, while a value lying above this line corresponds to defocusing and below to focusing. Experimental results are shown in Figure 30 for a probe beam that has a beam size of 35μm (FWHM). Strong focusing and defocusing is observed. The EIT window is also indicated showing that both effects occur within it.

The EIT/EIF window in the medium can be shown to exhibit the full range of physical optical effects expected of it. For example, if the waist of the coupling field is made small compared to that of the probe field, then diffraction rings appear on the probe field due to diffraction by the small circular aperture established by the coupling field. If the coupling field aperture is made slit-like by suitable focusing of this field, then the probe field exhibits the characteristic parallel linear fringes associated with a single slit. By suitable choice of parameters it is apparent that the medium can also be prepared by the coupling field to act as a waveguide for the probe field.

7 Inversionless Lasing.

A substantial literature has now developed on the associated topic of inversionless lasing, and more recently such effects have been reported for a number of different experiments.

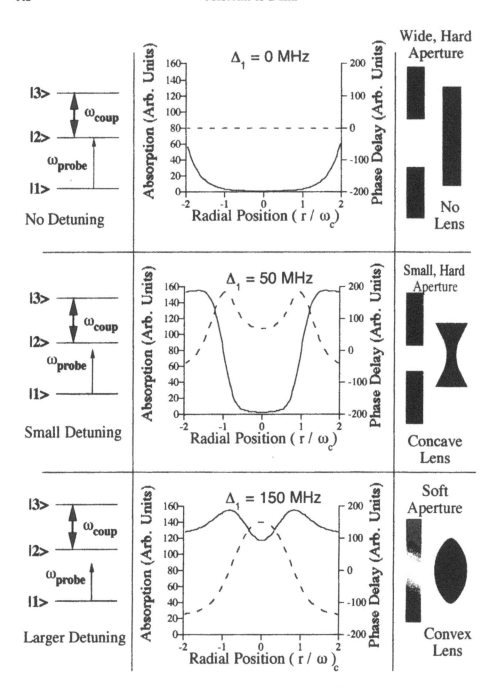

Figure 29. *Nature of the lens formed by electromagnetically induced focusing as the probe laser frequency is detuned from exact resonance with the transition from state 1 to state 2 (detuning is given by Δ_1). The solid curves describe absorption and the dashed curves dispersion (lensing effect).*

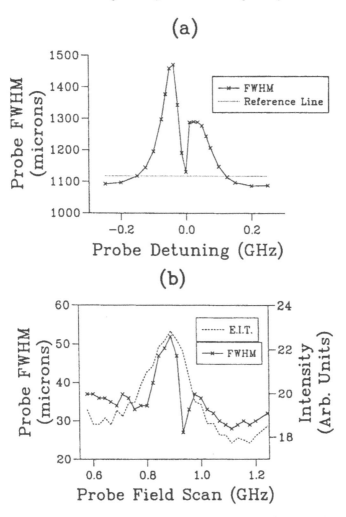

Figure 30. *Variation of the FWHM intensity of the probe field as a function of probe frequency detuning; (a) theoretical calculation, (b) experimental observations. In (b) solid line refers to the spot size, dashed line refers to the transmission of the vapour.*

It is not our intention here to review this area, but what we will do is to demonstrate that our simple one-atom model of the EIT process can be used to predict the occurrence of inversionless lasing in a simple three level system. Consider the level scheme shown in Figure 31a, and in particular for the case of an atom and field configuration with parameters as listed in the figure caption. If the atom is initially excited to state 1, the lower laser state, then the state amplitudes evolve as shown in Figure 31b. Little excitation occurs, because of the EIT effect, and the atom eventually decays away, largely as a result of the damping associated with state 1 (0.1 units). Over the lifetime of the atom, the contribution that it makes to the polarisation induced between state 1 and state 2 is +1.06 units, the positive sign corresponding to net absorption of the probe field. Consider now when the atom is initially excited to state 3, as illustrated

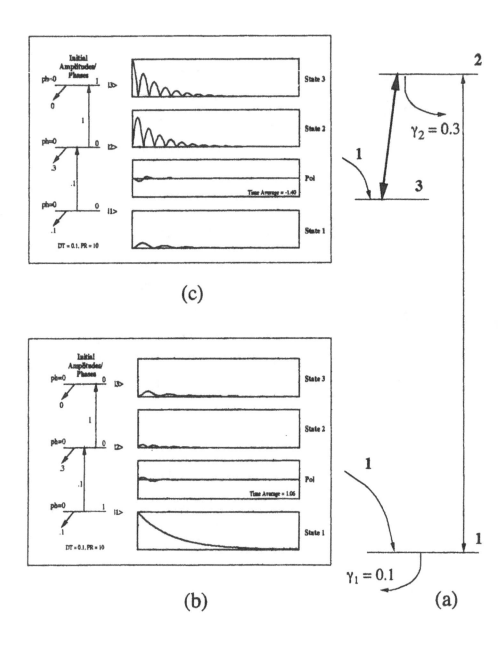

Figure 31. *Inversionless lasing in the three level scheme shown in (a). Evolution of state amplitudes when the atom is initially excited, (b) to state 1 , and (c) to state 3. Conditions: decay rate from level 2 is 0.3, coupling rate between states 3 and 2 is 1.0, and between states 2 and 1 is 0.1.*

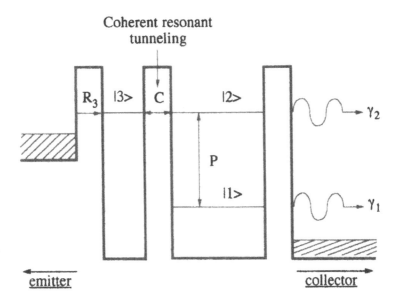

Figure 32. *Unipolar, double quantum well semiconductor laser structure for lasing without population inversion. Coherent resonant tunnelling couples state 3 to state 2. Lasing without inversion occurs between states 1 and 2. (Imamoglu and Ram 1994)*

in Figure 31c. The state amplitudes evolve as shown, the most pronounced effect being oscillation between states 3 and 2 due to the coupling field. Decay takes place from state 2, but also, as a result of the induced transition by the probe field, from state 1 as well. The atom overall decays away from the three level scheme rather faster because of the greater decay rate associated with state 2 compared to state 1. However, over its effective lifetime it contributes −1.40 units to the polarisation induced on the probe transition, the negative sign indicating net gain. If now we look at the whole medium made up of many such atoms excited randomly in time to states 1 and 3, but with the same excitation rate, then in so far as atoms excited to state 3 live overall for a shorter time than those excited to state 1, the average population in state 2 will be less than that in state 1; in other words a population inversion will not exist. However, it can be seen from the relative magnitudes of the two polarisations quoted above that there will be net gain (proportional to −1.40 + 1.06), even in the absence of a population inversion. Thus, the possibility of an inversionless laser is demonstrated in this configuration.

A version of this scheme proposed by Imamoglu and Ram (1994) for obtaining inversionless lasing within a semiconductor geometry is shown in Figure 32. A unipolar double-quantum well semiconductor laser structure is proposed in which coherent resonant tunnneling, rather than a coupling radiation field, couples levels 2 and 3.

Acknowledgements.

I would like to acknowledge with gratitude the many fruitful discussions I have had with both past and present members of the EIT group in St Andrews. These include Richard Moseley, Sara Shepherd, David Fulton and Bruce Sinclair. I am particularly grateful to David Fulton for allowing me to use one of his reports as the basis for the section on the use of the density matrix in EIT. In addition, I would like to thank Sara and David for preparing all the drawings in their final form.

Software

Please write to the author at School of Physics and Astronomy, University of St Andrews, North Haugh, St. Andrews, Fife, KY16 9SS, Scotland-UK (FAX: 44 1334 463104, e-mail: mhd@st-and.ac.uk) for further information on the PC-version of the three-level-atom/two-field computer programme.

References.

Arimondo E, and Orriols G, 1976, *Nuovo Cimento Lett* **17** 333.

Alezetta G, Gozzini A, Moi L, and Orriols G, 1976, *Nuovo Cimento B* **36** 209.

Alezetta G, Moi L, and Orriols G, 1979, *Nuovo Cimento B* **52** 209.

Boller K J, Imamoglu A, and Harris S E, 1991, *Phys Rev Lett* **66** 2593.

Fano U, 1961, *Phys Rev* **124** 1866.

Field J E, Hahn K H, and Harris S E, 1991, *Phys Rev Lett* **67** 3062.

Gray H R, Whitley R M, and Stroud Jr C R, 1978, *Optics Letters* **3** 218.

Imamoglu A and Ram R J, 1994, *Optics Letters* **19** 1744.

Moseley R R, Shepherd S, Fulton D J, Sinclair B D, and Dunn M H, 1995, *Phys Rev Lett* **74** 670.

Moseley R R, Shepherd S, Fulton D J, Sinclair B D, Dunn M H, 1996, *Phys Rev (to be published)*.

Scully M O, 1991, *Phys Rev Lett* **67** 1855.

Cooling and Trapping Neutral Atoms with Laser Light

Andreas Hemmerich

Universität München, Germany

1 Introduction

The study of the motion of neutral atoms under the action of light fields dates back to work of Albert Einstein published as early as 1916 (Einstein 1917). In this article Einstein described a sample of molecules subjected to a thermal light field. He found that the molecular dynamics is determined by the interplay of a frictional force (which cools the molecules) and a (heating) diffusive mechanism arising from the photonic nature of light. As a result,the molecules thermalise with the light field. The first use of the photon recoil for atomic beam deflection is due to Otto Frisch and his famous experiment carried out in the early thirties (Frisch 1933). However, only the advent of tunable narrow band lasers in the sixties has led us to exploit light forces for precision control of atomic motion. The renaissance of light forces began in the late seventies when Doppler-cooling was first proposed by Hänsch and Schalow (1975) for neutral atoms and also by Dehmelt and Wineland (1975) for ions. The development of laser cooling and trapping has enabled atomic physicists to produce intriguing novel types of ultra-cold atomic samples which display the lowest temperatures ever achieved in laboratories.

During the past few years various excellent and extensive lectures on the subject of laser cooling and trapping have been published by a number of researchers who have contributed significantly to the development of the field (Arimondo *et al.* 1992, Dalibard *et al.* 1992). These lectures given at Les Houches 1990 and Varenna 1991 together represent an extensive survey of fundamental concepts developed during the past two decades. Special issues of the Journal of the Optical Society of America have been dedicated to the field of laser cooling and trapping which give a broad collection of excellent reviews of recent achievements (J Opt.Soc.Am B2, Chu and Wiemann 1989).

A complete account of the field of laser cooling and trapping is far beyond the scope of this chapter. Even within the context of the selected materials completeness of presentation has not been attempted. Instead, the chapter is intended to be a brief introduction to a number of selected fundamental aspects of the physics of light forces which may serve as a first step towards the study of more complete and detailed surveys of the field. The average reader is assumed not to be an expert with regard to laser cooling and trapping. The selection of the covered topics consists mainly of a brief presentation of fundamental concepts (two-level atom, absorption and fluorescence spectra, radiation pressure, Doppler-cooling, light shifts, dipole forces, optical pumping, sub-Doppler cooling) which will be used to discuss the trapping of atoms in optical lattices where many of the basic physical mechanisms act together in a subtle interplay.

2 Classical description of light forces acting on an atom at rest

If an atom is subjected to a light field, it is not only the internal degrees of freedom that are affected. The exchange of energy between the light and the atom is accompanied by an exchange of momentum which affects the centre of mass momentum of the atom. Many aspects of the forces exerted on atoms in light fields can be described by a basically classical model in which the atom is treated as a classical object which can be polarised, and the light field is a classical electromagnetic wave: a quantum mechanical model of the internal degrees of freedom of the atom becomes necessary only if we want to specify the atomic susceptibility. In this section we will take such a classical approach which will allow us to identify two basic force contributions: 'radiation pressure' and 'dipole forces'.

Assume a charge e oscillating around the position $\mathbf{r}_0$, *i.e.* $\mathbf{r}(t) = \mathbf{r}_0 + e^{-1}\mathbf{P}(t)$ where $\mathbf{P}(t)$ is the corresponding dipole moment. If the dipole moment $\mathbf{P}(t)$ is subjected to an electro-magnetic radiation field, it will give rise to a time-averaged Lorentz force

$$\mathbf{F}_{\mathrm{L}} = \left\langle \frac{\partial}{\partial t}\mathbf{P} \times \mathbf{B} \right\rangle = -\left\langle \mathbf{P} \times \frac{\partial}{\partial t}\mathbf{B} \right\rangle = \langle \mathbf{P} \times (\boldsymbol{\nabla} \times \mathbf{E}) \rangle \tag{1}$$

and a time averaged Coulomb-force

$$\mathbf{F}_C = \langle e\mathbf{E}(\mathbf{r}(t), t) - e\mathbf{E}(\mathbf{r}_0, t) \rangle \approx \langle (\mathbf{P}\cdot\boldsymbol{\nabla})\mathbf{E} \rangle \tag{2}$$

where

$$\langle F \rangle \equiv \frac{1}{2T} \int_{-T}^{T} F(t)dt$$

denotes the time average over the oscillation period. (The second equality in Equation 1 follows via a partial integration.) The sum of the two force contributions is

$$\mathbf{F} = \mathbf{F}_{\mathrm{L}} + \mathbf{F}_C = \langle \boldsymbol{\nabla}_E(\mathbf{P}\cdot\mathbf{E}) \rangle$$

where ∇_E acts on $\mathbf{E}$ only. We now specialise to a harmonic monochromatic light field and assume that the polarisation is induced by the light field, i. e.

$$\mathbf{E}(\mathbf{r}, t) = \frac{1}{\sqrt{2}}(\mathbf{E}(\mathbf{r})e^{i\omega t} + \mathbf{E}^*(\mathbf{r})e^{-i\omega t}) \tag{3}$$

$$\mathbf{P}(t) = \frac{1}{\sqrt{2}}(\mathbf{P}e^{i\omega t} + \mathbf{P}^*e^{-i\omega t}) \tag{4}$$

The force is then given by

$$\mathbf{F} = \frac{1}{2}(\nabla_E(\mathbf{P}{\cdot}\mathbf{E}^*) + \nabla_E(\mathbf{E} \cdot \mathbf{P}^*)) \tag{5}$$

We express the complex polarisation $\mathbf{P}$ by means of the susceptibility tensor, *i.e.* $\mathbf{P} = \hat{\alpha}(\mathbf{E})\mathbf{E}$ where $\hat{\alpha}(\mathbf{E})$ is a complex $(3{\times}3)$ matrix and choose a basis such that $\hat{\alpha}(\mathbf{E})$ is diagonal with $\alpha_{\nu\nu} = \alpha_\nu + i\beta_\nu$. We may express the components of the electric field with regard to this basis $E_\nu(r) = \sqrt{\frac{I_\nu}{\epsilon_0}}e^{-i\psi_\nu}$ giving

$$\mathbf{F} = \frac{1}{2}\sum_{\nu=1}^{3}\alpha_\nu\nabla I_\nu + \sum_{\nu=1}^{3}\beta_\nu I_\nu\nabla\psi_\nu \tag{6}$$

A study of simple examples demonstrates that it is natural to call the first term of Equation 6 the 'dipole force' and the second term the 'radiation pressure'. Consider a linearly polarised light field $\mathbf{E} = \hat{\mathbf{z}}\sqrt{\frac{I}{\epsilon_0}}e^{-i\psi}$ and an isotropic susceptibility tensor $\alpha_{\nu\nu} = \alpha + i\beta$, $\nu = 1, 2, 3$. Equation 6 then simplifies to

$$\mathbf{F} = \mathbf{f}_{\text{dip}} + \mathbf{f}_{\text{rad}} = -\frac{1}{2}\alpha\nabla I + \beta I\nabla\psi \tag{7}$$

where I is the energy density and ψ is the local phase of the light field. In case of a plane travelling wave $\nabla I = 0$ and $\nabla\psi = \mathbf{k}$, where $\mathbf{k}$ is the wave vector of the light wave. Thus, only the first term of Equation 7 contributes, *i.e.* $\mathbf{F} = \mathbf{f}_{\text{rad}} = \beta I\mathbf{k}$. This force scales with the imaginary part of the susceptibility and is thus connected to relaxation by spontaneous emission. This suggests an interpretation of this force term as describing the phenomenon of radiation pressure. In contrast, for a standing wave we have $\nabla\psi = 0$ and $\nabla I \neq 0$ and the total force reads

$$\mathbf{F} = \mathbf{f}_{\text{dip}} = -\frac{1}{2}\alpha\nabla I.$$

which scales with the spatial gradient of the energy density and the reactive part of the susceptibility. This force is known as the dipole force because it is the induced dipole moment which is acted upon by an inhomogeneous electric field. We will discuss the physical origin of the two kinds of force contributions in Section 4 in more detail.

3 The two-level atom revisited

So far we have considered the forces exerted on polarisable, classical objects. We now treat the atoms in a simple quantum mechanical model which will let us express the

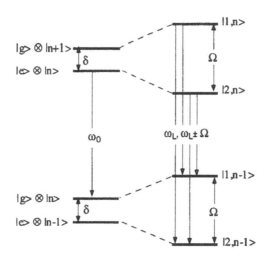

Figure 1. *Schematics of dressed atom. The case $\delta > 0$ is shown. In the absence of interaction the product states are grouped in doublets as depicted on the left side of the figure. When an interaction is present between the atom and the laser mode (right side) the states of each doublet are light-shifted*

atomic susceptibility in terms of physical parameters accessible in experiments. In this section we will briefly discuss atoms having only two levels at some optical energy separation. Such atoms may at first glance seem rather artificial; however, if we restrict ourselves to monochromatic light fields with spatially constant polarisation, then in many instances only two atomic states can be coupled by the light and we may thus neglect all others. We may then employ the force expression of Equation 7; what remains to be done is the calculation of the real and imaginary parts of the atomic susceptibility of a two-level atom. The two-level atom interacting with a monochromatic laser mode represents one of the most fundamental models (called the Jaynes-Cummings model) in quantum optics and has been extensively studied. There is no space here to repeat this discussion which can be found in many textbooks on quantum optics. I will merely summarise selected aspects in order to introduce some concepts and vocabulary which will turn out to be useful in the further discussion of light forces. More experienced readers may want to skip this section and proceed to Section 4.

Our object of interest is an atom interacting with a monochromatic laser mode. We are thus led first to construct the quantum state space of the combined system 'atom and laser mode' and then consider what happens when the two sub-systems interact. The state space of the combined system (called product space in the following) is most easily characterised by the picture given in Figure 1. We denote the atomic state by $|i\rangle$ where $i = g$ for the ground state and $i = e$ for the excited state. The photon states are denoted by $|n\rangle$, $n = 0, 1, 2, \ldots$. When no interaction is present the system is denoted by the product states $|i\rangle \otimes |n\rangle$. This is shown on the left of Figure 1, where positive detuning $\delta = \omega_L - \omega_0$ of the light frequency ω_L with respect to the atomic resonance ω_0 is assumed.

We now consider an interaction between both sub-systems via dipole coupling

$$W_L = -\hat{\boldsymbol{\mu}} \cdot \mathbf{E}_L(x).$$

Here $\hat{\boldsymbol{\mu}} = \boldsymbol{\mu}\hat{b} + \boldsymbol{\mu}^*\hat{b}^+$ is the atomic dipole operator with $\boldsymbol{\mu} = \langle g|\hat{\boldsymbol{\mu}}|e\rangle$, and the ground state projector $\hat{b} = |g\rangle\langle e|$, while

$$\mathbf{E}_L(x) = \frac{1}{\sqrt{2}}(\epsilon(x)\hat{a} + \epsilon(x)^*\hat{a}^+)$$

is the electric field operator with annihilation operator $\hat{a}$ and a normalised polarisation vector $\epsilon(x)$. The result of this interaction is a mixing of the product states $|i\rangle \otimes |n\rangle$ yielding new eigenstates $|1, n\rangle$, $|2, n\rangle$ and eigen-energies

$$E_1(n) = E_{g,n+1} + \Delta E$$
$$E_2(n) = E_{e,n} - \Delta E$$

shifted with respect to the unperturbed energies by an amount

$$\Delta E = \frac{\hbar\delta}{2}\left\{\sqrt{1 + (\omega_1/\delta)^2} - 1\right\} \tag{8}$$

which is called the light shift. The Rabi frequency ω_1 (which is a measure of the interaction strength and scales with the square root of the light intensity) is defined by the equation

$$\omega_1 e^{i\psi} = -\frac{\sqrt{2}}{\hbar}\sqrt{n+1}\,\boldsymbol{\mu}^*\cdot\epsilon(x)$$

The light-shift is an important phenomenon at the basis of many laser cooling mechanisms. In the presence of interaction, the combined system of an 'atom plus light-field' is often referred to as the 'dressed atom' (Cohen-Tannoudji and Reynaud 1977). In order to illustrate the significance of the dressed atom, let us briefly discuss how its fluorescence spectra and probe transmission spectra differ from those of a bare atom. For the bare atom (left in Figure 1) the atomic dipole operator has only one non-vanishing matrix element connecting adjacent state doublets, thus allowing fluorescence and probe absorption only at the Bohr frequency ω_0. In the case of the dressed atom, all matrix elements of the dipole operator (connecting adjacent doublets) are different from zero due to the mixing of the product states. In the fluorescence spectrum we correspondingly expect three emission lines at ω_L, $\omega_L \pm \Omega$ referred to as the Mollow-triplet in the literature (cf. right side of Figure 2; $\Omega = \sqrt{\delta^2 + \omega_1^2}$ is the off-resonant Rabi-frequency). Similarly, we find three frequency components in the probe transmission spectrum of the dressed atom. Two resonances occur at frequencies $\omega_L \pm \Omega$, one resulting in amplification and one resulting in absorption of the probe. This is because the transmitted probe intensity scales with the population difference of the states involved. For red detuning δ, the populations (depicted by the white circles on the left in Figure 2) are such that gain occurs at $\omega_L - \Omega$ and absorption occurs at $\omega_L + \Omega$. Higher order terms

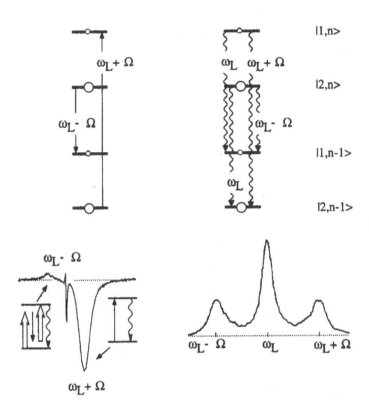

Figure 2. *Fluorescence - (right) and probe transmission (left) for a dressed atom (δ < 0). The probe transmission spectrum stems from atoms trapped in a magneto-optic trap. (cf. Grison et al 1991). The fluorescence spectrum showing the case of resonant excitation (δ = 0) is taken from Grove et al (1977). The filled circles depict the atomic population. The dotted lines in the spectra denote 100% transmission and zero fluorescence intensity respectively.*

with respect to the probe intensity also lead to a dispersive resonance at the laser frequency ω_L. However, the respective resonance on the left of Figure 2 is covered by a much larger contribution resulting from the multi-level structure of the cold magneto-optically trapped atoms from which the spectrum has been obtained.

In order to obtain expressions for the atomic susceptibility (which is needed to evaluate the force in Equation 7) we have to consider the time evolution of our system including the damping due to spontaneous decay. The two-level atom in a monochromatic laser field turns out to be equivalent to a damped driven spin 1/2 system. Therefore, its dynamics is described by a two-level optical Bloch-equation in analogy to the Bloch-equation used in NMR physics. A complete derivation of the two-level optical Bloch-equation can be found in most textbooks on quantum mechanics and will not be repeated here. I will merely comment on some of the steps to be taken and mention the final result. The starting point is the Heisenberg equation for the density operator of the combined system 'atom plus laser mode'. This equation is evaluated, e.g., in the basis of product states $|i\rangle \otimes |n\rangle$ and assuming the rotating wave approximation (*i.e.* keeping only the slowly oscillating terms, $\hat{a}\hat{b}^+$ and $\hat{a}^+\hat{b}$, in the interaction operator). Actually, we are only interested in the evolution of the atomic degrees of freedom and may thus trace over that part of the density matrix which describes the laser mode. We are then left with an evolution equation for the reduced density matrix s accounting only for the atomic degrees of freedom and we may define the optical Bloch vector

$$\begin{pmatrix} u \\ v \\ w \end{pmatrix} = \begin{pmatrix} \sigma_{eg} + \sigma_{ge} \\ i(\sigma_{eg} - \sigma_{ge}) \\ \sigma_{gg} - \sigma_{ee} \end{pmatrix} \tag{9}$$

where u and v denote the real and imaginary parts of the coherence σ_{eg} (which is proportional to the atomic susceptibility), and w is the inversion, *i.e.* the difference between populations in the excited state and ground state. The Bloch vector satisfies the Bloch-equation:

$$\frac{\partial}{\partial t} \begin{pmatrix} u \\ v \\ w \end{pmatrix} = \begin{pmatrix} -\omega_1 \\ 0 \\ \delta \end{pmatrix} \times \begin{pmatrix} u \\ v \\ w \end{pmatrix} - \frac{\Gamma}{2} \begin{pmatrix} 1 & 0 & 0 \\ 0 & 1 & 0 \\ 0 & 0 & 2 \end{pmatrix} \begin{pmatrix} u \\ v \\ w \end{pmatrix} - \begin{pmatrix} 0 \\ 0 \\ \Gamma \end{pmatrix} \tag{10}$$

Here Γ is the decay rate of the excited state. As can be seen directly from Equation 10, in the absence of the damping terms (*i.e.* when setting Γ to zero) the Bloch-vector maintains its length and precesses around the 'field'-vector. Damping leads to a decay of that motion and an approach towards a steady state value (reached within an atomic excited state lifetime Γ^{-1}) that we can easily calculate by setting the left side of Equation 10 to be zero. One finds that

$$\begin{pmatrix} u \\ v \\ w \end{pmatrix} = \frac{1}{1+s} \begin{pmatrix} -2s\frac{\delta}{\omega_1} \\ s\frac{\Gamma}{\omega_1} \\ 1 \end{pmatrix}, \qquad s = \frac{1}{2} \frac{\omega_1^2}{\delta^2 + \left(\frac{\Gamma}{2}\right)^2} \tag{11}$$

We immediately see that for increasing coupling strength (*i.e.* $s \gg 1$) the inversion w approaches zero, *i.e.* half of the atoms populate the excited state. From Equation 11 we may evaluate the polarisation $P = \text{trace}(\sigma\mu|g\rangle\langle e|) = \mu(u - iv)$ and find the steady state susceptibility of the two-level atom $\alpha + i\beta$ which, by means of Equation 7, allows us to calculate the light forces for a resting atom:

$$\alpha = \frac{\hbar c}{I_s} \left(\frac{\Gamma}{2}\right)^2 \frac{-\delta}{\delta^2 + (\frac{\Gamma}{2})^2} \frac{1}{1+s}, \qquad \beta = \frac{\hbar c}{I_s} \left(\frac{\Gamma}{2}\right)^2 \frac{\frac{\Gamma}{2}}{\delta^2 + (\frac{\Gamma}{2})^2} \frac{1}{1+s} \tag{12}$$

Here I_s is the saturation intensity which is a measure of the strength of the transition matrix element (*e.g.* for the Rubidium D2-line $I_s = 1.6\text{mW/cm}^2$, which is a typical value for a strong transition). Figure 3 shows the dependence of α and β on δ for small saturation $s \ll 1$. Both α and β display a resonance behaviour with a width given by Γ.

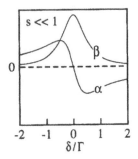

Figure 3. *The steady state susceptibility $\alpha + i\beta$ of a two-level atom.*

4   Radiation pressure

In this section we will discuss radiation pressure in two important examples of light fields, plane travelling and plane standing waves. We begin with a plane travelling wave $\mathbf{E} = \hat{\mathbf{z}}\sqrt{\frac{I}{\epsilon_0}}e^{-i\psi}$, *i.e.* the energy density I is constant and the local phase is $\psi = kx$. In this case only the second term on the right of Equation 7 contributes to the force, *i.e.* we have only radiation pressure in this example. The physical interpretation of radiation pressure is that atoms absorb photons from a light beam and thereby experience momentum kicks $\hbar k$. The absorbed energy is dissipated by some relaxation mechanism (which is spontaneous decay here). In this dissipation process no momentum is exchanged on average. This is true for spontaneous decay because photons are spontaneously emitted with an isotropic spatial distribution. Figure 4(a) summarises the physical mechanism. We may easily derive a quantitative expression from this figure. If we assume that a momentum $\hbar k$ is transferred per atomic lifetime Γ^{-1} and Π_e is the probability of the atom being excited, we expect the force to be

$$f_{\text{rad}} = \hbar k \Gamma \Pi_e, \qquad \Pi_e = \frac{\omega_1^2}{4\delta^2 + \Gamma^2 + 2\omega_1^2} \tag{13}$$

where Π_e is easily obtained from Equation 11. The maximum value for this force is $\frac{1}{2}\hbar k \Gamma$. This is in accordance with the result obtained if Equations 7 and 12 are combined. Because of the tiny atomic mass this force leads to extremely high values of the atomic acceleration, typically 10^6m/s^2.

Radiation pressure can be used to cool atoms. In order to see this, let us now consider the situation sketched in Figure 4(b). For an atom at rest we may expect that radiation pressure from both travelling waves cancels, as is seen directly from Equation 7. We will now go a step beyond Equation 7 and consider a moving atom. If the light frequency ω_L is tuned slightly below resonance (*e.g.* $\delta = -\Gamma/2$) a moving atom with velocity $v = \Gamma/2k$) will tune itself into resonance with the counter-propagating light beam by means of the Doppler-effect (*i.e.* $\delta_+ = 0$ and $\delta_- = -\Gamma$). This yields a net frictional force opposed to the direction of the atomic motion which leads to efficient cooling. The picture in Figure 4(b) lets us estimate the size of this force

$$f = (\Pi_e(\delta + kv) - \Pi_e(\delta - kv))\hbar k \Gamma = \gamma v + O(v^2) \tag{14}$$

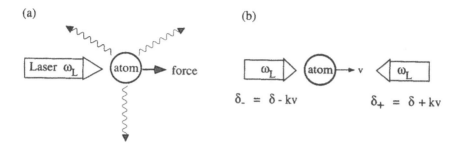

Figure 4. *(a): The physical mechanism of radiation pressure. (b): One-dimensional optical molasses. A moving atom interacting with light beams at effective detuning $\delta_\pm$.*

where

$$\gamma = \frac{\omega_1^2 \delta \hbar k^2 \Gamma}{\left(\delta^2 + (\Gamma/2)^2 + \omega_1^2/2\right)^2}$$

Here ω_1 is the Rabi-frequency corresponding to each travelling wave. The coefficient of friction γ is negative for negative δ and its magnitude acquires a maximum for $\delta = -\Gamma/2$ and $\omega_1 = \Gamma$ with $\gamma_{max} = -\frac{1}{2}\hbar k^2$.

The frictional force in Equation 14 leads to efficient cooling, *i.e.* a decrease of the mean kinetic energy of an atomic sample. At first glance one may expect that this continues until the kinetic energy transferred by a single photon recoil is of the order of the mean kinetic energy resulting in a steady state temperature, the recoil temperature, given by $k_B T_{rec} = (\hbar k)^2/m$ However, there is a diffusion mechanism that counteracts cooling leading to a steady state temperature which for many atoms is significantly larger then T_{rec}. This diffusion arises because of the discrete character of spontaneous emission which leads to a random walk of the atom in momentum space. Let us again consider the two travelling waves separately and then add their contributions to this random walk. Assuming that every lifetime, an atom has Π_e probability to be excited and to experience a momentum kick $\hbar k$ resulting from a spontaneous emission process, we can write the momentum accumulated through such processes at time t and the corresponding kinetic energy $E_{kin}(t)$:

$$\mathbf{P}(t) = \sum_{i=1}^{N=\Pi_e \Gamma t} \hbar \mathbf{k}_i$$

$$E_{kin}(t) = \frac{P(t)^2}{2m} = \frac{\hbar^2 k^2}{2m}\Pi_e \Gamma t \tag{15}$$

In taking the square of $\mathbf{P}(t)$ we have accounted for the randomness of the direction of the $\mathbf{k}_i$. The total increase in kinetic energy from both beams turns out to be D_1/m where m is the atomic mass and $D_1 = \hbar^2 k^2 \Pi_e \Gamma$ is the diffusion constant. There is a second mechanism of diffusion connected with the absorption of photons from the two light beams resulting from shot noise. The probability of absorbing a photon follows a Poisson distribution. Thus, the mean deviation of the number of absorbed photons at time t is the square root of the mean number of absorbed photons. The momentum accumulated due to shot noise in each beam is $P(t) = \hbar k \sqrt{\Pi_e \Gamma t}$. This results in a total increase of kinetic energy of D_2/m where $D_2 = \hbar^2 k^2 \Pi_e$ is the diffusion constant. The sum of the two diffusion constants is thus $D = D_1 + D_2 = 2\hbar^2 k^2 \Pi_e \Gamma$. In fact, the absorption statistics may be non-Poissonian at large light intensities where saturation plays a significant role. In this case the total diffusion can exceed the value estimated here. We can evaluate the mean kinetic energy, $\overline{E}_{kin}$, in steady state, *i.e.* when diffusion heating balances cooling:

$$\left(\frac{\partial}{\partial t}\right)_{fric} E_{kin} + \left(\frac{\partial}{\partial t}\right)_{diff} E_{kin} = 0 \qquad \Rightarrow \qquad \overline{E}_{kin} = \frac{D}{2\gamma} \qquad (16)$$

This allows us to derive the steady state temperature as

$$k_B T = \frac{1}{2}\hbar\Gamma\sqrt{1 + 2\left(\frac{\omega_1}{\Gamma}\right)^2} \qquad (17)$$

The steady state temperature in Equation (17) represents the minimum temperature that can be achieved with radiation pressure cooling and can be quite small. For example, for low saturation (*i.e.* $\omega_1 \ll \Gamma$) in the case of the Rubidium atom we find $T = 139\ \mu K$. You may be surprised that the temperature in Equation (17) acquires its minimum value at vanishing laser intensity ($\omega_1 = 0$). However, if the Rabi-frequency tends to zero, the time needed to reach the steady state temperature $(\Pi_e \Gamma)^{-1}$ approaches infinity.

Radiation pressure cooling, first proposed by Hänsch and Schawlow (1975), is also referred to as Doppler cooling (owing to the important role of the Doppler-effect) and its limit given by Equation (17) is referred to as the Doppler-limit. A superposition of three mutually orthogonal standing waves leads to a three-dimensional extension which is usually referred to as 'optical molasses' which was experimentally realised for the first time by Chu *et al.* (1985).

Doppler cooling has also been successfully used to decelerate and cool atomic beams using the experimental configuration sketched in Figure 4(a). Different techniques have been employed in order to keep the atoms in resonance during the deceleration process. One of them uses an inhomogeneous magnetic field along the atomic beam axis which tunes the atomic resonance frequency by means of the Zeeman-effect (Phillips and Metcalf 1982) (Zeeman-cooling). Another method is to actively change the light frequency synchronously with the deceleration process (Chirp technique) (Letokhov *et al.* 1976). The first method is a DC method; however, it uses a magnetic field gradient which needs carefully designed magnets, while the second method only leads to a pulsed beam of cold atoms but is more easily realised.

In Figure 5 a sketch of the chirping technique is depicted. A laser beam counterpropagates a beam of Rubidium atoms (Figure 5(a)). While the atoms are decelerated, the light frequency is increased in order to compensate for the decreased Dopplerdetuning (cooling chirp in (b)). After such a cooling chirp the cooling laser is shut off for a short time and a second light beam (active while the detection gate is high in (b))

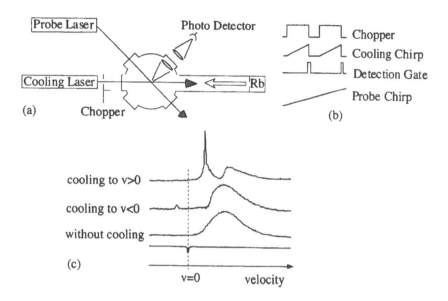

Figure 5. *Cooling an atomic beam using the chirp-technique. The data in (c) are taken from Hemmerich (1990).*

resonantly excites some atomic velocity class while the resulting fluorescence is observed (plotted on the y-axis in (c)) with a photo detector. For successive probe periods, the probe frequency is slightly altered (probe chirp in (b)) in order to measure the atomic velocity distribution. The highest frequency reached by the cooling chirp determines the final atomic velocity. The four traces in Figure 5(c) are (in descending order) cooling to positive and negative final velocities, the Maxwellian velocity distribution observed when no cooling chirp is applied and a reference trace which marks zero velocity.

5   Trapping atoms with radiation pressure

It also seemed clearly desirable to use the strong forces of radiation pressure to build atom traps. However, it soon turned out that just as electric charges cannot be trapped by static electric fields, radiation pressure is not appropriate for trapping two-level atoms. Note that even though long storage times can be realised, optical molasses only provide strong friction and thus do not represent a real trap. The reason that trapping is impossible is that under quite general conditions, the divergence of the radiation pressure force vanishes for resting atoms. One way to see this is to establish a connection between the radiation pressure force as given by Equation (13) and the Poynting vector describing the flow of electromagnetic energy. We consider an arbitrary monochromatic light field with spatially constant polarisation $\mathbf{E}(x) = \hat{\mathbf{e}}f(x)$ where $\hat{\mathbf{e}}$ is a complex unit vector, and $f(x)$ is a real function in three-dimensional space which can

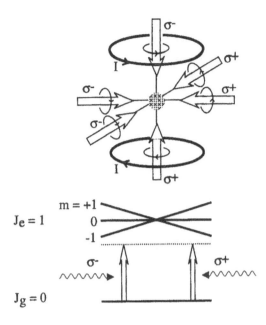

Figure 6. *Schematics of the magneto-optic trap.*

generally be written as

$$f(x) = \sqrt{\frac{I(x)}{\epsilon_0}}\,e^{-i\psi(x)}.$$

The restriction to spatially constant polarisation appears not to be a loss of generality as long as we consider two-level atoms. Using $\nabla\cdot\mathbf{E} = 0$, we can calculate the complex Poynting vector

$$\mathbf{S} = \frac{1}{\mu_0}\mathbf{E}\times\mathbf{B}^* = \frac{-i}{\mu_0\omega}\mathbf{E}\times(\nabla\times\mathbf{E}^*) = \frac{-i}{\mu_0\omega}f\nabla f^* = \frac{c^2}{\omega}(I\nabla\psi - \frac{i}{2}\nabla I).$$

In particular, in combination with Equation (7) we find that the radiation pressure force is $\mathbf{f}_{rad} = (\beta\omega/c^2)\mathrm{Re}(\mathbf{S})$. In combination with the fact that the real Poynting vector $\mathrm{Re}(\mathbf{S})$ has vanishing divergence, this is quite an interesting result because it tells us that $\nabla\mathbf{f}_{rad} = 0$ if the imaginary part of the atomic susceptibility β is spatially constant. As a consequence, for constant β we cannot use radiation pressure to trap atoms in three dimensions. This fact is sometimes called the optical Earnshaw theorem (Ashkin and Gordon 1983) in analogy with the Earnshaw theorem which makes a similar statement about trapping atoms with static electric fields.

There are a number of ways around the optical Earnshaw theorem, all of which rely on a spatially varying non-isotropic atomic susceptibility tensor obtained from a multi-level atomic transition. A very successful example is the magneto-optic trap (MOT) which uses a quadrupole magnetic field in order to spatially modulate the atomic resonance frequency and thus the imaginary part of the susceptibility tensor. This radiation pressure trap was first realised by Raab *et al.* (1987), building on a proposal by J. Dalibard. Consider an atom with a ground state of total angular momentum $J_g = 0$ and an excited state with $J_e = 1$ as depicted in Figure 6. Six laser beams with circular polarisation are arranged as shown in Figure 6 with a frequency detuning slightly below the atomic resonance frequency. In the trap centre, the magnetic field is zero and the radiation pressure from all beams adds up to zero (We neglect interference effects in our brief discussion here). When the atom is displaced from the trap centre, say to the right, the magnetic field tunes the $m=0 \rightarrow +1$ transition in to resonance and the atom predominantly interacts with the σ^+-beam which by means of radiation pressure redirects the atom towards the trap centre. The same holds for all other directions. The magneto-optic trap has been considerably improved since its first demonstration and has become a standard tool in many atomic physics laboratories. It was shown that the magneto-optic trap can be loaded from the low velocity tail of a thermal distribution of atoms in an atomic vapour at room temperature (Monroe *et al.* 1990). A standard magneto-optic trap allows trapping of typically 10^8 to10^{10} atoms with a density of 10^{11} atoms/cm^3. In slightly modified traps, densities above 10^{12} atoms/cm^3 have been observed (Gibble and Chu 1992). Steady state temperatures well below the Doppler-limit are possible due to cooling mechanisms connected with the atomic multi-level structure (an example is discussed in Section 7).

6 Dipole forces

In Section 5 we have concentrated on radiation pressure connected with the imaginary part of the atomic susceptibility. Here we are interested in the first force term in Equation 7 which scales with the real part α of the atomic susceptibility. We consider a light field with spatially varying intensity $\nabla I \neq 0$. We may insert α as given by Equation 12 into Equation 7 and find that

$$\mathbf{f}_{\text{dip}} = -\nabla \frac{\hbar \delta}{2} \ln(1 + s) \qquad s = \frac{1}{2} \frac{\omega_1^2}{\delta^2 + (\Gamma/2)^2}. \tag{18}$$

This force emerges as the gradient of a potential which we may call the optical potential. Let us specify our light field to be a one-dimensional plane standing wave of intensity $I = I_0 \cos^2(kx)$ so that $\nabla I = I_0 k \cos(kx - \pi/4)$. We then immediately see that, in the limit of large intensities and detunings ($\Gamma \ll \delta \ll \omega_1$), the force acquires a maximum strength $\frac{1}{2}\hbar k \delta$, which is a factor δ/Γ larger then the maximum possible strength of radiation pressure, *i.e.* it does not saturate with intensity.

We can easily identify the physical origin of the dipole force. The light field induces a dipole moment. This moment oscillates in phase with the driving field if its frequency is tuned below resonance and is thus driven towards high intensity regions. When the driving frequency is tuned above resonance the induced dipole is driven with a 180 degree phase lag and is therefore expelled from the high intensity regions.

This mechanism can be described particularly well in the dressed atom picture (cf. Figure 1) (Cohen-Tannoudji and Reynaud 1977). The standing wave induces light shifts

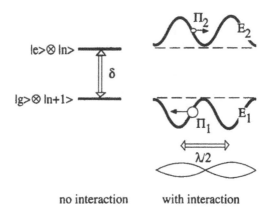

no interaction   with interaction

Figure 7. *The physical mechanism of the dipole force (cf. J. Dalibard and C. Cohen-Tannoudji 1985).*

depicted on the right of Figure 7. The total force suggested by this picture should be

$$\mathbf{f}_{dip} = \Pi_1 \nabla E_1 + \Pi_2 \nabla E_2 = (\Pi_2 - \Pi_1)\nabla\Omega$$

where Π_1 and Π_2 are the populations of the dressed states, E_1 and E_2 are the spatially light shifted energies of the dressed states, and $\Omega = \sqrt{\delta^2 + \omega_1^2}$ is the off-resonant Rabi-frequency. The inversion $\Pi_1 - \Pi_2$ of the dressed atom can be calculated by transforming the steady-state Bloch-vector of Equation 11 into the dressed state basis which yields the same result as given by Equation 18.

The dipole force can be used, *e.g.* to channel the atoms of an atomic beam into the valleys of the nodes or antinodes of an optical standing wave (Salomon *et al.* 1987, Hemmerich *et al.* 1991,1992). This is sketched in Figure 8(a). An atomic beam intersects an optical standing wave at 90 degrees. If the transverse kinetic energy is smaller than the potential depth, *i.e.* $\frac{1}{2}mV_{trans}^2 < \frac{\hbar\delta}{2}\ln(1+s)$, the atoms cannot climb the potential hills and are thus channeled into potential valleys, where they oscillate with a characteristic frequency. The atoms leave the standing wave either specularly reflected or with their incident k-vector unchanged because the interaction with the standing wave is purely elastic.

The channelling can be observed by exciting the atoms with a weak probe beam to a level which is not perturbed by the optical standing wave (cf. Figure 8(b)) (Hemmerich *et al.* 1991,1992). The resulting fluorescence is recorded versus the probe frequency in order to see spectra as shown in Figure 8(c). The fluorescence spectra show large inhomogeneous broadening because different atoms travel at different light intensities in the standing wave and thus contribute to the spectrum at resonance frequencies light-shifted by different amounts. When the atoms are channeled inside the nodes of the standing wave (which is the case for blue detuning δ of the standing wave frequency) the fluorescence spectrum is less broadened as compared to the case when channeling is suppressed by adjusting too high transverse velocities. The lowest trace in Figure 8(c) shows the fluorescence spectrum if no standing wave is applied.

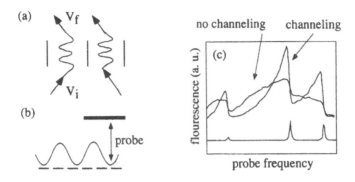

Figure 8. *Channeling of an atomic beam by dipole forces.*

7 Sub-doppler cooling: a semi-classical model

Around 1988 American researchers found in their experiments that the temperature in optical molasses was significantly lower than predicted by the theory of Doppler-cooling (Lett *et al.* 1988). A year later new cooling mechanisms were identified (Dalibard and Cohen-Tannoudji 1989) which turned out to be much more efficient then Doppler-cooling leading to temperatures two orders of magnitude below the Doppler limit. These new cooling mechanisms arise in optical fields which provide spatial gradients of the polarisation for atoms providing a manifold of degenerate Zeeman-levels. The susceptibility is not isotropic in this case and the force acting on a resting atom is then given by the tensorial expression in Equation 6 which is generally quite complex and not very intuitive.

The simplest and most instructive example is an atom providing a $(J = 1/2 \rightarrow 3/2)$ transition inside an optical standing wave composed of two counter-propagating travelling waves with orthogonal linear polarisations (this configuration is often designated as $lin \perp lin$) as depicted in Figure 9. The $lin \perp lin$ configuration can be decomposed into two circularly polarised standing waves with opposite helicity shifted with respect to each other by a quarter of the optical wavelength, *i.e.* the total polarisation alternates between clockwise circular (σ^+) and counterclockwise circular (σ^-). Noting that $\sigma^\pm$ light couples $(\Delta m = \pm 1)$ transitions and taking into account the different coupling strengths, due to different Clebsch-Gordon-coefficients, (1 and 1/3 as depicted in Figure 9a), we can plot the spatially varying light shifts of the two ground-state Zeemann components (cf. Figure 9b). Let us briefly consider an atom at some location where the light field is σ^+. If the atom is originally in the $m = -1/2$ groundstate, excitation and subsequent spontaneous decay leads back to $m = \pm 1/2$. Once the atom has been transferred to the $m = +1/2$ groundstate, further excitation can only reach the $m = +3/2$ excited state from where spontaneous decay always leads back to $m = +1/2$. Thus after a few cycles of excitation and spontaneous decay the entire population is optically pumped to $m = +1/2$. Similarly σ^- light transfers the atomic population into

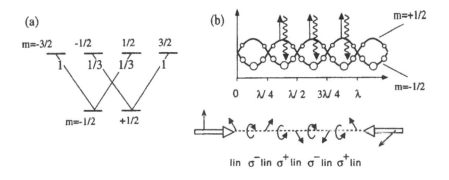

Figure 9. *Light-shifts for a $J = 1/2 \to 3/2$ transition in the $lin\perp lin$ -configuration.*

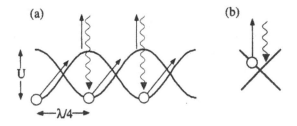

Figure 10. *(a) Model of Sub-Doppler cooling in the $lin\perp lin$-configuration. (b): Optical pumping yields fluctuations of the direction of the gradient force arising in the optical bipotential.*

the $m = -1/2$ ground state. The steady-state population is depicted by the open circles in Figure 9(b). Note that the lower lying state is always the most populated state.

Assume now an atom starting in some potential minimum and moving to the right in Figure 10(a). While the atom climbs up the potential well, kinetic energy is transferred into potential energy. The higher the atom climbs up, the higher the probability for an optical pumping process back into a potential valley. The spontaneous photon emitted in this process carries away the potential energy. Such cooling cycles repeatedly occur until the kinetic energy of the atom is not sufficient anymore to climb the potential well and the atom remains trapped. This cooling has been named Sisyphus-cooling because of the similarity of its cyclic mechanism to the unattainable task of the ancient Greek hero. We can roughly estimate the frictional force arising in Sisyphus-cooling in the regime of low saturation and large detuning ($s \ll 1$) where the saturation parameter s refers to the circularly polarised antinodes). In this case the optical pumping rate Γ_p for $\Delta m = \pm 1$ transitions (which are responsible for the cooling) approximately equals $\Gamma_s/2$ times $2/9$ (the factor $2/9$ is due to the respective Clebsch-Gordon coefficients) which is much smaller than the decay rate of the excited state Γ. The depth of the potential wells U is found to equal $\hbar\delta s/3$. The force should acquire a relative maximum if the atom moves a distance $\lambda/4$ during one optical pumping time, $\tau_p = \Gamma_p^{-1}$ *i.e.* the corresponding atomic velocity is

$$k v_{\max} \approx \Gamma_p, \qquad \Gamma_p = \frac{\Gamma s}{9} \tag{19}$$

Assume an atom moves at the speed $v_{\max}$. According to Figure 10(a), the energy W dissipated during the optical pumping time τ_p should be U. With the help of Equation 19 we find that

$$v_{\max}F = \frac{\partial W}{\partial t} = U\Gamma_p \qquad \Rightarrow \qquad F = \gamma v, \quad \gamma = -Uk^2\tau_p \tag{20}$$

The friction coefficient γ in Equation 20 is larger than the maximum friction coefficient for radiation pressure by a factor $2U/\hbar\Gamma_p = 6\delta/\Gamma \gg 1$ (cf. Equation14).

As for radiation pressure cooling we have to encounter diffusion due to the randomness of the direction of spontaneous emission and also due to the shot noise in the absorption processes (see Section 4). From these mechanisms we obtain a diffusion constant of the order $\hbar^2k^2\Gamma_p$. An additional source of momentum diffusion in Sysiphus-cooling is due to optical pumping induced instantaneous flips of the sign of the dipole force acting on atoms predominantly near a crossing of the two optical potentials (cf. Figure 10(b)). Let us estimate this, as it will turn out, dominant contribution to the diffusion by considering the momentum $P(t)$ accumulated at time t, if a randomly directed momentum kick $f\tau_p$ occurs during an optical pumping time τ_p where the force f equals the gradient of the optical potential well, *i.e.* $f = kU$:

$$P(t) = \sum_{i=1}^{N=\Gamma_p t} f_i \tau_p, \qquad |f_i| = f \tag{21}$$

By taking the square of Equation 21 and accounting for the randomness of the sign of f_i we may calculate the increase of kinetic energy

$$\left(\frac{\partial}{\partial t}\right)_{\text{diff}} E_{\text{kin}} = \frac{D}{m}, \qquad D = \frac{\tau_p f^2}{2} \tag{22}$$

The diffusion constant D is of the order $\hbar^2k^2\Gamma_p\delta^2/\Gamma^2$ and thus represents the dominant contribution in the case of low saturation (*i.e.* $\delta \gg \Gamma$). The steady state kinetic energy (when cooling compensates for heating due to momentum diffusion) is

$$\overline{E}_{\text{kin}} = \frac{D}{2\gamma} = U/2 \qquad \Rightarrow \qquad k_B T = U \tag{23}$$

From Equation 23, we see that the minimum temperature achieved with Sysiphus-cooling scales with the depth of the potential wells which, in principle, can be made arbitrarily small. However, our kinetic energy budget used to obtain Equation 20 is only correct under the basic assumption that the depth of the potential wells, U, is significantly larger than the energy E_{rec} transferred to an atom by the recoil of a single optical pumping photon. As U approaches E_{rec} the amount of kinetic energy dissipated per cooling cycle decreases significantly below the value assumed in Equation 20 and thus the cooling efficency is degraded and the steady state temperature will rise again. The minimum temperature T should occur for some optimal value $U = \epsilon E_{rec}$ where $\epsilon > 1$:

$$T = \epsilon T_{rec}/2, \qquad k_B T_{rec} = (\hbar k)^2/m \qquad (24)$$

The recoil temperature T_{rec} in Equation 24 is the temperature corresponding to the recoil exchanged in a single atom-photon interaction and can be quite low. For example, for Rubidium T_{rec} is only 370nK.

Furthermore, as we approach low temperatures, a basic condition for the derivation of Equation 22 is violated. We have assumed in this equation that all atoms experience the same friction coefficient, *i.e.* the mean velocity v_{rms} of our atomic sample should be smaller than v_{max}. We can easily evaluate this condition finding

$$v_{rms} = \sqrt{\frac{U}{m}} < v_{max} = \frac{\Gamma_p}{k} \qquad \Rightarrow \qquad \Omega_{vib}\tau_p < \sqrt{2}, \qquad \Omega_{vib} = k\sqrt{\frac{2U}{m}} \qquad (25)$$

where Ω_{vib} is the vibrational frequency corresponding to the harmonic part of the optical potential. According to Equation 25 our semiclassical model works as long as the time for a vibrational period $1/\Omega_{vib}$ of an atom inside the potential well of depth U is longer than the optical pumping time. When U tends to zero, $\Omega_{vib}\tau_p$ increases and will eventually exceed $\sqrt{2}$. At this point the friction coefficient will start to decrease and a further decrease of the temperature as predicted by Equation 23 is not to be expected. This limit for U roughly given by the condition $\Omega_{vib}\tau_p = \sqrt{2}$ would infer that $\epsilon = 18(\delta/\Gamma)^2$ in Equation 24. Since all considerations in this section were based on the low saturation assumption $\Gamma \ll \delta$, this shows, that long before we get even close to the recoil temperature, we enter the regime where the atom will oscillate many times in its potential well between successive optical pumping cycles ($\Omega_{vib}\tau_p > \sqrt{2}$). A more elaborate model of Sysiphus cooling is needed to describe the steady state temperature in this regime (Castin and Dalibard 1991).

8   Atoms trapped in optical lattices

The recoil limit at first glance appears to be a quite fundamental limit for any cooling mechanism that involves spontaneous emission. However, most interestingly, even this limit has been surpassed by light-induced cooling mechanisms that, in one way or another, prevent diffusion due to the photon recoil by entirely decoupling from the light field those atoms which are nearly at rest (Aspect *et al.* 1988, Kasevich and Chu 1992). In this chapter we will not follow these new developments, but turn our attention to some quite interesting aspects of atoms cooled to near the recoil limit. For such atoms the de Broglie wavelength is on the order of the optical wavelength and thus comparable to the width of the optical potential wells involved in the cooling mechanism illustrated in Figure 10. In this regime it is no longer justified to treat the atomic external degrees

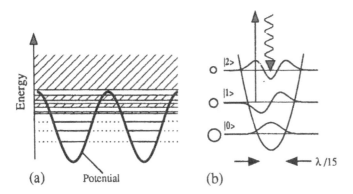

Figure 11. *Band-model for cold atoms moving in a perodic potential.*

of freedom classically. A quantum description of the position and momentum of the atomic centre of mass inside a periodic potential is required, similar to the description of an electron gas in a solid. Thus, we naturally arrive at a band model with discrete allowed energy bands separated by band gaps as depicted in Figure 11(a) (Castin and Dalibard 1991). The depth of the potential well U typically realized in experiments is on the order of a few MHz, while the typical band gap at the bottom of the potential well is on the order of 100kHz. The recoil temperature only corresponds to a few kHz showing that at such low temperatures we may expect only a few of the lowest states in the optical potential to be populated. The tunnelling rates for these low lying states are extremely small (Castin and Dalibard 1991) such that we may treat each potential valley as completely independent and thus may apply a harmonic oscillator model as sketched in Figure 11(b).

In order to find the temperature limit of Sisyphus-cooling we have to discuss it in terms of this oscillator model. In this picture, the cooling arises due to spontaneous Raman-transitions between different vibrational states (sketched in Figure 11(b) and in Figure 12(a), first and third detail). The relaxation rates of the vibrational levels via such Raman-transitions are proportional to the vibrational quantum number v and the optical pumping rate Γ_p for a free atom, multiplied by a small factor $\xi = E_{rec}/E_{vib}$. (sometimes called the Lamb-Dicke factor); where E_{rec} is the recoil energy and $E_{vib} = \hbar\Omega_{vib}$ is the energy separation between adjacent vibrational states (Courtois and Grynberg 1992) The factor $v\xi$ (similar to the Franck-Condon factor in molecular physics) comes into play because of the spatial confinement of the atomic centre of mass wave functions which decreases with increasing v: the wavefunctions of well confined levels have little overlap with those of other levels. A rate equation involving the relaxation rates of the vibrational levels yields their steady state populations. Typical vibrational temperatures for optimal conditions are ten times the recoil temperature. The relaxation rates for the low lying vibrational states are typically a few tens of kHz and thus are significantly smaller than $E_{vib}/\hbar$. Therefore, the vibrational

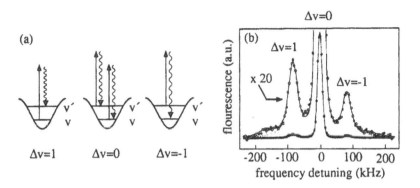

Figure 12. *Fluorescence of atoms cooled to near the recoil limit by Sysiphus-cooling. The spectrum has been taken from Jessen et al (1992).*

levels can be resolved, *e.g.* by probe transmission spectroscopy (Verkerk *et al.* 1992, Hemmerich and Hänsch 1993b), or by fluorescence spectroscopy (Jessen *et al.* 1992). In the fluorescence spectrum shown in Figure 12(b) we recognize a sharp central resonance due to transitions which do not alter the vibrational quantum number v (central feature of Figure 12(a)). These elastic processes are not suppressed by the Lamb-Dicke factor ξ and thus dominate the fluorescence. We also observe much smaller vibrational sidebands due to the spontaneous Raman transitions between adjacent vibrational levels which are responsible for the cooling mechanism. The asymmetry between the left and right sidebands results because a large fraction of atoms populates the vibrational ground state from where $\Delta v = -1$ processes are not possible.

The cooling and trapping of atoms in laser fields with polarisation gradients can be extended to two and three dimensions. By using appropriate light fields, two-dimensional (2D) and three-dimensional (3D) periodic arrays of microscopic light-traps can be realized (Hemmerich and Hänsch 1993b, Hemmerich *et al.* 1993, Grynberg *et al.* 1993) which have been called 'optical lattices'. The atomic vibrational motion in these traps is quantised and can be studied by means of spectroscopic methods in analogy with the one-dimensional case. A 2D example is realized by superimposing four plane travelling waves with parallel linear polarisations as depicted in Figure 13(a). Each pair of counter-propagating travelling waves form a 1D standing wave. These standing waves can oscillate with different relative phases leading to quite different 2D spatial patterns of the polarisation. If the relative time-phase difference ϕ is set to 90 degrees, a polarisation pattern is created in two dimensions (cf. Figure 13(b)) which is the natural extension of the 1D polarisation gradient used for Sysiphus-cooling in Figure 9(b). We may also proceed to a 3D extension by superimposing a third circularly polarised standing wave perpendicular to the drawing plane in Figure 13(a). If we choose the time-phase ψ of this third standing wave in order to obtain maximum constructive interference for circular polarisation (we define $\psi = 0$ in this case) we obtain a body-centered

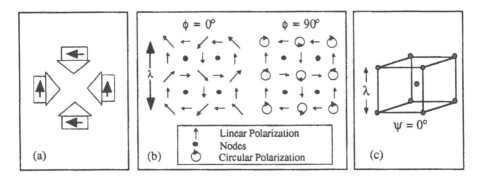

Figure 13. *Light-field geometries for 2D and 3D optical lattices.*

cubic lattice of circularly polarised antinodes as depicted in Figure 13(c) (Hemmerich *et al.* 1994b).

A different experimental solution to the creation of appropriate light fields has been suggested out by Grynberg *et al.* (1993). For a 2D optical lattice, three travelling waves are superimposed along the three bisectors of an equilateral triangle. A 3D lattice is made by directing travelling waves along the four surface normals of a tetrahedron. These experiments use fewer beams than the configurations shown in figure 13 at the expense of having fewer controllable parameters.

The vibrational quantum states can be experimentally observed by stimulated Raman spectroscopy. Assume that a weak probe laser beam is directed through the atomic sample and its transmission is recorded versus its frequency. When the probe frequency (ν_{probe}) differs from that of the lattice field (ν_{lattice}) by the energy separation between a pair of vibrational states, a Raman transition between these states can be excited (cf.Figure 14(a)). Such transitions can in principle occur in both directions, however, because lower states have higher thermal populations (as depicted by the open circles in Figure 14(a)), there is an excess of transitions from the lower to the upper state which leads to amplifying and absorbing resonances (Raman sidebands) in the probe transmission spectrum (cf. Figure 14(b)).

The photon picture shown in Figure 14(a) may be instructive in order to understand the energy budget of Raman transitions. However, a semi-classical picture appears to be more appropriate since all light fields are coherent fields with large mean photon numbers. In fact we can understand the excitation of atomic vibrations by considering the oscillating distortion of the optical potential introduced by the interference between the probe beam and the lattice field (Hemmerich and Hänsch 1993a). The spatial geometry of this distortion depends on the polarisation of the probe beam. In particular, the excitation rates of transitions involving an even or odd number of vibrational quanta result from contributions to the distortion of different parity (Hemmerich and Hänsch 1993a). The probe beam can only interfere with the 3D-field and thus distort the optical potential, if there exists a common polarisation component.

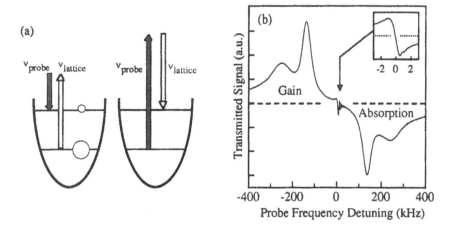

Figure 14. *(a): Raman transitions between vibrational levels of trapped atoms excited by a weak probe beam. When the probe frequency ν_{probe} is tuned below (above) the lattice frequency $\nu_{lattice}$ the probe is amplified (absorbed). (b): Probe transmission spectrum of a 3D optical lattice showing vibrational sidebands originating from Raman-transitions involving one and two vibrational quanta. The lattice geometry is that of Figure 13(c).*

The spectrum in Figure 14(b) also displays a very narrow dispersive four-wave mixing resonance which is resolved in detail in the upper right corner of Figure 14(b). For small angles enclosed by the probe beam and one of the trapping beams the linewidth of the central resonance scales linearly with this angle. Linewidths below 1kHz have been observed. A detailed discussion of this spectral feature (cf.Hemmerich *et al.* 1993) is beyond the scope of this chapter. Note, however, that quite generally, optical lattices have turned out to be interesting systems for studies in non-linear optics allowing the observation of multi-wave mixing processes (*e.g.* phase conjugation) or hyper-Raman transitions at low power levels accessible with diode lasers (Hemmerich *et al.* 1994a, 1994c, Lounis *et al.* 1993). Also, recently observed Bragg-scattering represents an interesting tool for extracting information about long range spatial order in optical lattices (Weidemüller *et al.* 1995).

A particularly exciting perspective of optical lattices is the achievement of atomic densities where the number of trapped atoms exceeds the number of lattice sites. In this regime, the lattice should acquire some solid state aspects and quantum statistics should play an important role for the system dynamics. Such hopes, however, are discouraged by the fact that the atomic density in near resonant optical fields is limited to the 10^{11}atoms/cm^3 level; this is well known, for example, in the case of the magneto-optical trap (MOT). This has also limited the atomic densities that can be obtained in optical lattices (typically 5% of the lattice sites can be occupied by an atom) because of inefficient loading (which so far has in all experiments involved a MOT) and at the same time because of the lattice itself. More efficient loading may be obtained by a

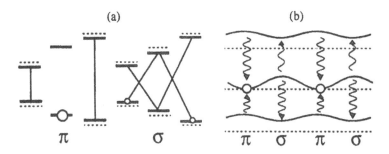

Figure 15. *(a) The dashed lines depict the Zeeman-components when only the magnetic field is present. In (b) only the groundstate is shown. The arrows indicate optical pumping processes (cf. Hemmerich et al (1995)).*

combination of multiple loading and refined magneto-optic trapping techniques (Gibble and Chu 1992, Ketterle *et al.* 1993), while the latter problem appears to be more fundamental. Different physical processes are responsible for these limitations. These all have the common feature that coupling the atoms to the light field modifies the field in a way that disturbs the trapping and cooling mechanisms (Dalibard 1988, Walker *et al.* 1990). As a consequence, atomic densities exceeding a few 10^{11} atoms/cm^3 appear difficult to reach in conventional optical lattices.

This has inspired research on new laser cooling and trapping schemes in which the elastic component of the fluorescence is strongly suppressed. At low fluorescence, the usual light-induced atom-atom interactions (Dalibard 1988, Walker *et al.* 1990) are expected to be suppressed. This may allow studies of novel atom-atom interactions which have not yet been considered. Moreover, in combination with appropriate loading techniques this may allow one to produce dense optical lattices in the future. A 1D example of a 'dark' optical lattice has been proposed by Grynberg and Courtois (1994). A 2D and 3D scheme has also been developed and demonstrated experimentally (Hemmerich *et al.* 1995). These novel cooling schemes use polarisation gradients which differ from those used in conventional Sysiphus-cooling and operate at blue detuning of the light field with respect to an atomic $(F \rightarrow F)$ transition or an $(F \rightarrow F - 1)$ transition (F is the total angular momentum). In addition a homogeneous magnetic field is employed.

Assume a light field composed of two polarisation components π and σ with field vectors parallel and orthogonal to the magnetic field, such that the nodes of π coincide with the antinodes of σ and vice versa. We consider the case when the light shifts are smaller than the Larmor frequency. At π-antinodes, all atoms are optically pumped to the $m_F = 0$ level where they decouple from the light field, whereas at σ-antinodes the atoms are pumped to the $m_F = \pm 1$ levels but remain strongly coupled (cf. Figure 15(a). For blue detuning of the light field, the groundstate lightshifts are positive and thus the decoupling of the $m_F = 0$ atoms occurs in a potential minimum. As illustrated in Figure 15(b), optical pumping cycles provide efficient cooling of the atoms into the

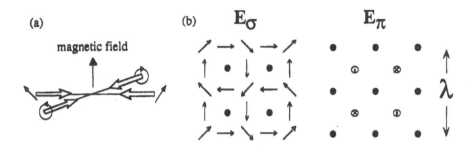

Figure 16. *Light field topography for a 2D dark optical lattice. The black dots depict nodes.*

$m_F = 0$ potential wells.

Let us briefly discuss the depopulation rates of vibrational levels in dark optical lattices. For the $m_F = 0$ potential, optical pumping processes leading back to the initial vibrational level without changing the value of m_F are not possible, because the coupling field $\mathbf{E}_\sigma$ has odd parity with respect to a nodal point. This is in contrast to the case of conventional optical lattices, where elastic processes exceed all other processes by more than an order of magnitude as seen in the fluorescence spectrum in Figure 12. The depopulation rates due to transitions changing the value of m_F have the same magnitude as those resulting from transitions which lead back to the initial $m_F = 0$ potential. This is also in contrast to conventional lattices where m_F-conserving transitions play the dominant role in the relaxation of vibrational populations.

It is not straight forward to use alternating π- and σ-components in order to produce 2D and 3D potential wells. A 2D solution is sketched in Figure 16. Four laser beams with polarisations as depicted in Figure 16(a) are superposed. If the time-phase ϕ difference between the two standing waves produced by each pair of counter-propagating beams is correctly adjusted, the total light field can be decomposed into the π- and σ-components shown in Figure 16(b) which display the correct topography for 2D potential wells. This light field has been employed in the experiment described in Hemmerich *et al.* (1995). In this experiment a cold sample of ^{87}Rb atoms is produced by a magneto-optic trap. For a short time the magneto-optic trap is disabled and a two-dimensional 'dark' lattice is created. In Figure 17(a) the fluorescence of the atoms during both periods is shown. A dramatic decrease of the fluorescence during the lattice phase is observed. When the magneto-optic trap is reactivated, the fluorescence reappears at an intensity exceeding the steady state value in the magneto-optic trap. When the lattice field is not present, the fluorescence also drops to zero, however, it takes more than ten milliseconds to recover when the magneto-optic trap is switched on and never exceeds the dotted line in Figure 17(a). Figure 17(b) shows a probe transmission spectrum of the dark optical lattice from Hemmerich *et al.* (1995). The $F = 1 \rightarrow 1$ component of the D1-transition of the ^{87}Rb isotope is used and the lattice light field has the topog-

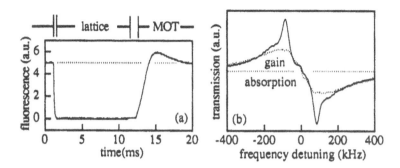

Figure 17. *Experimental observation in a dark optical lattice. For explanation see text.*

raphy shown in Figure 16. One clearly recognizes sharp resonances at 187kHz due to Raman transitions between adjacent vibrational states of trapped atoms (solid trace). These transitions are possible because the atoms are trapped in quantum states which are perfectly dark only at the centres of the microtraps. For comparison, when the time-phase difference is not correctly adjusted, the light field is not appropriate for 2D trapping and no vibrational resonances are observed (dotted trace).

9 Conclusion

The aim of this chapter was to point out that atoms subjected to laser light experience different kinds of strong forces which can be utilized in many different ways for the precision control of atomic motion. A number of fundamental concepts have been discussed here and supported by some experimental examples of possible applications. This chapter does not represent a review of the beautiful experimental and theoretical work in the field of laser cooling. I have omitted many fascinating aspects such as sub-recoil cooling, novel kinds of trapping techniques, atomic interferometry, and many others. Even with regard to the few topics which I have selected the presentation given here is very incomplete. Many important contributions from numerous research groups have not been discussed. In fact, most of the material in this chapter covers concepts and developments achieved all over the world during two decades, and I have used much of it without appropriate citation. I would like to thank Professor T W Hänsch for his support during many pleasant years of fruitful collaboration. I am grateful to the Munich laser cooling group and in particular to Matthias Weidemüller for many lively discussions and helpful remarks. Finally, this chapter has been improved by the unique remarks of Frank Sander.

References

Arimondo E, Phillips W D and Strumia F, eds., 1992, *Laser Manipulation of Atoms and Ions* in Proc of the Int Sch of Physics,'Enrico Fermi', North-Holland.

Ashkin A and Gordon J P, 1983, *Opt Lett* **8** 511.

Aspect A, Arimondo E, Kaiser R, Vansteenkiste N, and Cohen-Tannoudji, 1988, *Phys Rev Lett* **61** 826.

Castin Y and Dalibard J, 1991, *Europhys Lett* **14** 761.

Chu S, Hollberg L, Bjorkholm J E, Cable A,and Ashkin A,1985, *Phys Rev Lett* **55** 48.

Chu S, Wieman C, eds., 1989, Laser Cooling & Trapping, *J Opt Soc Am* **B6** special issue.

Cohen-Tannoudji C and Reynaud S, 1977, *J Phys B, Atom Molec Phys* **10** 345.

Courtois J and Grynberg G, 1992, *Phys Rev A* **46** 7060.

Dalibard J, 1988, *Opt Commun* **68** 203.

Dalibard J and Cohen-Tannoudji C, 1985, *J Opt Soc Am* **B2** 1707.

Dalibard J and Cohen-Tannoudji C, 1989, *J Opt Soc Am* **B6** 2023.

Dalibard J, Raimond J M and Zinn-Justin J, eds., 1992, *Fundamental Systems in Quantum Optics, Les Houches Session LII, 1990* North-Holland Physics Publishing.

Einstein A, 1917, *Phys Z* **18** 121.

Ertmer W, Blatt R, Hall J L, and Zhu M, 1985, *Phys Rev Lett* **54** 996.

Frisch R, 1933, *Phys Z* **86** 42.

Gibble K and Chu S, 1992, *Opt Lett* **17** 526.

Grison D, Lounis B, Salomon C, Courtois J, and Grynberg G, 1991, *Europhys. Lett* **15** 149.

Grove R, Wu F Y,and Ezekiel S, 1977, *Phys Rev A* **15** 227.

Grynberg G and Courtois J, 1994, *Europhys Lett* **27** 41.

Grynberg G, Lounis B, Verkerk P, Courtois J-Y, and Salomom C, 1993, *Phys Rev Lett* **70** 2249.

Hänsch T and Schawlow A, 1975, *Opt Commun* **13** 68.

Hemmerich A and Hänsch T, 1993a, *Phys Rev A* **48** R1753.

Hemmerich A and Hänsch T, 1993b, *Phys Rev Lett* **70** 410.

Hemmerich A, McIntyre D, Schropp D, Meschede D, and Hänsch T, 1990, *Opt Commun* **75** 118.

Hemmerich A, Schropp D, Esslinger T, and Hänsch T, 1992, it Europhys Lett **18** 391.

Hemmerich A, Schropp D, and Hänsch T 1991, *Phys Rev* **A44** 1910.

Hemmerich A, Weidemüller M, Esslinger T, Zimmermann C, and Hänsch T, 1995, *Phys Rev Lett* **75** 37.

Hemmerich A, Weidemüller M, and Hänsch T, 1994a, *Europhys Lett* **27** 427.

Hemmerich A, Weidemüller M, and Hänsch T, 1994b, *Laser Phys* **4** 884.

Hemmerich A, Zimmermann C, and Hänsch T, 1993, *Europhys Lett* **22** 89.

Hemmerich A, Zimmermann C,and Hänsch T, 1994c, *Phys Rev Lett* **72** 625.

Jessen P S, Gerz C, Lett P D, Phillips W D, Rolston S L, Spreeuw R J C, and Westbrook C, 1992, *Phys Rev Lett* **69** 49.

1985, *J Opt Soc Am* **B2** special issue.

Kasevich M and Chu S, 1992, *Phys Rev Lett* **69** 1741.

Ketterle W, Davis K B, Joffe M A, Martin A, Pritchard D E, 1993, *Phys Rev Lett* **70** 2253.

Letokhov V S, Minogin V, and Pavlik B, 1976, *Opt Commun* **19** 72.

Lett P D, Watts R N, Westbrook C I, Phillips W D, Gould P L, and Metcalf H J, 1988, *Phys Rev Lett* **61** 169.

Lounis B, Verkerk P, Courtois J-Y, Salomon C, and Grynberg G, 1993, *Europhys Lett* **21** 13.

Monroe C, Swann W, Robinson N, and Wieman C, 1990, *Phys Rev Lett* **65** 1571.

Phillips W D and Metcalf H J, 1982, *Phys Rev Lett* **49** 1149.

Phillips W D and Prodan J, 1984, *Prog Quantum Electron* **8** 231.

Raab E, Prentiss M, Cable A, Chu S, and Pritchard D E, 1987, *Phys Rev Lett* **59** 2631.

Salomon C, Dalibard J, Aspect A, Metcalf H, Cohen-Tannoudji C, 1987,
  Phys Rev Lett **59** 1659.

Verkerk P, Lounis B, Salomon C, Cohen-Tannoudji C, Courtis J-Y, and Grynberg G, 1992,
  Phys Rev Lett **68** 3861.

Walker T, Sesko D, and Wieman C, 1990, *Phys Rev Lett* **64** 408.

Weidemüller M, Hemmerich A, Esslinger T, and Hänsch T, 1995, *Laser Spectroscopy XII,
  Capri, Italy*..

Wineland D J and Dehmelt H G, 1975, *Bull Am Phys Soc* **20** 637.

Presentations by Participants

•1.3 μm diode pumped Nd:YLF additively modelocked laser
David Armstrong—(*University of Strathclyde*)

•Ultrasensitive measurements with short pulse ring lasers
Briggs Atherton—(*University of New Mexico*)

•Probe amplification in the presence of special density grating
Gianpaolo Barozzi—(*Universita di Milano, Institut Non Lineaire de Nice*)

•Efficient single frequency Q-switching via feedback controlled pre-lasing
Christoph Bollig—(*University of Southampton*)

•Hyper Rayleigh Scattering - OPO - OPA - with ns and fs pulses
Carlo Boutton—(*University of Leuven*)

•High power tapered semiconductor laser structures
Nicholas Brooks—(*University of Bath*)

•Soliton propagation in dispersion decreasing fiber
Eric Buckland—(*University of Rochester, Institute of Optics*)

•All solid-state femtosecond lasers
David Burns—(*University of St. Andrews*)

•High power, broadly tunable all-solid-state synch-pumped LBO OPO
Stuart Butterworth—(*University of Southampton*)

•The Cern Laser Ion Source
John Collier—(*Rutherford Appleton Laboratory*)

•Microchip Laser Studies
Richard Conroy—(*University of St. Andrews*)

•Application of radiative renormalisation to CARS spectroscopy
David Coppeta—(*Tufts University*)

•All-solid state femtosecond lasers
Matthew Critten—(*University of St. Andrews*)

•Ultrasensitive measurements with short pulse ring lasers
Scott Diddams—(*University of New Mexico*)

•Experimental studies in the role of EIT in four wave mixing schemes
Christopher Dorman—(*Imperial College*)

• **Laser performance and spectroscopic behaviour of Er^{3+} doped scandium silicates at 1.5 μm**
Livio Fornasiero—(*University of Hamburg*)

• **Noise fluctuations in stimulated Brillouin scattering in optical fibres at strong pump depletion**
Andrei Fotiadi—(*Laboratory for Quantum Electronics, A F Ioffe Physico-Tech Inst*)

• **Picosecond OPOs**
Steven French—(*University of St. Andrews*)

• **Electromagnetically induced transparency and focussing**
David Fulton—(*University of St. Andrews*)

• **Multi-millijoule, Q-switched, transeversely diode pumped laser with TEM$_{00}$ mode**
Efstratios Georgiou—(*Foundation for Research and Technology (F.O.R.T.H.)/I.E.S.L.*)

• **Confocal differential microscope using a single mode birefringent fibre**
Linas Giniunas—(*Vilnius University, Laser Research Centre*)

• **Laser beam quality and entropy, finding the limits of beam shaping**
Thomas Graf—(*University of Bern, Institute of Applied Physics*)

• **Short pulse generation in semiconductors**
Paul Gunning—(*BT Laboratories*)

• **State selective, high resolution two-electron two-photon laser-photo-detachment**
Gunnar Haeffler—(*Chalmers University of Technology*)

• **Low-voltage electro-optic modulation in BRAQWET structures**
Richard Hainzl—(*Institute of Optical Research, Stockholm*)

• **New developments of Cr^{4+} doped laser crystals**
Simone Hartung—(*University of Hamburg, Institut für Laser-Physik*)

• **Impact of spectral inverter fibre length**
Per Olof Hedekvist—(*Chalmers University of Technology*)

• **QPM in bulk LiNbO$_3$ for microchip lasers**
Peter Henriksson—(*Institute of Optical Research*)

• **Time-resolved measurements on 1.55μm in GaAsP/InP microcavity lasers**
Mathias Hilpert—(*Max-Planck-Institut für Festkörperforschung*)

• **The time evolution of the electric characteristics of a laser discharge through their waveforms of the voltage and the current**
Athanasios Ioannou—(*University of Patras*)

• **Quasi phase matching in organic waveguides**
Matthias Jaeger—(*CREOL, University of Central Florida*)

•**Upconversion processes and diode pumped laser experiments at 2.8μm**
Thomas Jensen—(*University of Hamburg, Institut für Laser-Physik*)

•**Frequency doubling of Nd-doped-solid-state lasers**
Tim Kellner—(*University of Hamburg, Institut für Laser-Physik*)

•**A study of coherent photon seeding applied to the NaCl:OH$^-$ colour centre laser**
Gordon Kennedy—(*University of St. Andrews*)

•**All solid-state femtosecond lasers**
Karen Lamb—(*University of St. Andrews*)

•**Development of a tunable laser based on copper(I) doped beta alumina**
Steve Lane—(*University of Manchester*)

•**Application of the ENSTA method to the characterisation of fs oscillators in the μJ region**
Hans Lange—(*Ecole Polytechnique, LOA - ENSTA*)

•**Double 1/4 discontinuities in DFB lasers**
Antonio Lucianetti—(*Politecnico di Milano*)

•**Two component superradiance**
Dmitry Mashkovsky—(*Physics Institute and Moscow State Technical University*)

•**Microchip Laser Studies**
David Matthews—(*University of St. Andrews*)

•**RTA based fs parametric oscillators**
Cate McGowan—(*University of St. Andrews*)

•**Laser requirements for spaceborne gravitational wave detectors**
Paul McNamara—(*University of Glasgow*)

•**Green Er^{3+}:LiYF$_4$ upconversion laser at room temperature, codoped with Y$_5^{3+}$ ions**
Patrick Moebert—(*University of Hamburg, Institut für Laser-Physik*)

•**Gain switched Ti:sapphire laser**
Gary Morrison—(*University of St. Andrews*)

•**Nonparaxial eigenmodes in nonlinear beam propagation**
Janet Noon—(*Imperial College*)

•**Photovoltaic nonlinearity and laser based on it**
Alexander Novikov—(*Institute of Physics, Kiev*)

•**QPM in bulk LiNbO$_3$ for microchip lasers**
Claes Magnus Olson—(*Institute of Optical Research, Stockholm*)

•**Signal restoration using the nonlinear optical loop mirror**
Bengt-Erik Olsson—(*Chalmers University of Technology*)

- The time evolution of the electric characteristics of a laser discharge through their wave forms of the voltage and the current
Giannis Parthenios—(*University of Patras*)

- Ultrafast carrier relaxation in low temperature grown InGaAs layers
Valdas Pasiskevicius—(*Semiconductor Physics Institute, Vilnius*)

- Spin relaxation in multiple quantum wells
Peggy Perozzo—(*Grand Valley State University*)

- Numerical results of split-step spectral method for propagation of a Gaussian beam in self focussing material
Monika Pietrzyk—(*Polish Academy of Sciences, Inst of Fundamental Tech Research*)

- Diode pumped Nd:YLF and nonlinear conversion schemes at high rep rates
Christian Rahlff—(*Technische Universitat Berlin*)

- Optical nonlinearities in CdS and CdSe Nanocrystallites
Philippe Riblet—(*University of Strasbourg, GONLO*)

- Traffic monitoring, the coherent laser radar approach
Alejandro Rodriguez—(*Polytechnic University of Catalonia*)

- Design of metamorphic buffer layers for 1.5 μm opto-electronic devices on GaAs
Ana Sacedon—(*Universidad Politécnica de Madrid, Ciudad Universitaria, E.T.S.I.T.*)

- Nonlinear optical properties of piezoelectric InGaAs/GaAs MQW p-i-n diodes
Jose Luis Sanchez-Rojas—(*Univ Politecnica de Madrid, Ciudad Univ, E.T.S.I.T.*)

- Ga:La:S a low phonon glass for long wavelength applications?
Thorsten Schweizer—(*University of Southampton, Optoelectronics Research Centre*)

- Bragg grating based single mode fibre lasers
Milan Sejka—(*Mikroelektronik Centret, Lyngby*)

- Photorefractive materials in the nonlinear regime
Eduardo Serrano Jerez—(*Heriot Watt University*)

- Electromagnetically induced transparency and focussing
Sara Shepherd—(*University of St. Andrews*)

- High order harmonic generation with Nd:glass Ti: sapphire CPA lasers
Georg Sommerer—(*Max Born Institut*)

- Continuous-wave OPOs
Tracy Stevenson—(*University of St. Andrews*)

- Cascaded nonlinearities in waveguides; Possibility of quasi-phase matched integrated optical parametric oscillators
Paper of Kacem El Hadi, presented by Michael Sundheimer—(*University of Nice*)

•**Kerr lens modelocked visible transitions of a Pr:YLF laser**
Jason Sutherland—(*Imperial College*)

•**Design of an optical parametric oscillator on behalf of a surface sum vibration spectrometer**
Teunis Tukker—(*FOM Institut, Utrecht*)

•**Four photon parametric oscillation in sodium vapour**
Virgilijus Vaicaitis—(*Vilnius University, Laser Research Centre*)

•**Linear and nonlinear optical properties of II-VI semiconductors studied with nanosecond lasers: ZnS as an example**
Jan Valenta—(*IPCMS - GONLO, Strasbourg*)

•**A study of coherent photon seeding applied to the NaCl:OH⁻ colour centre laser**
Gareth Valentine—(*St. Andrews University*)

•**Nonlinear refractive index near points of zero absorption and in the dead zone**
Arlene Wilson-Gordon—(*Bar-Ilan University*)

•**Nonlinear optical self-action and optical limiting**
Tiejun Xia—(*CREOL, University of Central Florida*)

•**CW frequency doubled Nd:YAG laser**
Carl Yelland—(*University of St. Andrews*)

Participants

•Azize Pelin Aksoy
Ankara Universitesi,
Fen Fakkultesi,
Elektronik Muhendisliji Bolumu,
06100 Tandogan,
Ankara, Turkey.

•David Armstrong
Strathclyde University,
Dept. of Physics & Applied Physics,
John Anderson Building,
107 Rotten Row,
Glasgow G4 ONG, UK.

•Briggs Atherton
University of New Mexico,
Dept. of Physics &Astronomy,
800 Yale Blvd. N.E.,
Albuquerque, New Mexico 87131,
U.S.A.

•Gianpaolo Barozzi
Institut NonLineaire de Nice,
1361 Route Des Lucioles,
06560 Valbonne, France.

•Gunnar Björk
Department of Electronics/FMI,
KTH Electrum 229,
S-164 40 Kista, Sweden.

•Christoph Bollig
Optoelectronics Research Centre,
University of Southampton,
Southampton SO17 1BJ, UK.

•Carlo Boutton
University of Leuven,
L.C.B.D.,Celestijnenlaan 200D,
3001 Heverlee, B3001 Leuven,
Belgium.

•Robert Boyd
University of Rochester,
The Institute of Optics,
Rochester, New York 14627,
U.S.A.

•Nicholas Brooks
University of Bath,
Electronic & Electrical Engineering,
Bath, Avon BA2 7AY, UK.

•Eric Buckland
University of Rochester,
Institute of Optics,
302 Wilmot Building,
Rochester, New York 14627
U.S.A.

•Stuart Butterworth
University of Southampton,
Optoelectronics Research Centre,
Southampton SO17 1BJ, UK.

•Alasdair Cameron
University of St Andrews,
School of Physics & Astronomy,
North Haugh, St Andrews,
Scotland, KY16 9SS, UK.

•Ali Cetin
Osmangazi University,
Fen edebiyat Fakültesi,
Fizik Bol., Camlik Kampusu,
26060/Eskisehir, Turkey.

•John Collier
Rutherford Appleton Labs,
Building R1,
Chilton, Didcot,
Oxon, OX11 OQX, UK.

•David Coppeta
Tufts University, EOTC,
4 Colby Street,
Medford,
Massachusetts 02155, U.S.A.

•Scott Diddams
University of New Mexico,
Department of Physics & Asronomy,
800 Yale Blvd. N.E.,
Albuquerque,
New Mexico 87131, U.S.A.

•Christopher Dorman
Imperial College, Physics Dept.,
LASP II,
London SW7 2BZ, UK.

•Malcolm Dunn
University of St Andrews,
School of Physics & Astronomy,
North Haugh, St Andrews,
Scotland, KY16 9SS, UK.

•Majid Ebrahimzadeh
University of St Andrews,
School of Physics & Astronomy,
North Haugh, St Andrews,
Scotland, KY16 9SS, UK.

•Roger Edwin
University of St Andrews,
School of Physics & Astronomy,
North Haugh, St Andrews,
Scotland, KY16 9SS, UK.

•TY Fan
MIT Lincoln Labs,
244 Wood Street, P.O. Box 73,
Lexington,
Massachusetts 02173, U.S.A.

•David Finlayson
University of St Andrews,
School of Physics & Astronomy,
North Haugh, St Andrews,
Scotland, KY16 9SS, UK.

•Livio Fornasiero
University of Hamburg,
Institute fur Laser-Physik,
Jungiusstr. 11,
20355 Hamburg, Germany.

•Andrei Fotiadi
A F Ioffe Physico Technical Institute,
Polytekhnicheskaya 26,
St Petersburg, 194021, Russia.

•James Fraser
University of Toronto,
Dept of Physics,
60 St. George Sreet,
Toronto, Ontario M5S 1A7, Canada.

•Graham Friel
University of Southampton,
Optoelectronics Research Centre,
Southampton SO17 1BJ, UK.

•Efstratios Georgiou
F.O.R.T.H./I.E.S.L.,
P.O. Box 1527,
Heraklio 71110,
Crete, Greece.

•Linas Giniunas
Vilnius University,
Laser Research Centre,
Sauletekio Ave 10,
Vilnius 2054, Lithuania.

•Philip Gorman
DRA Malvern, PEZIZ,
St Andrews Road, Malvern,
Worcs. WR14 3PS, UK.

•Thomas Graf
Institute of Applied Physics,
Sidlerstrasse 5,
CH-3012 Bern, Switzerland.

•Brett Guenther
2 Electrical Engineering East,
Penn State University, University Park,
Pennsylvania 16802, U.S.A.

•Paul Gunning
BT Labs, B55/131,
Ipswich IP5 7RE, UK.

•Gunnar Haeffler
Chalmers University of Technology,
Department of Physics,
Atomic Physics Group,
S-41296 Göteborg, Sweden.

•Richard Hainzl
Institute of Optical Research,
Lindstedtsvagen 24,
10044 Stockholm, Sweden.

●David Hanna
University of Southampton,
Optoelectronic Research Centre,
Highfield, Southampton,
Hamps. SO17 1BJ, UK.

●Simone Hartung
University of Hamburg,
Institut für Laser-Physik,
Jungiusstr. 11,
2035 Hamburg, Germany.

●Michael Hasselbeck
CREOL,
12424 Research Parkway,
Orlando, Florida 32826, U.S.A.

●Per Olof Hedekvist
Chalmers University of Technology,
Department of Optoelectronics,
S-41296 Gothenburg, Sweden.

●Andreas Hemmerich
University of Munich,
Department of Physics,
Schellingstr. 4,
D-80799 Munich, Germany.

●Peter Henriksson
Institute of Optical Research,
Lindstedtsvagen 24,
10044 Stockholm, Sweden.

●Mathias Hilpert
MPI fur Festkorperforschung,
Heisenbergstr. 1,
70569 Stuttgart, Germany.

●Christopher Howle
University of Strathclyde,
Physics Dept.,
John Anderson Bldg.,
Glasgow G40 1PN, UK.

●Günter Huber
University of Hamburg,
Institut für Angewandte Physik,
Jungiusstrasse 11,
D-20000, Hamburg 36, Germany.

●Athanasios Ioannou
University of Patras,
Department of Physics,
Patras 26500, Greece.

●Matthias Jaeger
CREOL, Suite 400,
12424 Research Parkway,
Orlando, Florida 32826, U.S.A.

●Thomas Jensen
Hamburg University,
Institute für Laser-Physik,
Jungiusstr. 11,
20355 Hamburg, Germany.

●Tim Kellner
Hamburg University,
Institute fur Laser-Physik,
Jungiusstr. 11,
20355 Hamburg, Germany.

●Christopher King
University of Toronto,
Physics Department,
60 St. George Street,
Toronto, Ontario M5S 1A7,
Canada.

●Karen Lamb
University of St Andrews,
School of Physics & Astronomy,
North Haugh, St Andrews,
Scotland, KY16 9SS, UK.

●Steve Lane
University of Manchester,
Laser Photonics, Dept of Physics,
Manchester M13 9PL, UK.

●Hans Lange
Ecole Polytechnique, LOA ENSTA,
Centre de l'Yvette,
91 120 Palaiseau, France.

●Antonio Lucianetti
Facolta di Fisica (Milano),
Via B. Palazzo 90,
24100 Bergamo, Italy.

●Pavel Mamyshev
AT&T Bell Laboratories,
Room 4C-310,
101 Crawfords Corner Road,
Holmdel,
New Jersey 07733-3030,
U.S.A.

•Dmitry Mashkovsky
General Physics Institute,
Vavilov Street 38, Box 117333,
Moscow, Russia.

•Paul McNamara
1 Torrin Road,
Summerston,
Glasgow G23 5HZ, UK.

•Alan Miller
University of St Andrews,
School of Physics & Astronomy,
North Haugh, St Andrews,
Scotland, KY16 9SS, UK.

•Patrick Möbert
University of Hamburg,
Institute fur Laser-Physik,
Jungiusstr. 11,
20355 Hamburg, Germany.

•André Mysyrowicz
Ecole Polytechnique,
Laboratoire d'Optique Appliquee,
ENSTA,
F-91120 Palaiseau, France.

•Janet Noon
LASP, Blackett Laboratory,
Imperial College,
Prince Consort Road,
London SW7 2BZ, UK.

•Alexander Novikov
Insitute of Physics,
Prospect Nauki. 46,
Kiev 252650, Ukraine.

•Magnus Olson
Institute of Optical Research,
KTH, Lindstedtsvagen 24,
Stockholm, Sweden.

•Bengt-Erik Olsson
Chalmers University,
Department of Optoelectronics,
412 96 Göteborg, Sweden.

•Miles Padgett
University of St Andrews,
School of Physics & Astronomy,
North Haugh, St Andrews,
Scotland, KY16 9SS, UK.

•Giannis Parthenios
University of Patras,
Department of Physics,
Patras 26500, Greece.

•Valdas Pasiskevicius
Semiconductor Physics Insitute,
Gostauto 11,
Vilnius 2600, Lithuania.

•Peggy Perozzo
Grand Valley State University,
Physics Department,
Allendale, Michigan, U.S.A.

•Monika Pietrzyk
Polish Academy of Sciences,
Institute of Fundamental Technological
Research,
Photonics Laboratory,
Swietokrzyska 21,
Warsaw, Poland.

•Christian Rahlff
University of St Andrews,
School of Physics & Astronomy,
North Haugh, St Andrews,
Scotland, KY16 9SS, UK.

•John Reintjes
NRL Code 6542,
4555 Overlook Avenue SW,
Washington, DC 20375-5000, U.S.A.

•Philippe Riblet
University of St Andrews,
School of Physics & Astronomy,
North Haugh, St Andrews,
Scotland, KY16 9SS, UK.

•Jean-Francois Ripoche
Ecole Polytechnique,
LOA ENSTA,
Centre de l'Yvette,
F-91120 Palaiseau, France.

•Alejandro Rodriguez-Gomez
Polytechnic University of Catalonia,
Campus Nord UPC, Edifici D-3,
C/Gran Capita,
s/n. 08034, Barcelona, Spain.

•Sean Ross
CREOL, 12424 Research Parkway,
Orlando, Florida 32826, U.S.A.

•Ana Sacedon
Universidad Politecnic de Madrid,
Ciudad Universitara,
Dpto. Ingenieria Electronica,
E.T.S.I.T.,
28040 Madrid, Spain.

•Jose Luis Sanchez-Rojas
Universidad Politecnic de Madrid,
Ciudad Universitara,
Dpto. Ingenieria Electronica,
E.T.S.I.T.,
28040 Madrid, Spain.

•Thorsten Schweizer
University of Southampton,
Optoelectronics Research Centre,
Southampton SO17 1BJ, UK.

•Milan Sejka
Mikroelectronik Centret,
DTU, Bldg. 345e,
2800 Lyngby, Denmark.

•Eduardo Serrano Jerez
Heriot Watt University,
Department of Physics,
Edinburgh EH14 4AS, UK.

•Wilson Sibbett
University of St Andrews,
School of Physics & Astronomy,
North Haugh, St Andrews,
Scotland, KY16 9SS, UK.

•Bruce Sinclair
University of St Andrews,
School of Physics & Astronomy,
North Haugh, St Andrews,
Scotland, KY16 9SS, UK.

•Georg Sommerer
Max Born Institut/B1,
Rudower Chausse 6,
D12489 Berlin, Germany.

•Michael Sundheimer
University of Nice,
Laboratoire de Physique
de la Matiere Condensee,
Sophia Antipolis, Parc Valrose,
06108 Nice, Cedex 2, France.

•Jason Sutherland
Imperial College,
Femtosecond Optics Group,
Blackett Laboratory,
Prince Consort Road,
London SW7 2BZ, UK.

•Tukker Teunis
University of Twente,
Department of Applied Physics,
P.O. Box 217,
7500 AE Enschede,
The Netherlands.

•Virgilijus Vaicaitis
Vilnius University,
Laser Research Centre,
Sauletekio Ave. 10,
Vilnius 2054, Lithuania.

•Jan Valenta
IPCMS - GONLO,
23 rue du Loess,
F-67037 Strasbourg Cedex, France.

•Eric VanStryland
CREOL/UCF,
12424 Research Parkway,
Orlando,Florida 32826, U.S.A.

•Ian White
University of Bath,
School of Physics, Claverton Down,
Bath, Avon BA2 7AY, UK.

•Arlene Wilson-Gordon
Bar-Ilan University,
Department of Chemistry,
Ramat Gan 52900, Israel.

•Tiejun Xia
CREOL, 12424 Research Parkway,
Orlando, Florida 32826, U.S.A.

Index